D1744351

Advances in Science, Technology & Innovation

IEREK Interdisciplinary Series for Sustainable Development

Advances in Science, Technology & Innovation (ASTI) is a series of peer-reviewed books based on important emerging research that redefines the current disciplinary boundaries in science, technology and innovation (STI) in order to develop integrated concepts for sustainable development. It not only discusses the progress made towards securing more resources, allocating smarter solutions, and rebalancing the relationship between nature and people, but also provides in-depth insights from comprehensive research that addresses the **17 sustainable development goals (SDGs)** as set out by the UN for 2030.

The series draws on the best research papers from various IEREK and other international conferences to promote the creation and development of viable solutions for a **sustainable future and a positive societal** transformation with the help of integrated and innovative science-based approaches. Including interdisciplinary contributions, it presents innovative approaches and highlights how they can best support both economic and sustainable development, through better use of data, more effective institutions, and global, local and individual action, for the welfare of all societies.

The series particularly features conceptual and empirical contributions from various interrelated fields of science, technology and innovation, with an emphasis on digital transformation, that focus on providing practical solutions to **ensure food, water and energy security to achieve the SDGs.** It also presents new case studies offering concrete examples of how to resolve sustainable urbanization and environmental issues in different regions of the world.

The series is intended for professionals in research and teaching, consultancies and industry, and government and international organizations. Published in collaboration with IEREK, the Springer ASTI series will acquaint readers with essential new studies in STI for sustainable development.

ASTI series has now been accepted for Scopus (September 2020). All content published in this series will start appearing on the Scopus site in early 2021.

More information about this series at https://link.springer.com/bookseries/15883

Cristina Piselli • Haşim Altan •
Osman Balaban • Peleg Kremer
Editors

Innovating Strategies and Solutions for Urban Performance and Regeneration

A culmination of selected research papers from the Second version of the international conference on Urban Regeneration and Sustainability (URS) held in collaboration with University of East London, UK (2020).

 Springer

Editors
Cristina Piselli 🆔
Department of Architecture (DIDA)
University of Florence
Florence, Italy

Osman Balaban 🆔
Department of City and Regional Planning
Middle East Technical University
Ankara, Turkey

Haşim Altan 🆔
Department of Architecture
Faculty of Design
Arkin University of Creative
Arts and Design (ARUCAD)
Kyrenia, Cyprus

Peleg Kremer 🆔
Department of Geography
and the Environment
Villanova University
Villanova, PA, USA

ISSN 2522-8714   ISSN 2522-8722   (electronic)
Advances in Science, Technology & Innovation
IEREK Interdisciplinary Series for Sustainable Development
ISBN 978-3-030-98189-1   ISBN 978-3-030-98187-7   (eBook)
https://doi.org/10.1007/978-3-030-98187-7

Max Timms

This Springer imprint is published by the registered company Springer Nature Switzerland AG
The registered company address is: Gewerbestrasse 11, 6330 Cham, Switzerland

Scientific Committee

Ahmed Abdel Ghaney Morsi, Helwan University, Cairo, Egypt
Aida Kesuma Azmin, International Islamic University Malaysia (IIUM), Kuala Lumpur, Malaysia
Anastasia Fotopoulou, Alma Mater Studiorum University of Bologna, Bologna, Italy
Anita Tatti, Polytechnic University of Milan, Milan, Italy
Anna Laura Pisello, University of Perugia, Perugia, Italy
Antonella Versaci, Kore University of Enna, Enna, Italy
Benedetta Pioppi, University of Perugia, Perugia, Italy
Chiara Chiatti, University of Perugia, Perugia, Italy
Çiğdem Varol, Gazi University, Ankara, Turkey
Claudia Fabiani, University of Perugia, Perugia, Italy
Cristina Piselli, University of Florence, Florence, Italy
Dimelli Despoina, Technical University of Crete (TUC), Chania, Greece
Duygu Cihanger M. Ribeiro, Middle East Technical University, Ankara, Turkey
Ekin Pınar, Middle East Technical University, Ankara, Turkey
Fabiana Frota de Albuquerque Landi, University of Perugia, Perugia, Italy
Hasim Altan, Arkin University of Creative Arts and Design, Kyrenia, Cyprus
Hossein Hosseiny, Washington University in St. Louis, Missouri, St. Luis, MO, USA
Ilaria Pigliautile, University of Perugia, Perugia, Italy
Ioannis Kousis, University of Perugia, Perugia, Italy
Kabindra Shakya, Villanova University, PA, USA
Klemens Laschefski, UFMG -Federal University of Minas Gerais, Belo Horizonte, Brazil
Marta Chàfer, University of Lleida, Lleida, Spain
Mattia Manni, University of Perugia, Perugia, Italy
Michelle Johnson, USDA Forest Service Northern Research Station, Bayside, NY, USA
Mona Helmy, The British University in Egypt (BUE), Cairo, Egypt
Mustafa Bayırbağ, Middle East Technical University, Ankara, Turkey
Nabil Mohareb, American University in Cairo, Egypt
Nihan Oya Memlük Çobanoğlu, Gazi University, Ankara, Turkey
Nil Uzun, Middle East Technical University, Ankara, Turkey
Olgu Çalışkan, Middle East Technical University, Ankara, Turkey
Osman Balaban, Middle East Technical University, Ankara, Turkey
Peleg Kremer, Villanova University, PA, USA
Raz Godelnik, The New School, New York, NY, USA
Sara Arko, IRI UL—Institute for Innovation and Development of University of Ljubljana, Ljubljana, Slovenia
Seri Park, Villanova University, PA, USA
Shrobona Karkun, Temple University, Philadelphia, PA, USA

Foreword

The world is becoming predominantly urban: around 55% of the world's total population live in cities today. This is an ongoing trend indicating that the world's cities will accommodate larger shares of the global population in the coming decades. However, this demographic trend does not show a uniform pattern across the globe. Majority of the future urban population growth will take place in the Global South. Cities in developing countries are now the hotspots of urban growth in demographic and economic terms, involving higher potential of transformation.

Increasing concentration of the global population in urban areas, especially in the Global South, comes at a significant cost. City administrations have to transform their cities in innovative ways in order not only to develop the urban infrastructure to accommodate the future population but also to renew and rehabilitate the old, decayed, or informal city quarters. Furthermore, this local transformation agenda has to respond to the constantly changing complexities of the current environmental crisis including climate change and COVID-19-like pandemics. Tackling such a multifaceted urban transition requires better urban regeneration policies and practices by relevant stakeholders, e.g., policymakers, city administrations, etc.

Accordingly, in the contemporary era of urbanization, where resources are decreased and challenges are multiplied, urban regeneration initiatives should deliver multiple benefits and generate win–win situations. On the one hand, regeneration initiatives need to rehabilitate and renew existing city parts in ways to create sustainable, healthy, and equitable living environments for the entire urban community. On the other hand, urban regeneration should improve urban performance so as to make urban economies more inclusive and greener, and urban services more affordable and accessible. This is easier said than done as urban regeneration initiatives usually face serious challenges in many parts of the world.

In this panorama, this book offers valuable insights into the above-mentioned debate based on a range of applications and case studies presented in key papers from the Second International Conference on Urban Regeneration and Sustainability (URS) of IEREK. Indeed, the book discusses the relevant aspects and potential solutions to manage the rapid growth of urbanization and the challenges of sustainable and effective urban regeneration. More in detail, the book is composed of three parts that progressively address the topic starting from discussions on planning, design, and management of urban regeneration, which is defined as a process and a strategy aiming to renew and transform public and private urban areas and buildings. Thereafter, established and innovative regeneration strategies to improve urban performance are presented when applied in different contexts and, therefore, tailored depending on the local contextual characteristics. Finally, the book aims at understanding and addressing the main challenges of urban regeneration towards sustainable and resilient urban development and transformation.

To this aim, this volume is meant to stimulate a discussion on key questions addressing challenges to urban regeneration and performance faced by cities based on research studies and real-case applications. Given the interdisciplinarity of the debate, one of the goals of primary importance is to understand the role of community engagement towards urban regeneration and sustainability. Which are the main social, environmental, and design challenges in urban planning and regeneration? Do developing countries show peculiar needs?

Can we identify the most effective strategies and processes to improve urban performance in new and existing urban environments? Do they significantly differ depending on the context of application? We hope that the readers of this volume will find significant insights into the answers of these questions.

Florence, Italy   Cristina Piselli
Ankara, Turkey   Osman Balaban

Introduction

With continued rapid growth of urban populations in many regions of the world, urbanization remains one of the main challenges faced by humanity today (Kuddus, 2020). Cities have been sites of major human, social, cultural, and economic development over centuries (Jacobs, 1961; McMichael, 2000). However, major challenges are still associated with the process of urbanization. These challenges include, among others, pollution (Han et al., 2018; Liang et al., 2019), inadequate infrastructure (Sohail et al., 2005), climate change (Srivastava, 2020), human health and wellbeing (McMichael, 2000), poverty and inequitable distribution of resources (Liddle, 2015), slums and squatters (Seto et al., 2013).

Urban areas are constantly changing. Change is inherent to the urbanization processes, as the constant flux of people, coupled with shifting social and cultural structures, translate into a physical adaptation of the urban landscape (Rodriguez and Juaristi, 2015). This change encompasses rapid, informal, and uncontrolled urban growth such as in urban slums and squatters across the global south, as well as the decay of urban centers, and cycles of boom and bust in the suburban development across the global north. Urban regeneration is the attempt to reverse urban decline by improving the physical structure and, more importantly, the economy of those urban areas (Rosentraub, 2015). There are many faces to urban regeneration, but generally it is successful only when a holistic concept of sustainability is integrated into the approach and applications of urban renewal (Boyle et al., 2018; Zheng et al., 2016).

In this book, urban regeneration is approached from a design-led and management perspective. The book provides many examples connecting the urban physical environment to the social, cultural, and economic conditions. It aims to improve our understanding of the challenges of urban regeneration in different contexts and contribute to enhancing urban regeneration programs. Chapters in this book illuminate the experiences of different localities facing current challenges due to their specific characteristics in geographic, historical legacies, environmental, economic, social, and cultural conditions. Sharing local interventions and responses to global issues such as climate change, urbanization, pollution will support the development of a connected global-regional-local understanding of sustainable urban regeneration. This book is divided into three parts. The first part focuses on urban regeneration from a design-led and management perspective. The second part introduces strategies to improve urban performance and the third part focuses on understanding the challenges of urban regeneration.

In detail, the first part addresses questions of "Design-Led Urban Regeneration and Management" and contains 7 chapters. The first four chapters present different approaches to urban design and regeneration through a series of local case studies. Chapter "Reconstruction of Resettlement Community from the Perspective of Rural Memory—A Case Study of Chun Xin Yuan in Shanghai" uses the theory of rural memory to propose strategies of community renewal and transformation for resettlement communities in China that are the product of rapid urbanization. Chapter "Temporary Urbanism and Community Engagement: A Case Study of Piazza Scaravilli in Bologna" explores the concept of temporary urbanism and the design of temporary interventions in unused urban spaces. Using a temporary installation at Piazza Scaravilli in Bologna, Italy, the authors examine its social impact through spatial, temporal, economic, and cultural contexts, and the possible solutions that temporary urbanism can provide for unused spaces. Chapter "Urban Circularity: City Planning Perspectives from the Regeneration of Amsterdam's Buiksloterham District" connects urban planning and design with the concept of the circular economy. Examining one of Amsterdam's post-industrial districts that is undergoing a transformation into a living example of a Circular City, the authors illuminate the intrinsic links between the organisation of urban resource flows and the production of space, and demonstrate how integrating circular economy principles into city planning requires behavioural adaptation and new technological systems. Chapter "Reinventing Bilbao, the Story of the Bilbao Effect" describes the urban regeneration effort of the city of Bilbao, Spain. The authors describe the key elements of the city's successful regeneration plan including a system-wide urban planning strategy, transportation system redesign, key infrastructure upgrades, new housing and affordable housing programs, and extensive job training programs in partnership with universities and local workers' unions, built upon its long history of technological innovation. The city leveraged many aspects of its local culture, including social resilience and the determination of Basque people.

While the first four chapters offer insights of urban redevelopment gained from observation and research of individual places, the latter three in this part address more general concepts of urban redevelopment and at a larger geographic scale. Chapter "The Waterfront Development in Europe: Between Planning and Urban Design Sustainability" centers on the redesign and regeneration of urban waterfronts across Europe. The authors use a collection of case studies of waterfront redevelopment of different types - from one-off sustainability projects to overhaul of urban transportation plans - to demonstrate that sustainable urban policies play a key role in regeneration, renewal, and innovation of urban natural infrastructure. Chapter "Transit-Oriented Development as a Tool of Urban Transformation Addressed at Urban Regeneration Processes" investigates the concept of mixed land use as a method to achieve vibrant, well connected, and sustainable urban patterns, and the relationship between transit-oriented development and regeneration through mixed land use development. Evaluating the European program "City Life", the authors use a multi-method approach to assess important aspects of mixed-use design and propose ways to enhance aspects of mixed-use urban redevelopment. In Chapter "The Concept of "Smart Density Planning" Principles for Livable and Sustainable Urban Transformation", the authors contrast the concept of "smart city" with a proposal for "smart density planning" to rethink urban transformations in terms of socio-spatial quality of life for inhabitants. Using the case of redevelopment of squatter developments in Turkey, the authors develop a framework of principles and indicators to assess "smart density planning" to support sustainable and livable urban transformation implementations.

The second part discusses potential strategies for urban performance improvement; thus, it has been given the title "Strategies to Improve Urban Performance" more generally, and consists of 8 chapters. The chapters in this part present several approaches to evaluating urban performance demonstrated through a series of local case studies in different contexts. Chapter "Livable Streets Determinants in Egypt: A Study on Streets' Physical Attributes in New Urban Communities" addresses the most integral parts of urban landscape and transit infrastructure within cities, namely streets. Indeed, they are widely known with their significance as a medium for urban transit and commute as well as part of the public realm where people usually gather to fulfill their social needs and aspirations. This chapter examines people's understanding of the physical attributes of Egypt's streetscapes by reviewing literature from 1980 to 2020 urban space architecture with the aim to explore how the provision, maintenance, and cleaning of basic facilities such as paving, roads, parking spaces, and traffic lights may affect streets' livability. Chapter "Using Cool Coating for Pavements, Asphalt, Façades and Building Roofs in the Urban Environment to Reduce the Summer Urban Heat Effect in Giza Square, Egypt" attempts to mitigate the urban heat island (UHI) effect using an elementary sustainable thermal-insulation panel from waste resources in informal settlements of Cairo, Egypt. The majority of housing projects in hot arid regions display very poor insulation due to building materials and low heat capacities, thus significantly increasing the UHI effect unless strategies against overheating in summer and overcooling in winter are developed.

Chapter "Towards Sustainability in Resettlement Plans, the Necessary Conditions and Their Interplay: The Case of Mumbai" describes the involuntary resettlement in development and infrastructure projects in Mumbai, India, while attempting to establish definite terms for sustainability of a resettlement plan where there are no well-defined units and analytical methods for impacts of resettlement on the affected families. The developing Mumbai city with fast-track involuntary resettlements is selected as a case study to demonstrate some of these issues that drastically uproot lives. Chapter "Urban Morphology, Environmental Performance and Energy Use: Holistic Transformation of Porto di Mare as Eco-District Via IMM" presents the process of local optimisation of Porto di Mare Eco-District masterplan in Milan, Italy through morphological and typological parameters to simulate alternative design scenarios and evaluate their performance using the visual programming interface of a Building Information Modeling (BIM) software. The main performance aspects considered are thermal loads of buildings, outdoor comfort, and energy use intensity. These challenges are addressed as part of this chapter with a strategy for the exploration of alternatives using existing energy modelling tools to show a replicable approach for similar problems and contexts.

Chapter "Can Skopje Museums Regenerate the Social and Urban Sustainability, Through "Social Friday Activity"?" focuses on the possibilities of revitalizing museums through social activities where the social, cultural, historical, arts and crafts, distinctive transformation, and occupation potential of the museums enable the variety of activities working as a catalyst for sustainable regeneration. By categorizing and prioritizing problems, based on opportunities and means of interventions, a way to globalization exists in a context as a linear process of local and global development, and interaction in every sphere raising awareness of "the back to the local community" creating sociality and urban sustainability. Chapter "Elementary Waste Insulation Panels in Hot Arid Regions" investigates whether locally available waste materials could be used to build reliable and replicable exterior retrofit insulation panels with focus on the panels' low thermal conductivity capabilities, cost effectiveness, ease of construction, buildability and acceptability by the locals, and capacity to fulfill a circular economy model.

Chapter "Comparative Review of Different Rating Systems Approach and Responses to Pandemic Situations" presents a comparative review of different rating systems' responses and approaches to pandemic situations to spot the shortcomings of the current sustainability rating systems in terms of protection against pandemics in the built environment. Six of the widely used rating systems have been studied to show the common aspects of a sustainable built environment and the credits/criteria aligned with COVID-19 protection precautions. Chapter "Multiplying Effects of Urban Innovation Districts. Geospatial Analysis Framework for Evaluating Innovation Performance Within Urban Environments" presents a novel database and analytical methodology to measure and describe the nonlinear benefits of geographic aggregation of knowledge-intensive activities within urban environments. The authors applied to the study of 50 notable innovation districts to benchmark them against a baseline of all districts in the United States. The analytical framework can then be applied to any geographical area to evaluate the economic performance of knowledge-intensive activities within urban environments. The work expands on general knowledge of how cities operate as complex systems and how they shape the collective knowhow of urban communities.

The third part, "Understanding Challenges of Urban Regeneration", consists of 8 chapters focusing on urban sustainability challenges in different contexts. In this part, we included many case studies at different scales - from individual spaces, e.g., cemeteries, brownfields and university campuses, to neighborhoods, cities, and countries. These case studies touch on the relationship between urban environmental, social, and ecological challenges and the planning and design processes.

In Chapter "Examination of the Population Density Impact on Major Air Pollutants: A Study in the Case of Germany", the authors examine the distribution of NO_2 and SO_2 air pollutants at the country-wide scale using Sentinel 5 remote images. They compute the relationship between population density and air pollution concentrations in Germany. Chapter

"A Comparison Between Italian and French Case Studies on Urban Regeneration" explores urban regeneration of some of the most problematic sites in urban areas, namely the brownfields. The chapter examines recently completed restoration projects in two types of brownfields in Italy and France. Using a socio-economic framework, the chapter highlights issues and strategies for sustainable urban regeneration and puts these spaces of decline at the centre of the future city. Chapter "Relevance of Urban Ecosystem Services for Sustaining Urban Ecology in Cities-A Case Study of Ahmedabad City" investigates the challenges related to urban expansion in the case of the rapidly growing metropolitan city of Ahmedabad, India. In this chapter authors connect long term survey and satellite imagery to determine the relationship between ecological concern, environmental degradation, and land use and cover change over a temporal scale of 25 years. Therefore, they formulate suggestions for a future policy framework for a city development plan. Chapter "Increase in Neighborhood Development and Urban Water Erosion in Tropical Cities from Central Brazil: The Case of Goiânia, Goiás" describes the interaction between water erosion processes and urban planning or lack thereof. Using the example of Goiânia, Goiás in Brazil, the authors demonstrate how lack of urban planning allowed for a social-environmental collision between the natural process of water-driven urban erosion and uncontrolled urban settlement. Chapter "The Dimension of Urban Morphology on Placemaking Theory: A Comparative Typomorphological Analysis of a Self-built Neighbourhood and Its Regeneration Project in Ankara" focuses on the morphology and typology of urban informal settlement in one neighborhood in Ankara, Turkey. The authors argue that self-organizing urbanity has distinguishable and recognizable design quality and the destruction of existing spatial patterns through regeneration projects can result in the loss of urban identity and environmental quality.

The final three chapters in the book address specific spaces within the urban fabric of unique importance and potential contribution in processes of urban regeneration. Chapter "Research on the Relationship Between Informal Learning and the Regeneration of Public Spaces on Chinese University Campuses: Taking Xian Jiaotong-Liverpool University as an Example" explores the relationship between informal learning in higher education and the use of public space to support and enhance the wellbeing of university students. The authors suggest that regeneration of public space can be important to different forms of informal learning. Chapter "Flora of Archaeological Landscape: Case Study of Arslantepe Mound and Its Territory" plays a key role in understanding archaeological landscapes. Floristic studies provide important information for archaeological research, for understanding the diet of societies and for determining agricultural activities of the periods. With the data obtained as a result of archaeobotanical studies in recent years, important findings regarding the general flora character of archaeological landscapes have been reached. In Chapter "Disused Urban Cemeteries: Unearthing User Experiences of Abney Park Cemetery", the authors similarly argue that disused urban cemeteries are an important and understudied urban green and open space. Using London's Abney Park Cemetery, UK, as case study, the authors assess the perception of users and find it highly valued as a place of respite and recreation. In all of these cases, deeper and more direct integration with urban planning for regeneration is needed.

The topics addressed in this book are interrelated and contribute to the understanding of urban regeneration and performance. Taken together, the chapters in this book suggest important relationships between urban planning and design, and success of sustainable urban regeneration programs, as follows:

– Proposing strategies for sustainable transformation of resettled communities for rapid urbanization. Multiple chapters emphasize and demonstrate the importance of the concept of temporary urbanism and the design of temporary interventions in unused urban spaces for enhancing social impact through spatial, temporal, economic and cultural contexts. Temporary urbanism offers agile and innovative urban interventions that can connect urban planning and design with different aspects of urban sustainability through experimentation and ongoing learning.

- The process of local optimization-master planning with morphological and typological parameters to simulate alternative design scenarios and evaluate urban performance using the visual programming interface of building information modelling.
- There are many urban spaces that provide opportunities for recreation, respite, and support to human wellbeing that are "hidden" or reside outside the scope of what is usually defined as urban green and open spaces, but still serve critical functions for urban residents. Many chapters in this book address such areas - unused urban cemeteries, abandoned docks and industrial areas, urban archeological sites. Conceptualizing such sites as part of the urban green space network, and developing specific interventions for temporary or long-term public use of such sites offers ways to expand sustainable urban land uses while preserving important social and cultural spaces.
- Aspects of urban sustainability are inherently linked and required for successful urban regeneration. When developing strategies for future cities, regeneration strategies cannot ignore social, economic, and environmental aspects of urban development. Generating more coherent frameworks for urban planning and design work on urban regeneration will allow for careful consideration of the sustainability of programs and interventions.
- Urban expansion in rapidly growing metropolitan areas, particularly in the developing world, presents unique challenges for urban regeneration that require careful and nuanced strategies. Urban informal settlements house some of the most vulnerable human populations on earth. Their physical structure and infrastructure hold an inherent logic that allows for a delicate balance of complex urban life. Interventions in these spaces should take into account the needs and perceptions of residents and attentively preserve the spaces and networks that provide structural support for communities in these spaces.
- Connecting long term survey and satellite imagery to determine the relationship between ecological concern, environmental degradation, land use and land cover change through modelling for a future policy framework for a city development and planning.

Different studies compiled in this book provide references for developing innovative strategies and solutions for urban performance and regeneration in different contexts around the world. Besides, this book could become a milestone for building better future and livable cities.

Kyrenia, Cyprus   Haşim Altan
Villanova, USA   Peleg Kremer

References

Boyle, L., Michell, K., & Viruly, F. (2018). A critique of the application of neighborhood sustainability assessment tools in urban regeneration. *Sustainability, 10*(4), 1005.

Han, L., Zhou, W., Li, W., & Qian, Y. (2018). Urbanization strategy and environmental changes: An insight with relationship between population change and fine particulate pollution. *Science of the total environment, 642*, 789–799.

Jacobs, J. (1961). The death and life of great American cities. Vintage.

Kuddus, M. A., Tynan, E., & McBryde, E. (2020). Urbanization: a problem for the rich and the poor?. *Public health reviews, 41*(1), 1–4.

Liang, W., & Yang, M. (2019). Urbanization, economic growth and environmental pollution: Evidence from China. *Sustainable Computing: Informatics and Systems, 21*, 1–9.

Liddle, B. (2017). Urbanization and inequality/poverty. *Urban Science, 1*(4), 35.

McMichael, A. J. (2000). The urban environment and health in a world of increasing globalization: issues for developing countries. *Bulletin of the World Health Organization, 78*, 1117–1126.

Rosentraub, M. S. (2015). Reversing Urban Decline, Why and How Sports, Entertainment, and Culture Turn Cities into Major League Winners, 2nd Edition, Routledge.

Zheng, H. W., Shen, G. Q. P., Song, Y., Sun, B., Hong, J., (2016). Neighborhood Sustainability in Urban Renewal: An Assessment Framework. *Environment and Planning B: Urban Analytics and City Science*, https://doi.org/10.1177/0265813516655547

Rodriguez, A., Juaristi, J. (2015). Transforming Cities: Opportunities and Challenges of Urban Regeneration in the Basque Country, Center for Basque Studies.

Seto, K. C., Parnell, S., & Elmqvist, T. (2013). A global outlook on urbanization. In Urbanization, biodiversity and ecosystem services: Challenges and opportunities (pp. 1–12). Dordrecht: Springer.

Sohail, M., Cavill, S., & Cotton, A. P. (2005). Sustainable operation and maintenance of urban infrastructure: Myth or reality?. *Journal of urban planning and development, 131*(1), 39–49.

Srivastava, R. K. (2020). Urbanization-Led Neo Risks and Vulnerabilities: A New Challenge. In Managing Urbanization, Climate Change and Disasters in South Asia (pp. 251–296). Singapore: Springer.

Acknowledgments

We would like to thank the authors of the research papers that were selected for addition in this book. We would also like to thank the reviewers who contributed with their knowledge and constructive feedback in hopes of ensuring the manuscript is of the best quality possible. A special thanks goes to the editors of this book for their foresight in organizing this volume and diligence in doing a professional job in editing it. Finally, we would like to express our appreciation to the IEREK team for supporting the publication of the best research papers submitted to the conference.

Contents

Design-Led Urban Regeneration and Management

Reconstruction of Resettlement Community from the Perspective of Rural Memory—A Case Study of Chun Xin Yuan in Shanghai

Xi Chen, Yan Hua, and Jun Wang

Abstract

Resettlement community is a kind of immigrant community which is a product of rapid urbanization in China. In these resettlement communities, although landless farmers live in the city, they lack community identity and sense of belonging, and still retain the rural living habits. Rural memory is the collective memory of rural culture and living habits. From the perspective of rural memory theory, this paper discusses the current situation and problems of resettlement communities in China, and the rural memory problems that need to be considered in the renewal and transformation of such communities. This paper takes Chun Xin Yuan community in Shanghai as the research object, observes the living habits and behaviors of the residents, and understands their demands on community environment and living facilities through field research and resident interviews. Then combined with the inheritance characteristics of rural memory, this paper puts forward the strategies of community renewal and transformation from four aspects on community space, life style, community management, and social culture. Through such renewal and transformation, the landless farmers can be integrated into urban life faster, and at the same time, the rural memory can be inherited.

Keywords

Rural memory • Resettlement community • Community renewal

1 Introduction and Basic Concepts

Since the reform and opening up, the speed of urbanization in China continues to accelerate. With the expansion of cities and large-scale land acquisition in rural areas, more and more farmers begin to live in various resettlement communities. Although the resettlement community provides living quarters for these farmers, the new environment still brings many problems to their life. When landless peasants enter the city, their original living habits and behavior will not continue. The change of living space will lead to the change or even break of their social communication circle. The scope of social communication will become fragmented because the original village residents are scattered into the new urban area (Wei, 2017). These memories from the countryside make them lack the sense of identity and hard to belong to the urban community.

In this context, exploring how to inject rural memory into the renewal and transformation of resettlement communities reflects the humanistic care for the relocated people. How to promote neighborhood relations, enhance the humanistic value of communities, and enhance the sense of community identity and belonging is also the key to the sustainable development of such communities. From the perspective of rural memory, this paper takes the renovation of Chun Xin Yuan in Shanghai as an example, analyzes the material and spiritual demands of landless farmers, explores the practical strategies of community renovation from the aspects of community space, life, management, and culture, so as to make them truly integrate into urban society.

X. Chen
Urban Planner, Shanghai Greenland Group, No.700, Da Pu Rd, Shanghai, China
e-mail: maigyscity@163.com

Y. Hua
East China University of Science and Technology, Shanghai, China
e-mail: 1303450502@qq.com

J. Wang (✉)
School of Art Design and Media, East China University of Science and Technology, No.130 MeiLong Rd, Shanghai, China
e-mail: denief@163.com

© The Author(s), under exclusive license to Springer Nature Switzerland AG 2022
C. Piselli et al. (eds.), *Innovating Strategies and Solutions for Urban Performance and Regeneration*,
Advances in Science, Technology & Innovation, https://doi.org/10.1007/978-3-030-98187-7_1

1.1 Rural Memory

Collective memory is a concept in the field of social psychology, which is different from personal memory (Morris, 2002). According to Jan Asman, "collective memory is something that people share, inherit and construct together in a group in modern society" (Yang, 2015). Rural memory is regarded as a specific regional memory of collective memory, which records the development and change of the whole village in the process of historical process. It is not only the common memory of villagers but also the link connecting the spiritual world and emotional world of villagers. There are various forms of expression of rural memory, including material ones, such as houses, temples, pictures, etc., and non-material ones, such as ballads, operas, crafts, technologies, etc. These elements all bear a certain rural memory. After generations of accumulation and inheritance, they form a kind of beautiful collective memory.

1.2 Resettlement Community

<The Land Administration Law of the People's Republic of China> (revised in 2004) stipulates that in case of land acquisition, compensation shall be given according to the original use of the land. In the process of urbanization, a large amount of land is needed for urban construction, the government collects farmers' fields, pays land compensation fees, and plans and builds communities for them. Such communities are resettlement communities (Zhang, 2019a). The resettlement community can be divided into rural resettlement community and urban resettlement community. This paper studies urban resettlement community, that is, the product of land acquisition caused by urban construction activities in the process of urbanization, with landless farmers as the main living body (Liu, 2017).

1.3 Community Renewal

In recent years, urban development has changed from incremental planning and construction to urban renewal and transformation, and the traditional urban renewal mode has also concentrated on small-scale incremental renewal. Community renewal is a small-scale way of mending and rejuvenating, which can bring vitality to the community. Community renewal mainly renews and reuses the public space, building facade and landscape, improves the community function, promotes the culture and the sustainable development of the community (Li, 2018). In <The Urban and Rural Planning Act of 1947>, Britain proposed public participation in urban planning. After the evolution of the times and the development of economic and social diversity, public participation gradually plays an important role in urban renewal. As the community involves many stakeholders, if it wants to get a good development, the public participation can not be separated (Lin, 2018).

2 Background

2.1 Research Summary

Most of the researches on new citizen groups abroad focus on the field of migration, and management of the resettlement community. From the existing research, mainly from the development policy, laws and regulations, spatial distribution, needs assessment, and so on. The United States is a multi-ethnic country composed of immigrants and their descendants. At the beginning of the twentieth century, the middle class sent immigrants to the community service center (or "resettlement center") which promoted the resettlement movement in the United States. It provided a series of help and services for immigrants, and promoted the social reform of American cities (Yang 2006); Singapore is also an immigration country which is distributed in different regions. Singapore government effectively integrates immigrants in space mainly through public housing policy (Sim et al., 2003); German refugee problem is an obvious long-term, complexity, and international problem. German government has issued different immigration policies and response measures at different stages, and finally accepts refugees in a completely open manner. Efforts have been made to rebuild and expand the old barracks, abandoned factories and other infrastructure to accommodate more refugees, providing high-level resettlement compensation for them (Wang, 2019).

Domestic research is mainly in the field of management and sociology, to explore the governance model of resettlement community and social integration of landless farmers. Yang Ying put forward the countermeasures to improve the community governance of landless farmers from three dimensions: the organizational system, the main cultivation and the foundation (Yang, 2012); Lu Ying gave countermeasures and feasible suggestions to the community public space governance based on the problems (Lu, 2017); Qian Quan clarified the "state city" in the social governance through the research on the practical experience of resettlement community, and the triple structure of "field society" reveals the multiple aspects of the modernization transformation of national grass-roots governance (Shan, 2019); Wei Lingqun used the professional methods of social work to deeply analyze the problems encountered by landless farmers, and put forward practical paths to solve the problem of urban integration of them in combination with his own experience (Wei, 2017); Chen Rui analyzed the problems existing in community governance, proposing to build a quasi-autonomous model under the multi-participation pattern, and explored countermeasures from the aspects of community social capital, community development vitality, improvement of community governance, and broad participation of residents, so as to further improve community governance (Zhang, 2019b).

Secondly, in terms of the culture of the resettlement community, researchers mainly discussed the cultural differences between urban and rural areas and the causes of cultural conflicts and solutions. Cheng Yu started with two kinds of culture in urban and rural areas, analyzeing the differences and conflicts between urban and rural cultures, and puts forward countermeasures for cultural integration from the aspects of life style, psychology, and values (Cheng, 2012); Shan Jing took the resettlement community in Chongqing University City as an example, studying the cultural conflicts in the resettlement community (Shan, 2019); Jiang Yan proposed that the resettlement community must gradually return to community construction, cultivate and improve the citizenization of landless farmers, and promote the conversion of landless farmers to citizens (Jiang, 2012); Li He summarized the cultural adaptation of landless farmers and analyzed the main reasons for the cultural adaptation of them from the perspective of Cultural Anthropology (Li, 2014).

2.2 The Disappearance of Rural Memory

The rapid development of Chinese urbanization has led to the growing gap between urban and rural areas, forming a dual economic structure. Comparative labor productivity,

binary comparative coefficient, and binary contrast coefficient are the important indicators to measure the strength of urban–rural dual economic structure.

The binary comparative coefficient is the ratio of agricultural comparative labor productivity to non-agricultural comparative labor productivity (Li, 2011). Suppose the binary comparative coefficient is R1, the binary contrast coefficient is R2, the proportion of agricultural added value is G1, the proportion of non-agricultural added value is G2, the proportion of agricultural employment is E1, and the proportion of non-agricultural employment is E2. According to formula: $R1 = \frac{G1/E1}{G2/E2}$;

$$R2 = \frac{/G1 - E1/ + /G2 - E2/}{2}$$

and use the data in <Chinese Statistical Yearbook (1978–2017)> . The closer the binary comparative coefficient is to 0, on the contrary, the closer the binary contrast coefficient is to 1, the stronger the binary economic structure is. The binary comparative coefficient of developing countries is generally 0.31–0.45, and that of developed countries is generally 0.52–0.86 (Li, 2011). However, Chinese urban–rural dual comparative coefficient fluctuates in the range of 0.14–0.26, indicating that Chinese urban–rural gap is large (Fig. 1).

In China, there are economic and cultural gap between urban and rural. Urban culture pays attention to real life, higher openness, and effective management, while rural culture has the characteristics of simplicity, local, closed, and conservative (Cheng, 2012). When landless farmers enter the city, it means that their culture has changed. Under

Fig. 1 The strengthen of dual economic structure of urban–rural in China

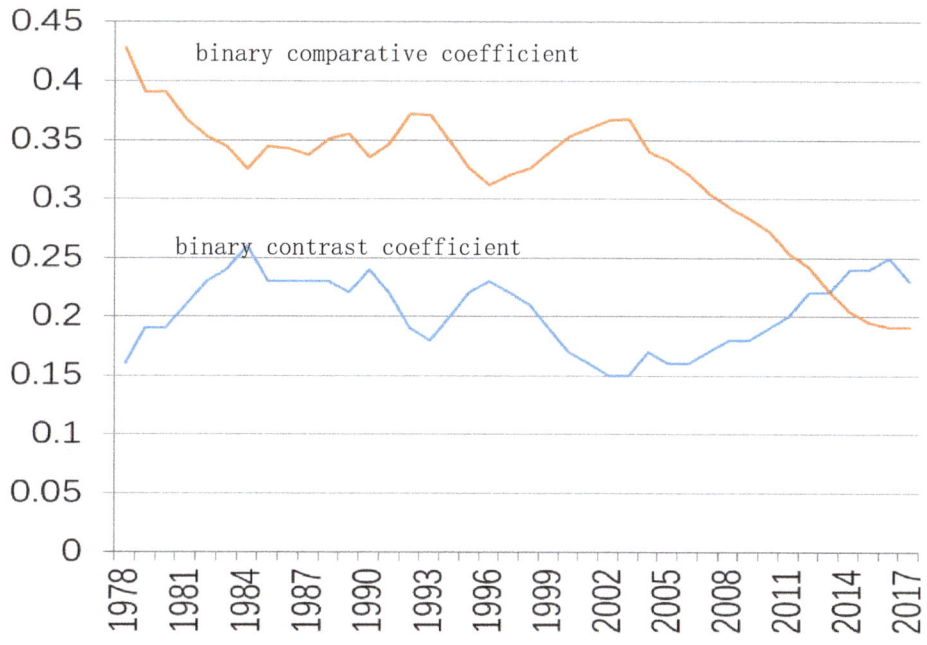

the strong collision of the inherent rural culture and the emerging urban culture, their living habits and behavior will face great changes. The new culture regulates their behavior, and also begins to change their original ideas, to match and adapt to urban life, and finally to adapt to urban culture. Cultural adaptation is a process in which different cultures change their original nature and patterns through long-term contact, connection, and adjustment. It is a new comprehensive process. In the process of cultural adaptation, in the conflict between rural culture and urban culture, the rural memory of landless farmers is constantly disappeared.

3 Renewal Practice of Chun Xin Yuan in Shanghai

3.1 Overview of Chun Xin Yuan

Chun Xin Yuan is located in Minhang District, Shanghai. It was the outcome of the expansion of Shanghai central city in 2000. The original Ji Xin village was expropriated and the resettlement community is constructed nearby. Chun Xin Yuan is divided into two phases, with a total of 680 households, of which 540 are farmers' relocated households, with a total number of about 2000 residents. There is an public activity room for the elders and a community club, as well as a fitness venue and central green space. With the urbanization of the surrounding areas, various types of commercial residential projects have been built around, and a series of public service facilities such as supermarkets, kindergartens, primary schools, public transportation, and subway stations are gathered within the 15-min walking range. The location of Chun Xin Yuan is becoming more and more superior.

3.2 The Path of Renewal and Transformation of Chun Xin Yuan

3.2.1 Residents' Needs

In order to understand the needs of community residents, we interviewed 106 residents of different age, about 5% of all residents. Among them, 62% are aged 60–80, 23% are aged 40–60, 10% are aged 20–40, and 5% are under 20. After the interview, we mainly got the following feedback from the residents.

Firstly, in terms of residential space, people of different ages put forward different needs. Among them, most of the elderly still retain their original rural living habits. They like to sit at the entrance of the corridor during the day, enjoy the cool in summer and the sun in winter. At the same time, they can choose vegetables, read books, chat, and so on. Therefore, the elderly hope to set up more spaces in the residential

area that can allow people to sit down and rest. These spaces can also block the wind and rain as a place for the elderly to communicate. Middle-aged people are more concerned about parking, greening, and outdoor drying space. Among them, due to the low proportion of parking spaces, the parking contradiction is very prominent, and the phenomenon of parking occupying sidewalks, green belts, and public spaces is very common. The problem of drying in public space has also been mentioned more. Many residents in Chun Xin Yuan maintain the habit of drying clothes in rural areas in outdoor venues so that on good days, the trees, shrubs, and fitness equipment in the residential area are covered with quilts. For young people, the biggest problem is the lack of activity space and facilities, and the lack of cultural exhibition space.

Secondly, it is the problem of community integration. Since most of the residents in the resettlement community are villagers from the same village, they are familiar with each other. With the passage of time, a lot of new residents have moved into the community, which makes the community form a different circle of communication. In some matters related to the interests of residents, the voices of different circles are not same, which leads to the disharmony of the community.

Third, the confusion of community management. Like most urban communities, Chun Xin Yuan is jointly managed by the Residents' Committee and the House Owners' committee, but there is also a Villagers' Committee in such resettlement community. This will lead to the problem of multi-management. The functions of the three parties are not clear enough, and the authority and scope of management are also crossed, which also leads to the prevarication and wrangle of some things.

Fourth, the demand for community facilities. Because the buildings in the community are multi-story, there is no elevator installed at the beginning. At present, the majority of the residents are the elderly, they need elevators and also need open space to do some fitting.

3.2.2 The Way to Implant Rural Memory

The common rural memory makes the villagers more closely connected and enhances the cohesion and identity of the villagers (Zhang, 2017). From the survey, most of the demands has an obvious rural mark, residents of Chun Xin Yuan have a common rural memory. Therefore, in the renovation of the community, we should not only solve the problems raised by the residents but also implant rural memory in these solutions. Rural memory exists in the behavior and life of villagers. At the same time, this kind of rural memory does not exist alone, but penetrates into all aspects of community renewal and transformation.

In terms of space improvement, firstly, the fragmented space in the community will be integrated and transformed into fitness and rehabilitation square, vestibule and backyard, drying flat, own vegetable garden, rest oxygen bar, etc. And then a series of paths will be integrated and connected to form a leisure Ring Road within the community. Secondly, optimize the parking system, control, and guide the parking of bicycles. And improve the road space environment, enhance residents' sense of security. Finally, the utilization rate of public space should be considered for the elders.

In the aspect of community life, the design should increase the function of matching the needs of all kinds of people and forms a diversified and complex functional layout structure. In order to solve the problem that it is difficult for residents to buy vegetables, a fixed-point market can be set up in the open space between Chun Xin Yuan and Lian Hua New Village. Residents can be employed to clean up and manage the market environment. Considering that the residents like to chat at the corridor, on the premise of not affecting the traffic, use the space in front of the house to set up seats and sunshade umbrella for the residents to rest, show the original living habits, and reshape and continue the rural culture.

In terms of management, it needs to clarify the position of different committees, ensure the transparency and accessibility of management, and improve residents' participation in community affairs.

In terms of facilities, the community should popularize community culture and education, hold some cultural activities, convey the civilized rules of urbanization, garbage classification, environmental protection, civic morality, etc. The cultural activities organized by the resettlement community can not only enhance the public communication of the new community residents, promote the construction of the residents' social relationship network, but also help to continue the collective memory (Jiang, 2012). To hold traditional rural cultural activities helps to create landscape sketches with residents' rural memory, to enhance residents' sense of community belonging.

4 The Practical Strategy of the Resettlement Community Renewal from the Perspective of Rural Memory

Landless farmers miss rural life, which does not mean that they want to return to their previous life. The renewal and transformation of the resettlement community should be different from the general community, not limited to the planning of space and landscape, but from the space level, life level, management level, and cultural level, integrate into the rural memory, and reshape the identity and sense of belonging to the community. From the practice of the renewal of Chun Xin Yuan, we can sort out the renewal

strategies that can be adopted by the resettlement community in these aspects.

4.1 Community Space

The space concept of ancient Chinese is a kind of psychological view, which is a high level concept about the universe, nature, society, and life (Bai, 2009). Space is closely related to nature, society, and people. Community space is not only the place of community production and life but also community culture, spirit, and environment.

The first is how to increase public space. On the one hand, we should respect the living habits of landless farmers, such as stacking farm tools and sundries, like to gather and chat in the corridor. At the same time, according to the behavior activities and transformation needs of the elderly, children and other different groups, the fragmented space should be fully integrated to provide residents with public space to meet various functions. At the same time, landscape sculpture, landscape sketch, and landscape wall are put in to sketch the rural impression and continue the rural memory.

Secondly, the improvement of traffic space. For parking problems, we can try to adopt time-sharing parking management; or after consultation with the urban traffic management department, we can designate night parking spaces on both sides of the peripheral roads for the residents; third, we can build three-dimensional parking facilities in the community to increase the supply of parking spaces. At the same time, we will open up a centralized parking area for mopeds and bicycles, and assign the management staff.

The third aspect of space improvement is the greening landscape. The community tree species should be the local tree species that are easy to survive, especially some fruit trees, such as orange tree, fig, persimmon tree, to continue the memory of the village. In the community, there are also some spaces for residents to rest and exchange, such as the small open space in front of the house and behind the house, the small pavilion beside the fitness facilities, etc., which can let residents sit and chat, read, and rest, and reproduce typical rural life scenes.

4.2 Community Life

In <Communication and Space>, Jan Gail divides activities into three types: necessity activities, selective activities, and social activities. He pointed out that "social activities are passive that depend on the participation of people (Jan, 2002)". For community residents, social activities are inevitable and the most extensive in life. To build a vibrant community, residents must have more activities and rich experience.

In the <Humanized City>, Jan Gail put forward that "attractive activities largely determine whether urban space has vitality and vitality (Jan, 2010)". This requires the community to give people more possibilities for communication, create a platform, and enhance the vitality of the community. The community can hold all kinds of collective activities regularly, such as traditional festival activities, handmade activities, cultural lectures, etc., so that the residents can understand each other and get closer in the activities.

In the countryside, the settlement style residence makes the communication between villagers very convenient, and the neighborhood relationship naturally formed. Now, the residential form of the building has hindered the communication of residents (Jan, 2002). This requires building a proper size and well-designed public space in the community to enhance and maintain neighborhood relations. For example, the rest area at the entrance of the corridor and the community vegetable field.

Then the facilities for the elderly, children, and other specific groups is in need. In particular, the elderly resettlement community, elevator, barrier-free passage, elderly fitness facilities, elderly activity room, daycare room and other facilities, can increase the community care for the elderly. Similarly, special consideration should be given to children's playground, day nursery, and other facilities.

4.3 Community Management

The residents in resettlement community are complex and has various backgrounds. Community management needs public participation, innovative mechanism, and joint construction of sustainable community.

First of all, it needs to clarify the responsibilities and obligations of management subjects. For example, there are usually village committee and neighborhood committee in the resettlement community, so rights and responsibilities of these two subjects needs to be clarified. The village committee is about to continue the functions under the original rural system, such as the issuance of resettlement subsidies, medical reimbursement, etc.; the neighborhood committee should regularly carry out collective activities, strengthen the contact and interaction between residents. The neighborhood committee also could reshape the "acquaintance society", build the trust foundation, and strengthen the residents' recognition, trust and support (Song et al., 2013).

4.4 Community Culture

The original human value system of rural areas for the relocation of farmers has collapsed, while the formation of urban community order under the modern pattern is not a one-day success (Zhao, 2019). For example, airing quilts on the square fitness equipment, planting vegetables on the community green space, accumulating sundries in the corridor, etc. These are the living habits left by the rural culture, which have certain conflicts with the modern urban culture. We can not blindly pursue urban culture and use it to regulate and restrict the behavior of landless farmers, but we should keep good rural culture and integrate rural culture with urban culture.

First of all, we should form a new cultural identity on the basis of continuing the rural memory. Community organizations build cultural lecture halls and places for cultural activities, provide exhibition places for rural traditional culture, and enhance residents' sense of identity and belonging to the community. Only when human development is realized, can the sustainable prosperity of culture and the continuous progress of community be realized (Huang & Zhou, 2018).

Then, it is to explore the rural culture that can be inherited. To explore the talented people in the original countryside, such as craftsmen, craftsmen, old cadres, etc., and organize them to carry out professional training for community residents so that the traditional non-material arts and crafts can continue. At the same time, make these community talents become the business card of the community, and enhance the residents' community pride.

5 Conclusion

With the acceleration of China's urbanization process, a large number of villages are disappearing, and more and more farmers live in resettlement settlements. Resettlement settlement is a special community, which not only covers the basic elements of modern urban community but also has the characteristics of some rural communities. This paper analyzes the causes and characteristics of resettlement settlements, discusses their existing difficulties, and puts forward the strategy of introducing rural memory into the renewal and transformation of resettlement settlements. Specifically, taking Shanghai Chun Xin Yuan as the research object, through the investigation of the residential area and the interview with residents, this paper summarizes the problems faced by Chun Xin Yuan at present, and puts forward the specific strategies for the renewal and transformation of Chun Xin Yuan residential area and urbanization transformation according to the needs of residents.

In terms of space, it is proposed to improve the space environment from the aspects of public space, traffic space, and landscape greening, create a human and attractive residential area, and enhance the residents' sense of community identity and belonging. In terms of facilities, respect residents' willingness to continue rural production and lifestyle, increase places for neighborhood communication, set up residential vegetable gardens, care for the elderly, children and other special groups, and provide humanized design. In terms of management, on the basis of clarifying the responsibilities and obligations of different management subjects, advocate the establishment of various community autonomous organizations, take residents as the main body, innovate the mode of co construction, CO governance and sharing, and realize the modernization of community governance. In terms of culture, we should continue and inherit rural traditional culture, respect, and introduce the guiding role of rural talents and talents in the community, cultivate and resettle the community's own characteristic culture, alleviate the conflict between urban culture and rural culture, and build a new humanistic value system.

The purpose of the research on the renewal, transformation and transformation strategies of resettlement settlements is to promote community construction, especially to enable farmers who have lost their land to integrate into urban society faster and better. At the same time, it is also of great value to retain and inherit some rural traditional culture, production, and lifestyle in this process. It can enrich the cultural gene of urban society and communicate the past and future.

Acknowledgements This research is supported by China National Social Science Fund, No. 19BJY063 and The IV Peak Plateau Discipline of Shanghai Design.

References

Bai, J. (2009). *Perspective analysis of the differences between Chinese and Western Classical Gardens. School of architecture.* Master dissertation, Tianjin University. http://www.cnki.net.

Cheng, Y. (2012). *Study on cultural conflict and integration in urban and rural development.* Master dissertation, Southwest Petroleum University. https://kns.cnki.net/kcms/detail/detail.aspx?dbcode=CMFD&dbname=CMFD201301&filename=1012516246. nh&uniplatform=NZKPT&v=LEHVkx4yr0F3nUr6nuV5KRFhfKI gA8yNL33Ej56qbpu2vuc5qWYfmophrplGm0q_.

Huang, L., & Zhou, M. (2018). Study on urban community renewal strategy in the context of cultural rejuvenation. *Western Journal of Human Settlements, 33*(04), 1–7.

Jan, G. (2002). *Communication and space.* China Construction Industry Press.

Jan, G. (2010). *Humanized city.* China Construction Industry Press.

Jiang, Y. (2012). Study on the deep influence of cultural construction of resettlement community on the citizenization of landless farmers. *Journal of Yunnan University of administration, 2*(12), 115–117. https://kns.cnki.net/KNS8/Detail?sfield=fn&QueryID= 59&CurRec=1&recid=&FileName=YNXY201202030&DbName= CJFD2012&DbCode=CJFD&yx=Y&pr=CJFX2012;&URLID=53. 1134.D.20120301.1110.035.

Li, Y. (2011). China's dual economic structure: characteristics, evolution and adjustment. *Rural Economy, 9,* 11–18. https://kns. cnki.net/KNS8/Detail?sfield=fn&QueryID=64&CurRec=1&recid=

&FileName=NCJJ201109020&DbName=CJFD2011&DbCode= CJFD&yx=&pr=&URLID.

Li, H. (2014). *A study on the centralized living and cultural adaptation of the landless farmers in the suburbs: a case study of Xinglong Village, Liaohe Town, Tongliao Economic Development Zone.* Master dissertation, Inner Mongolia Normal University. https://kns. cnki.net/kcms/detail/detail.aspx?dbcode=CMFD&dbname=CMFD 201501&filename=1014348859.nh&uniplatform=NZKPT&v=1d2 IAj2hV0ltGRZqq1SoWQbOiCeybuWfpGlyHFWcPTv-tsqrjzddq YzTwquJsI9F.

Li, W. J. (2018). *Discussion on sustainable development strategy of traditional Indonesian community under mixed community governance mode—Taking tongkol dalam krapu lodan as an example.* Master dissertation, Suzhou University of Science and Technology. https://kns.cnki.net/kcms/detail/detail.aspx?dbcode=CMFD&db name=CMFD202001&filename=1019191942.nh&uniplatform= NZKPT&v=8rdYUdN8h3rJ4zT2dZuNfaNuYTjauCcewkKTfEoX 1jV5elyvBN02aW6jScMNMrXK.

Lin, X. L. (2018). *Development, effectiveness and practice of participatory planning in traditional community renewa—a case study of Shapowei, Xiamen City.* Master dissertation, Xiamen University. https://kns.cnki.net/kcms/detail/detail.aspx?dbcode= CMFD&dbname=CMFD201902&filename=1018195568.nh&uni platform=NZKPT&v=OJsrwAkPEJsxUJdtxgdKkWwLP2L79ZVJ t1pUjNwBG_mam0ZmNp6bbmYAPEK06ydX.

Liu, J. (2017). *Spatial resettlement and social integration of landless farmers in big cities: a case study of Nanjing landless farmers' resettlement area.* Master dissertation, Southeast University. https:// kns.cnki.net/kcms/detail/detail.aspx?dbcode=CMFD&dbname= CMFD201801&filename=1018037810.nh&uniplatform=NZKPT& v=_wF1AN_Zz3hfNQTOZHVThnpu09CQRg2DyxHCXaMtFG LkZJNYERuVBRvsXggbbojh.

Lu, Y. (2017). *Research on the Governance Dilemma and Countermeasures of public space in relocation oriented community—based on the survey of Wuxi communitys.* Master dissertation, Nanjing University of Technology. https://kns.cnki.net/KNS8/Detail?sfield= fn&QueryID=16&CurRec=1&FileName=1017055413. nh&DbName=CMFD201702&DbCode=CMFD.

Morris, H. (2002). *On collective memory.* Shanghai People's Publishing House.

Shan, J. (2019). *Study on cultural conflicts and Countermeasures of resettlement communities in Chongqing—a case study of resettlement communities in University City.* Master dissertation, Chongqing Normal University. https://kns.cnki.net/KNS8/Detail?sfield= fn&QueryID=41&CurRec=1&FileName=1011103479. nh&DbName=CMFD2011&DbCode=CMFD.

Sim, L. L., Yu, S. M. & Han, S. S. (2003). Public housing and ethnic integration in Singapore. *Habitat International* 27(2), 293–307. https://doi.org/10.1016/S0197-3975(02)00050-4.

Song, H., Zhang, X., & Su N. (2013). Innovation of the new type of community management from agriculture to Africa: a case study of Tongle community in Chengdu. *Theoretical Exploration, 5,* 85–89. https://kns.cnki.net/kcms/detail/detail.aspx?dbcode=CJFD&db name=CJFD2013&filename=LLTS201303020&uniplatform=NZK PT&v=8YsBDkCsnWh3d7ksvSGgPkgAPBGXSSvFoR9Zhv8c Lz31TPZ45Z4vek2_KpcceszC.

Wang, X. (2019). *A study of refugee problems in Germany.* Master dissertation, Shandong University. https://kns.cnki.net/kcms/detail/ detail.aspx?dbcode=CMFD&dbname=CMFD201902&filename= 1019138020.nh&uniplatform=NZKPT&v=z7Lij158NpNn5Q9K R6ymPuwMGxJLVhrZzgtju6823HvI8JZLoUjCZWZ9nvTWqz2z.

Wei, L. Q. (2017). *Study on urban integration of landless farmers in Community Governance: a case study of FX community in Hefei.* Master dissertation, Anhui University. https://kns.cnki.net/kcms/ detail/detail.aspx?dbcode=CMFD&dbname=CMFD201902&file

name=1019142518.nh&uniplatform=NZKPT&v=DGd1BW3l
AtvpzEBOA-COP-fJP2DyH1u_NvsSYA6zEA6AydzadvlO6Oo
HS1WuwIqx.

Yang, C. J. (2006). Immigration and urban reform in the early 20th
century in the United States. *American Studies*, 6,133-149. https://
kns.cnki.net/kcms/detail/detail.aspx?dbcode=CJFD&dbname=
CJFDLAST2017&filename=MGYJ201606009&uniplatform=
NZKPT&v=T4J69EJJnwxm6NtD7jbErUEYg70tqxmwefofO02tO
D2-LIvjMhFespEEl9ZJvMoZ.

Yang, Y. (2012). *On the governance of suburban landless farmers'
communities*. Master dissertation, Suzhou University. https://kns.
cnki.net/kcms/detail/detail.aspx?dbcode=CMFD&dbname=CMFD
201301&filename=1013120351.nh&uniplatform=NZKPT&v=iiV
Cl0_7VTzeDmOw7hmagqLiotqflyBDXfHMSWh6t1aG9fK4pi
Oipc77ux9uAiyw.

Yang, A. (2015). *Cultural memory*. Peking University Press.

Zhang, S. X. (2017). *Study on the inheritance path of rural memory in
the context of Urbanization—Taking y village as an example*.
Master dissertation, Anhui University. https://kns.cnki.net/KNS8/
Detail?sfield=fn&QueryID=51&CurRec=1&FileName=101718
5744.nh&DbName=CMFD201702&DbCode=CMFD.

Zhang, C. P. (2019a). *A study on the welfare changes of landless
farmers before and after resettlement: a case study of Huangdao
District, Qingdao*. Master dissertation, Yantai University. https://
kns.cnki.net/kcms/detail/detail.aspx?dbcode=CMFD&dbname=
CMFD2019a02&filename=1019656397.nh&uniplatform=
NZKPT&v=sDWfNl9y2p1fqutzqX4KTF5XDvx3VnEm6vm
C52326iZy-PD1_BY5DYUHbwm64ifv.

Zhang, R. (2019b). *Problems and Countermeasures of community
governance in urban relocation: a case study of C community in
Hefei*. Master dissertation, Anhui University. https://kns.cnki.net/
kcms/detail/detail.aspx?dbcode=CMFD&dbname=CMFD20
19b02&filename=1019142448.nh&uniplatform=NZKPT&v=
VmyvxOQhW1SU-J3igVQaXsiiOsgpYkRxlaLUB4WqV9wq
DK125Z_0b52GonZsF7Lh.

Zhao, Y. (2019). *Practical research on social work participating in
community governance of landless farmers—Taking m village
community as an example*. Master dissertation, Henan University of
Finance, Economics and Law. https://kns.cnki.net/KNS8/Detail?
sfield=fn&QueryID=60&CurRec=1&FileName=1019842291.
nh&DbName=CMFD201902&DbCode=CMFD.

Temporary Urbanism and Community Engagement: A Case Study of Piazza Scaravilli in Bologna

Yreilyn Cartagena

Abstract

In the contemporary urban context, the concept of Temporary Urbanism is debated due to its different names, definitions, and heterogeneity. This type of intervention makes use of empty spaces—private, public or open—for a short period of time. This article explores the importance of citizen participation, associated challenges and opportunities, and focuses on the construction process of a temporary "SLAB" intervention erected in the summer of 2018 in Bologna, Italy. An understanding of the level of community engagement achieved, and the social and physical effects of a city's life are established. The urban intervention was introduced into an existing installation, at Piazza Scaravilli, as an urban transformation in response to its needs. An ethnographic study approach was conducted at Piazza Scaravilli on the urban interaction before and during construction. This study demonstrates this installation's influence, focusing specifically on the social impact on the spatial, temporal, economic and cultural contexts, and the possible solutions that temporary urbanism can provide for unused spaces.

Keywords

Urban identity • City transformation • Temporary urbanism • Community engagement

1 Introduction

Temporary urbanism is the organic evolution of "temporary uses" (Talen, 2014). It is defined as a small-scaled intervention constructed for a short period of time, for public or private areas without a specific use.

Today, this trend is known as "everyday urbanism", "loose spaces", "guerrilla urbanism", "do-it-yourself urbanism" or "ad hoc urbanism" (Benner, 2013). The term "everyday urbanism", specifically emerged in the 90 s as a way to encourage informal engagement in social and cultural spaces. These events paved the way for new categories such as "Guerrilla Gardening" which are based on ecology and sustainability (Adams & Hardman, 2014). As a result, temporary urbanism came from the idea of experimental interventions, where the need for change in our cities left room for similar activities which adapt to the needs and desires of citizens (Haydn & Temel, 2006).

The projects are internationally recognised. However, it is more common in today's contemporary society that temporary proposals are presented as short-term solutions for public spaces where fundamental changes can be seen in the urban landscape. It is important to note how the integration of community participation in temporary urbanism can help strengthen a city's identity; a process where the citizens are in control of the changes that are being made and therefore assume leading roles.

This case study is based on a variety of temporary actions held at the Piazza Scaravilli, Bologna, Italy. Located in Via Zamboni, The Pizza Scaravilli was used as an experimentation site. Its implementation of a variety of temporary activities transformed the urban integration. Currently, the historical centre has had a substantial impact on the development of the city, which is organised by the Bologna City Council, and is known as ROCK PROJECT and UniBo.

This article's focus is on in-depth investigation of the interventions located in Piazza Scaravilli providing an understanding of the temporary uses and the influence of community participation in Urban Planning. The evidence presented in the article are supported by the data taken from relevant articles based on temporary urbanism and community participation. In addition, included are the visual examples of methods, which help establish a critical framework that addresses the goals of this research.

Y. Cartagena (✉)
University of Huddersfield, Huddersfield, England
e-mail: yreilyn.cartagenadelgado@hud.ac.uk

© The Author(s), under exclusive license to Springer Nature Switzerland AG 2022
C. Piselli et al. (eds.), *Innovating Strategies and Solutions for Urban Performance and Regeneration*,
Advances in Science, Technology & Innovation, https://doi.org/10.1007/978-3-030-98187-7_2

Therefore, in order to understand the benefits and consequences of temporary urbanism, ethnographical methodologies are used as an approach to investigate the area, noting participants' behaviours on a daily basis.

In that sense, this method helped to establish an understanding of the city–citizen relationship in Piazza Scaravilli in addition to the socio-cultural and economic influence that could be generated by temporary urban planning.

This paper highlights the limitations and benefits of temporary urbanism. The resilience demonstrated by the citizens within the experimentation site, proves that with each global misfortune experienced, public spaces regain their strength as protagonists of our cities. Today, communities, neighbourhoods and sectors in deplorable conditions have come back to life, thanks to communities or social groups that took over empty or abandoned spaces, squares or parking lots, and converted them to usable public areas through temporary urbanism. Communities are collectively reclaiming their environments. It is true that, together with other disciplines, temporary urbanism provides the opportunity to create scenarios where people's living conditions can improve and enhance community engagement.

2 Temporary Urbanism, the City and the Community

The term "temporary" is understood as a finite moment with a defined beginning and end (Bishop & Williams, 2012) It is followed by the concept of "urbanism", which is defined as the combination of design and planning, taking into account different aspects of the city (Chase et al., 2008). Since the beginning of history, temporary urbanism, originally known as "temporary uses", had an important impact on cities. Its influence remained unnoticed until the modern age during which different terms emerged and notorious changes were reflected in cities (Oswalt et al., 2014). Figure 1 shows the first temporary interventions in Europe and America that gave rise to this type of urbanism.

At present, "temporary urbanism" is understood as the application of experimental actions, or activities, to test strategies in an attempt to transform the city. Those actions are first studied through investigation of the process of the space, the culture, inhabitants and user needs to respond appropriately to each aspect through a temporal approach (Bishop, 2019).

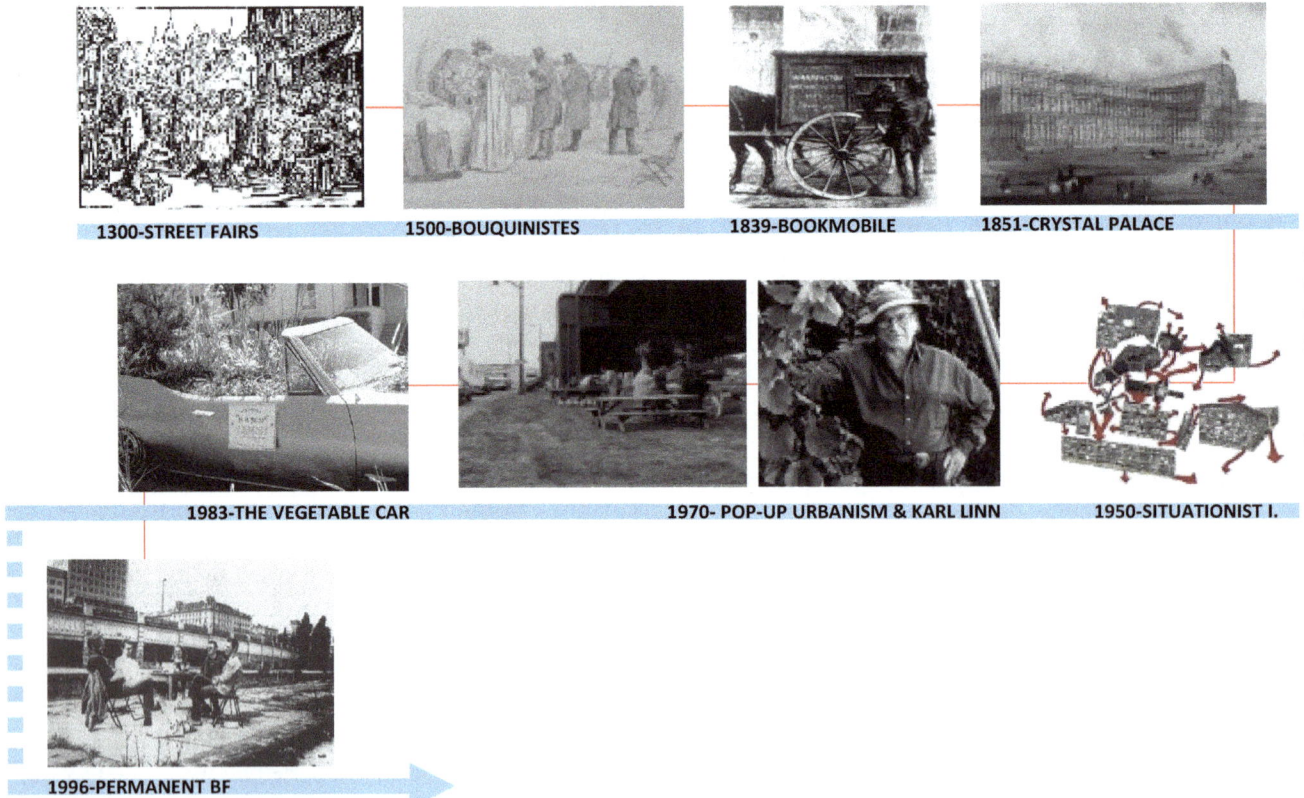

Fig. 1 Temporary urbanism timeline

2.1 The Importance of Community Engagement: Challenges and Opportunities

In recent years, the temporary use of spaces has become a significant trend, playing an essential role in exploring identity, innovation and the study of disused spaces. This phenomenon represents an urban development opportunity for urban planners, designers and the creative industry (Oswalt et al., 2014); one where the shape, spatial identity and idea of creating a neoliberal urban development through interdisciplinary discourses are adopted and named a "magical" trend (Ferreri, 2015).

As much as the understanding of the city is vital for the designer, the temporary revitalization of urban gaps is also a part of that participatory conception. The need to empower the neighbours to act as promoters of projects in their coexistent space or, at least, make them feel "creators of place" is also of significance (Rueda, 2018).

An example of the theory explained is the Open Sky Museum, an artistic intervention located in San Miguel—Chile, proposed by citizens to enhance the identity of the neighbourhood using narratives that represent it (Contreras & Garcia DelaVega, 2019).

Figure 2b shows a sample of the murals and graffiti made by national and international artists. The initiative emerged to restore the San Miguel area, taking inspiration from the Bicentennial of Chile, financed by the National Fund for Cultural Development and the Arts Bicentennial of 2010, "FONDART". The museum began to come to life through a series of participatory activities involving the community, where the main attractions were the walls of the buildings found on the main avenue, which can be seen in Fig. 2a (Rodríguez, 2017).

According to Bishop and Williams, the implementation of temporary interventions in the city cannot occur without the community's participation. The author states that as a result of the urban planning in San Miguel, the heritage created between the artists and the residents of the San Miguel community transformed a humble and destroyed neighbourhood into an arts complex with the ability to attract tourists from all parts of the world (Contreras & Garcia DelaVega, 2019). Consequently, the regeneration of the buildings was achieved, in part, and attracted the state government's attention to the improvement of public spaces (Urzúa Martínez, 2016). Nonetheless, working together with a local community can be challenging, especially when citizens are trying to decide on how best to represent the neighbourhood through imagery. This explains why the first proposal was rejected by the community (Rodríguez, 2017). Several elections held, however, it was recorded that there was a lower percentage of participants who engaged with the artist's initial proposals. The sense of permanence was also one of the challenges to overcome when carrying out this project. This case study is an example where community participation has substantially helped shape the way in which their neighbourhood is represented and how their cultural identity is viewed by others (Andres et al., 2019). Temporary urbanism is generally related to areas with poor conditions, humble communities or projects without sufficient financial funds and with time constraints as its strategies stemmed from an idea of solving problems immediately. Theories of resilience (Katz, 2004) expose groups that work together to implement renewal strategies in their communities independently from any government entity (Keck & Sakdapolrak, 2013). The purpose of community resilience aims to find its strengths, understand where and what its challenges are, and eliminate its weaknesses. It is here that these citizen groups understand the power and prominence they have in their community. In that manner, many branches arise in temporary urbanism, each

Fig. 2 **a** Community participation (Rodríguez, 2017). **b** Chilean Workers Mural (Rodríguez, 2017)

one with its goals, opportunities and proposed solutions for each conundrum and context, reflecting the needs of the spaces (Lydon & Garcia, 2015).

3 Introduction to Piazza Scaravilli

A historic road, known as Via Zamboni, is located in the centre of Bologna and the University, and is heavily used and noticed by the population. This road hosts various forms of culture and entertainment activities, such as the Teatro Comunale Palazzo Poggi, among other historical venues around the University of Bologna. Figure 3a represents the central axis, capturing the essence of coexistence between students and citizens. Different uses, needs, styles and representations exist simultaneously, however, the different road uses bring about varied criticisms.

Piazza Scaravilli, located on Via Zamboni, has undergone significant changes after the area was transformed from a parking space into a public space. The site hosts different activities daily, contributing enormously to Via Zamboni. Figure 4 shows Piazza Scaravilli before the transition into a public space.

3.1 Malerbe: "From a Parking Space to a Public Space"

By: ROCK, University of Bologna, Centro Antartide, Beyond Architecture Group "BAG" and EU Community and the Municipality.

Personal Communication: Professor Danila Longo, Jul 2018—University of Bologna

The ROCK project carefully considers the transformation, or creation of public spaces, to increase and improve the value of the city and develop cultural collaborations in a sustainable municipality. Together with other collaborators, ROCK Bologna has decided to create a virtual public space

Fig. 4 Parking (Google_earth, 2015)

in Piazza Scaravilli. The selected site takes advantage of the attractive area, being the only space in Bologna with four sides of porticoes near the main street, making it a valuable architectural heritage.

As a result, Malerbe was born: "From a parking space to a public space". The project was originally proposed by students, where phrases such as "Let me breathe" and "do not park" became the focus of the space. Constructed in 2016 by students, with the help of a cooperative "Centro Antartide", the project took the form of a co-design workshop, a "Utopia Concreta", with the participation of Paolo Robazza "BAG", the architect.

The project consisted of a temporary garden seating, built with recycled materials and weeds known as "Malerbe"; a wild garden that prevents people from stealing it. By

Fig. 3 **a** Via Zamboni (BO2ND, 2018). **b** Via Zamboni

developing the Piazza, the community gained a free public space to socialise, relax and host public events, with the goal of attracting visitors of varying ages and backgrounds. Piazza Scaravilli has hosted prestigious events such as the "ROCK Arena" or the "Bologna Design Week".

Despite its significant progress, there was an additional need to improve the space and increase community participation, which is why U-LAB "ROCK Living Lab" worked in the Piazza to define the priorities and potential of the place. Consequently, an initiative to create a Co-designing workshop, known as the "Utopia Concreta - Malerbe # 2, was born (Fig. 5).

Personal Communication: Architect. Paolo Robazza, Jul 2018—BAG.

The idea was born during the summer of 2017 when the plaza hosted numerous events and the organisers realised that instead of designing a platform and seating area for each event, it would be more beneficial to request a proposal for a "temporary intervention". The process was carried out in the form of a co-designing workshop called "Utopia Concreta - Malerbe # 2", in collaboration with Architect Paolo Robazza from "BAG" the University of Bologna and ROCK Bologna.

The main goal in Piazza Scaravilli was to use the area as a "permanent laboratory" and the host of temporary activities for the transformation of the area. The temporary structure represents the last phase of the Utopia project and the name "*SLAB*" means, "*a type of powerful wave generated from the sea's deep waters, breaking into shallow water*". It hosts part of the performance review that animates the summer in prominent places in the university. By hosting this new space, the university area has become one of the prominent places in the centre calling for interaction among the community.

The second part is explained in detail in Sect. 5 with an ethnographic documentation of the construction process of *SLAB* (Fig. 6).

4 Methodology

As (Rockwell, 1980) indicates, an essential characteristic in the ethnographic application is to understand the spaces from the same point of view of the community as they experience them in their everyday lives and describe them in a socio-cultural sense through the interaction between the citizens spaces within a specific population (Geertz, 1974).

According to (Werner & Schoepfle, 1987), this methodology (Fig. 7), as part of an urban study, focuses on the process of directing observations and describes three types of techniques (Angrosino & Rosenberg, 2011):

The first is a descriptive observation in which you observe anything and everything. Assuming you ignore everything, the downside of this model is that it can lead also to the gathering of irrelevant information.

The second type—focused observation—emphasises interview-based observations, in which the opinions of the participants guide the researcher's decisions on what to observe.

The third type of "observation", considered more systematic by Angrosino & Deperez, is selective observation, in which the researcher focuses on different kinds of activities to help delineate the differences in these activities. According to (Mariampolski, 2006), the information collected from this methodological approach is undoubtedly effective as its investigation reveals much more information than what could

Fig. 5 **a** Pilot \"Do not park\" (Longo, 2016). **b** Malerbe Pilot Action (Longo, 2016)

Fig. 6 **a** Malerbe (Longo, 2017). **b** Co-design workshop (Longo, 2017)

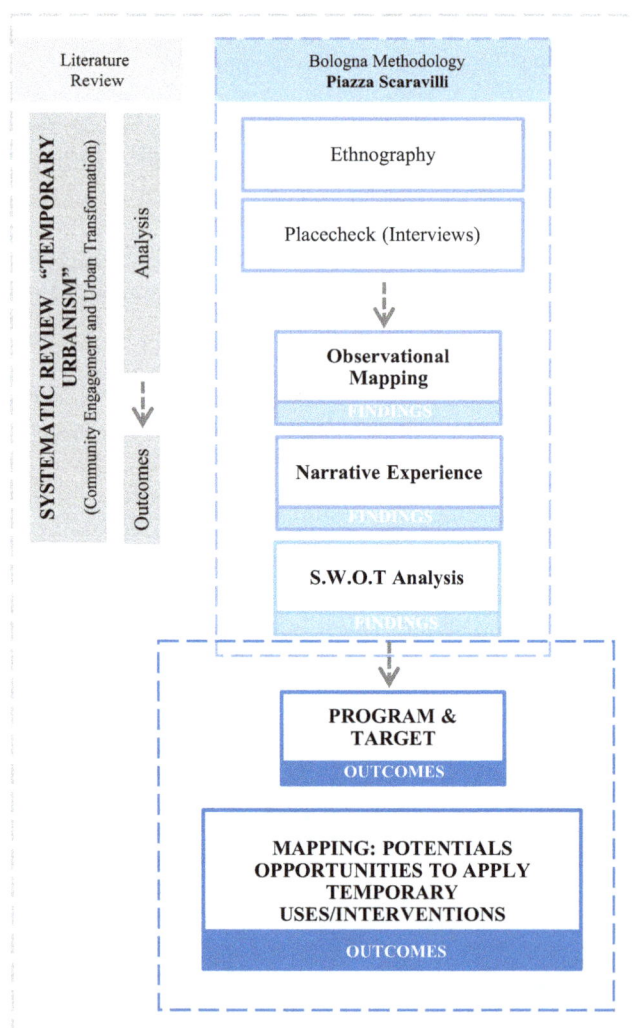

Fig. 7 Designed methodolgy

be obtained through documented sources or other types of research. On a contradicting note, (Klima, 2019) states that ethnography is limited in application to bigger communities and is only useful to smaller areas with a behaviour and

structure that is easy to understand and describe. Similarly, the observer's prejudices and experiences could influence their ethnographic descriptions or affect the community's behaviour forcing them not to act naturally. Therefore, objectivity will always be an existing theme among the limitations of ethnography (Boellstorff et al., 2012).

Moreover, (Banks & Vokes, 2010) state that utilising more images in the ethnographical approach can help add to the context of the study and complement the researcher's textual descriptions. Photographic ethnography presents the participant with specific situations and moments, and the combination of these together with the scenography may help communicate and express certain circumstances (Brisset Martín, 1999). However, ethnography, in the urban landscape, offers active participation, visual crossing techniques and interviews to understand and reformulate the way to observe the city (Duneier et al., 2014).

5 Case Study: "SLAB" Assembly Process—Part II

The case study goes into the construction of "SLAB", a temporary installation in Piazza Scaravilli. The workshop was held in July 2018 and was directed by Architect Paolo Robazza and his team from "BAG", in collaboration with a group of students and teachers from the University of Bologna.

A considerable amount of the work was observational and focused on the urban analysis, activities and community documentation before and during the "SLAB" construction process, all the while capturing the community's engagement and responses. The objectives are to use the area as a laboratory to host activities, take advantage of its privileged location and improve its quality by integrating a socially diverse society.

Ethnography (Table 1) was chosen as a research method as this approach allows for exploration through anecdotal living experiences. This created the potential to address observational concerns and detect the impact that this type of

Table 1 Ethnography approach in Piazza Scaravilli

Component	Variables of analysis
Users: Bolgna Students Local Community	Recognise the expectations of the community and its belonging to the city Understand socio-economic and cultural dynamics of the population involved Identify the habitat needs of the community
Area: Piazza Scaravilli	Study the accessibility and integration of the specific sector in the city Explore the insertion of temporary interventions Investigate the agents involved Add the position and participation of the community

urbanism makes on the city, the impressions and the community's behaviours.

In this context, it was not only interesting to examine the influence of student participation but also the thoughts and intentions of the local community. On the other hand, this also helps to understand the relationship between the temporary, cultural and architectural interventions along Via Zamboni.

Data Collection: From Monday, July 2 to Friday, July 6, 2018: The construction process recorded each day through interviews and observation techniques for data collection as recommended by (Kawulich, 2005). The techniques were applied as follows:

In phase 1, daily photographic documentation was conducted, creating a series of photographs that portrayed and documented the creation of the space. See Fig. 8 for day one and two of the workshop.

Secondly, the direct observation technique was undertaken under the Piazza porticoes. Photographs were taken from a variety of perspectives to avoid interruptions and allow for direct participation to capture all visual angles. Figure 9 shows a series of words and symbols that narrate the situation on day three.

Finally, I was fortunate to be a participating observer. For 2 h, I was present during this assembly process this has made my experience radically different when compared to the other techniques because the observer perception was united with the feeling of belonging. In particular, in the field of community work, you will understand leadership skills, communication, working well in a team and finding solutions to problems, in bringing the best to the city. The integration between the student community and the emotions that are evoked by each piece assembled give shape to that wooden Wave called "SLAB". See links to a video

Day 1

Day 2

Fig. 8 Photographic narrative

Fig. 9 Narrative experience

University Life

Travel Corridor

Identity "Place"

Community Area   Homeless

summarising the construction process of the project here: https://youtu.be/NSq_5TVL9bo.

5.1 Observational Summary

Results from the case study

The analysis helped establish an understanding of the situation and the impact of the temporary interventions in the Piazza, whilst taking the community's experiences and opinions into consideration.

Phase 1 was divided into three parts and is explained in detail in Sect. 4. The language barrier signified a major limitation during interviews conducted and in communicating with the participants. However, the volunteers were very interested in collaborating. The following analysis (see Table 2) utilises the SWOT Analysis technique and applies it to the Piazza Scaravilli case to understand and evaluate the place. Humphrey explains through this exercise how potential opportunities are enhanced when you have a complete outlook on future weaknesses and threats. In this case, Piazza Scaravilli is given an advantage by its location in terms of accessibility and positioning between two essential axes of Bologna's centre, generating a meeting point for the community. The presence of art and the university environment keeps the area alive most of the time. On the other hand, existing weaknesses, such as crime and lack of green spaces in good condition, predict future threats in the area (see Appendix A).

Table 2 S.W.O.T analysis

Bologna "Via Zamboni"	
Strength	Weakness
Permeability between the university campus and the centre Public spaces in use Proximity to the city centre The arcades and the passages Artistic value Cultural value Appreciation of medieval architecture	Areas with poor lighting The cycle line does not respect the pedestrians Lack of the private sector engagement Public spaces used as garbage area The need for green areas
Opportunities	Threats
Strong creative activity Big potentiality in open spaces Potential to develop urban areas with community participatory process Opportunity to increase the footfall-taking advance of urban traditional activities	Poor lighting will generate crime, low footfall at night and more homeless Lack of green areas in the centre may force people to do not stay in open areas in the hot weather Informal commercial zones

Phase 2 of the ethnographic study was designed as a semi-structured interview by way of a place check. A place check is a simple way of assessing what a place and its inhabitants are trying to convey (Cowan, 2000). The objective was to ascertain the citizens' needs and their views on the current situation of their city, its problems and shortcomings.

In summary, 20 interviews indicated that 70% of interviewees between 18 and 35 years old support the participatory process. In comparison, almost 80% interviewees aged 35–60 believe that the local government and the university should take responsibility. In its entirety, the community asks for more green spaces and equipped public spaces that host enjoyable public events for all ages. Finally, it was interesting to observe how the population living on the outskirts of the centre began to appreciate and think about how they could creatively utilise empty or disused spaces within their community. They began associating these empty spaces with temporary urban planning and exploring how they could improve the appearance of their neighbourhoods (see appendix B).

To sum up, the different observation phases helped explain the area's current situation and provided before and after perspectives of the "Malerbe" intervention in the square. Following the methodology, attention then turns to the response of the citizens', their opinions and behaviours within ethnographic explorations while getting to know, exploring, and enjoying the site and experiencing local activities.

The conception of projects such as "Malerbe" and "SLAB" demonstrate that they have the potential to bring life and create an engaging space that had previously only been used as a parking lot, regenerating the area and offer Via Zamboni and the community a permeable, healthy and adaptable public space. This intervention introduced the idea of citizen participation (see Appendix B).

6 Conclusions

Through the observation and analysis of current urban characteristics, Piazza Scaravilli and Via Zamboni aim to capture the attention of the university area's diverse population; to welcome it, to challenge it, and to host various forms of culture and entertainment. The reflection is drawn upon the significant change that took place when the parking

lot was transformed into an open community area. The area now hosts different daily activities and contribute to Via Zamboni massively. Yet, there are also negative aspects to the study such as follows: (a) some seating areas attract homeless people in the Piazza, thus increasing insecurity. (b) Empty spaces along Via Zamboni (c) Absence of community participation (d) lack of green spaces (e) Loss of identity (f)Economy (g)Regeneration of public spaces.

This study is of relevance for temporary urbanism as it makes a remarkable contribution through dense descriptions of specific experiences and activities and insight into concerns within communities. Through a primary investigation, this study was able to verify situations in current contexts. By utilising proposals and hypotheses found in contemporary literary writings, this study has established a better understanding of the potential of temporary urbanism. After observing daily customs, needs, and behaviours, it was easy to understand the city as a place full of culture, art, and tradition. Simultaneously, to analyse it from an urban and social point of view: it was necessary to investigate spatial, temporal, economic and cultural contexts as they are key to the transformation and evolution of a city. Thus, the collected data and the theoretical knowledge on temporary urbanism are reflected in Fig. 10 where a social mapping of the area indicating uses, community, and social impact come together to generate a diagram that adds to the application of temporary interventions for the city and community.

The diagram in Fig. 11 explains how temporary urbanism, when studying the city, can be adapted to each situation, thus allowing community integration and ideas, an exchange of ideas to address the needs of the city from the perspective of the community.

Cities worldwide have been promoting temporary interventions that offer quick and effective alternatives for empty or unused public spaces. Nevertheless, today, and more than ever, the concept of "temporary" has gained momentum. In considering the COVID-19 pandemic and its ongoing global impact, Social Distancing has forced us to stay at home while the economy has forced us to leave it. Therefore, even the "new normal" encompasses a temporary trend that is adaptable to the needs of society.

In summary, the future of cities is not an independent or a sole objective for governments. Cities and citizens need to evolve ecologically, while taking into consideration the associated strengths and weaknesses, demographic factors and the essential needs to begin the transformation of a city.

Fig. 10 Program and target

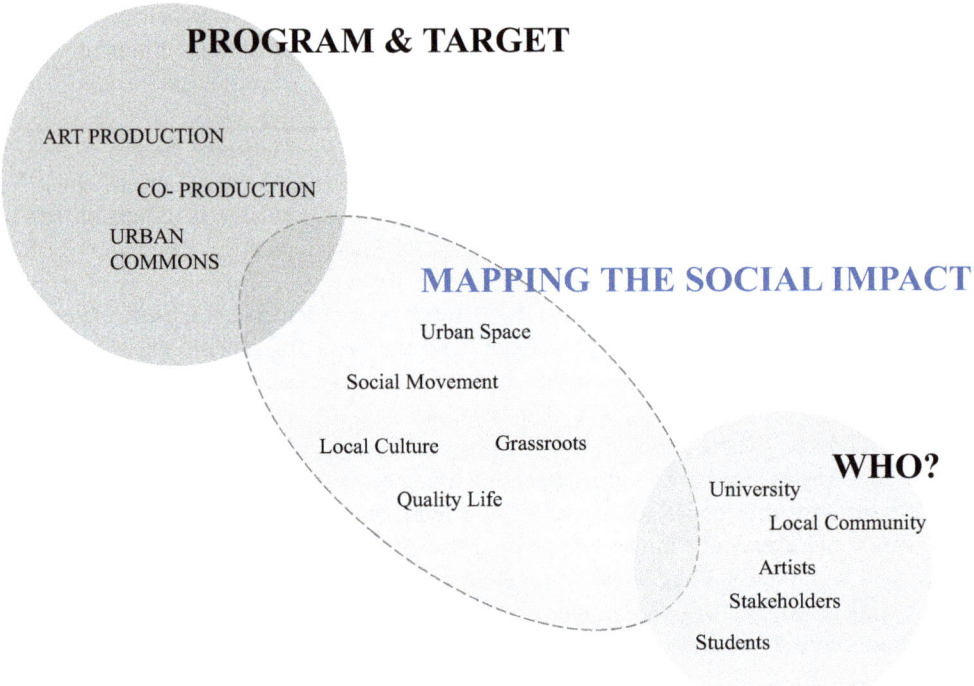

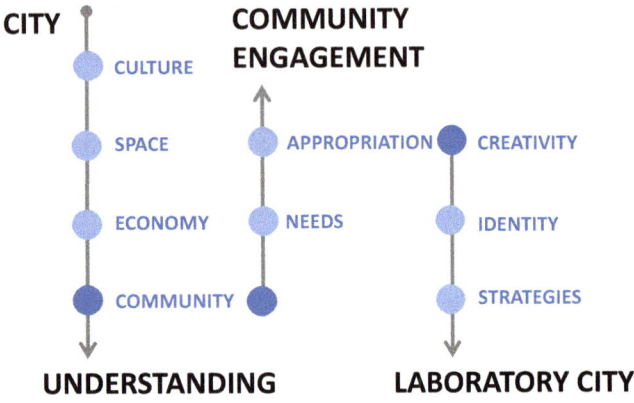

Fig. 11 Mapping temporary opportunities and engagement

Acknowledgements The author wishes to acknowledge Professor Danila Longo and Architect Paolo Robazza and many of the student contributors mentioned in this research. I thank them for their collaboration and participation in the data collection process. I would also like to thank my Ph.D. supervisor, Dr. Ioanni Delsante, for his guidance and support in the development of this research.

Appendix A

A.1 Comments collected during interviews or engaged conversation with the local community:

Personal Communication: Emanuel Bombardini, Jul 2018

I considered the new design of the Piazza as an evolution in Via Zamboni; the piazza is now an important meeting place for students, artist's organisation and the community, we organise several free activities here.

Personal Communication: Sandra Lucchini, Jul 2018

Good job from the university; this is what Bologna needs, free seating spaces.

Personal Communication: Political Sciences Student-Mario Mirabile, Jul 2018

I am a participant in this project to contribute something to the city before my graduation
   Bologna has a low level of inhabitants in summer and needs free social spaces so that this intervention will help as a meeting point for many, regardless of social class.

Personal Communication: Architecture Student- Eleonora Savini, Participant in workshop 1 and 2, Jul 2018

At the state one of these projects I had a great experience in team working with the design process

This summer, looking at the success of the current intervention on Piazza, as people are enjoying it every day, the SLAB pavilion will have a good response from the community.

Appendix B

B.1 Placecheck Piazza Scaravilli

Date: 02 Jul 2018 (Morning) 03 Jul 2018 (Afternoon) 05 Jul 2018 (Afternoon)

Participants: Local Community: 10 citizens—Students: 10 citizens

Questions:

What do you like or dislike about Malerbe?
Did the intervention produce any impact on the quality of community's life?
Will you participate in the application of any urban intervention?
Do you have any suggestion of how to improve the piazza?

See Fig. 12 and Table 3.

Fig. 12 Placecheck Results

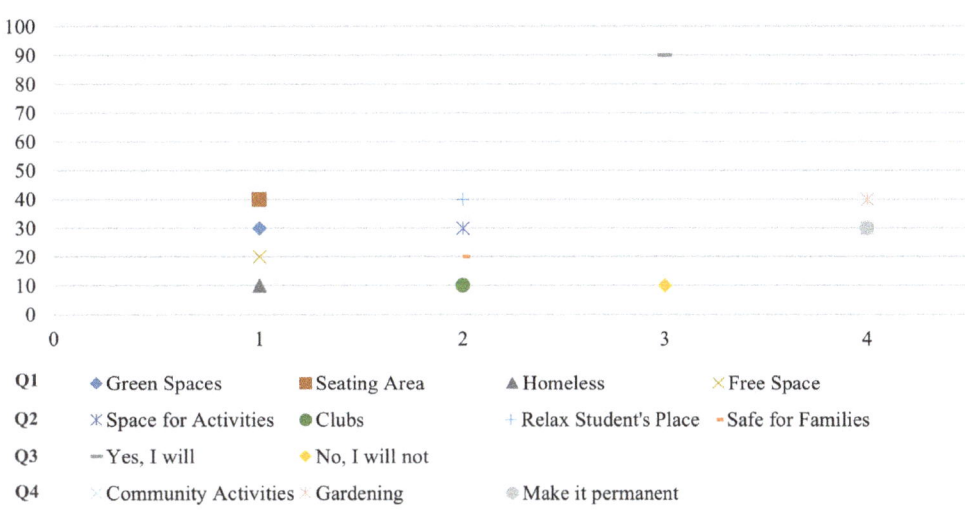

P. Scaravilli's Chart

Q1	◆ Green Spaces	■ Seating Area	▲ Homeless	✕ Free Space
Q2	✳ Space for Activities	● Clubs	✛ Relax Student's Place	− Safe for Families
Q3	− Yes, I will	◆ No, I will not		
Q4	Community Activities	Gardening	⬤ Make it permanent	

Table 3 Placecheck Q&A

Question 1	Question 2
Green Spaces: "Via Zamboni needs more vegetation, the University did a good job putting some flowers at Scaravilli" Seating Areas: "The right choice, before this space used to be a university parking, now as you can see this is full of people, better use of the space" Homeless: "This is a good idea; the only negative point is the homeless people who are coming every night" Free Space: "As students, we need free-seating (public seating) places like this one, we now come here every day for lunch"	Space for Activities: "Yes, the piazza is now hosting plenty of activities" Clubs: "University clubs started to meet here because is better, free and easy for different club's activities" Relax Student place: "I like it very much, we came here just for chill after class" Safe for families: "It a safe place for families to come with their children and let them play"
Question 3	Question 4
Yes: "Yes, I see this process very interesting. (Interview during SLAB build process)" No: "No, but the idea of public spaces like these is good"	Community Activities: "The city should organise free activities for the community (children, oldies or parents) at the moment the majority of the events here are for students or young people" Gardening: "The idea of add vegetation together with seating was good but, do the council also organise gardening activities with us" Make it permanent: "We all like this place, and think that it should be permanent"

References

Adams, D., & Hardman, M. (2014). Observing guerrillas in the wild: Reinterpreting practices of urban guerrilla gardening. https://doi.org/10.1177/0042098013497410

Andres, L., Bakare, H., Bryson, J. R., Khaemba, W., Melgaço, L., & Mwaniki, G. R. (2019). Planning, temporary urbanism and citizen-led alternative-substitute place-making in the Global South. *Regional Studies*, 1–11. https://doi.org/10.1080/00343404.2019.1665645

Angrosino, M., & Rosenberg, J. (2011). Observations on observation. *The Sage handbook of qualitative research*, 467–478.

Banks, M., & Vokes, R. (2010). Introduction: Anthropology, photography and the archive. *History and Anthropology, 21*(4), 337–349.

Benner, S. M. (2013). Tactical urbanism: From civil disobedience to civic improvement. Retrieved from http://hdl.handle.net/2152/23995

Bishop, P., & Williams, L. (2012). The temporary city Routledge.

Bishop, P. (2019). Urban design in the fragmented city.

Boellstorff, T., Nardi, B., Pearce, C., & Taylor, T. L. (2012). *Ethnography and virtual worlds: A handbook of method.* Princeton University Press.

Brisset Martín, D. E. (1999). Acerca de la fotografía etnográfica. Gazeta de Antropología Nº 15. Retrieved from http://hdl.handle.net/10481/7534

Chase, J., Crawford, M., & Kaliski, J. (2008). *Everyday urbanism.* Monacelli Press.

Contreras, D., & Garcia DelaVega, A. (2019). Museo A Cielo Abierto En San Miguel Como Experiencia De Paisaje Musealizado. Diferents. Revista de museus, 98–111. https://doi.org/10.6035/Diferents.2019.4.7

Cowan. (2000). Placecheck: A user's guide. Urban Design

Duneier, M., Kasinitz, P., & Murphy, A. (2014). *The urban ethnography reader.* Oxford University Press.

Ferreri, M. (2015). The seductions of temporary urbanism. *Ephemera, 15*(1), 181.

Geertz, C. (1974). "From the native's point of view": On the nature of anthropological understanding. *Bulletin of the American academy of arts and sciences*, 26–45.

Haydn, F., & Temel, R. (2006). Temporary urban spaces: concepts for the use of city spaces.

Katz, C. (2004). *Growing up global: Economic restructuring and children's everyday lives.* U of Minnesota Press.

Kawulich, B. B. (2005). *Participant observation as a data collection method.* Paper presented at the Forum qualitative sozialforschung/forum: Qualitative social research.

Keck, M., & Sakdapolrak, P. (2013). What is social resilience? Lessons learned and ways forward. *Erdkunde, 67*, 5–18. https://doi.org/10.3112/erdkunde.2013.01.02

Klima, A. (2019). *Ethnography# 9.* Duke University Press.

Lydon, M., & Garcia, A. (2015). Tactical urbanism: Short-term action for long-term change.

Mariampolski, H. (2006). *Ethnography for marketers: A guide to consumer immersion.* Sage.

Notes for an Urban Community Work of Art., (2017).

The Open Sky Street Art Museum in San Miguel.

Oswalt, P., Overmeyer, K., & Misselwitz, P. (2014). Urban catalyst: the power of temporary use.

Rockwell, E. (1980). Etnografía y teoría en la investigación educativa. *Revista Dialogando, 29.*

Rueda, E. T. (2018). Urbanismo Adaptativo.

Talen, E. (2014). Do-it-yourself urbanism. *Journal of Planning History, 14*(2), 135–148. https://doi.org/10.1177/1538513214549325

Urzúa Martínez, S. (2016). Museo a Cielo Abierto de San Miguel: Transformar y re-crear el espacio público. INVITRO. Hábitat residencial y territorio. Blog del Instituto de la Vivienda de la Universidad de Chile.

Werner, O., & Schoepfle, G. M. (1987). *Systematic fieldwork. graph. Darst. 2. Ethnographic analysis and data management. 1987. 355 S. 2. Ethnographic analysis and data management. 1987. 355 S.* Sage.

Urban Circularity: City Planning Perspectives from the Regeneration of Amsterdam's Buiksloterham District

Georg Hubmann, Theresa Lohse, and Jonas Plenge

Abstract

This article connects city planning with the concept of the Circular Economy (CE). It analyses the case of 'Circular Buiksloterham', one of Amsterdam's post-industrial districts that is transforming into a living example of a Circular City. Firstly, we relate the conceptual background of CE with urban planning and design. Then, we develop an analytical framework for circular urban developments including the following categories: operational implementations, spatial design, social integration and policy integration. In discussions on the interlinkages between circularity and urban planning, we identified and addressed an arguable lack of implementation and little attention for the built environment as gaps in the literature. Based on the case analysis, qualitative interviews and document analyses, we argue that there is an intrinsic link between the organisation of urban resource flows and the production of space. The findings demonstrate that the marriage of circular principles and city planning implies both behavioural adaptations and new technological systems. Furthermore, we identified that making the step up in scale from experimental zones to an entire 'circular district' blurs the holistic nature of CE.

Keywords

Circular economy • Circular city • Resource flows • Urban regeneration • Buiksloterham

G. Hubmann (✉) · T. Lohse · J. Plenge
Chair for Urban Design and Sustainable Urban Planning, Faculty of Architecture, Technical University Berlin, Berlin, Germany
e-mail: g.hubmann@tu-berlin.de

1 Introduction

Urbanisation and climate change are two of the major concerns for city administrations today. Between 1950 and 2000, the global population living in urban areas grew from 30 to 50% and keeps on growing strongly with a prediction of around 68% by 2050 (UN DESA, 2018). At the same time, environmental degradation and the increase of extreme weather events caused by climate change have become key challenges that impact ecosystems, economies and communities around the world (IPCC, 2018; Lehmann et al., 2018; Trenberth et al., 2015). Cities find themselves in a complex dilemma. On the one hand, they are the focal points of development; on the other hand, cities consume up to 80% of total global energy production and account for 71–76% of the world's CO_2 emissions (Hoornweg et al., 2011, pp. 207–211; Marcotullio et al., 2013, p. 622; Satterthwaite, 2008, p. 543). How cities are designed and how they operate significantly affect direct and indirect Greenhouse Gas (GHG) emissions. Especially urban form, infrastructure and supply systems are critical factors because of their strong link to the throughput of materials and energy, waste generation and system efficiencies of a city (Seto et al., 2014, p. 927).

In recent years, the concept of Circular Economy (CE) has gained popularity among urban practitioners, politicians and scholars. CE is seen as a way to help solve cities' complex sustainability challenges by remodelling their urban metabolisms and by applying looping actions to material flows. Yet, there is no clear evidence whether CE initiatives, as they are increasingly implemented by cities and formulated as urban policies, can provide more sustainable results. What is more, the relationship between CE and its biophysical limits including system-wide thinking on entropy and the laws of thermodynamics as well as the altering of materials over time remain unclear (Calisto Friant et al., 2020, p. 4). However, according to the International Resource Panel Report 'Weight of Cities', optimising systems and creating cross-sector synergies between buildings,

C. Piselli et al. (eds.), *Innovating Strategies and Solutions for Urban Performance and Regeneration*,
Advances in Science, Technology & Innovation, https://doi.org/10.1007/978-3-030-98187-7_3

mobility, energy and urban design can reduce GHG emissions and resource use by up to 55 per cent (IRP, 2018). But critical scholarship has pointed to considerable implementation gaps in how such ideas are operationalised in practice (Monstadt & Coutard, 2019, p. 12; Williams, 2019; Williams, 2013), indicating that 'circular urban developments' tend to remain largely on a rhetoric strategy level. At the same time, other authors have noticed a lack of attention for the 'meso-scale' (buildings and built environment) that is between the macro-scale (cities or eco-parks) and micro-scale (manufactured products or construction materials) (Appendino et al., 2019, p. 3; Pomponi & Moncaster, 2017, pp. 3–5). Additionally, the assessment of urban planning strategies in connection with CE are largely missing (Petit-Boix & Leipold, 2018, p. 1276).

Addressing the above-mentioned gaps in the literature, this research analyses the transformation process of Buiksloterham (BSH), an urban neighbourhood in the north of Amsterdam that was launched as the Netherlands' first Living Lab for smart, circular and bio-based urban development in 2015. The two research questions are: How does the notion of CE influence city planning at district scale? How can circularity principles from a Living Lab approach be mainstreamed in urban planning? The purpose of this paper is to take a spatio-temporal perspective during the analysis to investigate the relationship between CE and city planning after 5 years of implementation in the selected case. The goal is to identify levers and barriers of a 'circular urban planning approach' and to give an indication whether the practices of the experimental Living Lab have the potential to be mainstreamed in city planning. Cities are viewed as complex dynamic systems in a continuous state of change, which is reflected in their size, social structures, economic systems, geopolitical settings and the evolution of technology (Kennedy et al., 2007, p. 44). City planning is a way to govern these dynamics with urban planning providing for a spatial structure of activities at city or regional level (Hall & Tewdwr-Jones, 2019, p. 3) and urban design as the process of giving form, shape and character to groups of buildings, to whole neighbourhoods, and to a city, bringing together place-making, environmental stewardship, social equity and economic viability (Raven et al., 2018, p. 142). We argue that there is an intrinsic link between the organisation of urban resource flows and the production of space. Thus, the way resources are managed in a CE does not only effect the urban metabolism but also impact spatial configurations, which are usually under the jurisdiction of spatial planning. Methodologically, this article starts with a qualitative literature review to determine principles of circularity in city planning. Then, an analytical framework is designed that takes into consideration the systemic multi-criteria logic of CE in urban planning. This framework is then used to carry out the analysis by taking a spatio-temporal

perspective. In order to monitor the spatial regeneration process of the area, qualitative data was reviewed that is relevant for two points in time; in the case of 'Circular Buiksloterham' the reference years are 2015 and 2020. The empirical analysis relies mainly on three sources: (a) the planning document entitled 'Circular Buiksloterham—Transitioning Amsterdam to a Circular City' (Manifesto),[1] (b) an investment memorandum with the title 'Investeringsnota Buiksloterham 2019' (Ivn.B.2019),[2] and (c) three semi-structured interviews with relevant people involved in the development process that were conducted by the authors between November 2019 and June 2020.[3] The paper is structured as follows: first, we trace the trajectory of the CE concept and link this to urban planning and design. Then, we introduce our analytical framework capable of detecting key features of a 'circular urban development' over time. Next, we apply this framework and present empirical results before we discuss the outcomes and give a conclusion.

2 Principles of Circularity in Urban Planning and Design

The concept of CE and its underlying principles have various conceptual touching points with city planning, understood as both the practices of urban planning and urban design. Despite being multidimensional and transdisciplinary by nature, the common denominator of city planning is its spatial logic. However, there is a tendency to look at innovation processes without reference to the spatial context (Monstadt, 2009, p. 1924). We argue that in the midst of the global resource crisis, there is a gap to close regarding the inclusion of CE in city planning efforts because decisions on urban form and infrastructural systems have long-term consequences and strongly affect a city's capacity to address resilience and sustainability.

---

[1] This planning document was developed as a vision and ambition document in 2014 and signed by 20 partners in March 2015.
[2] Published in late 2019, this recalibrated version of the 2006 development plan lays out the programmatic, urban planning, environmental, civil engineering and financial framework for the building stock establishment and the land transfer from the municipality to third-party financiers. It is also the development framework for private developers aimed at transforming their own land on their properties or leases.
[3] The first interviewee (Interviewee 1) is a member of the Smart City Amsterdam programme. The second interviewee (Interviewee 2) was a project leader at Waternet from 2014 to 2016. This is Amsterdam's water utility company that was mandated by the municipality with the execution of all water-related tasks in BSH (Gemeente Amsterdam, 2019, p. 73). Interviewee 3 is a representative of Amsterdam's municipality 'Gemeente Amsterdam' working in the team of sustainability advisors for Amsterdam Noord.

A growing number of scholars, politicians and practitioners across different fields are engaging with the concept of CE as an answer to the fossil-based, waste-generating and unsustainable linear economy model (Reike et al., 2018, p. 249; Sillanpää & Ncibi, 2019, p. 26). CE is not new: early thinking on cyclical processes or closing loops dates back to the eighteenth and nineteenth centuries, but integrative waste management approaches, which mark the beginning of the current day discussion of CE, only emerged in the 1970s (Korhonen et al., 2018, p. 545; Murray et al., 2017, pp. 372–373; Reike et al., 2018, p. 248). There are several concepts that laid the way for how CE is perceived today. It is noteworthy that among them are some scholars with a spatial background: the Operating Manual for Spaceship Earth by architect R. Buckminster Fuller (1969), Regenerative Design by landscape architect J. T. Lyle (1970s), and Performance Economy by architect and industrial analyst Stahel (1976) (Ellen MacArthur Foundation, 2013, pp. 30–31; Homrich et al., 2018, p. 527; Sillanpää & Ncibi, 2019, pp. 16–18; Winans et al., 2017, pp. 825–826). Despite its academic origin, contemporary notions of the CE approach have been largely brought forward by practitioners. At the same time, scholarly research is in its infancy and there is substantial conceptual unclarity and fragmentation (Korhonen et al., 2018; Lieder & Rashid, 2016). Blomsma and Brennan who frame CE as an 'umbrella concept' divide its development into three stages, where the latest period starting in 2013 is characterised by a substantial need for theoretical or paradigmatic clarity (Blomsma & Brennan, 2017, pp. 607–610).

The dichotomy between practitioners and academics also unfolds regarding the definition of CE.[4] While a widely cited definition by the Ellen MacArthur Foundation (EMF) has an implicit focus on business models, a more comprehensive definition by van Buren, Demmers, van der Heijden, & Witlox, 2016, p. 3 includes the following dimensions of CE: the 3R framework,[5] waste hierarchies, a systems perspective, environmental quality, economic prosperity and social equity (Kirchherr et al., 2017, p. 228). Using definitions without waste hierarchies might not have adequate impacts on the status quo, e.g. increasing recycling rates instead of initiating a more holistic transformation (Kirchherr et al., 2017, p. 229). Another blurry characteristic of the CE concept is its relationship with sustainable development. While CE is mainly focused on the narrow goal of closing loops giving attention to resource inputs or waste outputs,

sustainable development has—similar to city planning—open-ended goals that focus on multiple dimensions surrounding environmental, social, and economic sustainability. Thus, CE is viewed as just a condition for sustainability or a beneficial relation (Geissdoerfer et al., 2017, pp. 762–768). But in the CE literature, there is agreement about missing emphasis on the social dimension while links between environmental and economic issues are pointed out in several publications (Homrich et al., 2018, p. 534; Murray et al., 2017, p. 369; Sauvé et al., 2016, p. 54). Yet, stressing only one or two categories of the sustainability triple bottom line might, for example, lead to a lack of social considerations in implementing CE (Kirchherr et al., 2017, p. 227). It is also unclear how the concept of CE will lead to greater social equality, defined as inter- and intra-generational equity, diversity or equality of social opportunity (Murray et al., 2017, p. 376). The lack of social aspects in the CE concept is thus identified as a conceptual shortcoming that might be problematic for the application to city planning, which fully embraces the three pillars of sustainable development.

There is consensus in the literature about at least three characteristics of CE: (1) the idea of closed loops, (2) decoupling resource use from economic growth and (3) being a radical concept with far-reaching implications for human systems. In general terms, an economy that operates in a circular way should not have negative effects on the environment; rather, the damage done in resource acquisition is restored while as little waste as possible is generated in the production process and during the life of a product (Murray et al., 2017, p. 371). CE enables thinking in cycles and aims at keeping the valuation of materials in closed loops instead of having an open-ended conception of value chains. This can be realised by having an understanding of the relevant 'biological' and 'technical' loops and including at least the notions of input reduction, reuse, and recycling (Bocken et al., 2016, p. 308; Homrich et al., 2018, p. 526; Winans et al., 2017, p. 825). In other words, virgin material or energy inputs to the system and waste as well as emission outputs from the system should be reduced (Korhonen et al., 2018, p. 544). The second common understanding of the CE concept is to allow for natural resource use while reducing pollution or avoiding resource constraints but at the same time sustaining economic growth. CE is widely described as an alternative model of production and consumption as well as a growth strategy that allows for the 'decoupling' of resource use from economic growth, and in doing so, contributing to sustainable development (European Commission, 2020, p. 2; Geissdoerfer et al., 2017; Ghisellini et al., 2016, p. 24). 'Decoupling' means that an economy can grow without increasing the pressure on the environment. Thus, the same economic input is generated based on using less material, energy, water and land resources, while at the same time delinking environmental deterioration. The third

---

[4] Due to the maximum allowance of words for this publication, there is not enough room for stating the two definitions.
[5] The three Rs mean reduce, reuse and recycle and are part of a framework called waste management hierarchy that indicates an order of preference for action to reduce and manage waste. It is usually presented in the form of a pyramid.

characteristic suggests that a true CE would imply fundamental systemic change including new concepts of systems, economy, value, production and consumption leading to sustainable development of the economy, environment and society (Wu, 2005). However, only 40% of the analysed definitions by Kirchherr et al. conceptualise CE from a holistic systems perspective (Kirchherr et al., 2017, p. 229). In its origins, CE was considered as an approach for waste management but recent notions conceptualise it as "a broader and much more comprehensive look at the design of radically alternative solutions, over the entire life cycle of any process as well as at the interaction between the process and the environment and the economy in which it is embedded (…)" (Ghisellini et al., 2016, p. 12). Thus, CE has the potential to be a model for entirely new living and economic configurations, which overcome business-as-usual resource management.

To conclude, there are various authors who highlight the importance of CE for city planning but further research is necessary to clarify this relationship. Despite the critical importance of land as a source of biomass and energy, for example (Hubacek & Van Den Bergh, 2006, p. 23), and space as interlinked with the main arguments for CE including environmental, economic, social and geostrategic improvements (Van den Berghe & Vos, 2019, p. 2), " [i]t is still unclear how land use can be integrated into CE-related initiatives, design, and evaluation" (Winans et al., 2017, p. 830). However, city planning is considered as a key for moving towards more circular cities because it affects relevant urban strategies such as mobility or construction (Petit-Boix & Leipold, 2018, p. 1279), can provide for the geographical–spatial proximity necessary for industries and businesses to close energy and material loops (Winans et al., 2017, p. 830), and can—through its systemic perspective—promote regulations for stimulating reuse, recovery, repair and the maintenance of existing resources (Girard & Nocca, 2019, p. 34). Agudelo-Vera et al. even argue that some form of urban planning has always been present in the relationship between resource management, urbanisation and technological development (Agudelo-Vera et al., 2011, p. 2302). Still, dealing with resource flows has barely found its way into urban planning literature. Two exceptions are I. McHarg's Design With Nature and M. Mostafavi's Ecological Urbanism. The former aims to protect the values of natural processes and uses matter and cycles as inspiration for sustainable landscape design (McHarg, 1969, p. 55) while the latter argues for overcoming the inherent conflictual conditions between ecology and urbanism (Mostafavi, 2010, p. 3). The premise for bridging resource flows with urban planning and design, however, is an interdisciplinary understanding of the relationships between socio-economic and environmental dynamics and the built environment (Remøy et al., 2019, p. 2).

Recent examples make clear that spatial factors influence the viability of more complex, circular, infrastructural systems, as demonstrated by the district Hammarby Sjöstad in Sweden (Williams, 2013, p. 3) or a case from the Finnish pulp and paper industries (Ghisellini et al., 2016, p. 26). Another example is Saint-Vincent-de-Paul, the transformation of a former hospital in the centre of Paris, which enabled by its strategic location, implemented CE goals, such as the storage and reuse of building components as well as the revitalisation through experimental projects (Appendino et al., 2019, p. 11). In the context of city planning and CE, specific emphasis should be given to the role of people and, in particular, low-income households since mainly a socio-technical approach has been used as of yet (Pomponi & Moncaster, 2017, p. 17). Thus, introducing CE for city planning means to develop areas based on the principle of decentralised, interconnected, polycentric circular urban systems. This requires a re-evaluation of traditional values of the disciplines urban planning and urban design including the scalability of solutions, the approach to infrastructure, the creation of interrelated networks and the role of public space. It would also require an integrated vision and management of the many existing planning tools at the municipal level and appropriate policy instruments (Girard & Nocca, 2019, p. 38; van der Leer et al., 2018, p. 301; Winans et al., 2017, p. 830).

3 Analytical Framework for Circular Urban Developments (AFCUD)

To the knowledge of the authors, there exist at least three evaluation frameworks for CE in urban planning that are designed for specific research goals: the Tetrahedron—a Discourse-Institutional Analytical Framework (Van den Berghe & Vos, 2019, p. 6); the Integrated Evaluation Framework for the Circular City (Girard & Nocca, 2019, p. 41); and the V-H CE Evaluation Framework for Urban Planning (van der Leer et al., 2018, p. 301). These tools make clear that a multi-scale, systemic approach is needed for analysing resource-conscious urban planning. When applied, information can be obtained on the complex interlinkages between transformation processes and resource flows (Voskamp et al., 2018, p. 524). Typically, within CE research there are three key aspects that define different overarching research directions: the technical requirements for closing loops, socio-technical criteria and socio-ecological criteria (Barrera, 2016, pp. 20–24). We enlarge this view by introducing the SETS model that includes social, ecological and technological systems and their dynamic interactions as starting points (Urban Systems Lab, 2020). In addition, the following four multi-scale systemic categories were defined to particularly frame the

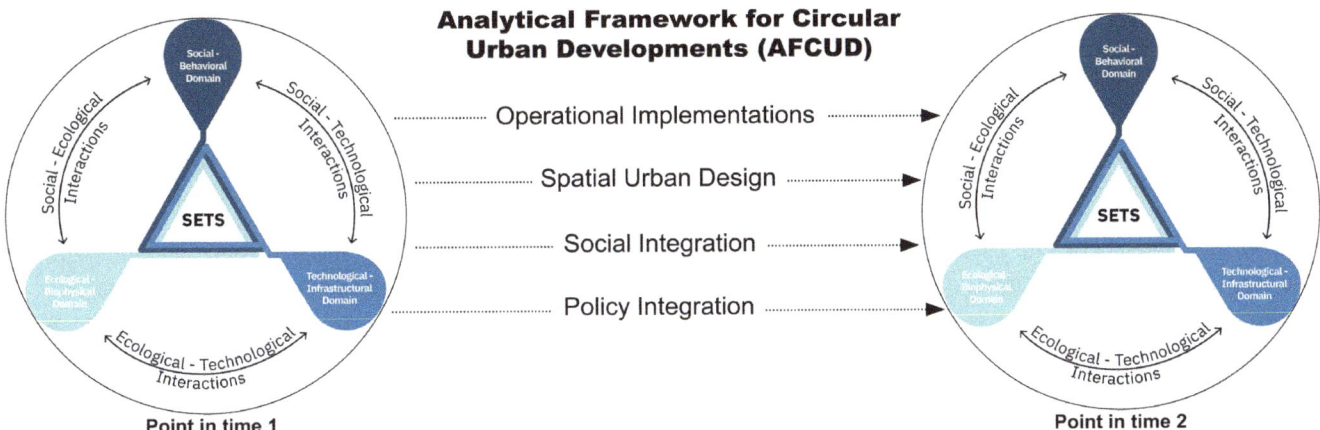

Fig. 1 Analysis of a CE related urban transformation process through the selection of two points in time complemented with four cross-cutting categories

evaluation towards circularity and to be able to focus on the transforming space as well as the political ecology of the processual interactions in reference to the SETS model: operational implementations, spatial urban design, social integration and policy integration (See Fig. 1). 'Operational implementations' refers to the technical requirements for closing loops and the day-to-day components that are relevant in an applied CE. 'Spatial urban design' scrutinises the production of urban space within a given development. 'Social integration' refers to the type of inclusion of local citizens in the redevelopment processes including those who live, work and use the neighbourhood with a specific focus on those who are part of low-income housing schemes. Finally, 'policy integration' is defined as a formalised common set of goals and rules that promote cross-functional communication, collaboration and optimisation that ends ideally in a more or less holistic governance of the systems and subsystems involved.

4 Buiksloterham: Transformation in Progress

The district of BSH is slowly transforming into a revitalised mixed-use area following the ambition to become a fully circular neighbourhood. BSH is a former industrial port area located in the north of Amsterdam with a size of around 100 hectares. In the early twentieth century, it was home to large private productive industries, for example, Fokker Aviation and a waste and power plant that operated until 1993. In the 1980s, the municipality re-examined the area's zones and functions towards a mixed-use residential and working district. In the course of a spatial vision and land development declaration in 1998, the municipality purchased a large share of the industrial property in BSH (Metabolic, 2015, p.86). In 2006, the municipality of Amsterdam took the decision to

redevelop the BSH district and published the first version of an investment decision document for the area. As a consequence of the financial crisis and the lack of commercial real estate development interest, the municipality launched a call for tender for the former DeCeuvel shipyard in 2010 (Gladek, et al., 2014, p. 90). Targeted were creative and sustainable proposals for a self-build temporary use. Following the well-received winning proposition for a contaminated soil regeneration through phytoremediation vegetation, the municipality articulated their interest in a vision for 'Circular Buiksloterham'. The planning framework for a circular district was then developed in 2014 as a collaboration between the municipality (Gemeente Amsterdam), an urban design and architecture office, academics as well as urban and cultural organisations (Table 1). The project was termed 'Living Lab Plan' with the underlying idea that individuals, households, collectives and private businesses build their own homes side by side—driven by the fundamental idea of closing urban material cycles (van Bueren & Steen, 2017, p. 18). The following sections focus on the recent transformation of the district and will evaluate the urban regeneration of the district between two points in time (2015, 2020) and along the defined categories in the analytical framework considering social, ecological and technological perspectives.

4.1 Operational Implementations

Material Use: As key circular operation, building materials and their recycling methods are envisioned to be closed loops in the new building developments of the district. The Manifesto states an ambition of nearly 100% material recovery and reuse for any new builds in the area of BSH and demands a material passport "to record material properties and origin" (Metabolic, 2015). These two ambitions are readopted in the Ivn.B.2019 but the building material reuse is further specified for 50% recycled materials and 30% renewable materials; and 80% of both in public space interventions. One of the first successful implementations of recycled materials was completed in the mobility hub space of the 'Schoonschip' scheme by using recycled baked bricks (Gemeente Amsterdam, 2019, p. 68).

Waste: To enhance a full waste recycling plan at urban scale, the overall strategy targets the apartment level and user-behaviour as well as the overall district level. Different strategies were envisaged for further development in both source documents. As for the Manifesto, waste separation is laid out for buildings and public spaces with source separation via a colour-coded bin system, containers on ground floors, and through vacuum systems that transport waste underground to a central facility (Metabolic, 2015). An explicit technical proposal for waste recycling at the user end is a sink drainage macerator that grinds food and organic waste in apartment kitchens. The biomass is centrally collected and used for green energy generation. In the recent Ivn.B.2019, these two strategies are detailed resulting in a waste separation agenda based on six waste fractions for both commercial and residential. The collection will be underground and laid out for efficient use of space at district level, e.g. keeping short distances between the end user and container locations. This solution, which is proposed by the municipality, is only assigned for households, yet larger companies need to directly contract waste disposal businesses that implicate longer hauls (Interviewee 3, personal communication, June 5, 2020).

Water: the analysis of all three sources made clear that the technical circular objectives concerning water management mostly proceeded to a successful implementation at a small to medium scale. The Manifesto plots two primary aims: first, natural water management and the smart use of water and second, efficient treatment of wastewater for nutrient recovery and micropollutant removal (Metabolic, 2015). This results in the definition of separate urine collection, water saving measures, rain water collection and natural water buffering, resource recovery and entire micropollutant removal (Metabolic, 2015). The companies Waternet and DeAlliante started to realise a black and grey water separation system for about 100 households in the course of the Living Lab project of 'Schoonschip' (See Fig. 2). In this system, a neighbourhood biorefinery recovers minerals from grey water that can be reused as fertiliser and biogas is produced from the black water that is further used as energy. The black water is collected from the toilets through a vacuum sewer system (Interviewee 1 & 2, personal communication, November 6, 2019). Alongside heat generation from grey water recovery, the Ivn.B.2019 suggests vacuum sewage and wastewater separation for three plots with over 600 dwellings including 'Schoonschip' and optionally some connections to large mixed-use plots (Gemeente Amsterdam, 2019). In practice, the vacuum toilet flush and its increased noise level was received divergently throughout the resident groups. The acceptance in social housing properties is very low compared to the free market properties and Living Lab projects (Interviewee 3, personal communication, June 5, 2020).

Energy: With respect to energy-related strategies, it is a general goal of the City of Amsterdam to reduce CO_2 emissions by 55% until 2030 compared to 1990 and to become fully supplied with renewables by 2050. The primary approach for this goal is the production of local energy—ideally directly on the plots. In the Ivn.B.2019, it is stated that this is obligatory for all energy except district heating. The following strategies for this goal's accomplishment are proposed for private land: (a) using roof and façade areas to generate electricity through photovoltaics and (b) using surface water for cooling and/or regeneration. For public space, the goal is to become energy-neutral and provide sustainable energy from solar panels (Gemeente Amsterdam, 2019). In

Fig. 2 Different housing projects in Buiksloterham: floating architecture of the Schoonschip project (left), elevated walkways and reused boats of DeCeuvel (centre) and the beginnings of the Cityplot project (right). *Sources* author's own

comparison, the Manifesto constitutes an estimated energy supply decline of 60% financed by the Circular Investment Fund and maps out pilots for a parallel AC/DC smart grid (Metabolic, 2015). Selected were the two Living Lab projects 'DeCeuvel' and 'Schoonschip' (See Fig. 2) that are equipped with a smart grid and are envisioned to be scaled via tenders on municipal, leasehold or private plots in BSH, as indicated in the investment memorandum. 'DeCeuvel' established an energy token for all locally generated renewable energy, which is tracked in a blockchain and a money token to trade the produced energy with local goods in the community (Metabolic, 2018, p. 1). Following in the footsteps of 'DeCeuvel' and 'Schoonschip', two additional plots will implement a smart grid in the upcoming years and establish a local energy market for residents. This way, users will be able to trade energy that can not be used sustainably by themselves (Gemeente Amsterdam, 2019).

4.2 Social Integration

As part of the assessment in preparation for the Manifesto in 2014, stakeholder analyses and interviews were conducted to assign interested parties to engagement groups. The most significant declaration hereof states: "it is impossible to create any kind of meaningful or lasting change without broadly engaging the parties within a particular context who either: (1) have decision making power or (2) are subject to the results of decisions made" (Metabolic, 2015). The project coordinator Metabolic and Stadslab (local communication platform) held stakeholder meetings at regular intervals, especially with residents. A member intensively involved in these meetings, however, strikingly expressed that the participants' attention and willingness to invest time was most challenging during the process but, over time, resident groups were increasingly taking part in participation formats. Yet, there is a tendency that people with a 'green living mindset' have been moving to BSH (Interviewee 1, personal communication, November 6, 2019). In contrast, residents of the social housing schemes who were assigned to a home in the area and did not specifically apply for living there are generally less engaged in the circular neighbourhood development. A simple formula for the future of social integration to develop an urban circular economy was voiced by one interviewee: "In the end, it's not about what we have achieved so far, it's about making circular thinking mainstream" (Interviewee 1, personal communication, November 6, 2019). The Manifesto set out a Neighbourhood Action Plan as an active participation method that translates high-level goals into everyday actions (Metabolic, 2015). This was introduced together with an award economy on community and Living Lab level where people can trade their self-building and other skills in exchange for money (Metabolic, 2015).

In preparation for the recent investment memorandum, the municipality (Gemeente Amsterdam) hosted several stakeholder meetings during which the details of the area development were left aside but instead the focus was on participatory instruments and developing common visions. Additionally, walk-in evenings with the purpose of relationship management between residents, entrepreneurs and developing parties were held monthly with most questions forwarded to the municipality (Gemeente Amsterdam, 2019). Nonetheless, the investment document records that public participation will only be administered by the municipality for public space with exact topics defined for wilful engagement. For instance, this includes existing road refurbishment as well as park, playground and sport facility design (Gemeente Amsterdam, 2019). For all other plots, the level and method of participation needs to be proposed by the developer who applies for a building permit. Overall, the envisioned participation is based on a new policy framework called 'Democratisation' that frames Amsterdam's principles and action streams with regard to increasing the control and ownership of residents over their neighbourhood. It is the goal to decide up front which instrument of participation will be chosen and the intention is to clearly communicate when, why and which form of cooperation is intended (Gemeente Amsterdam, 2019).

4.3 Urban Spatial Design

For the development of BSH, a mixed working and living structure at district level is foreseen. In contrast to the Manifesto, where an equal distribution between production and living was proclaimed (Metabolic, 2015), a minimum of 20% of the programme is reserved for business activities and a maximum of 70% for housing according to the Ivn.B.2019 (Gemeente Amsterdam, 2019). This shift might indicate the great pressure on the housing market Amsterdam is currently experiencing. According to the municipality of Amsterdam, a productive city district contributes to circularity in the way that production and consumption flows are allocated closer to each other (Gemeente Amsterdam, 2019). The programmatic mixture gives rise to the need for flexible and adaptive buildings that can react to programmatic change. Therefore, rules for the development of building plots were introduced (Gemeente Amsterdam, 2019). Also, special zones for the experimental development of new building typologies were declared (Gemeente Amsterdam, 2019). The aerial views depicted in Fig. 3 clearly reveal the development progress of the two mentioned experimental zones 'DeCeuvel' and 'Schoonschip' over recent years. Furthermore, development and construction on the site of the 'Cityplot' has started. After the experimental zones, this is the first larger scale area

Fig. 3 Aerial views of BSH at different points in time indicating the progress of development

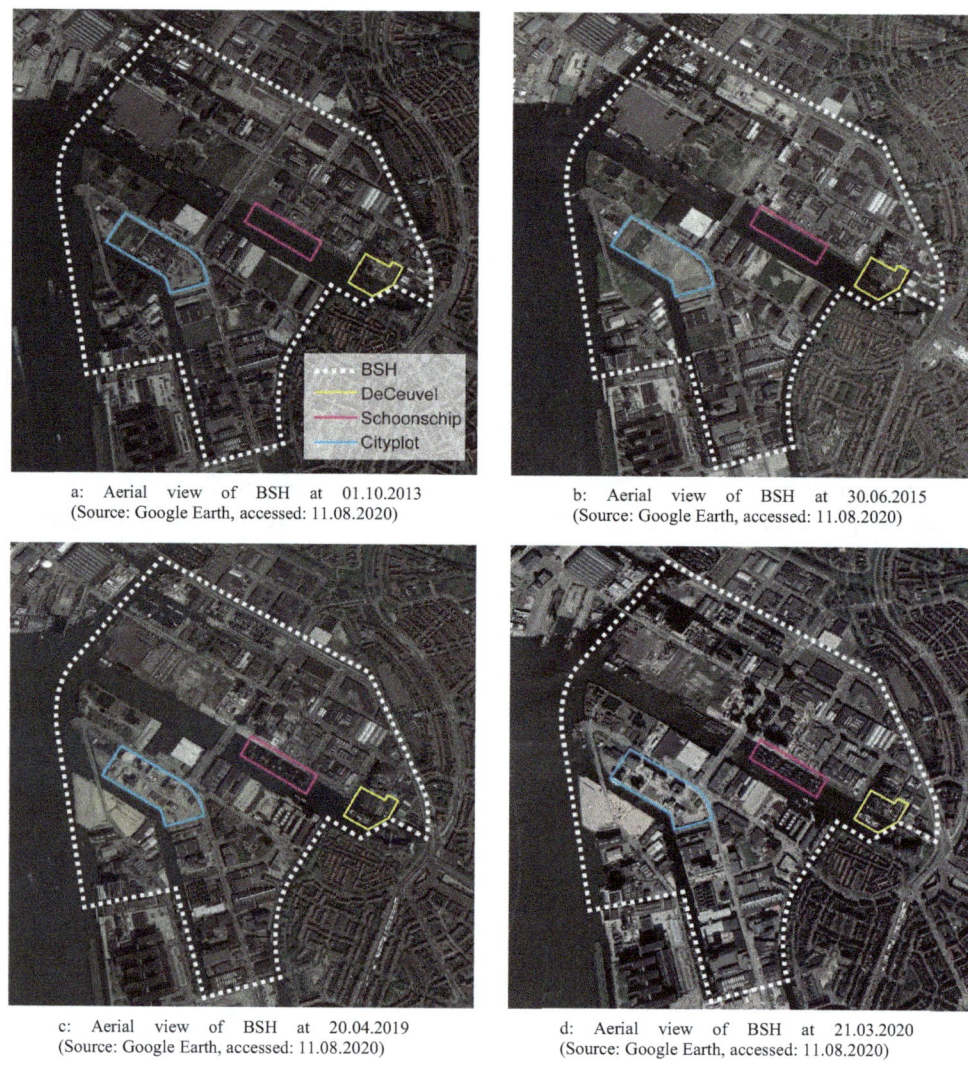

a: Aerial view of BSH at 01.10.2013
(Source: Google Earth, accessed: 11.08.2020)

b: Aerial view of BSH at 30.06.2015
(Source: Google Earth, accessed: 11.08.2020)

c: Aerial view of BSH at 20.04.2019
(Source: Google Earth, accessed: 11.08.2020)

d: Aerial view of BSH at 21.03.2020
(Source: Google Earth, accessed: 11.08.2020)

that is supposed to be developed following the guidelines of the Inv.B.2019.

Mobility: One of the operational fields that to a large extent interferes with aspects of spatial and urban design is mobility. The Manifesto demands zero-emission mobility in the area of BSH by reducing the overall mobility and considering concepts such as shared mobility and mobility hubs with the opportunity to host innovative modes of transport as well as elevated bike and pedestrian paths (Metabolic, 2015, p. 32, 39,164). Importance is given to the integration of the waterways (River IJ, Johan van Hasseltkanaal) as a hub not only for passenger transport but also for any other kind of infrastructural connection and logistics (Gemeente Amsterdam, 2019; Metabolic, 2015). Resembling partly the proposal in the Manifesto, the Ivn.B.2019 sets a general plan of parallel roads and canals that give access to the plots, which —depending on their depth—are interconnected by informal green pathways. Most of the plots allow public access to the waterfront (Gemeente Amsterdam, 2019). It can be seen as a

spatial decision to ensure water accessibility for a later implementation of technical interventions that interact with water areas (Gemeente Amsterdam, 2019). As for circular mobility, the Ivn.B.2019 plans a bicycle network embedded at city scale, the encouragement of subsystems for (electric) cars, and a combination of bicycle and car parking. Restrictions for motorised through-traffic are additionally planned to reduce any further emissions (Gemeente Amsterdam, 2019). The ambition to establish an integrated mobility system was mentioned as one of two successfully started projects touching on the circular planning mindset. (Interviewee 1, personal communication, November 6, 2019).

Green Areas: Another aspect that affects both spatial planning and the circular constitution of a city district is the realisation of green areas and vegetation. According to the vision of 2015, street vegetation should increase ecological and economic value and urban greenery is intended to relieve local sewage systems (Metabolic, 2015). Spatially disconnected and inaccessible green areas are supposed to be integrated into the city structure (Metabolic, 2015). The current zoning plan reacts to these requirements. The vision of having a "green neighbourhood bustling with activity" (Metabolic, 2015) is continued in the Ivn.B.2019. At district level, at least 22 m wide green corridors are placed to connect canals and streets to guarantee the buffering of water and heat (Gemeente Amsterdam, 2019). Additionally, river and canal banks will be reserved for greenery and public parks (Gemeente Amsterdam, 2019). At plot scale, greenery is to be included and integrated and will be part of tender negotiations. Especially on deep parcels, green courtyards are essential to support circular goals of rainwater reuse and increased biodiversity.

4.4 Policy Integration

The intention of establishing a circular city district requires a stable policy framework that exceeds the given laws without losing its original intention (Metabolic, 2015). The Manifesto lays out a policy regarding the Living Lab idea, executed, e.g. in projects like 'DeCeuvel' and 'Schoonschip' (See Fig. 2) in form of a 'Neighbourhood Action Plan'. This required, among others, the declaration of 'special physical zones' for urban experiments in order to be able to continuously experiment within the given laws (Metabolic, 2015). Furthermore, detailed guidelines for housing developers concerning energy demand and the choice of building materials are postulated (Metabolic, 2015). The Manifesto proposes loosening restrictions in current laws to make an integration of technical interventions feasible, for example, the reuse of grey water (Metabolic, 2015). In practice, the process of adapting the circular approach to current rules is depicted by an involved planner. He describes the encounter of barriers set by legislation as a habitual thing when accessing a new innovative idea. However, translating solutions into existing legislation settings often turned out to be a tiring and time-consuming process (Interviewee 2, personal communication, November 14, 2019). By signing the Manifesto in 2015, the municipality declared the 'Circular Buiksloterham' approach an inherent part of all future developments in the area (Gemeente Amsterdam, 2019). To emphasise that the ambitions displayed in the Manifesto of 2015 will be complied with, the essential points are partly translated into a set of 'circular rules', valid from 2019 onwards. But the municipality aims at a different level of policy integration due to its impetus of setting a framework for developing an entire city district rather than only focusing on experimental Living Labs. Circular rules are applied at two levels: firstly, the so-called 'starting points' are defined as minimum requirements to be met; secondly, 'ambitions' can be discussed within the tendering process of

a new development project (Gemeente Amsterdam, 2019). According to a representative of the municipality, the mentioned framework for BSH has higher ambitions towards circularity than other districts. The level of innovation included in the circular urban development idea and formalised in both the Manifesto and Ivn.B.2019 is a requirement when approaching a new development (Interviewee 3, personal communication, June 5, 2020). Negotiations are conducted with and are depending on the influence of the sustainability advisor responsible for the area (Interviewee 3, personal communication, June 5, 2020). The suggested special physical zones can be re-encountered in the spatial programme proposed by the municipality that foresees experimental forms of housing as well as the experimental development of production (Gemeente Amsterdam, 2019). Exemplary is the approval of the 'Schoonschip' project through an amendment plan. By making use of the option of deviation offered in the zoning plan, an agreement could be made and innovative ideas were realised (Gemeente Amsterdam, 2019). The set of rules described as circular mainly focuses on the implementation of circular aspects at an operational level, comparable to the guidelines for (housing) developers and the technical interventions in the Manifesto.

5 Discussion

In this article, we analysed the urban regeneration of 'Circular Buiksloterham' with a spatio-temporal approach taking into account a time span of 5 years to understand more closely how the concept of CE is interrelated with city planning. First of all, the city of Amsterdam is a key driver in pushing the development of BSH towards a resource conscious neighbourhood. This can be noticed, for example, in more ambitious regulations than in other districts, the financial commitment and the establishment of 'special physical zones' for experimentation with existing laws. Additionally, the aerial images show a certain level of progress regarding the implementation. Besides greenery for climatic functions and recovery strategies for operational elements (energy, water and materials), aspects of circularity were also considered in the spatial design of the district. All the plots have proximity to the water and to the streets for the potential implementation of technical interventions (e.g. traffic or recycling infrastructure) and waste hubs were installed at an early stage. Thus, by considering the results of the analysis and against doubts from the literature review, a circular urban development is taking place. The proposed and partly implemented solutions in BSH go beyond the rhetorical level and the spatial dimension is included in the innovation process. However, it is interesting to see that the target for the share of housing in the area increased from 50 to 70% during the observed period between 2015 and 2020, whereas the production/business share declined from 50% to a minimum of 20%. This might be connected with the city-wide economic pressure to build more social housing and opens up simple technical questions for a 'full circularity'. For example, will it be possible to generate enough electricity or heat on-site with a significantly lower amount of industry in the area? Another striking aspect is the introduction of starting points and optional ambitions for developers. For example, in terms of material use this means that all the timber used for construction must comply with forest management standards but the application of 50% recycled materials and 30% renewable materials is optional. This might be another drawback for realising the ambition of a fully circular neighbourhood. Given these points, our general impression is that 'Circular Buiksloterham' is at a stage where it is scaling from small Living Lab zones to larger developments such as 'Cityplot' with 550 dwellings (See Fig. 2), and eventually to the entire district. This seems to result in a relaxation of the requirements and in a new development logic that is no longer targeted at specific user groups but instead entails planning for a more general public while at the same time answering to economic pressures.

The modifications made at the planning level over time suggest a less holistic implementation of CE principles than originally proposed. Yet, conceptually speaking, there is agreement in the literature that CE is a radical concept with far-reaching implications for human systems. When the Manifesto was published, a holistic approach was proposed including reward schemes and rules for social behaviour. The IncB.2019's ambition and view on circularity follow the Manifesto but the definition of the actual rules reveal trade-offs. At this point, the nature of the CE approach shifts to a rather operational, technically driven concept. Although the renewable energy reward system established at 'DeCeuvel' shows that under 'urban laboratory conditions' a closed regenerative system can lead to a strong alternative. In fact, urban experiments on a small scale can be successfully operated and legislative hurdles can be cleared but the development process is time and energy consuming (Interviewee 2, personal communication, November 19, 2019). And since the next step of the development is to scale up the size of plots and the number of inhabitants in order to address the housing shortage, a transition of the good practices collected at the Living Labs seems to fall flat. Thus, due to the scale of implementation, the more radical theoretical notion of CE including behaviour and system changes conflicts with the new plan for Buiksloterham that is predominantly relying on technical and operational solutions (See Appendix).

Taking a closer look at the social dimension reveals that the suggestions in the literature review about missing emphasis on social considerations while implementing CE-related urban functions are reasonable. In the case of Buiksloterham, the two pioneer projects 'DeCeuvel' and 'Schoonschip' (See Fig. 2) show that the idea of Urban Circularity attracts 'avantgarde' thinkers who constitute a small section of society (Interviewee 3, personal communication, June 5, 2020). In general, the target group of the two Living Lab projects were young, well-educated people including entrepreneurs who contributed with their own ideas about the design of the buildings via a participation process. Sharing the same convictions and beliefs around closing loops makes them invest time, money and energy, taking into account the barriers and associated frustrations (Interviewee 2, personal communication, November 14, 2019). Thus, it is of great importance to address the question of social acceptance for new systemic solutions that include technical artefacts that are not trivial but necessary due to the requirements of circularity. This is exemplified by the introduction of a vacuum toilet system, which is a component of a loop-closing technical intervention. The increased noise levels, the somewhat unusual design of the toilets and the strict guidelines for what to put into this system were not well received by social housing tenants while it was successfully tested in the experimental environment at 'DeCeuvel'. Thus, new systems that are necessary or planned in order to close loops at neighbourhood scale need to be clearly articulated and widely tested in order to raise acceptance for the required changes. Otherwise, this can lead to refusal among future residents because the acceptance of a new technology or a change in behaviour is not scaleable but instead requires a step-by-step adaptation process.

Lastly, the proposed analytical framework (AFCUD) was useful to structure the research process and to generate conclusions regarding the relationship between CE and city planning. Interestingly, as also the reviewed literature demonstrates, the ecological perspective was weakly covered during our analysis. This was first and foremost due to the lack of data, however analysing Urban Circularity from an ecological perspective holds potential for future research.

6 Conclusion

The case of 'Circular Buiksloterham' indicates a strong relationship between CE and city planning. It became clear that the central component of city planning, its spatial logic, is coupled with resource-oriented fields such as energy, water, waste and materials. These are important for establishing a CE while every systemic intervention within those fields generates spatial conditions and dependencies. The development rules of BSH clearly include ideas that are derived from resource management and the notion of circularity is indeed part of all the analysed research categories. At the same time, there are two well established Living Labs that have been implemented in recent years taking into consideration circularity measures on both construction and behavioural levels. But making the step from those experimental zones to an entire 'circular district' was identified as the key challenge in Buiksloterham. Many prerequisites, such as a joined-up regulatory framework as well as technical and political levers, are in place but there is increasing economic pressure to build affordable housing and the institutional capacity for scaling up the Living Labs including lengthy coordination processes is lacking. Instead, the guiding principles were relaxed to make way for large-scale housing developments—maybe still more ambitious regarding circularity than other projects but no longer following the holistic idea of CE. Scaling up as the next step in the development process also reveals a double social problem when combining principles of CE and city planning at district scale. On the one hand, there is a need to accept new technological solutions necessary for Urban Circularity and on the other hand, the appropriation of a new reality is necessary that includes the adaptation of behaviour and an openness for cultural change. With this research, we addressed two research questions. How the ideas of CE impact city planning at neighbourhood scale was answered

throughout the paper with the example of BSH. Whether the Living Lab approach of 'Circular Buiksloterham' can be mainstreamed in city planning is more difficult to answer. Our opinion is that there is a lot of potential in the areas of energy, waste, water reuse, biodiversity and ecosystem services but the case of BSH showed that the closer the realisation of a 'fully circular neighbourhood' is, the less radical the proposed solutions will be. Further research is necessary to address the 'problem of scale' within the idea of Urban Circularity and if and how the marriage of circularity and urban planning contributes to environmental goals such as climate-neutrality.

Acknowledgements We wish to thank our three interview partners for their valuable insights into the planning process of 'Circular Buiksloterham', two anonymous reviewers who provided useful comments as well as Annett Bochmann, Isabel Ordoñez and Vera van Maaren for comments on earlier versions of this paper.

Appendix

See Table 1.

Table 1 Circular rules (Aadapted from Gemeente Amsterdam, 2019, p. 79)

Subject (Definition)	Rules
General	**Starting Points** The developer subscribes to the circular ambitions letter of BSH and takes the included principles and rules into account during the plot development
Sustainable Mobility Mobility that contributes to a clean, safe and accessible neighbourhood	**Starting Points** – Sufficient space for parking bikes – At least 25% of the parking spaces in parking facilities should be equipped with electric charging points. Prepare the remaining 75% of parking spaces for electric recharging by using sheathing tubes and cable ducts. A checklist is available on request – Smart and sustainable construction logistics **Ambitions** – Application of subsystem for (electric) (container) bicycles or (electric) cars on the plot – Linking local sustainable energy generation to energy storage in electric cars
Energy The energetic quality of the programme and its functions	**Starting Points** Energy-neutral or energy-generating new building with an EPC $\leq$ 0,00 (in accordance with NEN 7120) – When determining the EPC, renewable energy is generated on location/plot (with the exception of district heat and cold) – The application of the EMG (NEN 7125)—sustainable energy generation remotely—is excluded – Renewable energy generation on location must be incorporated into the architectural design **Ambitions** – Application of smart grid system on the plot, possibly linked to electric cars charging points. The various systems will be headed towards a smart grid at district level

(continued)

Table 1 (continued)

Subject (Definition)	Rules
Durable heat and cold supply The way in which the programme is provided with space heating, hot tap water and cold supply	**Starting Points** – A connection to natural gas is not possible – There is an obligation to connect to the municipal grid 'Westpoort heat network' – The developer/initiator can create an exemption or apply for exemption from the connection obligation if the development is sufficiently sustainable – The application of conventional cooling or heating systems is not a problem – Boilers, electric boilers or conventional cooling machines are not allowed
Commodities and materials The application of raw materials and materials on the development plot	**Starting Points** – The development has an MPG $\leq$ 0.5 (excluding input PV panels) – All the timber used must demonstrably comply with the principles of sustainable forest management (FSC label or equivalent) – Application of material passport **Ambitions** – Apply 50% recycled material – and 30% use of renewable materials
Waste collection The way household waste is collected	**Starting Points** – Facilitate separate waste collection on the plot—in at least the following six fractions: glass, paper, PMD (plastic/tin/metal/beverage cartons), textiles, GFE (vegetable, fruit and vegetable/food waste) and residual waste
Green and Biodiversity The amount of green space, the quality of green spaces	**Starting Points** – Within a development unit, an area that is comparable to at least 40% of the surface area of the development unit should consist of greenery. This includes green roofs, green skins or greening of the ground level that is not built on. For the roof, a minimum substrate thickness of 10 cm is required – Biodiversity: minimum application of three measures from the nature-inclusive building manual
Water: Rainproof and reusing water The way in which measures are taken at plot level that contribute to climate mitigation and water reuse	**Starting Points** – The minimum water storage on the lot is the total lot area x 60 mm – 60 mm precipitation (=60 L per sqm) is retained for more than 24 h – The water is drained off with a maximum rate of 2.5 L/sqm/hour – Collected rainwater collected should be reused for the irrigation of green spaces on the plot **Ambitions** – Rainproof for 90 mm precipitation – Reuse of water for toilet flushing or possible other functions
Flexible and future-oriented buildings	**Starting point** – Separation of carrier and installation is mandatory for a flexible building layout that possibly can change in the future

References

Agudelo-Vera, C. M., Mels, A. R., Keesman, K. J., & Rijnaarts, H. H. M. (2011). Resource management as a key factor for sustainable urban planning. *Journal of Environmental Management, 92*(10), 2295–2303. https://doi.org/10.1016/j.jenvman.2011.05.016.

Appendino, F., Roux, C., Saadé, M., & Peuportier, B. (2019). Circular economy in urban projects : a case studies analysis of current practices and tools. *2019 Aesop*, (Ademe 2014).

Barrera, P. P. (2016). *Circular Buiksloterham: Assessing Circular Urban Development in Amsterdam North*. https://doi.org/10.13140/RG.2.1.2926.5120.

Blomsma, F., & Brennan, G. (2017). The emergence of circular economy: A new framing around prolonging resource productivity. *Journal of Industrial Ecology, 21*(3), 603–614. https://doi.org/10.1111/jiec.12603.

Bocken, N. M. P., de Pauw, I., Bakker, C., & van der Grinten, B. (2016). Product design and business model strategies for a circular economy. *Journal of Industrial and Production Engineering, 33*(5), 308–320. https://doi.org/10.1080/21681015.2016.1172124.

Calisto Friant, M., Vermeulen, W. J. V., & Salomone, R. (2020). A typology of circular economy discourses: Navigating the diverse visions of a contested paradigm. *Resources, Conservation and Recycling, 161*(November 2019), 104917. https://doi.org/10.1016/j.resconrec.2020.104917.

Ellen MacArthur Foundation. (2013). *Towards the Circular Economy: Opportunities for the consumer goods sector*. Retrieved from https://doi.org/10.1162/108819806775545321.

European Commission. (2020). *Circular Economy Action Plan for a cleaner and more competitive Europe. #EUGreenDeal*.

Geissdoerfer, M., Savaget, P., Bocken, N. M. P., & Hultink, E. J. (2017). The circular economy—a new sustainability paradigm? *Journal of Cleaner Production, 143*(0), 757–768. https://doi.org/10.1016/j.jclepro.2016.12.048.

Gemeente Amsterdam. (2019). *Investeringsnota Buiksloterham 2019*. Retrieved from https://www.amsterdam.nl/projecten/buiksloterham/plan-publ-buiksl/.

Ghisellini, P., Cialani, C., & Ulgiati, S. (2016). A review on circular economy: The expected transition to a balanced interplay of environmental and economic systems. *Journal of Cleaner Production*, 114, 11–32. Retrieved from https://doi.org/10.1016/j.jclepro.2015.09.007.

Girard, L. F., & Nocca, F. (2019). Moving towards the circular economy/city model: Which tools for operationalizing this model? *Sustainability (Switzerland)*, 11(22), 1–48. https://doi.org/10.3390/su11226253.

Gladek, E., Van Odijk, S., Theuws, P., & Herder, A. (2014). *Circular Buiksloterham*. De Alliantie, Waternet, Ontwikkelingsbedrijf Gemeente Amsterdam.

Hall, P., & Tewdwr-Jones, M. (2019). *Urban and regional planning*. Urban and Regional Planning. https://doi.org/10.4324/9781351261883.

Homrich, A. S., Galvão, G., Abadia, L. G., & Carvalho, M. M. (2018). The circular economy umbrella: Trends and gaps on integrating pathways. *Journal of Cleaner Production, 175*, 525–543. https://doi.org/10.1016/j.jclepro.2017.11.064.

Hoornweg, D., Sugar, L., & Gómez, C. L. T. (2011). Cities and greenhouse gas emissions: moving forward. *Environment and Urbanization, 23*(1), 207–227. https://doi.org/10.1177/0956247810392270.

Hubacek, K., & Van Den Bergh, J. C. J. M. (2006). Changing concepts of 'land' in economic theory: From single to multi-disciplinary approaches. *Ecological Economics, 56*(1), 5–27. https://doi.org/10.1016/j.ecolecon.2005.03.033.

IPCC (2018). Summary for Policymakers. In: Global warming of 1.5° C. An IPCC Special Report on the impacts of global warming of 1.5°C above pre-industrial levels and related global greenhouse gas emission pathways, in the context of strengthening the global response to the threat of climate change, sustainable development, and efforts to eradicate poverty. In V. Masson-Delmotte, P. Zhai, H. O. Pörtner, D. Roberts, J. Skea, P. R. Shukla, A. Pirani, W. Moufouma-Okia, C. Péan, R. Pidcock, S. Connors, J. B. R. Matthews, Y. Chen, X. Zhou, M. I. Gomis, E. Lonnoy, T. Maycock, M. Tignor, T. Water eld (Eds.), World Meteorological Organization, Geneva, Switzerland, p. 32.

IRP (2018). The Weight of Cities: Resource Requirements of Future Urbanization. In M. Swilling, M. Hajer, T. Baynes, J. Bergesen, F. Labbé, J. K. Musango, A. Ramaswami, B. Robinson, S. Salat, S. Suh, P. Currie, A. Fang, A. Hanson, K. Kruit, M. Reiner, S. Smit, S. Tabory, S. (Eds.) *A Report by the International Resource Panel*. Nairobi, Kenya: United Nations Environment Programme.

Kennedy, C., Cuddihy, J., & Engel-yan, J. (2007). The Changing Metabolism of Cities. *Industrial Ecology, 11*(2).

Kirchherr, J., Reike, D., & Hekkert, M. (2017). Conceptualizing the circular economy: An analysis of 114 definitions. *Resources, Conservation and Recycling, 127* (September), 221–232. https://doi.org/10.1016/j.resconrec.2017.09.005.

Korhonen, J., Nuur, C., Feldmann, A., & Birkie, S. E. (2018). Circular economy as an essentially contested concept. *Journal of Cleaner Production, 175*, 544–552. https://doi.org/10.1016/j.jclepro.2017.12.111.

Lehmann, J., Mempel, F., & Coumou, D. (2018). Increased occurrence of record-wet and record-dry months reflect changes in mean rainfall. *Geophysical Research Letters, 45*(24), 13,468–13,476. https://doi.org/10.1029/2018GL079439.

Lieder, M., & Rashid, A. (2016). Towards circular economy implementation: A comprehensive review in context of manufacturing industry. *Journal of Cleaner Production, 115*, 36–51. Retrieved from https://doi.org/10.1016/j.jclepro.2015.12.042.

Marcotullio, P. J., Sarzynski, A., Albrecht, J., Schulz, N., & Garcia, J. (2013). The geography of global urban greenhouse gas emissions: An exploratory analysis. *Climatic Change, 121*(4), 621–634. Retrieved from https://doi.org/10.1007/s10584-013-0977-z.

McHarg, I. (1969). *Design with Nature*. University of Pennsylvania.

Metabolic. (2015). *Circular Buiksloterham: Transitioning Amsterdam to a Circular City*. Retrieved from https://www.metabolic.nl/publications/circular-buiksloterham-roadmap-amsterdams-first-circular-neighborhood/.

Metabolic. (2018). *DeCeuvel Sustainability Engergy Flows*. Retrieved from https://deceuvel.nl/wp-content/uploads/2018/06/DeCeuvel_Sustainability_Energy_Flows.pdf.

Monstadt, J. (2009). Conceptualizing the political ecology of urban infrastructures: Insights from technology and urban studies. *Environment and Planning A, 41*(8), 1924–1942. Retrieved from https://doi.org/10.1068/a4145.

Monstadt, J., & Coutard, O. (2019). Cities in an era of interfacing infrastructures: Politics and spatialities of the urban nexus. *Urban Studies, 56*(11), 2191–2206. Retrieved from https://doi.org/10.1177/0042098019833907.

Mostafavi, M. (2010). Why ecological urbanism? Why now? *Harvard Design Magazine, 32*, 1–12.

Murray, A., Skene, K., & Haynes, K. (2017). The circular economy: an interdisciplinary exploration of the concept and application in a global context. *Journal of Business Ethics, 140*(3), 369–380. Retrieved from https://doi.org/10.1007/s10551-015-2693-2.

Petit-Boix, A., & Leipold, S. (2018). Circular economy in cities: Reviewing how environmental research aligns with local practices. *Journal of Cleaner Production, 195*, 1270–1281. Retrieved from https://doi.org/10.1016/j.jclepro.2018.05.281.

Pomponi, F., & Moncaster, A. (2017). Circular economy for the built environment: A research framework. *Journal of Cleaner Production, 143*, 710–718. Retrieved from https://doi.org/10.1016/j.jclepro.2016.12.055.

Raven, J., Stone, B., Mills, G., Towers, J., Katzschner, L., Leone, M. F., … Rudd, A. (2018). *Urban Planning and Urban Design. Climate Change and Cities: Second Assessment Report of the Urban Climate Change Research Network*. Retrieved from https://doi.org/10.1017/9781316563878.012.

Reike, D., Vermeulen, W. J. V., & Witjes, S. (2018). The circular economy: New or Refurbished as CE 3.0?—exploring controversies in the conceptualization of the circular economy through a focus on history and resource value retention options. *Resources, Conservation and Recycling, 135*(February 2017), 246–264. https://doi.org/10.1016/j.resconrec.2017.08.027.

Remøy, H., Wandl, A., Ceric, D., & van Timmeren, A. (2019). Facilitating circular economy in urban planning. *Urban Planning, 4*(3), 1–4. https://doi.org/10.17645/up.v4i3.2484.

Satterthwaite, D. (2008). Cities' contribution to global warming: Notes on the allocation of greenhouse gas emissions. *Environment and Urbanization, 20*(2), 539–549. Retrieved from https://doi.org/10.1177/0956247808096127

Sauvé, S., Bernard, S., & Sloan, P. (2016). Environmental sciences, sustainable development and circular economy: Alternative concepts for trans-disciplinary research. *Environmental Development, 17*, 48–56. Retrieved from https://doi.org/10.1016/j.envdev.2015.09.002.

Seto, K. C., Dhakal, S., Bigio, A., Blanco, H., Delgado, G. C., Dewar, D., Huang, L., Inaba, A., Kansal, A., Lwasa, S., McMahon, J. E., Müller, D. B., Murakami, J., Nagendra, H., & Ramaswami, A. (2014). Human settlements, infrastructure and spatial planning. In: O. Edenhofer, R. Pichs-Madruga, Y. Sokona, E. Farahani, S. Kadner, K. Seyboth, A. Adler, I. Baum, S. Brunner, P. Eickemeier, B. Kriemann, J. Savolainen, S. Schlömer, C. von Stechow, T. Zwickel and J.C. Minx (Eds.), *Climate Change: Mitigation of Climate Change. Contribution of Working Group III to the Fifth*

Assessment Report of the Intergovernmental Panel on Climate Change. Cambridge, United Kingdom and New York, NY, USA: Cambridge University Press.

Sillanpää, M., & Ncibi, C. (2019). *Getting hold of the circular economy concept. The Circular Economy*. Retrieved from https://doi.org/10.1016/b978-0-12-815267-6.00001-3.

Trenberth, K. E., Fasullo, J. T., & Shepherd, T. G. (2015). Attribution of climate extreme events. Nature Climate Change. *Nature Climate Change*.

UN DESA. (2018). *Revision of World Urbanization Prospects [2018 revision]*. Retrieved July 6, 2020, from https://www.un.org/development/desa/publications/2018-revision-of-world-urbanization-prospects.html.

Urban Systems Lab (2020). *Social, ecological, and technological systems (SETS)*. Retrieved July 10, 2020, from http://urbansystemslab.com/about.

van Buren, N., Demmers, M., van der Heijden, R., & Witlox, F. (2016). Towards a circular economy: The role of Dutch logistics industries and governments. *Sustainability (Switzerland), 8*(7), 1–17. https://doi.org/10.3390/su8070647.

Van Bueren, E., Steen, K. (2017). *Urban Living Lab*. Retrieved from https://www.ams-institute.org/news/urban-living-labs-living-lab-way-working/.

Van den Berghe, K., & Vos, M. (2019). Circular area design or circular area functioning? A discourse-institutional analysis of circular area developments in Amsterdam and Utrecht, The Netherlands. *Sustainability (Switzerland), 11*(18), 1–20. https://doi.org/10.3390/su11184875.

van der Leer, J., van Timmeren, A., & Wandl, A. (2018). Social-ecological-technical systems in urban planning for a circular economy: an opportunity for horizontal integration. *Architectural Science Review, 61*(5), 298–304. https://doi.org/10.1080/00038628.2018.1505598.

Voskamp, I. M., Spiller, M., Stremke, S., Bregt, A. K., Vreugdenhil, C., & Rijnaarts, H. H. M. (2018). Space-time information analysis for resource-conscious urban planning and design: A stakeholder based identification of urban metabolism data gaps. *Resources, Conservation and Recycling, 128*, 516–525. https://doi.org/10.1016/j.resconrec.2016.08.026.

Williams, J. (2013). The role of planning in delivering low-carbon urban infrastructure. *Environment and Planning B: Planning and Design, 40*(4), 683–706. https://doi.org/10.1068/b38180.

Williams, J. (2019). Circular cities: Challenges to implementing looping actions. *Sustainability (Switzerland), 11*(2). https://doi.org/10.3390/su11020423.

Winans, K., Kendall, A., & Deng, H. (2017). The history and current applications of the circular economy concept. *Renewable and Sustainable Energy Reviews, 68*(October 2015), 825–833. https://doi.org/10.1016/j.rser.2016.09.123.

Wu, J. S. (2005). *New Circular Economy*. Tsinghua University Press.

Reinventing Bilbao, the Story of the Bilbao Effect

Marc Wouters

Abstract

The City of Bilbao, Spain, garnered worldwide recognition for its economic regeneration, known as "The Bilbao Effect". It suffered a total economic collapse in the 1980s after a downturn of its industrial economic base and a devastating flood. This paper examines the methods The City of Bilbao used to forge its path back to international success, and ultimately win The European City of the Year Award in 2018. While Bilbao is known for its world-famous Guggenheim Museum, the critical elements of the regeneration strategy are less well known. They include its system-wide urban planning strategy, its housing programs, its job training programs, the leveraging of its local culture, banking system, long history of technical ingenuity, and the determination of the Basque people. This paper is a companion to the documentary film "Reinventing Bilbao, the Story of the Bilbao Effect" by Manta Wake, LLC. Bilbao collapsed due to a significant reduction of its shipbuilding of industry, a dramatic flood of the city's historic center, and high levels of pollution. The strategy to recover and reinvent the city included three components: conversion of the economy from an industrial economy to a professional service economy, cleaning the environmental problems, and transforming their former industrial brownfield sites into new walkable mixed-use districts that would host the new economy. One of the most visible projects is the redevelopment of Abandoibarra District, a large abandoned industrial brownfield site adjacent to the city center. The new walkable mixed-use development on this site is a focal point of the regeneration plan. It includes new housing, offices, university facilities, cultural facilities, retail, and parks. One of the last major buildings constructed on the site is The Guggenheim Museum by architect Frank Gehry. According to the city's Director of Urban Planning, the worldwide success of the museum would not be as great if it were not for the entire urban and economic planning effort the city had established prior to the construction of the museum. The city chose to invest in architectural design to continue its tradition of high-quality culture. The city's transportation system improvements were designed to connect the Basque Region to the city's economic center. Improvements included a new subway system and light rail system, a new airport designed by Santiago Calatrava, easier pedestrian street crossing points, and new bike lanes. The accomplishments of the "Bilbao Effect" include a reduction of the city's unemployment rate to amongst the lowest in Spain. New businesses have opened in the city center including offices, research firms, and major retailers. Tourism has increased over tenfold. National and international conferences have taken place in the city regularly. The city also reduced its deficits to near zero and the Basque Region has re-emerged as a major source of economic growth in Spain.

Terms:

Brownfield sites	Abandoned industrial yards.
Greenfield sites	Previously undeveloped land.

1 Introduction

The 800-year-old City of Bilbao, Spain has garnered worldwide recognition for its economic regeneration, known as "The Bilbao Effect". Its economy collapsed after the downturn of its industrial economic base and a devastating flood. Its residents had no choice but to reinvent their economy and the city's image. Bilbao's highly successful resurgence comes from a system-wide regeneration plan that took two decades to develop and execute. The city used

M. Wouters (✉)
Marc Wouters | Studios, New York, USA
e-mail: marc@mwouters.com

urban planning strategies to develop new mixed-use districts, rebuild its transportation system, and upgrade key infrastructure. It also created new housing and affordable housing. The city created extensive job training programs in partnership with universities and local workers unions and built upon its long history of technological innovation. The city leveraged many aspects of its local culture, including social resilience and the determination of the Basque people. The long history of banking allowed them to create innovative financing strategies. Together this series of strategies created economic growth, stabilized the city's population, and positioned the city to serve its future generations.

2   Background, the Collapse of Bilbao

The City of Bilbao is the capital of the Basque Region of Spain. It is located on the north coast of Spain near the Bay of Biscay. Founded June 15, 1300, it has a long history of shipping, trade, and production of iron and rioja wine. Industrial shipbuilding and trade blossomed in the 1800s. Trade supported the growth of local banking institutions. These banks grew to worldwide recognition. The city built several significant classical buildings including the grand beaux-arts Teatro Arriaga. By 1970, Bilbao constructed 4.7% of the world's ships (Arana, 2017). The city became home to international companies such as BBVA, Banco Bilbao Vizcaya Argentaria S.A., and Iberdrola Energy. The city's population in 2018 was approximately 350,000 and the region's population includes approximately 1 million (Alchetron, 2018).

The city's economic collapse occurred in the early 1980s. Its shipbuilding economy experienced growing competition from overseas. Bilbao's Director of Urban Planning, Asier Abaunza Robles (Abaunza Robles, 2017), stated that the large shipbuilding yards closed due to lack of work. Resident M. Iturregi recalls that local workers marched in protest of the closing of the shipyards. In some districts, the unemployment rate reached 30% (Areso, 2017), and youth unemployment rose to 50%. Approximately 60,000 jobs were lost (Power, 2016). The city's docks along its main river were vacant, and the cranes that loaded the ships lay still. On August 26, 1983, a massive flood struck the city's historic center. The ground floors of most buildings in the historic center, the Caja Vieja, were destroyed. The flood damaged businesses and residential areas. The flood also exposed the low quality of housing conditions of many residents. Prior to the flood, high levels of uncontrolled dumping from manufacturing destroyed the city's river. The Nervión River was filled with solid waste, raw sewage, and toxic chemicals (Urkijo, 2017). One resident, T. Busto (Busto, 2017), recounted that she was advised to leave the city because the living conditions were unhealthy. After the flood, the city's population declined from over 400,000 to below 350,000.

3   The Recovery Strategy

The city undertook an extensive plan to regenerate itself. The strategy to recover and reinvent the city included three components: conversion of the economy from an industrial economy to a professional service economy, cleaning the environmental problems, and transforming the vacant industrial brownfield sites into new walkable mixed-use districts that could host the new economy (Abaunza Robles, 2017).

For the first component, the city decided to expand its economy to include technology services, biotech, robotics, tourism, and more. The conversion of the industrial economy to a service economy required retraining of the existing workforce, and also attracting a young professional workforce that was educated with the newest skills. To grow local talent, the city provided support to its schools (Areso, 2017). Support was given to the University of Deusto to expand training opportunities. Support included new university facilities, a new library building, partnerships between private companies, and university programs. The area's main worker association, the Mondragon Workers Cooperative, was founded in 1943 to train area workers. Starting in the 1990s, it created programs including its own university to retrain union members in a variety of new technological skills and help them transition to a new economy (Power, A. Bilbao City Story, 2016).

The second component, the cleaning of the environment, was viewed as an essential step to improving the new economy. The city reasoned that new hi-tech workers and businesses would not move to Bilbao if the main areas were polluted. To clean the river, the city invested over 1 billion Euros to rebuild its waste and sewer system (Areso, 2017). The Galindo Waste Water Treatment Plant (EDAR) was opened on the left bank of the Nervión River in 1991. The funding was provided through the Bilbao Rai, 2000 public–private investment organization (European Commission, 2004). The replacement of sewer lines took almost 20 years to complete. Manufacturing plants that emitted toxins were closed or redesigned. According to local guide, M Urkijo (2017), the river was the heart of the city, and cleaning the river was one of the first widely visible steps that marked a clear turning point in the city's direction.

4   The Abandoibarra District

The third component was the building of new mixed-use urban areas that would house the new economy. According to the city's head of urban planning, these were built on the brownfield sites of the abandoned industrial yards. The city did not build on any previously undeveloped, greenfield sites, but instead chose to redevelop the vacated industrial sites that were closest to the city center (Abaunza Robles, 2017). The most visible development was the Abandiobarra District, located directly adjacent to the historic city center (Fig. 1). The new walkable mixed-use development was built on the site of the former dockyards and was the main focal point of the regeneration plan. The new development included new housing, offices, university facilities, hotels, parks, and cultural facilities. Some of the principal buildings include The University of Deusto Library, The 2000 seat Euskalduna concert and convention hall built in 1999, The Maritime Museum constructed in 2001, the 8000 m^2 Basque Public University Auditorium built in 2010, the 22,000 m^2 Zubiarte Retail Centre built in 2004 (Bilbao Rai, 2000), and the new Guggenheim Museum (Fig. 2). The buildings were governed by a consistent set of height limits and street setbacks. Parking was hidden below grade and included a large 800-car garage below the new Plaza Euskadi. The new district was laid out with avenues of buildings and walkable streets (Fig. 3). New avenues connected to nearby existing avenues, while some new retail streets featured arcaded sidewalks. The emphasis on walkable design included continuous sidewalks, well-designed crosswalks, and parks and paths integrated between the buildings. All the needs of daily life including office, retail, residential, and recreation were located within close walking distance of each other, and the entire Abandoibarra District was located with walking distance of the historic city center and the city's downtown core. This proximity fostered economic growth across several city districts.

The city chose to invest in high-quality architectural design to continue its long tradition of well-crafted cultural edifices. Architecture for the new buildings was commissioned from a wide variety of internationally known architects. One of the last major buildings constructed on the site was The Guggenheim Museum by architect Frank Gehry, and it was considered one of the great architectural works of the twentieth century. The only tall building in the plan, the new 500,000 m^2 41-story headquarters tower for Iberdorla Energy was designed by Pelli and Partners, and was situated in the center of the new district. Raphael Moneo designed the University of Deusto's new library. A new retail complex was designed by RAMSA, and a new mixed-use building was designed by Robert Krier (Fig. 4). Near the district, there are other notable projects including the new Puente Zubizuri pedestrian bridge designed by Santiago Calatrava Architects. It connects to a grand stair that leads to the main urban area. A pair of residential high rises, designed by Arata Izosaki are set on either side of the grand stair and created a grand framed portal to the city.

The new Abandoibarra District includes a new waterfront esplanade. The esplanade includes several elements that recall the city's shipping past. Large light fixtures along the walkway are shaped to recall the prows of the ships that once occupied the area (Fig. 5). Some of the lifting cranes of the old docks, as well as anchor chains and anchors, have been left in place and incorporated into the outdoor landscape. Two old dry docks were kept and now hold a museum of Bilbao's historic boats. The Euskalduna Palace Conference Center and Concert Hall on the promenade was clad in corten steel to recall the metal cladding of the ships.

Fig. 1 Bilbao with Abandoibarra District

Fig. 2 Abandoibarra District aerial

Fig. 3 Abandoibarra District residential

Fig. 4 Abandoibarra District with buildings designed by Pelli and Krier

Fig. 5 Abandoibarra Esplanade

The director of city planning explained that this first new district for the city was critical to the city's success. It had to successfully attract and house the new economy, and it needed to demonstrate to the residents of Bilbao that the regeneration of districts could be done properly. Before the building of the Abandoibarra District, many residents were reluctant to see renovations in the many other neighborhoods of the city. Following the successful construction of Abandiobarra, many residents were requesting modifications to their neighborhoods.

Prior to the construction of the Abandoibarra District, the city built a new deepwater port on the Bay of Biscay. A modern port that could handle large ships was seen as an essential piece of infrastructure, necessary to maintain Bilbao's competitive role in international trade (Abaunza Robles, 2017). The new port replaced the old port that was previously located at Abandoibarra. The new port was 770 acres and transferred 6.4 million tonnes of cargo in 2016 (Bilbao Port, 2018). The port allowed for bigger ships that were an essential part of the modern international cargo industry.

5 Transportation

The city determined that its new economy would require a complete rethinking of its transit system. According to the city's director of urban planning, a new subway, and light rail system was viewed as a way to connect residents of the surrounding region to the new economic activity of the revitalized city center. The subway was constructed to reach more distant communities, and the light rail lines supplemented local connections (Fig. 6). Some individuals were opposed to building the subway system before the economic resurgence occurred (Arana, 2017), but the city opened the system in an early phase of the regeneration plan in November 1995 so that it would be ready to support the beginning of economic growth. The stations were designed by Foster + Partners and the curved glass canopies of the station entrances were designed to resemble local shellfish (BIA Press, 2014) (Fig. 7). To improve international business connections and support tourists, the city also built a new airport designed by architect Santiago Calatrava.

Photo Credit: Marc Wouters c 2018

Fig. 6 New light rail system

Fig. 7 New subway station

The new Metro system opened 1995

Photo Credit: Marc Wouters c 2018

The city also improved pedestrian routes by adding two new pedestrian bridges over the main river in the center of town. Before the bridges, many residents took small boats to travel across the river (Busto, 2017). The new pedestrian bridges greatly shorten the walking distances between various parts of the city. One bridge is situated to provide a direct connection between the University of Deusto and the new employment opportunities in the Abandoibarra District.

On city streets, crosswalks, and major pedestrian areas were improved to ease pedestrian and bike circulation. Neck downs were inserted at many street intersections to reduce the pedestrian crossing distance on streets and increase pedestrian safety. Some areas, such as the Don Diego Lopez Haroko Kale Avenue, were reconstructed with curbless sidewalks so that pedestrians would not be hampered by stepping over curbs when crossing a street. Lighted bollards at crosswalks of Ercilla Kalea Avenue include red and green 'walk/don't walk' signals to increase pedestrian safety. Bike improvements include new dedicated bike lanes in some parts of the city.

6 Affordable Housing

Affordable housing was a significant part of the plan. New affordable housing was seen as a necessary way to attract young professionals, as well as a way to support a growing senior population. The housing was built with public/private partnerships. City-owned industrial land was cleaned and made available to use for some of the new affordable housing projects. According to local guide Urkijo, The Saralegi Affordable Housing Community in the neighborhood of San Francisco was built on the former grounds of iron mines (Fig. 8). Brick furnaces from the old iron

Fig. 8 The Saralegi Affordable Housing Area, with a preserved historic iron furnace, and urban planner and author Marc Wouters

smelters were kept as historic landmarks for the new community. The housing was often sold at affordable prices to qualified residents and allowed the residents to build long-term equity (Urkijo, 2017).

7 Governance

Bilbao had to create a new organizational structure to implement its vision. The city created a new collaborative organizational system, which consisted of an alliance between the Bilbao City government, the Bilbao Regional government, The Bizkaia Province, and the national government of Spain (Abaunza Robles, 2017). All four levels of government created a system to work together. It also created a local advisory council known, as Bilbao Metropoli 30 comprised of government, university experts, leaders of the private sector, and community representatives. The organization operated as a think tank to devise growth strategies and promote local culture. The city also established a public–private redevelopment company called Bilbao Ría 2000, which oversaw the redevelopment of public land by private development teams. The organization facilitated the redevelopment of vacated industrial sites, generated revenue from those investments, and reinvested the revenues into other city growth projects. The Basque region of Spain also had full authority to collect and invest taxes within its region

and had a strong ability to direct funds to local needs (Power, 2016).

The city's long history of banking facilitated its capabilities to structure financing for the new projects. The overall plan cost approximately 6 Billion Euros (Areso, 2017). The city used low interested loans from the EU investment bank to finance the construction of the subway. The reconstruction of the sewer system was financed by a surcharge on water bills and public funds. The cost of the new deepwater port, 700M Euros, was funded by port activities and the national Spanish Ministry.

8 Social Resilience

The city's rich historic culture was essential to the implementation of the plan. Before the economic collapse, the residents had spent decades building a strong social and cultural network. The social network allowed residents to provide valuable support to each other during the economic collapse. It also enabled them to organize recovery plans, and implement them. The social network was supported by the design of older public spaces that were shaped to encourage the informal gathering of residents. These included the arcaded Plaza Nueva, the many pocket plazas, and local landmarks (Fig. 9). The local culture and after-work social rituals strengthened these social bonds. Venues such

Fig. 9 Bilbao Plaza Nueva with series of social establishments along the arcade

as food establishments were designed to encourage people to mix and socialize. These venues were arranged in a series along pedestrian streets and plazas that allowed residents to mingle up and down the street and meet with a wide range of friends and acquaintances. This regular social ritual existed for generations and facilitated strong social bonds and social resilience. Food portions, seating arrangements, portable plate sizes, were all designed so that people could mingle in a crowd and chat with each other (Urkijo, 2017). Many residents have attested to the importance of the social network and the corresponding pride residents take in their city.

9 The Guggenheim Museum, Bilbao

During the city's long-term urban planning process, it identified a strategically significant site for a major museum. The site was located in the new Abandiobarra District, within walking distance of the historic city center, at a highly visible site on the river, and adjacent to one of the new light rail lines. The city created a partnership with the Guggenheim Museum and held an international competition for the design of the building. The design by architect Frank Gehry was selected (Moore, 2017). The building was composed of curving shapes that were clad in shining titanium (Fig. 10).

The architect designed the building to reflect its cultural context: "….the river, the sea, the boats coming up the channel. It was a boat." (Gehry) (Moore, 2017). The city and Basque Region paid for the building and the Guggenheim assembled the art collection. When the museum opened in 1997, it garnered worldwide attention. It was featured in a wide variety of publications, and television media. According to the World Architecture Survey of 2010, the building was regarded as one of the most significant pieces of architecture of the twentieth century (Tyrnauer, 2010).

The international press often gave credit to the museum for attracting new businesses to the city, but the museum was one of the last projects constructed during the reinvention of Bilbao. The city's director of urban planning says the worldwide success of the museum and Bilbao would not be as great if it were not for the entire urban and economic planning effort the city had established before the construction of the museum. He compared the city master plan to a cake, and that the museum was like the cherry on the cake. He says many other cities commissioned individual signature buildings hoping to repeat the "Bilbao Effect", but they did not attain as great an economic success because they did not establish a broader comprehensive regeneration urban plan before designing the signature buildings (Abaunza Robles, 2017).

Fig. 10 The Guggenheim Museum in the new Abandoibarra District of Bilbao

10 Future Projects

The city's next major district is the peninsula of Zorrotza-urre. The peninsula reaches into the city's river and houses a mix of older shoreline communities and large industrial sites. The extensive shoreline hosts several older shipping-focused industries. The underutilized industrial lands are to be remediated and converted into new housing, affordable housing, a new educational campus, and new technology-based businesses. One goal of the plan is to foster relationships between the new educational campus and the new technology-focused businesses. Some portions of the community, which have historic character, will be retained. The plan includes filling large portions of the peninsula with earth to create higher ground that is resistant to flooding (Abaunza Robles, 2017). Plans include separating the peninsula to make it an island. New bridges to the peninsula have already been constructed.

11 Result: "The Bilbao Effect"

The accomplishment of the "Bilbao Effect" is the rebirth of the city's economy and culture. New businesses have opened in the city center including offices, research firms, and major retailers. The city's unemployment rate dropped from almost 30% in the 1980s to 11% in 2004 (Power,

2016), amongst the lowest in Spain. Major international retailers have opened stores in the city center. Tourism has increased from 24,000 visitors per year in the late 1990s to over 600,000 visitors per year in 2018. National and international conferences take place in the city regularly drawing approximately 100,000 visitors a year. Many of these conferences take place at the new Euskalduna Conference Centre. Job growth in the city and surrounding metropolitan area is estimated at 113,000 jobs between 1995 and 2005. Job growth occurred in the professional and service sectors (Power, 2016).

Older buildings have been restored to their original grandeur (Fig. 11). New residents have moved into Bilbao and housing stock has improved. The city used profits from its investment projects to reduce its debt. Its debts were reduced from approximately 200 M euros in 1995 to near zero in 2008. The Basque Region has re-emerged as a major source of economic growth in Spain (Wallander, 2014).

Some residents have observations on the success of the plan. A series of residents who were interviewed agree that the city is a much better place and they like the overall improvements to the city (Busto & Iturregi, 2017). There is a long history of pride in the city that many residents discuss, and they feel they can have pride again. Yet there are concerns expressed by some residents. One resident expressed concern that too much focus had been placed on tourism, and hoped that the city would maintain its long professional

Fig. 11 Bilbao downtown historic buildings restored

standing in banking and innovation. They also suggested that maybe more of the old shipyard buildings could have been saved and repurposed for new uses. One resident M. Iturregi (2017) says the change in work schedules and the rise of social apps has changed the after-work social rituals and reduced the frequency people go out after work. The social network that had been built over the decades is starting to fray. He says people don't know each other within the community the way they used to.

There is criticism of the architectural approach as well. One local architect describes the planning strategy and the architectural strategy as having two opposing approaches (Nicholson Viar, 2017). The urban plan is conceived to unify and integrate a variety of uses, whereas the architectural strategy appears to strive for unique pieces that, in some cases, detract from the whole. Unlike the older parts of the city center, which have a cohesive appearance in the rows of buildings, each new international architect brought their own personal style to Bilbao. The styles are very idiosyncratic to each architect. The wide variety of aesthetics of the new architecture does not provide a cohesive appearance to the new districts.

The city has received several international awards for its work. These include the Lee Kuan Yew World City Prize in 2010, the World Mayor Prize to Inaki Azkuna in 2012 from the City Mayor's Foundation, and the European City of the Year Award in 2018 from the Academy of Urbanism, Europe. Individual architectural projects have also received countless design awards.

12   Conclusion

The City of Bilbao, Spain used urban planning as an integral part of its regeneration strategy. After the city's economic collapse in the late 1980s, the city used the urban planning of new districts, transportation systems, new city infrastructure, and public spaces to foster the growth of new businesses. Bilbao is an example of the strong interrelationship between high-quality design and economic success. The core principle of pedestrian-oriented urban planning guided most of their projects including the layout of the new mixed-use district of Abandoibarra, the riverfront esplanades, new pedestrian bridges, and transit systems in walkable neighborhoods. The close proximity of residents, businesses, universities, and cultural venues is made possible by the principle of compact mixed-use development and allows economic growth to spread across the city. Rather than build new developments on the outskirts of the city, Bilbao chose to repurpose the vacant shipyards near the center. By locating mixed-use development near the center of the city, they were able to create an economic district that was within walking distance of many parts of the city.

The decades-old social network in Bilbao may be the most essential ingredient in the making of the "Bilbao Effect". These networks allow residents to offer each other support during challenging times, allow them to collaborate on regeneration plans, and provide a communication network for implementing the plan. Many residents testify to the importance of these relationships. Historic Bilbao's many varied public spaces, cultural venues, and social rituals support these social networks.

Other key aspects of the plan follow the theme of increasing connectivity. These include improvements to its transportation system such as the new subway system and light rail line. The subway was constructed during the early phases of the overall plan before economic growth occurred so that it was ready to support the early growth. The subway was able to tie many residents who lived in more distant communities to the new economic opportunities, and it brought these communities together.

The strategy for new buildings included specific sites for affordable housing so that young professionals would find the city desirable to live in. The affordable housing also accommodated a growing senior population. The housing was planned as part of an overall economic growth strategy to attract young workers. It was not added as an afterthought. To continue the city's tradition of high-quality design, the city focused on the cultural contribution of each new building. It selected high-quality design firms for each of its major new buildings.

Finally, one major misconception of the "Bilbao Effect" is that it was caused by one structure, The Guggenheim Museum. The building is one of the most significant buildings of the twentieth century, and its notoriety brought media attention to Bilbao. But the museum alone did not create the overall success. The city's intricate and broad planning program, which focused on many parts of the city, is the foundation of Bilbao's success. The museum was integrated into an overall city-wide plan. The overall plan created a large series of economic benefits ranging from the growth of new types of industries, population growth, and high-quality living conditions.

References

Abaunza Robles, A. (2017). Interview, August 23, 2017.

Alchetron. (2018). *Bilbao general data*. https://alchetron.com/Bilbao

Arana, K. L. (2017). Behind the Bilbao effect: An overnight success in 20 years. *MAS Context, 30–31*, 26–33

Areso, I. (2017). Bilbao's strategic evolution: Metamorphosis of the industrial city. *MAS Context, 30*, 126–147.

BIA Press. (2014, September 24). *Lord Foster receives inaugural BIA award in Bilbao*. https://biarchitecture.org/

Bilbao Port. (2018). *Bilbao Port authority website*. https://www.bilbaoport.eus/en/

Bilbao Rai. (2000). *Zubiarte Shopping Center.* https://www.bilbaoria2000.org/en/actions/abandoibarra/zubiarte-shopping-centre/

Busto, T. (2017). Interview September 7, 2017.

Busto, T., & Iturregi, M (2017). Interviews, Sept 7, 2017.

European Commission. (2004, January 1). *Bilbao to have clean river, regional policy Bilbao.* https://ec.europa.eu/

Iturregi, M. (2017). Interview, Sept 7, 2017.

Moore, R. (2017 October 1). The Bilbao effect, how Frank Gehry's Guggenheim started a global craze. *The Guardian.* https://www.theguardian.com/artanddesign/2017/oct/01/bilbao-effect-frank-gehry-guggenheim-global-craze

Nicholson Viar, A. (2017). Interview Sept 7, 2017.

Power, A. (2016). Bilbao city story. *CASEreport101*, May 2016.

Tyrnauer, M. (2010, June 30). *Architecture in the age of Gehry.* Vanity Fair, https://www.vanityfair.com/culture/2010/08/architecture-survey-201008

Urkijo, M. (2017). Interview, August 22, 2017.

Wallander, A. (2014, May 7). *An economic exception: The Basque Country,* Project for Democratic Union.

Marc Wouters RA, LEED AP is director of Marc Wouters | Studios in New York, which focuses on urban planning, architecture, urban regeneration, climate adaptation of urban areas, sustainable city design, and documentary films on various issues facing communities. Mr. Wouters directed the documentary film, "Reinventing Bilbao", a companion to this paper. Projects led by Mr. Wouters have received AIA, ULI, and CNU national awards. Projects include work in the US, Europe, Canada, and in Latin America. Regeneration projects are located the City of Baltimore, the City of Saskatoon, in Puerto Rico following Hurricane Maria, in Washington, DC, several small cities, and more.

The Waterfront Development in Europe: Between Planning and Urban Design Sustainability

Donatella Cialdea and Chiara Pompei

Abstract

This paper analyzes the different possibilities of acting on natural infrastructures in order to guarantee a safe environment experience for citizens, especially in congested urban centers close to rivers. The paper identifies the urban policies and funding through which urban regeneration comes to life for waterfront development. In particular, different kinds of case studies are presented in two blocks. The first one acts on sustainable projects not linked by a general strategy, but which produce positive effects on the city in terms of safety. The second block collects experiences linked to general strategies that secure and encourage slow mobility, in order to regenerate the entire city network. The waterfront development can be implemented in several ways. Nevertheless, the collection of the case studies demonstrates that sustainable urban policies play a key role to regenerate, renew and innovate the natural infrastructures and that sustainable urban design has the tools to transform the routes in roots, turning the habit of passing through a city onto the practice of living through the city.

Keywords

Natural infrastructure • Waterfront • River • Planning and urban design • Sustainability

D. Cialdea (✉)
University of Molise, Campobasso, Italy
e-mail: cialdea@unimol.it

C. Pompei
Sapienza University, Rome, Italy

1 Introduction

This paper analyzes the different possibilities of acting on waterfront development in order to guarantee a safe city environment experience for citizens, especially in urban centers close to rivers. In these cases, there are various declinations of urban routes that aim not only to offer a safe crossing of the city but above all a new way of living it. The paper seeks to understand and identify projects, urban policies and funding through which these paths come to life, with strengths and weaknesses, and the new possibilities of action along unexplored paths. The waterfront, the object of this study, is a place that represents the historical memory of the cities' development. All the major cities have sprung up along sea and river networks for their utility in agriculture, food, connection with the rest of the territory, trade, spaces of worship, meeting and reflection, recreation and enjoyment, and also have been neglected by these very useful patterns (Cialdea, 2020; Jacobs, 1961; Sairinen & Kumpulainen, 2006). The attention to rivers as integral elements of city life has undergone many changes over time, without however losing sight of their commercial role. Waterfront sites had a decay period in the post-industrial age, as many industrial sites have been decommissioned for a change in more profitable economic activities, production technologies, environmental regulations for the aquatic environment protection and a shift of production sites to Third Countries (Carta, 2007; Carta & Ronsisvalle, 2016; Marshall, 2016). It is precisely when the waterfront lost this commercial and industrial role that different functions arose for their new use and a different function. In particular, the theme of waterfront transformation, together with the brownfield regeneration process, is actually linked first of all to the combination City-Port. Several port cities can be identified: the medieval port city, characterized by a strong functional and spatial relationship between port and city; the nineteenth-century port city characterized by the introduction of several significant technological innovations; the modern and industrial port city characterized by a clear spatial separation between city and port; the contemporary and post-industrial port city, characterized by the phenomenon of waterfronts redevelopment and by the rediscovery/reconquest of their public civic dimension (Cialdea, 2019a; Clemente, 2011; Hoyle & Pinder, 1992; Vincenti, 2011). The factors that have contributed to the resurgence of waterfront development, both fluvial and coastal, are different. The first is that the abandoned waterfront led to depressed land values, attractive redevelopment schemes and cheaper available land. The second is that beginning in the 1970s and 1980s, environmental regulations and remediation to guarantee cleaner water and land made the land appealing. The third one is the naissance of Preservationists' Movements that took to preserving historic abandoned industrial structures along waterfronts and riverfronts. The fourth factor is the leadership citizen activism in reclaiming "lost" waterfronts and historic regions. The fifth is that with residential developments with supporting services, waterfronts have become prime real estate (Fisher et al., 2004). In particular, the contemporary waterfronts recovery and regeneration processes also arise for the international competitiveness of the cities in terms of productive investments, new governance, evaluation and management strategies of the territory, and enhancement of share capital. Therefore, the process of urban transformation and requalification of the waterfront creates competitive opportunities, capable of attracting productive investments, cultural events, increasing the tourist offer and the positive image of the city. It is a network of places, functions and flows, an osmotic interface, an intersection of infrastructural beams, a creative synthesis of space and community, an active place of production and commerce, a fruitful synthesis of identity and perspectives and a place where people can be involved in the analysis, planning and design of the river public dimension (Carta & Ronsisvalle, 2016; Cialdea, 2019b; Cialdea & Pompei, 2020; Van Der Knaap & Pinder, 1992; Vincenti, 2011). The waterfront, therefore, both coastal and river, take on a very important catalyst role to revitalize and transform the cities affected by deindustrialization into cultural centers for a renewed relationship with water and new urban spaces of multiple functions. Another essential aspect to consider when talking about the waterfront is the river or the sea that it overlooks. The ecological aspect of the river/sea as a whole must be considered, as it is made up of water which, as Garett Eckbo says, "is integral to life processes, to the natural shape of earth and landscape" (Eckbo, 1990) and therefore it needs to understand how the waterfront transformations affect this network, what impact they have. Disciplinary advancements in the field of environment and ecology have been made significantly over the past two decades, especially at the European level. These highlight one of the essential issues facing river and coastal cities: the risk of flooding. In fact,

due to climate change, the more frequent occurrence of flood phenomena and low-flow emergencies have directed attention to the necessity of adopting urban river spaces. The first directive has been the Habitats Directive 92/43/EEC that ensures "the conservation of a wide range of rare, threatened or endemic animal and plant species" (European Commission, 1992). Subsequently, the Directive 2000/60/EC, for the redevelopment of surface and ground waters, establishes a framework for community action in water management, identifying the river basin as the correct territorial reference. It prioritizes ecological objectives and requires to "protect, enhance and restore all bodies of surface water" (European Parliament & Council, 2000). Then the Directive 2007/60/EC on flood risk was issued for watercourses and coastlines at risk from flooding "to map the flood extent and assets and humans at risk in these areas and to take adequate and coordinated measures to reduce this flood risk" (European Parliament & Council, 2007). With urban regeneration and sustainable development, there is the need to achieve and synthesize all the cultural, economic and ecological elements of the cities. So, in relation to the water-city, due to the multifarious nature of urban rivers and coasts, the waterfronts become an interdisciplinary challenge field (Cialdea & Quercio, 2017; Fusco Girard et al., 2004; Prominski et al., 2017; United Nations, 2017).

Therefore, when it comes to waterfront development, planning and design must guarantee in a sustainable, integrated and interscaling way the safety of citizens through the implementation of interventions, projects, strategies and policies that transform the waterfront into a place of life, commerce, job and leisure, to contribute to the improvement of the quality of urban life in social, economic and cultural terms.

2 The Methodological Approach and Samples Selection

The work is part of a research, conducted for some years by the authors, on the theme of rivers and their relationship with the city (Cialdea & Pompei, 2018a, b, 2019a, b, 2020). Therefore, in this work some cases are shown, with the aim of researching elements and tools that achieve objectives linked to sustainability (criterion of representativeness), with attention to the 11th (Sustainable cities and communities) and the 13th (Climate action) Sustainable Development Goals of the Agenda 2030 (United Nations, 2015). For the samples, reading sheets have been created, to allow more direct comparison (opportunities and limitations of the policies adopted).

The reading sheet created in this work is divided into the first part of general data, where basic information is collected to locate the intervention geographically. Then, there is an interpretative-reading part of the urban center evolution in relation to the river that crosses it. Finally, there is a third synthetic-interpretative part divided into inputs, represented by the boundary conditions, and outputs, represented by the results—which will be analyzed in the discussion section. In general, the inputs were the guidelines for the selection of the samples and they consists of five. The first one is the presence of river regeneration projects implemented or partially implemented, thus excluding cases in which there is an unfulfilled forecast. The second input is the position of the river in relation to the city center. Case studies were selected in which the river centrally crosses the urban center or laps the municipal border, positioning itself however in close relationship with the urban center of the city. The third input is the type of the river and the riverfront. Case studies have been chosen in which the riverfront has urban or mixed features since it is in these situations that the river and its banks, as well as ecological elements of a wider network, are configured as real public urban spaces. The fourth input is the type of the area where the project is carried out, and mainly external or central areas with medium and high-density areas have been chosen. The fifth input, on the other hand, concerns the type of policies that have made it possible to implement river regeneration, encouraging municipal or regional/national policies. These boundary conditions seek to identify selective elements to describe the importance that the proximity to the riverways represents since by lapping or crossing the urban space they represent a plus in terms of cultural and touristic attractiveness for the urban center itself.

The collection of images and plans of the realized projects has been developed in order to highlight the sustainability issues with a graphic structure useful for comparison, especially through the design tools part of the reading sheet.

By analyzing different situations, in our ongoing research, it was possible to identify two large blocks of interventions that have in common, in addition to the elements mentioned above, the will to be planning and urban design sustainable and it allowed us to select four emblematic case studies, which highlight current trends in the design and sustainable planning of riverfronts. The first acts on riverfronts projects which are linked by local planning policies and decisions and the second block collects the experiences linked to national and regional sustainable and economic planning policies.

2.1 The First Block: Projects Linked to Local Planning Policies

The first block of interventions collects the experiences of riverfronts projects which are not linked to a national sustainable planning strategy but to strong municipal policies

for the development of river and port areas that produce positive effects on the city in terms of safety and urban design.

2.1.1 Germany

The first experience is in Germany. Here, widespread flooding in 2002 caused many damage costs to river areas and mark a reorientation toward an integrated flood risk management system in the whole of Germany. In particular, many cities have set in motion projects for the safety of the city and riverfronts, even after other widespread flooding events (Thieken et al., 2016). Among these is Hamburg. Hamburg is located in the north-west of Germany and is located at the point where the Alster and Bille rivers flow into the Elbe. It is crossed by a dense network of canals called Fleete and the city center surrounds two artificial lakes formed by Alster: Lake Binnenalster (the part of the river that was inside the ancient city walls) and Lake Außenalster (external Alster). It is a city-state, the second-most populous city in Germany and here there is the second largest port in the European Union. In 2002, Hamburg decided to strengthen its flood protection infrastructure and enhance the Elbe river as a public city network. The Elbe is a navigable basin that touches the cities of Berlin, Hamburg, Prague, Dresden and Leipzig (International Commission for the Protection of the Elbe River, 2016; Vitillo, 1997). In particular, the most representative interventions for the development of the riverfront are those of the port. The first analysis of the port urban peripheries regeneration was commissioned by the Hamburg architect, Professor Volkwin Marg. The study presented in 1996 clarifies many of the development issues that are the urban structure and the principle of mixed uses. Then, in 1997, Henning Voscherau presented "Vision HafenCity" to the public as an opportunity for the inner city to regain its waterfront (Clemente, 2011; Mazzoleni, 2013). In the same context, an interesting project to regain the waterfront is represented by the Promenade Niederhafen. The Promenade Niederhafen is located in the north-west of HafenCity, in the same area as the flood protection project which saw a barrier, of 7.20 m above sea level, built between 1964 and 1968 on the banks of the river. In the last decades following inspections of Niederhafen's existing flood barrier in 2006, the municipality of Hamburg determined that supporting elements of the existing structure were overburdened and its foundations needed significant reinforcement. The city of Hamburg organized a competition to design the existing flood barrier, and Zaha Hadid Architects' project was selected in 2006. The redevelopment project aims to reconnect the 625 m river walk with the urban fabric surrounding, acting as a popular river walkway and at the same time creating connections with the nearest neighborhoods. The linear structure is 8.60 m above sea level in the eastern part and 8.90 m above sea level in the western part, to protect the city from the highest winter storms and extreme high tides. The river promenade is divided into different spatial qualities. The west zone—at a larger scale—offers wide views downstream of all shipping activity. The east one creates a long ramp alongside the amphitheater, leading visitors down to the water's edge. The walk is located at the top of the barrier and develops like a staircase-amphitheater from which to observe the port and the river with broad riverbank steps, moored ships, floating jetties, dike steps and promenades, superdikes, and marinas. The riverside promenade's width offers generous public spaces for different city users. Shops and public utilities are also accommodated within the structure at street level facing the city (Benelli, 2019; Castro, 2019; Prominski et al., 2017; Zaha Hadid Architects, 2019) (Fig. 1).

2.1.2 The Netherlands

The second experience is in The Netherlands. The Netherlands means lower countries in reference to its low altitude and flat topography, with about 50% exceeding 1 m above sea level and almost 17% falling below sea level. Most of the areas below sea level are polders. It is crossed by several rivers, there are many lakes and has an extensive system of waters within canals, many of which are navigable. The region is crossed by flood phenomena for which the entire planning, both national, and above all local, is solid in tackling the flood protection system (Voorendt, 2015). Among the most interesting waterfront redevelopment interventions, both naturalistic and urban, there are those relating to the IJssel river, a branch of the Rhine that is characterized along its entire course by alluvial landscapes in the agricultural expanses and urbanized waterfront when it

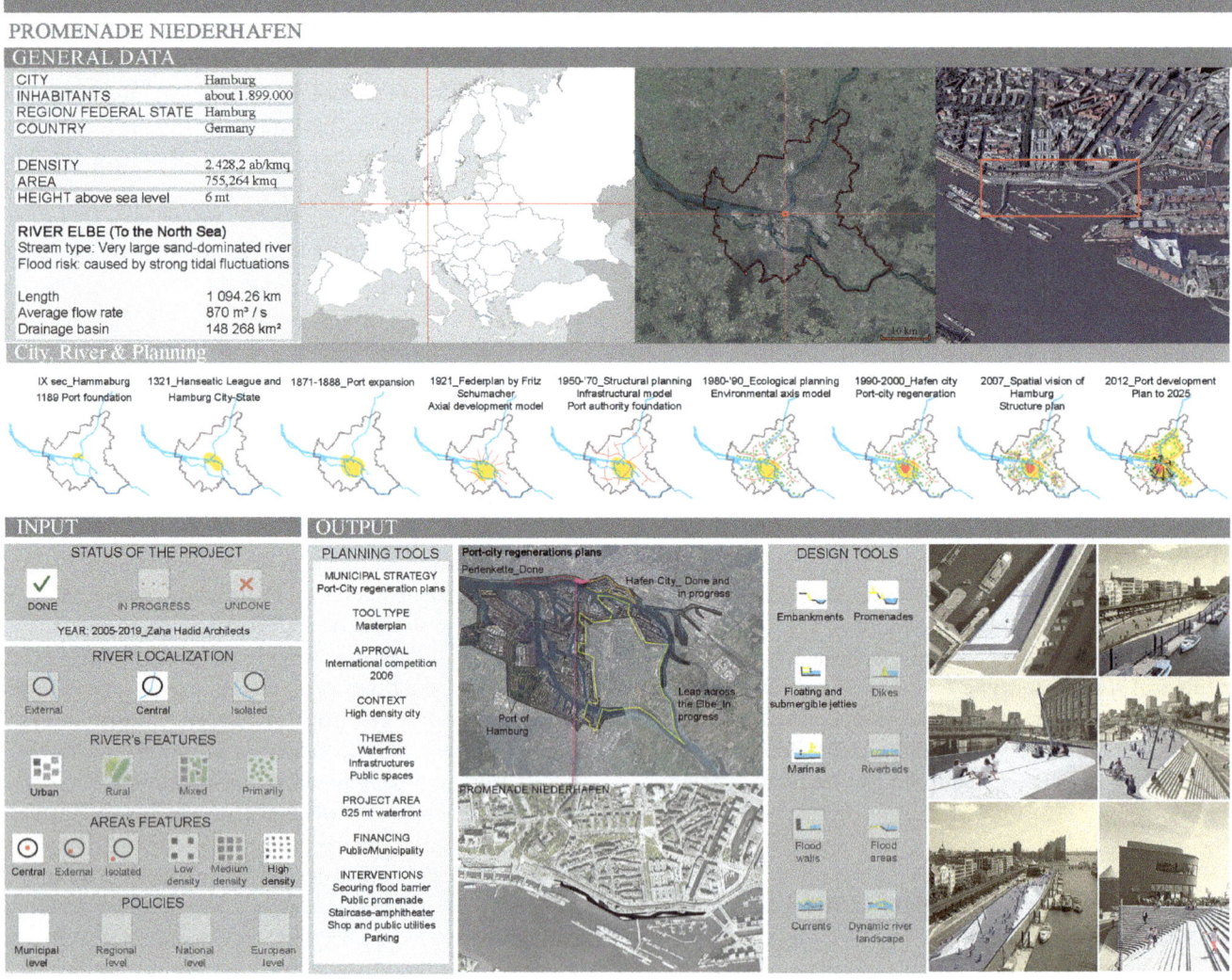

Fig. 1 Promenade Niederhafen Project: **General Data Section**: (*Source* Google Earth 2020); **City, River and Planning Section**: Historical Evolution; **Input–Output Section**: Project Analysis (*Source* Google Earth 2020, https://www.zaha-hadid.com/architecture/hamburg-river-promenade/, Authors own elaboration, 2020)

crosses the cities. Among the waterfront projects of the IJssel, there are those of the city of Doesburg, which has variations in water levels up to 5 m and which is surrounded by anti-flood barriers arranged in a radial shape for flood protection. The selected project area is located on this barrier between the city center and the river. In the past, there was also a moat which was then filled and industrial activities were carried out on it, thus losing the relationship with the river. Following the disposal of these areas, it was planned to locate residential functions here and re-establish the relationship with the river. It is called the IJsselhade Residential Area and was commissioned by the municipality of Doesburg in collaboration with Watershap Rijn en Ijssel and Johan Matser Projectontwikkeling BV of Hilversum. The project, an area of 1.5 hectares for a budget of about 8 million euros, consists of two parts, an architectural one for the construction of new residences, entrusted to the architect Adolf Nicolini, and a landscaping one, entrusted to the OKRA landscape Studio of Utrecht. Between 1997 and 2005, about 150 housing units and a hotel were built on the anti-flood barrier, 6 m over the river level, which reconnects to the lower level by a staircase, at the end of which the old moat was also restored. The city and river are linked to each other through a framework bridge consisting of two arched tubes 18 m in length. The development leads to the creation of an inner and an outer riverfront on the IJssel. Like the IJssel quay, the former moat parts are treated as waterfronts. The wall represents a water defense and has water and electricity for ships and on this barrier the promenade was then designed to admire the landscape. The highlights of the riverfront are formed by the grandstand stairs with natural seating at the port basin and a peninsula, where there is the "Doesburg panorama", an observation tower on a pontoon on the peninsula that has yet to be developed. The promenade has two levels, one floodable and the other dedicated to the public space, with a series of canals and bridges connecting the various parts. The layout of the riverfront is contemporary, with reference to the historical city center (Knuijt, 2002; OKRA, 2010) (Fig. 2).

2.2 The Second Block: Projects Linked to National and Regional Policies

The second block of interventions collects the experiences of general strategies and policies at the national or regional level that include encouraging the sustainable regeneration of the entire city network and the economic development of an attractive city.

2.2.1 Portugal

The first experience is in Portugal. Here in 2001, the POLIS program (*Programa de Requalificação Urbana e Valorização Ambiental das Cidades*/urban redevelopment and environmental valorization of cities) was launched, with the main objective "(…) to improve the quality of life in the cities, through interventions in urban and environmental aspects, improving attractiveness and competitiveness of urban centers that have an important role in structuring the national urban system (…)" (MAOT, 2000). Among the intervention strategies, there is the creation of new urban centralities within metropolitan areas, the regeneration of abandoned industrial areas, the urban and environmental regeneration of riverfront and waterfront, and the enhancement of factors generating new identities, with a forecast investment of 800 million euros. In 1999, at the wish of the Minister of the Environment, a working group was established that defines the *Cidades program* to concretize what has been defined at the national and regional economic strategic level. The POLIS program was designed taking into account the deep changes in Portugal in the last three decades reflected with great intensity in spatial planning, particularly in the urban structure resulting from the economic development, based on industrialization. It is very interesting to underline that this program is a management process of collaboration between the State and municipal administrations. The cities that participate are called *Polis*, which become joint-stock companies with 60% of capital belonging to the state and 40% to municipalities. In most cases, the interventions were carried out within 38 consolidated cities, leaving the suburbs to play a marginal role (Fedeli, 2006; Partidário & Correia, 2004; Pelucca, 2010). Among the most extensive waterfront projects are those built in the city and the metropolitan area of Porto—among the most industrialized cities in the country of around 1,700,000 inhabitants, which is also called the Northern Capital. The Porto Metropolitan Area (AMP) is made up of 16 municipalities, occupies an area of 1.885,10 km^2 and is located on the northern bank of the Duoro River—a riverfront of approximately 35 km, not far from the Atlantic Ocean. In terms of territorial macro-structure typology, the AMP is classified as an urban region, a territorial entity with a strong population density, businesses and services, which has undergone intense transformation processes and strong territorial interactions. Thanks to the POLIS program, various plans and projects are implemented, both in Porto and in the nearby city of Gondomar. In Porto, with the POLIS Program in 2001 interventions on the waterfront are implemented in *Vila Nova de Gaia*. This one represents a vast urban and environmental redevelopment operation and the main objectives include the recovery of the Douro riverfront through the design of a new road section capable of making the car flow compatible with the needs of pedestrians, cyclists, fishing-bulls and tourists; the urban recovery of the fishing village of the Afurada through the expansion and redevelopment of the port (Pelucca, 2010; Vincenti, 2011).

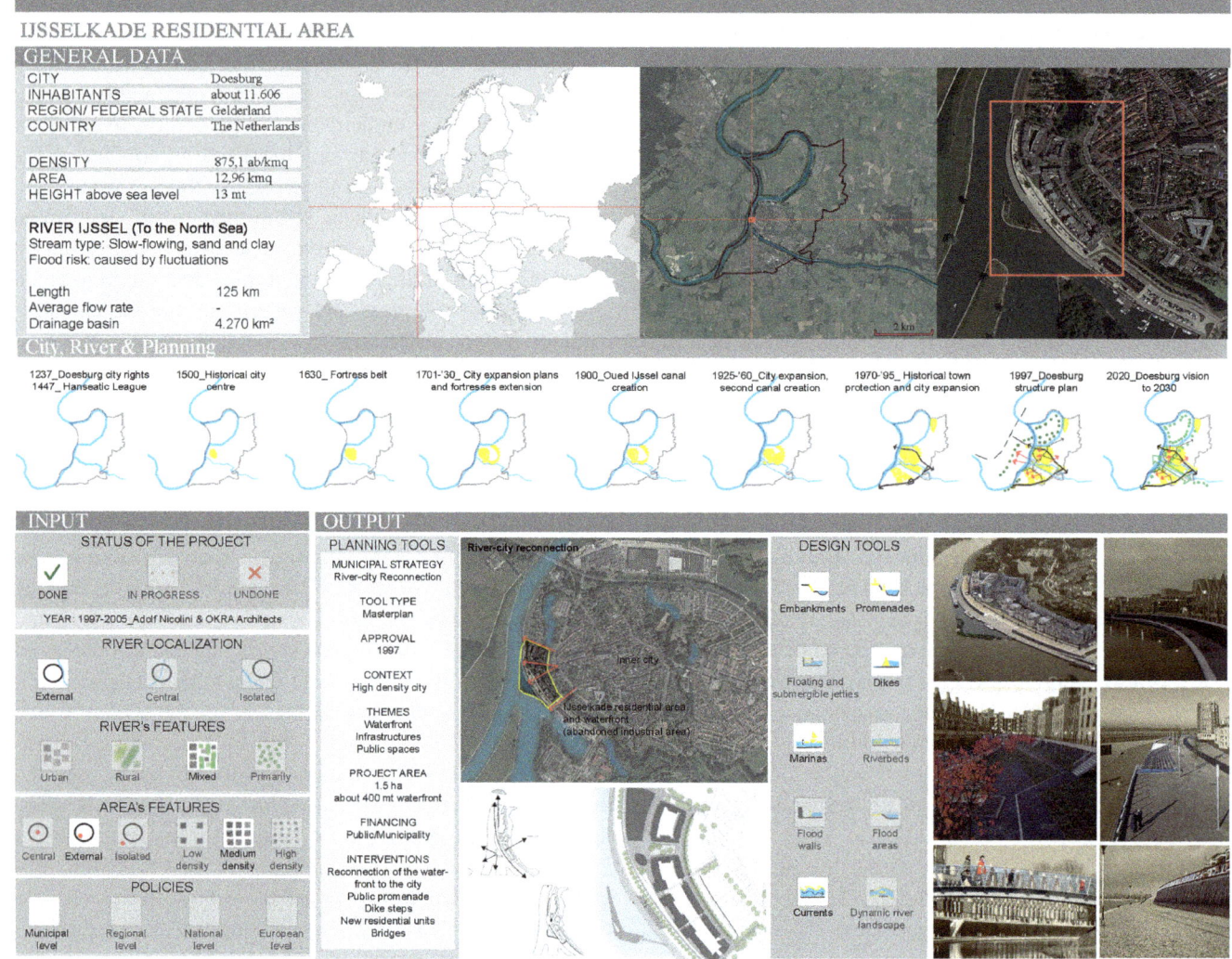

Fig. 2 IJsselkade residential area Project: **General Data Section**: (*Source* Google Earth 2020); **City, River and Planning Section**: Historical Evolution; **Input–Output Section**: Project Analysis (*Source* Google Earth 2020, http://landezine.com, Authors own elaboration, 2020)

The Municipality of Gondomar instead is to the East of Porto. Comparatively to the other municipalities of the Porto metropolitan area, Gondomar tries to redevelop the 4.5 km riverfront of the Douro in order not to lose its continuity with the nearby Porto. The problems encountered by the municipality concerned the stronger and stronger urban growth, the pressure generated by the leisure phenomenon and the private ownership of the land. This riverfront is one of Gondomar's major environmental opportunities to improve the quality of life. Aware of these opportunities and threats, the municipality defined a Strategic Plan covering 4.5 km of this margin, which was executed in 1999/2000 and which had as main objectives to control the urban sprawl, requalify the area in a sustainable way by saving environmental qualities and improve at the same time the economic and social level of the local inhabitants. The riverfront requalification projects are promenades, both naturalistic and urban, the

rehabilitation of Gramido Historic House, the construction of Gramido Nautical Centre and Gramido car parking, bridges, floating jetties, green corridors and little park and naturalistic rest areas (Camara Municipal de Gondomar, 2016; Flores, 2012; Marcolin et al., 2015). The project thus gave Gondomar an attractive and dynamic new pedestrian promenades, panoramas and paths (Fig. 3).

2.2.2 The United Kingdom

The second experience is located in the United Kingdom. Here the water development operations are being carried out according to regional public and private policies and investments to obtain places of strong economic attraction and prime properties, with the conversion of large areas waterfront, along the sea, rivers and canals, in new city centers inspired by art and culture, innovation and technology. The roots of this operation are to be found in the

Fig. 3 Gondomar Polis Project: **General Data Section**: (*Source* Google Earth 2020); **City, River and Planning Section**: Historical Evolution; **Input–Output Section**: Project Analysis (*Source* Google Earth Pro 2020, Authors own elaboration, 2020)

English planning system (European Communities, 2000). After several reforms, regional planning is responsible for economic development as well as for the issues of requalification and competitiveness (Spaan and De Wolff, 2005; Cuturi, 2006). To ensure the development of cities defined as metropolitan areas and new regional poles, the operation that is defined as "Waters" is applied to various English cities including Liverpool and Manchester, the two English regional poles economically stronger as indicated in the UK Regional Scheme Strategy to 2021 (Government Office For the North West, 2008). The program relies on different development companies, guided by Knight Knox Unique Property Consultancy company. It offers delivering prime rental accommodation to high-growth cities, primarily in the north of the UK, with leading developers, and it specializes in land acquisition, design, marketing, financing, lettings, property management and sales. Liverpool is the second

leading center in the north-west of England and it is fundamental to the economic growth of the City Region. It forms a strategic hub with retail floor space, leisure, cultural and tourist facilities. Within the 2015's Liverpool City Region Devolution Agreement, Liverpool City Region has control over transport, skills, business support and other areas, and the Liverpool City Region Combined Authority is established. The Liverpool City Region is crossed by the River Mersey developed as a key asset for tourism and trade. For this reason, the government recognizes that the River Mersey and Liverpool Bay area can drive growth within the Northern Powerhouse (GOV.UK, 2015; Liverpool City Region, 2016, 2019; Peel Land & Property, 2016). Then the Liverpool City Region Local Enterprise Partnership (LEP) has been constituted and working on behalf of the Liverpool City Region Combined Authority has produced the "Building our Future: Liverpool City Region Growth

Strategy" to deliver greater economic growth and prosperity for residents. The Liverpool City Region LEP's Growth Plan also sets out a long-term ambition for the City Region, with a Strategic Economic Plan (SEP). They identify five strategic projects with maximum impact that will tie together various elements of the Growth Plan's approach to deliver sustainable growth (Liverpool City Region Local Enterprise Partnership, 2016). Along the Mersey riverfront two kinds of projects are being realized: the "Liverpool Waters" in the northern part, by Chapman Taylor architect, and the "Liverpool Docks" in the southern part, by BACA architects. In particular, the project "Liverpool Waters" will transform the city's northern docks along the Mersey River into a prime and investment attractive quarter. The project is going to regenerate a 60-hectare site for a mixed-use waterfront quarter. Together with Wirral Waters on the West bank of the Mersey, the project will benefit the structure and economy of Liverpool City Region and the LEP (Liverpool City Council, 2002, 2010, 2017, 2018; Peel Land & Property, 2016). The project, after several analyses, underlined Liverpool as a place of cultural and social heritage and landscaping, ecological and economic interest. The project site is located on the eastern bank of the River Mersey, in the north of Liverpool's Pier Head. The urban design objectives take into account the need to create a place with identity, where public and private spaces are distinguished and where there is variety and choice. Most part of the land use will be used for commercial offices, residential and tourism purposes. The area development considers five neighborhoods: Princes Docks, Central Docks, Clarence Docks, Northern Docks and King Edward Triangle. Princes Dock is characterized by an existing hotel, office and parking uses. The project will activate additional residential and hotel land uses, with cafes, restaurants and promenades spaces. The Central Docks are focused to create open spaces for business, entertainment and leisure uses around the Leeds and Liverpool Canal extension. Here the cruise liner will be the new landmark. Clarence Docks are characterized by the presence of two World Heritage sites, the Clarence Graving Docks and the Victoria Clock Tower, and it will be a residential area with night-time activities. The Northern Docks is going to form a medium-rise residential area with a strong visual connection with the riverfront. The King Edward Triangle is a transition area from the City center and the riverfront, and it is dominated by the Shanghai Tower and will be characterized by primary tall buildings cluster (Liverpool Waters, 2011). All the interventions focus on realizing different design elements, such as streets, promenades, marinas, new residences, water spaces, squares, parks and gardens to define a public framework to link all the neighborhoods and transform the riverfront into a key asset to attract business investments on a regional context (Fig. 4).

3 Discussion

Each city has benefited from the waterfront regeneration in its own way. The analysis has tried to underline the positive and negative outcomes of different planning and design tools, to respond to the 11th and 13th Sustainable Development Goals and their targets, which are related to Sustainable cities and communities and Climate action. In these goals, there is the aim to individuate policies, strategies or actions that can achieve the sustainable transformation of the cities in balance with the environment and all the society. For this purpose, the third synthetic-interpretative part of the reading chart allowed the comparison through the inputs, represented by the boundary conditions (physical and political), and the outputs, represented by the results (the projects and the spatial transformations), which can be divided into the economic, social and environmental dimensions.

As far as the general considerations represent, a common analysis basis and feature, positive aspects have been underlined in the planning, economic, social and environmental dimensions that can be achieved with this kind of waterfront regeneration processes.

– *The inter-scalar capacity of intervention.* All the waterfront projects are part of a large area of intervention, in which the waterfront plays a key role in the enhancement of the urban quality of life. In Hamburg, the pedestrian paths and the embankments are both for the entire stream of the Elbe river and the city public spaces. In Doesburg the embankments affect the entire course of the IJssel river. In Gondomar, the POLIS program, in particular, follows four lines of intervention integrated with the metropolitan system of Porto, pursuing urban redevelopment and environmental enhancement operations, as well as cultural heritage, re-housing areas and measures to improve the urban and environmental conditions of the city. In Liverpool, the Mersey waterfront represents a great asset for regional and international tourism and trade.

– *Investments attractiveness.* Promenade Niederhafen has given a high-value attractiveness to the entire city with the provision of contemporary public space in the inner part of Hamburg. In Gondomar, the project board of directors is made up of public figures, but private companies are also involved on an international competition basis to guarantee the quality of the projects and respect for the process times. Liverpool Waters has a strong focus on physical and economic requalification through specific programs to promote partnership between the public, private and voluntary sectors. Furthermore, the policies pursue an improvement in the quality of life of the

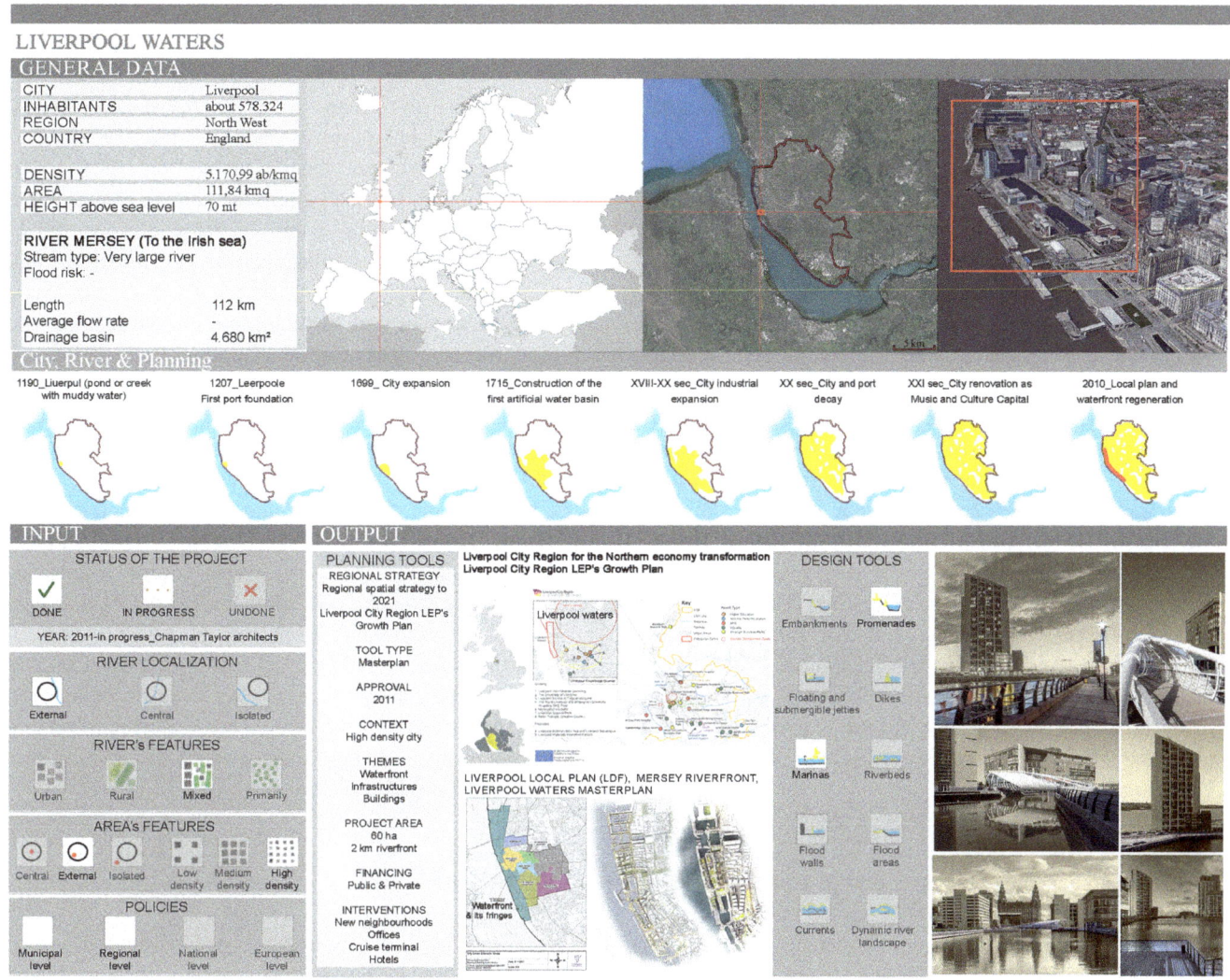

Fig. 4 Liverpool Waters Project: **General Data Section**: (*Source* Google Earth 2020); **City, River and Planning Section**: Historical Evolution; **Input–Output Section**: Project Analysis (*Source* Google Earth Pro 2020, https://liverpoolwaters.co.uk/, Authors own elaboration, 2020)

residents, in terms of crime, traffic and environmental pollution, raising standards for residence and education, and offering economic opportunities.

– *Social gathering in public spaces*. Promenade Niederhafen allowed the creation of promenades and new public and panoramic spaces for citizens. The community of Doesburg has acquired a new large open public space, a reconnection with its territory and a rediscovery in a contemporary key of its relationship with its historical origins, with the creation of panoramic promenades and new public spaces. Gondomar has expanded much more in the north part far from the river with panoramic naturalistic promenades and new public spaces through the conservation of more significant historic buildings and commercial, touristic and service activities. Liverpool Waters gives public spaces and promenades to citizens

along the Mersey river which was once a private port space.

– *Environmental improvement and mitigation of climate change*. Promenade Niederhafen allowed the embankments to be secured with the strengthening of engineering work for the protection from the Elbe river flood risks as well as in the case of IJssel river in Doesburg. In Gondomar, the planning contained within itself sustainable planning and design guidelines, in particular with the aim of strengthening the relationship between city and river and the link between urban activities and cultural and economic dynamization of the riverfront. In Liverpool, the Mersey transformation in environmental performance aims to commit to the cleanest river standard by 2030 and a discharge-free Mersey by 2040.

Negative aspects can be defined along with all these pro- cesses, and divided into the planning, economic, social and environmental dimensions. They share problems related to the integration between natural infrastructures and historical cities linked to brownfield regeneration.

- *Influence of private stakeholders in the planning trans- formation*. In all the case studies, the private stakeholders (investors or property owners) influenced through plan- ning, application or financial resources during the project implementation. In this case, the public stakeholders have been able to seize the private income, to achieve also collective outcomes. In particular, in Liverpool, the real estate company Peel Land & Property decides to submit to the Liverpool City Council an outline planning appli- cation for the development of the riverfront of its prop- erty. The Liverpool City council had already approved the Unitary Development Plan in 2002, but it imple- mented and modified it.
- *The city branding consumerism model*. The great eco- nomic investments of huge and international private companies lead the public administrations to invent a model of city that if not managed for the public interests risks to jeopardizing the city and the planning processes to allow the market processes, as in Liverpool or Hamburg.
- *Gentrification*. The waterfront regeneration of these cases, except for Gondomar, is often linked to the phe- nomenon of gentrification, in terms of concentration of social categories or development oriented towards prime functions, which expel cultural and recreational activities related to the medium-level community or even the res- idents themselves. In fact, the regeneration increases the surrounding living costs, privileging the "prime" society.
- *Decrease of the river's naturalistic features*. In Hamburg, Doesburg and Liverpool the design tools for the river- banks securing transformed the riverbed into high engi- neering embankments, removing the naturalistic features of the river.

In general, several considerations have to be introduced, which are the background of these positive or negative outcomes.

For planning tools, which are the first type of outputs, the most used tools for waterfront regeneration are national or regional policies, local plans, economic development strategies and master plans. In this case studies analysis, it has emerged that the inner difference between the national planning system of the European states considered is not a limitation for waterfront regeneration. There are no more efficient or less efficient policies or strategies. Despite their differences—local for Germany and The Netherlands,

national for Portugal and regional for the United Kingdom— all the policies have achieved waterfront regeneration. The difference itself is a great opportunity to understand how it is possible to invest, in different ways, for the same purposes. Nevertheless, it is important to underline that the achieve- ment has been possible, thanks to the presence of an up-to-date strong legislative system that allows planning and design transformations, disposing of financial resources or partnership opportunities, throughout the planning and design level processes.

Among the design tools, which represent the second type of output various types of intervention are highlighted, ranging from the creation of dams or barriers for the safety of riverbanks, to the creation of paths and pedestrian areas designed to rediscover and enhance urban and landscape resources, as well as to satisfy a request by citizens for spaces reserved for pedestrians and cyclists. In particular, these interventions lead to the creation of embankment walls, promenades, dikes, floodwalls, currents, marinas, public spaces as floating and submergible jetties and cur- rents. Therefore, there is a high risk of loss in terms of naturalistic features, instead of valorizing the river itself. In all the case studies, the design interventions have been related to the securing of the riverbanks—from the flooding and the pollution in all the samples—and the cultural and social reconnection with the river.

4 Conclusion

The discussion tries to demonstrate how urban policies for the territory allow to reconfigure, regenerate, renew and innovate natural infrastructures as urban spaces or networks for safer urban spaces. The collection of case studies high- lights various guiding trends underway in the development of river and coastal fronts. The paper underlines several trends as the redevelopment for safety and protection from flooding and fruition by citizens (Germany), the reconnec- tion between the city center and the abandoned industrial riverfront (The Netherlands), the redevelopment to enhance the historical and environmental resources through the riverfront (Portugal) and the development of real estate riverfront operations to produce economic attractiveness for the prime market, touristic and commercial uses (United Kingdom). In each case, there are positive effects, such as the creation of new public spaces and the securing of the riverbanks, but there are negative effects too because there is a high risk of gentrification, especially in the structured economic riverfront operations. This kind of comparison can help to produce interventions evaluation not only by iden- tifying possible inputs–outputs combinations but also by identifying the conditions that produce negative effects, trying to minimize them. So, in ongoing research, evaluation parameters can be implemented both from a professional point of view and from the point of view of citizens who benefit from new projects. In the first case, these parameters should consider the effectiveness of the interventions in relation to the procedures, funding and techniques used, and in the second case, the appreciation of citizens for the functionality, fruition and perceived beauty of the interventions must be considered. In fact, the action on the infrastructures should have the objective to transform the routes in roots, the aim to modify the habit of passing through a city in the practice of living through the city.

References

Benelli, A. (2019). *Niederhafen River Promenade.* Area Review online. https://www.area-arch.it

Camara Municipal de Gondomar. (2016). *Estudo estratégico para definição das linhas orientadoras de "Gondomar 2020".* https://www.cm-gondomar.pt/

Carta, M. (2007). *Creative City. Dynamics, innovations, actions.* List.

Carta, M., & Ronsisvalle, D. (2016). *The Fluid City paradigm: Waterfront regeneration as an urban renewal strategy.* Springer Verlag.

Castro, F. (2019). *Niederhafen River Promenade.* ArchDaily Architecture Website. https://www.archdaily.com

Cialdea, D. (2019a). Spatial evolutions between identity values and settlements changes. Territorial analyses oriented to the landscape regeneration. In C. Gargiulo & C. Zoppi (Eds.), *Planning, Nature and Ecosystem Services* (pp. 10–19). fedOAPress. https://doi.org/10.6093/978-88-6887-054-6

Cialdea, D. (2019b). *The Tourism connected to the Waterways/Il turismo dell'acqua.* Artigrafichelaregione.

Cialdea, D. (2020). Landscape features of costal waterfronts: Historical aspects and planning issues. *Sustainability, 12,* 1–22. https://doi.org/10.3390/su12062378

Cialdea, D., & Pompei, C. (2018a). Landscape urbanism's interpretative models. A new vision for the Tiber river. In A. Leone & C. Gargiulo (a cura di), *Environmental and territorial modelling for planning and design* (pp. 57–68). Napoli: fedOAPress. ISBN: 978-88-6887-048-5. https://doi.org/10.6093/978-88-6887-048-5

Cialdea, D., & Pompei, C. (2018b). Paesaggio e spazio pubblico. Una proposta per il nuovo Contratto di Fiume Medio-Basso Tevere/Landscape and public space. The new proposal for the Middle-Low Tiber River Contract. In L. Ricci, A. Battisti, V. Cristallo, & C. Ravagnan (a cura di), *Costruire lo spazio pubblico tra storia, cultura e natura/Building the public space between history, culture and nature* (pp. 185–189). ROMA: INU Edizioni. ISBN: 978-88-7603-195-3

Cialdea, D., & Pompei, C. (2019a). Smart city and alluvial park: The role of the "urban green" in the water management through historical and natural values. In A. Gospodini (Ed.), *Proceedings of the international conference on changing cities IV: Spatial, design, landscape & socioeconomic dimensions* (pp. 757–769). THESSALY: University of Thessaly. ISBN: 978-960-99226-9-2. https://iris.uniroma1.it/retrieve/handle/11573/1346213/1322388/Cialdea_Smart-city_2019a.pdf

Cialdea, D., & Pompei, C. (2019b). The past and future of the Tiber River: Urban and infrastructure transformation. In E. G. Arun, & F. Akoz (Eds.), *Proceedings of structural engineers world congress on architecture and structure* (pp. 3–14). Istanbul: maya basin yayin mat. tic. ltd. şti. ISBN: 978-605-62703-8-3

Cialdea, D., & Pompei, C. (2020). The territorial framework of the river courses: a new methodology in evolving perspectives. *European Planning Studies.* ISSN: 0965-4313.https://doi.org/10.1080/09654313.2020.1747401

Cialdea, D., & Quercio, N. (2017). Natural spaces river land in the urban context area. In A. Gospodini (Ed.), *Changing Cities III: spatial, design, landscape & socio-economic dimensions* (pp. 62–78). Grafima Publications. Syros-Delos-Mykonos Islands, Greece.

Clemente, M. (2011). *Città dal mare. L'arte di navigare e l'arte di costruire le città.* Editoriale Scientifica.

Cuturi, C. (2006). *Strategie integrate di riqualificazione urbana e sviluppo locale nel Regno Unito e nella Repubblica d'Irlanda. Politiche, strumenti, prospettive di valutazione* [Doctoral dissertation, Università degli Studi di Napoli "Federico II"]. http://www.fedoa.unina.it/1431/1/Cuturi_Metodi_Valutazione.pdf

Eckbo, G. (1990). Water in total landscape design. In K. Tsuru (Ed), *Elements and total concept of waterscape design.* Mimeograph.

European Commission. (1992). *Council Directive of 21 May 1992 on the conservation of natural habitats and of wild fauna and flora (92/43/EEC).* OJ No. L 206/7 July 22 1992. https://eur-lex.europa.eu/legal-content/EN/TXT/?uri=celex%3A31992L0043

European Communities. (2000). *The EU compendium of spatial planning systems and policies.* Luxembourg: Regional Development Studies. http://aei.pitt.edu/99144/1/28G.pdf

European Parliament and Council. (2000). *Directive of 23 October 2000 establishing a framework for Community action in the field of water policy (2000/60/EC).* OJ No. L 327/1 December 22/2000. https://eur-lex.europa.eu/legal-content/EN/TXT/?uri=celex:32000L0060

European Parliament and Council. (2007). *Directive of the of 23 october 2007 on the assessment and management of flood risks (2007/60/EC).* OJ No. L 288/27 November 6/2007. https://www.eumonitor.eu/9353000/1/j9vvik7m1c3gyxp/vitgbgimtmez

Fedeli, V. (2006). *Rethinking European spatial policy as a hologram: Actions.* Routledge.

Fisher, B., Benson, B., & Institute, U. L. (2004). *Remaking the urban waterfront.* Urban Land Institute.

Flores, J. (2012). *Gondomar Polis Project -Sustainable Requalification of Gondomar's River Douro Waterfront.* Comunicação apresentada no 5th European Conference on Sustainable Cities & Towns» - Session A5- Sustainable (re-)designing of districts, promovido pelo ICLEI – 21 a 24 de Março, Sevilha. 2007. https://www.academia.edu/799966/Gondomar_Polis_Project_Sustainable_Requalification_of_Gondomar_s_River_Douro_Waterfront

Fusco Girard, L., Forte, B., Cerreta, M., De Toro, P., & Forte, F. (2004). *The human sustainable city: Challenges and perspectives from the habitat agenda.* Ashgate Pub Ltd.

GOV.UK (2015). *Liverpool City Region devolution agreement.* https://www.gov.uk

Government Office For the North West. (2008). *The North West of England plan regional spatial strategy to 2021.* http://www.gov.uk

International Commission for the Protection of the Elbe River. (2016). *The Elbe river and its basin.* https://www.ikse-mkol.org/

Jacobs, J. (1961). *Life and death of great American cities.* Vintage.

Knuijt, M. (2002). IJsselkade Doesburg – Ijssel Waterfront in Doesburg. *Topos, 2002*(39), 19–23.

Hoyle, B., & Pinder, D. (1992). *European Port cities in transition.* Belhaven Press.

Liverpool City Council. (2002). *Liverpool unitary development plan.* https://liverpool.gov.uk/

Liverpool City Council. (2010). *Core strategy revised preferred options.* https://liverpool.gov.uk/

Liverpool City Council. (2017). *Liverpool Maritime Mercantile city world heritage site management plan 2017–2024*. https://liverpool.gov.uk/

Liverpool City Council. (2018). *Liverpool local plan 2013-2033*. https://liverpool.gov.uk/

Liverpool City Region. (2016). *The state of Liverpool city region report: Making the most of devolution*. https://www.liverpoolcityregion-ca.gov.uk/

Liverpool City Region. (2019). *Liverpool city region spatial planning statement of common ground*. https://www.liverpoolcityregion-ca.gov.uk/

Liverpool City Region Local Enterprise Partnership. (2016). *Building our future: Liverpool city region growth strategy*. https://www.liverpoollep.org/

Liverpool Waters. (2011). *Liverpool waters—Design and access statement*. https://issuu.com/christopherm.gibson; https://liverpoolwaters.co.uk/

MAOT (Ministério do Ambiente, do Ordenamento do Território e do Desenvolvimento Regional). (2000). *Programa Polis. Programa de Requalificação Urbana e Valorização Ambiental de Cidades*. Lisboa. https://dre.pt/pesquisa/-/search/274224/details/maximized

Marcolin, P., Flores, J., & Cortesão, J. (2015). Outline for a human-based assessment methodology of urban projects. The case of Polis Gondomar. *(GOT), n.º 7 (junho). Centro de Estudos de Geografia e Ordenamento do Território* (pp. 183–211). https://doi.org/10.17127/got/2015.7.008

Marshall, R. (2016). *Waterfronts in post-industrial cities*. Taylor & Francis.

Mazzoleni, C. (2013). Amburgo, HafenCity. Rinnovamento della città e governo urbano. *Imprese & città, 2*, 138–155. https://www.milomb.camcom.it/c/document_library/get_file?uuid=e3216bf5-1726-4315-b5b3-674e99a75ad3&groupId=10157

OKRA. (2010). *River Ijssel*. Landezine. http://landezine.com/index.php/2010/11/quay-and-bridges-on-the-river-ijssel-by-okra-landscape-architecture/

Partidário, M. R., & Correia, F. N. (2004). POLIS—The Portuguese programme on urban environment. a contribution to the discussion on European urban policy. *European Urban Studies, 12*(3), 409–424.

Peel Land & Property. (2016). *Peel in the Northern Powerhouse. The Peel Group partnering for growth, prosperity and a lasting social legacy*. https://www.peel.co.uk/

Pelucca, B. (2010). *Progetto e rinnovo urbano nella città contemporanea: il caso del Portogallo*. EdA Esempi di Architettura.

Prominski, M., Stokman, A., Zeller, S., Stimberg, D., & Voermanek, H. (2017). *River.Space.Design: Planning strategies, methods and projects for urban streams*. Birkhauser Architecture.

Sairinen, R., & Kumpulainen, S. (2006). Assessing social impacts in urban waterfront regeneration. *Environmental Impact Assessment Review, 26*, 120–135. https://www.researchgate.net/publication/222548733_Assessing_social_impacts_in_urban_waterfront_regeneration

Spaans, M., & De Wolff, H. (2005). Changing spatial planning systems and the role of the regional government level; Comparing the Netherlands, Flanders and England. *ERSA Conference in Amsterdam*.

Thieken, A. H., Kienzler, S., Kreibich, H., Kuhlcke, C., Kunz, M., Mühr, B., Müller, M., Otto, A., Petrow, T., Pisi, S., & Schröter, K. (2016). Review of the flood risk management system in Germany after the major flood in 2013. *Ecology and Society, 21*(2), 51. https://doi.org/10.5751/ES-08547-210251

United Nations. (2015). *Transforming our world: the 2030 Agenda for Sustainable Development*. United Nations. https://sdgs.un.org/2030agenda

United Nations. (2017). *New Urban Agenda*. Habitat III Secretariat. https://habitat3.org/the-new-urban-agenda/

Van Der Knaap, B., & Pinder, D. (1992). Revitalising the European Waterfront: Policy evolution and planning issues. In A. Carpenter & R. Lozano (Eds.), *European Port cities in transition* (pp. 155–175). Belhaven Press.

Vincenti, T. (2011). Realtà urbane a confronto. Il caso del waterfront di Porto: un percorso valutativo ex post. *XL Incontro di Studio del Ce.S.E.T.* (pp. 299–323). Firenze University Press www.fupress.com/ceset; https://core.ac.uk/download/pdf/228540579.pdf

Vitillo, P. (1997). Amburgo Metropoli Verde. *Territorio, 4*, 71–83.

Voorendt, M. Z. (2015). *Examples of multifunctional flood defences working report*. Delft University of Technology, Department of Hydraulic Engineering.

Zaha Hadid Architects. (2019). *Niederhafen River Promenade*. https://www.zaha-hadid.com/architecture/hamburg-river-promenade/

Transit-Oriented Development as a Tool of Urban Transformation Addressed at Urban Regeneration Processes

Cristina I. Covelli

Abstract

Transit-oriented development is a theoretical and practical framework that represents an important tool for municipalities, private actors and citizens to enhance urban transformation. Furthermore, there has been a growing interest to understand and implement this concept within urban planning practice. Mixed land use is an element that influences the transit-oriented development level of an urban area; therefore, it is important to further understand the role of this element and find paths to strengthen it and use it for urban regeneration processes. The main objective of the principle of mixed land use is to create more vibrant, well-connected and sustainable urban patterns. Adding to that, urban structures with a high mixed land use percentage provide citizens with a sufficient amount of services within walking distance and therefore decrease car dependency structures. However, areas with low mixed land use patterns will reproduce less dense, car-dependent urban areas. This research intends to provide tools to measure and assess mixed land use in an urban context so that it could be applied in urban regeneration processes. This research wishes to expand the scope of previously done studies in the sense that it will evaluate six parameters within mixed land use and at the same time expand the knowledge of the influence of TOD in urban regeneration projects. The study case that was assessed was Citylife, which is one of the biggest urban generation projects in Europe. The parameters measured were complementary uses, access to local services, access to parks and playgrounds, affordable housing, housing preservation, and business and services preservation. The methods that were used to perform the evaluation of these parameters are mapping, using geoinformation systems, specifically Qgis. Other methods used were literature review and observations. The conclusions describe the weakest and strongest parameters that were evaluated and the way in which Citylife could increase mixed land use patterns.

Keywords

Mixed land use • Urban regeneration • Complementary uses • Affordable housing • Access to local services • Housing preservation • Transit-oriented development • Mixed urban patterns

1 Research Framework

The goal of this research was to perform a mixed land use assessment of Citylife using the transit-oriented development theoretical and practical framework to highlight the importance of encouraging higher mixed land use patterns in urban regeneration processes. This research aims at understanding the implementation of the transit-oriented development mix principles in the area around an urban regeneration project in Italy and contribute to aim at a more sustainable growth pattern. Also, this research contributes to expanding the knowledge about how transit-oriented development is being implemented in Italy's urban development and land planning policies.

More in detail, this research will help to understand how TOD principles could be effectively implemented in a specific area, analyse the results and propose improvements to further contribute to transport planning and urban planning synergies and therefore thrive for planning methodologies that contribute to aim at more sustainable cities.

To develop the proposed goal, the following objectives were formulated:

(1) Implement the transit-oriented development standard 3.0, specifically the variable mix around the Citylife urban regeneration project.

C. I. Covelli (✉)
KTH, Stockholm, Sweden
e-mail: covellig.cristina@gmail.com

(2) Analyse the results of the assessment done on the area of study.

(3) Propose enhancements to the area of study to aim at a more sustainable neighbourhood aligned with TOD principles.

Within the first objective, the assessment evaluation method was implemented within a 500 m radius surrounding the area of the Citylife urban regeneration project. The mix variable with its six subcategories included in the TOD standard 3.0 was evaluated using spatial analysis tools as well as observations, satellite images, pictures and an interview. Regarding the second objective, the evaluation was applied, and the classification of the chosen area was made accordingly to the three different categories defined by the TOD standard: Gold, Silver and Bronze. To achieve this objective, the results were analysed throughout the document and then summarized in the conclusion section. In this case the observations and literature review were used to understand the results.

The third objective was achieved through the suggestion of improvements to the variables that did not achieve the highest score. To accomplish this objective, spatial analysis was performed in order to physically place the areas in which the proposed improvements could be implemented. Within the performance of the assessment, it was also suggested how the mix variable could reach 100% of the points assigned to each subcategory.

2 Research Methodology

The methods used in this analysis were a combination of qualitative and quantitative research methods. The quantitative method used was mapping. Geographical information systems were used to score the variables that resulted from the TOD literature review. Each of the variables analysed was assessed using Qgis software. The qualitative methods were observations and a literature review which were useful to understand and interpret the spatial study that was performed. The literature review was focused on the transit-oriented development theoretical framework which intended to fulfil a comprehensive mixed land use assessment of Citylife. And the observations were performed in the area of study to understand the interaction between the dynamic and static elements of the environment.

More in detail, regarding the first method used which was mapping, geographical information systems were a useful tool to relate geographical information with variables within urban development. In this case, the Qgis software was used to score the variable of mix that resulted from the TOD literature review. The subindicators used on this scoring system included complementary uses, access to local

services, access to parks and playgrounds, affordable housing, housing preservation, and business and services preservation. Each of the subcategories analysed was assessed using the Qgis software, complemented also using statistical tools to retrieve more useful information to understand the spatial results. The statistical analysis was used to build up the ratios needed to unveil the relationship between the analysed variables. The majority of the ratios were shown as a percentage. The ratios that were expressed in percentage were complementary uses and affordable housing.

The second method was observations. A direct on-site observation was performed, specifically, unstructured observations were used to walk around the chosen area of study and analyse how easy was it to walk and get from the transit station Tre Torri to the commercial areas next to it as well as the residential areas.

The third method used was online satellite imagery. The method was used to analyse variables using satellite pictures of the area and sample areas. Specifically, this method was useful to analyse the transformation and urban regeneration that took part during the process in the periods before and after the construction of the Citylife project.

The fourth method was a literature review. This qualitative method allowed further analysis of the spatial information constructed using the first method. It was important to make a comprehensive literature review to understand and analyse the TOD variables and parameters.

The most important steps followed during the research were first to make a comprehensive literature review and research. After that, it was necessary to make sufficient data collection to measure all six subcategories. Also, structure the study visits which took place during the period from January to April 2020. Then, the data was processed, and most results were produced. After gathering all the information that was collected the overall evaluation and recommendations were made. The last part of the research focuses on the final reflections and conclusions.

3 Transit-Oriented Development and Mixed Land Use in Citylife

Transit-oriented development of a theoretical and practical framework encourages sustainable mobility patterns in the sense that it intends to diversify land use. With the help of this tool, private actors, public organizations, citizens and NGOs can aim at making cities more accessible, safer and more attractive. Urban regeneration projects constitute an important tool for urban planners to use urban land in the most efficient way possible. Subsequently using the TOD framework in urban regeneration processes can unveil guidelines to enhance urban development in a sustainable way. According to Bertolini, transit-oriented development (TOD) is a concept that aims at building sustainable mobility patterns based on high synergies between land and transport planning. The main essence of this concept is to encourage sustainable means of transportation, build denser and more mixed neighbourhoods and shift away from car-dependent urban areas. This concept has evolved, including more variables that add more complex and integral solutions to shift away from car-dependent urban structures (Curtis et al., 2009).

Areas around a transit station have the highest opportunities to develop high TODness levels. In particular, in Milan, areas around railway stations represent core zones where synergies between land and transport planning could be achieved. Furthermore, within the TOD concept, transit stations represent the node area where mixed land use patterns can be implemented, as well as enhancing walking and cycling networks. Also, within TOD, areas around transit stations should be dense, compact and with less space possible destined for motor vehicles (Liu et al., 2016).

Furthermore, Cervero includes other dimensions within TOD. This theoretician emphasizes the importance of smaller units of analysis such as blocks. He argues that the way blocks are structured in future urban areas is key to determining the accessibility of pedestrians and bicycle users, and how easy it would be for those actors to navigate the road network. Adding to that, the shape of the urban structure can influence car use. He states that shorter distances for pedestrians and adequate networks for bicycle users along with an appropriate mix land use can shape cities into a more sustainable future. To go over the main points of transit-oriented development, Cervero summarizes that TOD policies: "promote more walking and transit riding and less driving: pedestrian-friendly designs such as safe and attractive sidewalks; small city blocks and a highly connected grid like street network; mixed land uses that place many destinations close to each other, including small storefront ground-floor retail in commercial districts; sufficiently high densities to justify high-quality and frequent public transport services; and community hubs and civic places that promote social interaction and a sense of belonging" (Cervero, 1993).

Spatially on the urban structure, the TOD model has been explained extensively by Peter Calthorpe. Researchers Mingqiao Zou et al. explained the TOD morphological model based on Calthorpe's illustrations. Figure 1 shows a TOD area based on elements such as public transportation station, core commercial, office/employment area, TOD residential area, secondary area and public open space (Zou et al., 2014).

This model explains that within the TOD structure the areas closer to the station are the ones that should have higher development, specifically concentrated within a 500 m radius. Land use intensity starts decreasing from 600 to 1000 m away from the transit station. This model also contributes to understanding the importance of mixed land use within this concept. There is a high mix between residential, commercial, offices and transportations infrastructure. All these elements aim at reducing travel demand (Zou et al., 2014).

This demonstrates that mixed land use can be considered an important factor to aim at TOD development patterns, as it has been proved that higher mixed land use patterns discourage the use of motorized vehicles to transport and encourage walking and cycling.

Affordable housing is one of the topics that have been widely related to TOD. In this regard, the research developed by Pal explores the notion of developing affordable housing

Fig. 1 TOD model based on Calthorpe (*Source* Zou et al., 2014)

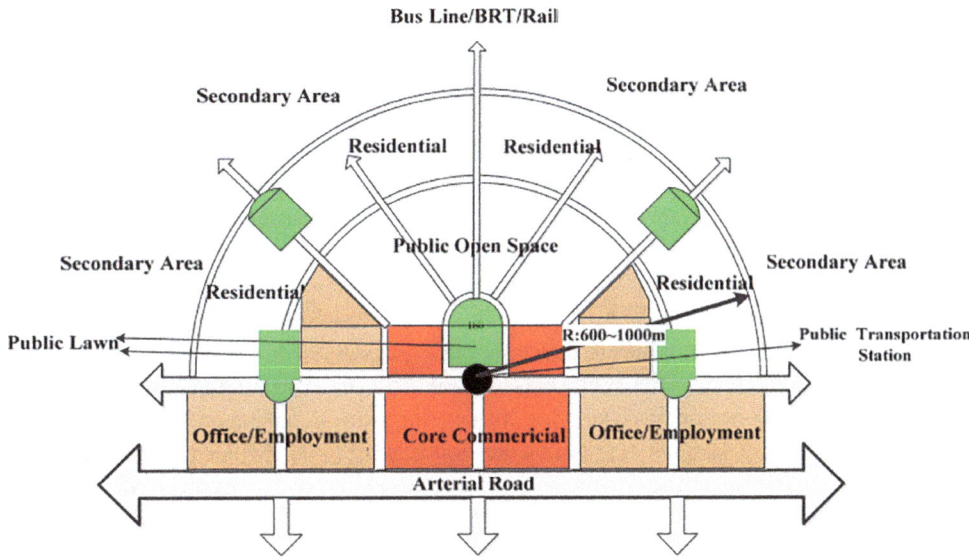

projects close to transit stations. Specifically, railway stations so that citizens would have access to public transportation options. In that way "easy access to multiple modes of public transit would increase a household's budget for housing by eliminating the financial burden of owning a car" (Pal, 2018).

Likewise, Pengjun et al. studied the relationship between the citizen's patterns to choose the residential location and the housing prices. They stated that "When it comes to residential location choice, land use policy should be assisted by housing policy to encourage passengers to choose to live near metro station areas. The high housing prices nearby transit stations may force people, in particular, low-income people, to live far away from metro station areas" (Pengjun, 2018). In this sense, promoting affordable housing residential projects can contribute to TOD principles by aiming at a social mix by the income level of the area and increasing synergies between housing policies and transportation planning.

Transit-oriented development is a concept that contributes to the design of less car-dependent societies and urban areas. Within the theoretical framework of transit-oriented development, the Institute for Transportation and Development Policy (ITDP) formulated a standard that can be used to measure the TODness of a study area. The ITDP is an organization that promotes the use and implementation of TOD policies and best practices to actively reduce car use in cities. This tool within TOD proposes to implement practical measurement of parameters in already built cities as well as in future undeveloped areas to discourage car use and promote sustainable mobility alternatives. This tool is called the TOD standard 3.0. The following section (Sect. 5) will describe the components, variables and parameters used by this TOD standard.

3.1 The Citylife Case

Milan planned several urban regeneration projects in the vision 2030 of the comprehensive plan PGT (Piano di governo del territorio, 2019). This research will study one of those projects, specifically Citylife, which is planned to be one of the biggest urban regeneration projects in Europe. The TOD standard 3.0 was applied in the area around Citylife to practically assess the TODness level, specifically the mix standard was evaluated. Areas around a transit station have the highest opportunities to develop high TODness levels, and in this case Citylife represents a study case in which public transportation was key for its development. Areas around railway stations represent core zones where high synergies between land and transport planning could be achieved. Furthermore, within the TOD concept, transit stations represent the node area where mixed land use patterns can be implemented. Also, within TOD developments, areas around transit stations should be dense, compact and with less space possible destined for motor vehicles (Liu et al., 2016).

In 2015, the metro station "Tre Torri" was built as part of the expansion of subway line 5, which aimed to connect the new inhabitants and visitors of the area. According to the TOD model, the urban environment should be based on elements such as public transportation stations, core commercial areas, office/employment areas, TOD residential areas, secondary areas and public open spaces (Zou et al., 2014). This theory explains that within the TOD structure the areas closer to the public transport stations are the ones that should have higher development, specifically concentrated within a 500 m radius. There is a high mix between residential, commercial, offices and transportations infrastructure. All these elements aim at reducing travel demand (Zou et al., 2014). Within TOD there is a strong link between mixed land use and public transport ridership. According to Sarkar et al. "land-use mix was found to be strongly associated with the choice preference of public transport. The coefficients were positive for public and nonmotorized modes, which imply that the trip makers residing in the areas with mixed land use prefer public and nonmotorized travel modes" (Sarkar & Chunchu, 2016). This demonstrates that mixed land use can be considered an important factor to aim at TOD development patterns, as it has been proved that higher mixed land use patterns discourage the use of motorized vehicles to transport and encourage walking and cycling.

In this case, Citylife urban structure represents an area with intense use combining high rise residential, commercial, offices and recreational uses. Higher intensity use areas next to transit stations are also known as urban cores, and urban cores have different types and vary according to their function "there is a desirability core (Stojanovski, 2013), the most desirable TOD zone in the center. The morphological effect on architecture and cityscapes is visible firstly in the TOD CORE zone. The TOD CORE zones are tentative and question the traditional urban design heuristics about walking distances" (Stojanovski, 2015). These core zones areas around a transit station which are reachable within walking distance stand as a valuable element to foster a sustainable urban land use pattern. Under this description Citylife could be considered as a TOD core which enables residents and visitors to be served by a wide array of services within walking distance and foster a high quality of life, decreasing car-dependent patterns.

Milan planning authorities, private developers and citizens involved in the formulation and development of the Citylife project conceived the urban regeneration of an area

Fig. 2 Milan conference centre hall in 2021 (left); current Citylife project (2020) (*Source* Google Earth)

that was previously destined as a conference centre. The objective was to include Citylife area into the urban tissue as a vibrant and active part of the city, as stated by Citylife official website: "CityLife is the urban redevelopment project of the Portello district of Milan. This area was previously occupied by the Fiera Milano City and to give new life to the neighbourhood, by creating a space that would enhance the entire city, international architects were used to design the spaces" (Taccioli, 2019). Also, according to the Guiding architects studio "City life went from the former trade fair area of Milan to a new multipurpose district, with three futuristic towers in the middle, residential blocks designed for a new way to live in the city and the first urban shopping district in Italy" (Guiding architects, 2020). Figure 2 shows the physical transformation from the conference centre into the mixed-use new project called Citylife.

Planning documents such as the Milan PGT contain strategies that plan to enhance connectivity, innovation, social equity and regional cohesiveness (PGT, 2019). The specific objectives are to encourage: a connected, metropolitan and global city, a city of attractive and inclusive opportunities; a green, liveable and resilient city and a city that regenerates itself (PGT, 2019). Within the objective of urban regeneration, there are several strategies that include redevelopment and efficiency of urban land use on private and public lots. Also, it considers the equalization and transfer of building rights, recovery of abandoned and disused buildings, among others.

3.2 Evaluating Mixed Land Use in Citylife

The fifth principle of the TOD standard evaluation system was applied in Citylife project. More in detail, the Mix TOD principle is subdivided into six subcategories. The first is complementary uses, then access to local services, access to parks and playgrounds, and the fourth one is affordable housing. The fifth one is house preservation, and the sixth one is business and services preservation. The general objective of this parameter is to plan urban spaces adapted for a mix of income, demographics and uses. The first subcategory, complementary uses, focuses on quantifying the residential and nonresidential uses in the same area. More points are given to areas that have a higher percentage of mixed land use. The method consists of finding the mix ratio between the variables of residential and nonresidential use. For example, when the same type of land use has more than 80% of the total floor area the evaluation parameter attributes 0 points. And on the contrary, when the total of the floor area has a predominant use of 50–60%, the parameter attributes 8 points which is the highest (ITDP, 2017).

The second parameter is access to local services. This parameter measures the percentage of buildings that have proximity to the following amenities: primary schools, healthcare service or pharmacy, and source of fresh food (ITDP, 2017). In order to evaluate this parameter, the entrances of the residential buildings in the area are mapped and overlayed with a 300 m distance to the entrances of the local services. More in detail, higher points are given to developments in which the residential buildings have a walking distance to the entrance of three local service types. On the contrary, fewer points are given to developments in which the residential buildings have close access to two or one type of local service. In this case, the three residential projects in Citylife were mapped along with the education facilities, primary and secondary schools, healthcare services and pharmacies, and supermarkets and restaurants (ITDP, 2017).

The third subcriterion is access to parks and playgrounds which measures the percentage of buildings located within a walking distance to areas such as parks or playgrounds (ITDP, 2017). To evaluate this parameter it was necessary to map the three residential complexes in Citylife along with the parks and playgrounds in the area of the project. The evaluation system gives a higher score to the projects in which at least 80% of the residential buildings are within a walking distance of 500 m of a park or a playground. It is important to highlight that the ITDP scoring system considers only the "eligible parks and playgrounds" and defines them as "at least 300 m^2 in area and publicly accessible 15 h or more per day. If the park or playground has shared use as school yard or physical education facility, school time can be deducted from the opening hours" (ITDP, 2017).

The fourth subcriterion is affordable housing, and this parameter measures the percentage of the housing areas that are considered within the criteria of social housing. Affordable housing is considered as "housing rent which is below 30% of the mean income in the relevant income category" (ITDP, 2017). Within this section, it is required to make a ratio between the residential housing and the affordable housing area. Higher scores are given to areas that have a higher percentage of affordable housing. The fifth subcriterion is housing preservation. This parameter intends to measure the relocation percentage of previous residential buildings that were on the lot in which the new project is built. The evaluation system gives a higher score to those projects in which the previous residential areas were relocated within a walking distance of 500 m away from the original location. In order to perform the evaluation of this subcriterion, it was necessary to find out if there were previous residential areas in Citylife, which was accomplished by a mapping time-lapse using Google Earth. Also, it was complemented with research on the previous land use of the area of study.

The last subcriterion is called business and services preservation. This subcriterion quantifies the percentage of local services that were relocated within a walking distance from the original location. To follow the evaluation process, it was necessary to identify the previous commercial uses of the area of study. In order to perform this evaluation, the previous commercial buildings were mapped. This

procedure was complemented by contacting the current shopping centre located in the area of study and requesting information about possible previous commercial activities.

4 Results of Mix Principle Evaluation of Citylife

After implementing the evaluation system, it was found that the best-ranked parameters within mixed land use in this area were complementary uses, access to local services and access to parks and playgrounds. However, there are several actions that could be implemented in the study area in order to improve the mixed land use level. Specifically, affordable housing and business and services preservation. Citylife obtained 17 points out of 25 possible. Table 1 shows in detail the results of the evaluation of all subcriteria considered in the evaluation of the Mix TOD principle within the Citylife development. Results will also be shown more in detail by each subcriteria.

4.1 Complementary Uses

The complementary uses subcategory aims to estimate the percentage of each land use category and evaluate its overall weight. With the purpose of finding complementary uses percentage, it was necessary to make a general land use map of the study area. It is required to find the total of the square meters on the project and then classify it according to the use. It was found that the total area of this project was 365.000 m^2 (Systematica.net, 2019). Through the revision of the detailed plans, it was found that 51% of the land is destined for residential use, whereas 34% for office space, 8.7% is commercial, 4.1% for events and 1.7% is destined for parking spaces. According to the scoring system, if the predominant use of the block is between 50 and 60%, it is awarded 8 points out of 8. In this case, Table 2 summarizes the results of the percentage of land use. This table shows that the area studied has optimal land use diversification. This subcategory obtained the maximum score. It could be recommended that other urban regeneration projects would adopt similar percentages of land use, and therefore endure a high land use diversification in the city.

4.2 Access to Local Services

To evaluate the access to local services subcategory, the residential buildings of *Citylife* were mapped. Also, three different types of local services were mapped and then the distance between the two layers was quantified and analysed. The first type of service was food, so all supermarkets, small

Table 1 Mix standard results of Citylife

Mix				
Objective A. Opportunities and services are within a short walking distance of where people live and work, and the public space is activated over extended hours				
5.A.1 Complementary Uses	Residential and nonresidential uses within the same or adjacent blocks	Max score 8	Obtained score 8	Predominant use is 51%
5.A.2 Access to Local Services	Percentage of buildings that are within walking distance of an elementary or primary school, a healthcare service or pharmacy, and a source of fresh food	3	3	80% of buildings have access to two types of services
5.A.3 Access to Parks and Playgrounds	Percentage of buildings located within a 500-m walking distance of a park or playground	1	1	100% of the buildings were located within 300 square meters from a playground, park or green area
Objective B. Diverse demographics and income ranges are included among local residents				
5.B.1 Affordable Housing	Percentage of total residential units provided as affordable housing	8	1	0% of the housing market has a selling price below 30% of the mean of the selling price
5.B.2 Housing Preservation	Percentage of households living on site before the project that is maintained or relocated within walking distance	3	3	No relocation; there were no any previous housing projects
5.B.3 Business and Services Preservation	Percentage of pre-existing local resident–serving businesses and services on the project site that are maintained on site or relocated within walking distance	2	1	0% of the local businesses were relocated within a 500 m radius
Total score: 17/25				

Table 2 Land use Citylife

Land use	Square meters	Percentage	Project
Residential	188,000	51%	Residenza Hadid Residenza Libeskind
Commercial	32.000	8.7%	Shopping district
Offices	124,000	34%	Tre Torri Citylife; Torre Hadid:177 m, 39 floors. Torre isozaki: 209 m, 46 floors. Torre libeskind: 175 m, 28 floors
Events	15,000	4.1%	Pallazzo scintille
Green areas	173,000 (6,000 flower field)		Green áreas. Gli Orti Fioriti di CityLife
Parking	7,000	1.7%	Underground parking spaces
	Total: 365,000 square meters		

stores, bakeries, pubs, cafes, wine shops, fast food services and restaurants were mapped. After that, the second type of service was mapped including all the schools, libraries and kindergartens in the area. Lastly, type 3 of local services was mapped, identifying all hospitals, dentists, veterinaries and pharmacies (ITDP, 2017). After the overlaying analysis, it

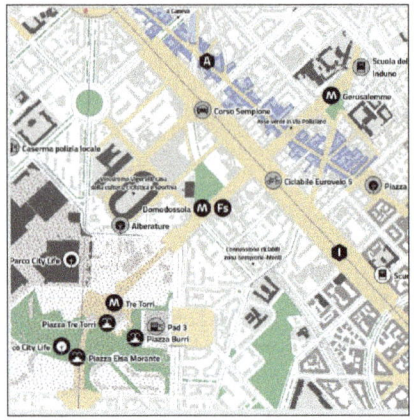

"Urban regeneration Will be at the center of future developments" Strategic plan of Milan

Fig. 3 Milan PGT Citylife urban regeneration spaces (*Source* Milano's PGT, 2009)

Types of local services

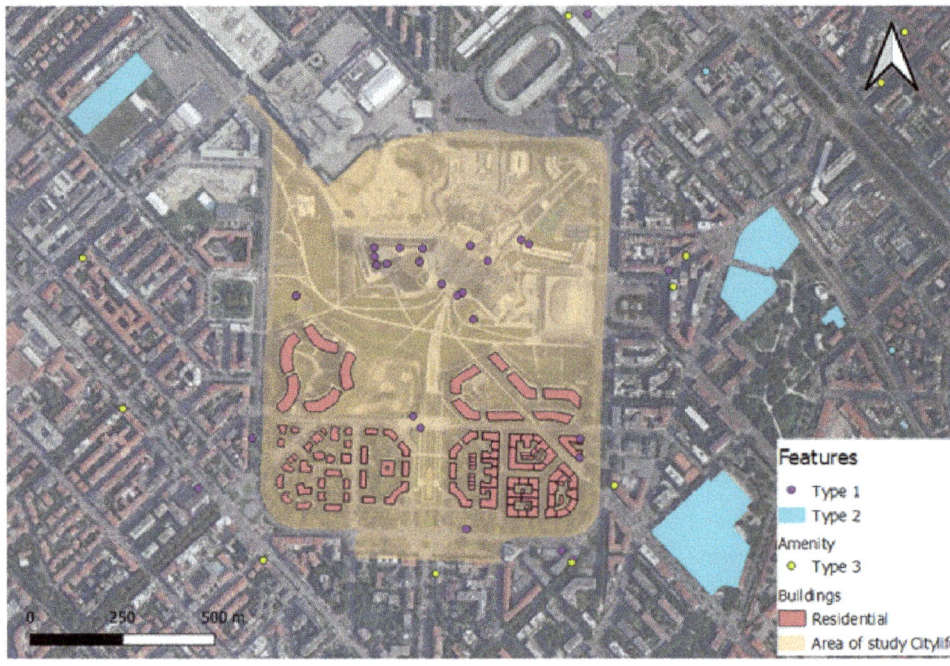

Type 1: supermarkets, wine stores, restaurants, bakery, fast food, café, pub

Type 2: Schools, libraries, kindergarten.

Type 3: Dentist, pharmacies, hospitals, veterinary.

Fig. 4 Types of local services (*Source* own work, 2020)

was found that 100% of the buildings in the area have access to at least two different types of services. According to the scoring system, the study area is awarded three points. Figure 3 illustrates the Qgis result (see Fig. 4). Even though this parameter got a sufficient score, there is space for improvement. The types of services offered in the area could be more diversified and accessible to the residential areas in a higher percentage than 80%. It is important that services related to education are provided in a more extensive way in urban regeneration projects so that citizens would lower car use for the shortest trips and would use sustainable transport mobility alternatives.

4.3 Access to Parks and Playgrounds

This subcategory evaluates and classifies the proximity between residential buildings and eligible green areas. Eligible green areas are parks or playgrounds bigger than 300 square meters. In order to perform this evaluation, it was necessary to identify and map the residential buildings in *Citylife* as well as the eligible green areas and measure the distance between them. The evaluation revealed that 100% of the residential buildings in the area are located within 500 m of an eligible park, playground or green area. In this case, it was attributed 1 point which is the highest score possible (ITDP, 2017). The evaluation of this subcategory received the highest score, so there are no further suggestions in this regard in *Citylife*. Figure 4 shows the Qgis result (see Fig. 5). This area has 173,000 sqm of a public park inside the area of the project (Distribuzione moderna, 2020).

4.4 Affordable Housing

This parameter measures the percentage of affordable housing within the area of study. In Italy there are three types of affordable housing projects: subsidized, assisted and agreed housing (Wang, 2019). According to Wang, Italy is lacking affordable housing policies compared to other European countries: "The public housing stock in Milan is just 5% of all the dwellings, similar with the average percentage of Italy, which is 4% (Pittini et al. 2015). This number is very low compared with other countries, such as Netherlands (36%), Great Britain (22%), and France (20%), which shows Italy is actually a country lacking of social housing stock" (Wang, 2019). Under this analysis the concept of subsidized housing will be used. Following the notion of subsidized housing, it was found that 0% of the project was destined for affordable housing units.

Citylife has two residential projects designed by the architect Libeskind (see Fig. 6b) and the other one by Zaha Hadid (see Fig. 6a). Libeskind project is worth "150-million-dollars (…) These residences are quite exclusive and costly apartments for wealthy people, indeed"

(Inexhibit, 2020). In fact, the last three apartments are being sold at 9,400 euro per square meter on average, whilst the average price for a square meter in city life is 6,450 euro (Citylife, 2020). Similarly, the residential complex designed by Zaha Hadid cost about $3,500 per square foot and has double-height penthouses, called sky villas, provided with generous-sized panoramic terraces (Inexhibit, 2020). According to immobiliare.it, in Milan the housing price on average is 4,179 euro per square meter which makes Citylife 120% more expensive (Immobiliare.it, 2020). *Citylife* is located in the highest price range in the city. Figure 7 illustrates the housing price per square meter according to the geographical area of the city of Milan (see Fig. 7).

These conditions award the area 1 point out of 8 possible points. The ITDP (Institute for Transportation and Development Policy) stresses the importance of the role of affordable housing policies to effectively promote social mix and social wellbeing. As this subcategory got the lowest score possible, it is important to reflect on the inclusion of socio-economic diversity within urban regeneration projects in Milan. The Mix TOD evaluation systems encourage social mix and therefore it envisions the inclusion of at least 15% of the total project of affordable housing that has at least a 30% lower market value. In order to achieve that, it is imperative that diverse types of actors like private developers, civil organizations and municipal governmental agencies partner up to formulate projects where a wider range of market prices are available for citizens that have lower income.

Fig. 5 Access to parks and playgrounds (*Source* own work, 2020)

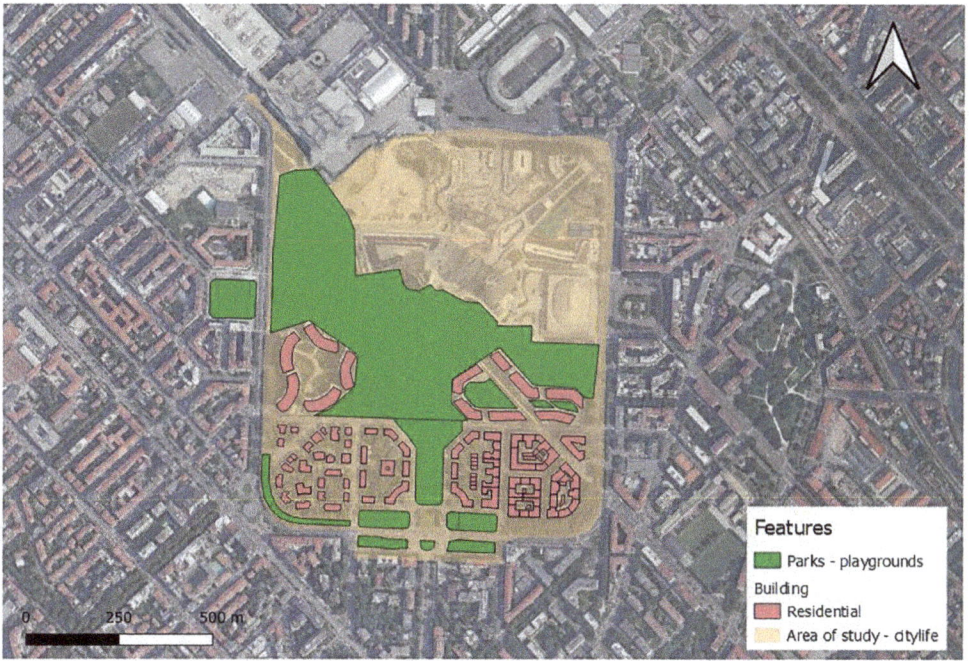

Access to parks and playgrounds

Fig. 6 Residential areas **a** Zaha Hadid **b** Libeskind (*Source* Citylife's website, 2020)

Fig. 7 Residential price per square meter Milan February 2020 (*Source* Immmobiliare.it, 2020)

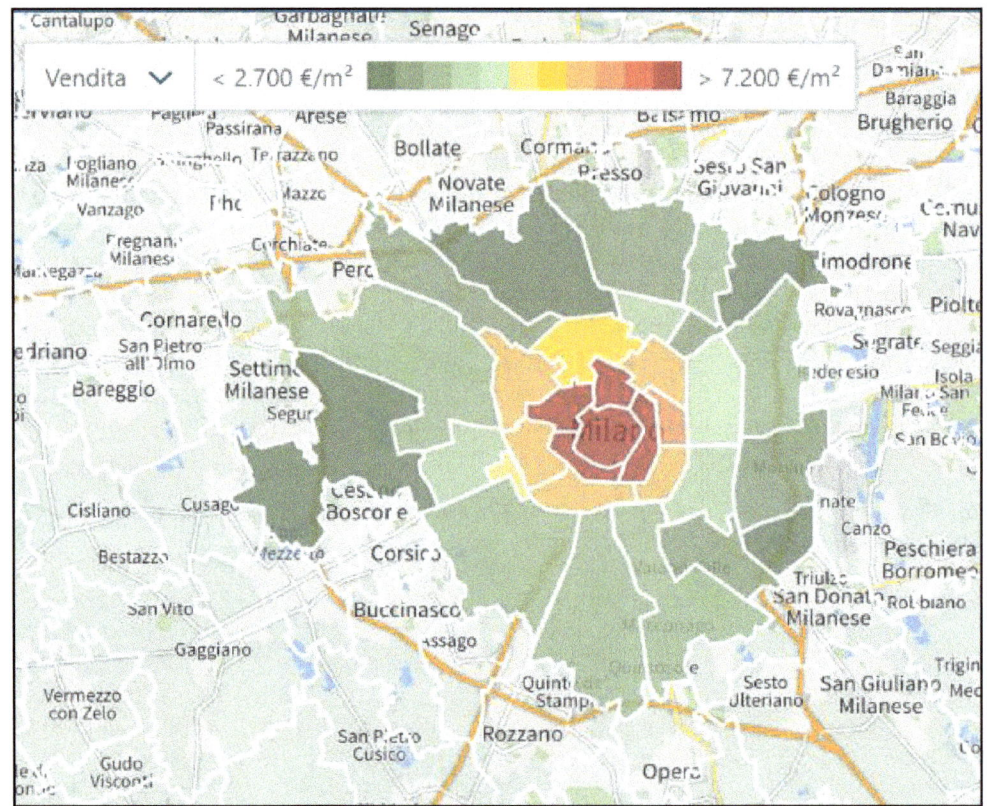

4.5 Housing Preservation

The housing preservation subcategory evaluates the percentage of households that were relocated within a walking distance from the original location of the site. This subcategory gives a higher score to the projects in which households are relocated within 250 m of walking distance from the lot where they were living previously. Projects that relocate people within a 500 m walking distance are given 2 points. If the project considers relocating less than 100% of the households within a 500 m radius, it receives 0 points

(ITDP, 2017). On the site where Citylife took place, there was a conference hall with 20 exhibition halls, with a total volume of about 2,500,000 m³ (Citylife, 2020). The previous structure was demolished in 2007, so the lot could be used for urban regeneration development; nevertheless, 120 trees that were on the lot were relocated and recovered into different parks in the city of Milan (Citylife, 2020). Since there were not previously built residential areas there were no housing relocations. In this case the study area gets a total of 3 points. In this respect, it is possible to suggest that future developments relocate 100% of the households living in the area previous to the project within a 250 m radius (ITDP, 2017). Within this parameter, Citylife got the highest score by default and there are no further recommendations to be made.

4.6 Business and Services Preservation

This subcategory evaluates the percentage of business and services that were relocated within a 500 m radius from the original location. The score system gives a maximum of 2 points to those projects that relocate previous businesses and services within a 500 m radius. In this case, through an inquiry to Maria Antonietta Morello, the press office contact of the Citylife shopping district, it was possible to find out that there were not any commercial activities in place before the arrival of the project (Morello, 2020). However, through a mapping analysis of the area, it was found that inside the lot of the conference hall there was a Hotel called Fiera Congress and a restaurant called Spizzico (Google earth, 2020). The hotel was relocated along with the conference hall to the periphery of the city, 8.5 km away from the original location. On the other hand, the restaurant was not relocated. In this case, the area is awarded a total of 1 point as the previous commercial facilities were not relocated within a walking distance. Regarding this subcategory, it is important to point out that even though there was not a strong commercial activity before the construction of the Citylife development, the relocation of the previous commercial uses should have been made in the area close to the original location. Relocation of commercial activities is important to preserve the socio-economic conditions of the place and include the old tenants in possible new developments.

5 Conclusions

Within the ITDP standard that evaluates the TODness level, Mix stands as the variable with the highest weight on the scoring system. This research evaluated the Mix TODness level of an urban regeneration project in Milan. The result awarded Citylife a total score of 17 out of 25 possible points. The overall result is sufficient to comply with a silver classification according to TOD standards; nevertheless, there are specific parameters that could improve. More in detail, Citylife got the highest score in the parameter of access to parks and playgrounds as 100% of the buildings were located within a 300 m^2 radius of a playground, park or green area. Nevertheless, it is important to highlight that plain green areas should not be considered necessarily as parks or playgrounds are also important to consider the quality of the space. Parks or playgrounds should be set as a

space where people can enjoy nature, but also that certain urban furniture and facilities need to be placed in order to make the space attractive for children and adults.

Also, Citylife got the highest score awarded for the complementary use parameter. This analysis revealed that land use in Citylife has a high diversification. The assessment performed in the area of study demonstrates that most of the development has low predominant land use, which provides diversified land uses for inhabitants and visitors. The most predominant land use was residential, which represents 51% of the total area, followed by commercial which stands for 34% of the area. These results are aligned with the ideal distribution of land use patterns according to the TODness level, where the predominant land use ideally should not be more than 50–60% of the total area of the project. This result could be an important guideline to be incorporated in future developments and further promote mixed land patterns in urban regeneration processes where Citylife could be taken as a positive example.

Likewise, the parameter of access to local services got a high result on this assessment. The results regarding the access to local services revealed that 100% of the buildings in the area have access to at least three types of local services. Also, in this regard, Citylife could be set as a positive example for urban regeneration developments. It is important to highlight the importance of the proximity of residential buildings to certain services such as healthcare facilities, pharmacies, education and fresh food supplies. Specifically, projects should provide access to at least three types of services to 80% of the residential buildings or higher (ITDP, 2017). Also, by default, Citylife had the best possible score on the parameter that measures housing preservation due to the inexistence of previous residential activities on the lot that was used to build the project. In this respect, it could be highlighted that for future urban regeneration developments it is important to relocate 100% of the households within a 250 m radius of the previous location.

On the other hand, there are certain aspects that could be enhanced within the Mix TODness level in Citylife. Citylife did not relocate all previous commercial uses, so in this case the assessment was awarded 1 point, which is the lowest score possible. Within the parameter of affordable housing, it was found that 0% of the housing market has a selling price that is below 30% of the mean of the selling price of the area nor does it belong to the category of subsidized housing or affordable housing. In order to aim at the highest score, Citylife would need to increase at least 20% of housing projects that have a selling price 30% below the mean of the area or that could be categorized as affordable housing according to Italian standards. It is relevant to point out that in order to achieve this objective it is imperative to cooperate between the private, public developers and other housing market stakeholders in order to diversify the housing offer and make it accessible for people with different levels of income. Likewise, it is important to stress the fact that the affordable housing concept is not universal. Each country sets an affordable housing standard that best fits its socioeconomic system. The ITDP recommendation is to follow the standard that establishes affordable housing as household units worth 30% less than the mean of the housing selling price.

The conditions described earlier pose a major challenge for Milan's case, as land prices and construction costs in housing projects are more expensive and scarcer than in other areas of Italy; in fact, Milan has the highest land value in Italy (Global property guide, 2020). Nevertheless, urban regeneration projects regardless of the land value could offer a percentage of affordable housing, encourage social mix and increase the quality of life of people with low income. In this sense, as the selling price of a housing unit determines the social mix of an area, if only the richest segment of the population can buy the housing units in one specific area the social mix will most likely be low. Therefore, it is important to diversify the housing price offers. In brief, Milan could have higher housing affordability offers, especially in urban regeneration projects, which is a goal that can be addressed by involving private and public stakeholders along with civil organizations involved in urban redevelopment from the very early stages.

Overall, the Mix TOD principle gives a higher score to those areas with more diversified land uses and with a higher social mix. This principle stands for the importance of providing different urban services to citizens within the same area so that it would result in a more active and vibrant space as well as achieve higher diversity in income and demographics. Within this analysis, there are subcategories that represent a greater challenge for new urban regeneration developments, such as the potential threat to economical profit caused by a high affordable housing percentage. Also, it could be challenging to provide a land use pattern where the predominant land use does not exceed 60% of the development. Adding to that, in Milan, urban practitioners have the challenge of planning urban regeneration projects next to qualifiable green areas that meet diversified land prices. But to the same extent, diversified land use and socially mixed projects could offer wide opportunities for public and private actors to provide different land patterns and less segregated neighbourhoods.

Out of the six parameters studied, three stood out as the strongest: Complementary uses, access to parks and playgrounds and access to local services. However, one parameter could not be fully assessed as there were not any housing relocations that took place on this project. On the other hand, the parameter of affordable housing and commercial relocation got the weakest score. Applying the TOD

assessment in Milan was relevant to understand the factors that can be improved in the area in order to aim at a higher Mix TODness level in future urban regeneration projects. There are some parameters that have a higher weight on the Mix TOD level. Complementary uses and affordable housing are the parameters that most influence the Mix TOD level. Since affordable housing is the weakest qualified parameter in this project and at the same time one of the most influential within the TODness level, this could be a parameter that should be prioritized in future regeneration projects in Milan. Encouraging social mix in an urban regeneration project contributes immensely to reaching a higher TODness level. This research study investigated the existing land use diversity in an urban regeneration project in Milan in order to identify the strongest and weakest parameters with the ultimate purpose of enhancing livability and increasing TODness levels in future projects. *Citylife* project has several strong parameters but also opportunities for improvement.

References

Beske, J., & Dixon, D. (2018). *Suburban remix creating the next generation of urban places*. Island press.

Calthorpe, P. (1993). *The next american Metropolis: Ecology, community, and the American dream*. Princeton Architectural press.

Cervero, R., Guerra, E., & Al, S. (2017). *Beyond mobility: Planning cities for people and places*.

Cervero, R. (1993). *Transit-supportive development in the United States: Experiences and prospects*. UC Berkeley.

City life S.p.A. (2020). Ultimi tre appartamenti gia finiti. Retrieved from: https://www.city-life.it/it/ultimi-tre-appartamenti-gia-finiti

Curtis, C., Renne, J., & Bertolini, L. (2009). *Transit oriented development making it happen*. Ashgate publishing limited.

Distribuzione moderna. (2020). Citylife un intervento urbano da 336mila metri quadri. Retrieved from: https://distribuzionemoderna.info/real-estate/citylife-un-intervento-urbano-da-366mila-metri-quadrati

Dittmar, H., & Ohland, G. (2004). *The new transit town best practices in transit oriented development*. Island press.

Furlan, R., & AlMohannadi, M., (2016). Light rail transit and land use in Qatar. *Archnet JAR*.

Global property (2020). Price history. Retrieved from: https://www.globalpropertyguide.com/Europe/Italy/Price-History

Google earth. (2020). *Comparison between 2001 and 2020 in Citylife lot*. Retrieved from: 45°28′42.18″ N 9°09′27.68″ E

Guiding architects (2020). Contemporary urban developments a rebirth of Milan. Retrieved from: http://www.guiding-architects.net/tours/contemporary-urban-developments-a-rebirth-of-milan/

Immobiliare italiana (2020). Immobiliare.it

Inexhibit. (2020). Daniel Libeskind case study. Retrieved from: https://www.inexhibit.com/case-studies/daniel-libeskind-citylife-residences-milan/

ITDP (2017). TOD standard (3rd uppl.). New York.

Liu, C., Erdogan, S., Ma, T., & Ducca, F., (2016). How to increase rail ridership in Maryland: Direct ridership models for policy guidance. *KTH*.

Morello, M. A. (2020). Interview hel on march 17th 2020.

Ming, W., ZhangbYu, H., & Changa, J. (2016). Alternative transit-oriented development evaluation in sustainable built environment planning. *Transportation Research Procedia*, 3220–3232.

Papa, E., & Bertolini, L. (2015). Accessibility and transit-oriented development in european metropolitan areas. *Journal of Transport Geography*, 70–83.

Pengjun, Z. (2018). Suburbanization, land use of TOD and lifestyle mobility in the suburbs. *JTLU*, 195–215.

Piano di governo del territorio. (2019). PGT. Retrieved from: https://www.comune.milano.it/aree-tematiche/urbanistica-ed-edilizia/pgt-approvato-e-vigente-milano-2030

Sarkar, P., & Chunchu, M. (2016). Quantification and analysis of land-use effects on travel behavior in smaller indian cities: Case study of Agartala. *American Society of Civil Engineers*.

Sohonia, A., Thomasa, M., & Rao, K. (2016). *Application of the concept of transit oriented development to a suburban neighborhood*. Elsevier.

Stojanovski, T. (2013). *Public transportation systems for urban planners and designers: The urban morphology of public transportation systems*.

Stojanovski, T. (2014). Transit-oriented development (TOD): Analyzing urban development and research gate.

Stojanovski, T. (2015). The morphological effect of public transportation systems on cities: Urban analysis of transit-oriented development (TOD) in Swedish cities.

Systematica. (2019). Citylife development. Retrieved from: http://www.systematica.net/project/citylife-development/.

Taccioli, L. (2019). Citylife il quartiere piu modern ditalia. Retrieved from: https://www.lorenzotaccioli.it/citylife-il-quartiere-piu-moderno-ditalia/#le-residenze-hadid-di-city-life

Transportation research board. (2004). *Transit-oriented development in the United States*. Transportation research board.

University of California. (2007). Histories of transit-oriented development: Perspectives on the development of the TOD concept real estate and transit, urban and social movements, concept protagonist.

Wang, S. (2019). *The regeneration of social housing estates in Italy*. Retrieved from: file:///C:/Users/Cristina/Downloads/2017_10_WANG%20(1).pdf

Zou, M., Lin, X., Mao, C., Ke, Z., & Li, M. (2014). Review on the theory and planning principle of transit-oriented development. *KTH*.

The Concept of "Smart Density Planning" Principles for Livable and Sustainable Urban Transformation

Hatice Kalfaoglu Hatipoglu and Seher B. Mahmut

Abstract

The aim of this study is to contribute to sustainable and livable environments with the approach of "smart density planning", which ensures a smart and planned growth with an acceptable density taking into consideration the efficient distribution of public spaces and services in urban regeneration areas in Turkey. There is a rapid urban transformation especially after the 2000s in Turkey in order to redevelop squatter settlements that have been built in order to respond to housing needs in rapid urbanization in cities unofficially and building areas that require improvement for structural resilience. However, there are significant problems confronting the process of regeneration. Squatter settlements have been demolished in order to create modern urban areas with better "physical conditions". Multi-story housing blocks are the dominant shape of a new style of housing. Their scale and high number of flats provide shelter for more people on limited land and cause a high-pressure density which ignores the livability and socio-spatial needs of the inhabitants. Since housing is a condition more than a shelter, creating livable residential areas is a crucial issue. The paper first discusses the meaning, context, relation and importance of these terms in order. Moreover, a critical review of the "smart city" has been conducted in relation to the presented "smart density planning" concept in order to rethink urban transformations in terms of socio-spatial quality of inhabitants. The authors introduce this concept of "smart density planning", which refers to the logical distribution of facilities and public spaces for overall inhabitants in efficient planning. Moreover, the proposed concept defines a morphology that takes into account the human scale and optimization of activity patterns in terms of smart planning, rather than a well-known definition of "smart cities" that is minimized to technological developments. Consequently, a framework has been composed to define the principles/indicators of "smart density planning" to ensure sustainable and livable urban transformation implementations in Turkey. These indicators assay how smart planning can be achieved in terms of density-based planning. This analytical framework also acts as a guideline to lead future neighborhood designs in Turkey.

Keywords

Urban Regeneration • Density • Smart Planning • Livability • Housing Quality

1 Introduction

Cities are densely populated areas by many people from different social and cultural backgrounds. This high density has complex spatial relationships supported by the opportunities provided by high technology. Urban spaces where residential, commercial, industrial and green areas are intertwined face some physical and social problems as a result of the high density of the cities. In order to respond to the need for accommodation in the urbanization process, unofficial squatter areas emerged in many cities of Turkey. Since the 2000s these squatter areas and several old districts have been transformed into high-rise/density concrete apartment blocks with the aim of providing better physical conditions. However, there are significant problems that cities are confronting in the process of regeneration such as disregarding socio-spatial quality, well-being of the people, logical distribution and ratio of facilities and public spaces. The design of these areas determines the livability and

H. Kalfaoglu Hatipoglu (✉)
Assoc. Prof., Department of Architecture, Ankara Yildirim Beyazit University, Ankara, Turkey
e-mail: hhatipoglu@ybu.edu.tr

S. B. Mahmut
Department of Architecture, Ankara Yildirim Beyazit University, Ankara, Turkey

sustainability of our environments. The multi-story housing blocks provide shelter for the people ignoring the reality that housing is more than a simple condition of physical protection. Housing is a spatial environment that has cultural, economic, psychological and social dimensions (Payne, 1977). Home is anchored with its special environment and humans interact with this environment (Angerbauer, 2001). Urban design is the art of shaping the interaction between people and places, environment and urban form, nature and built fabric, and has a great effect on the development of successful and livable cities (Campbell & Cowan, 1999).

These urban transformations in the form of mass housing have satisfied the quantity of housing but created many problems with regard to having serious quality insufficiencies and a great impact on urban patterns (Tekeli, 2010; Şenyel, 2006). Consequently, the solutions brought with these transformations to eliminate the housing problems could not demonstrate real success in terms of ensuring sustainable and livable environments due to the ignoring socio-spatial dimensions of the urban environment (Sezer, 2009).

Moreover, the dominant morphology in these high-dense areas is the high-rise due to the high amount of land value and the unearned income. Since there are a lot of disadvantages of high-rise buildings that have negative influences on livability and sustainability, it is crucial that these people have enough public spaces such as facilities, greenery, playgrounds, schools etc. and a well-designed infrastructure such as roads, parking places and technical equipment. Most of the urban transformations, which have been realized in squatter areas, have residences with low or middle income due to their previous reputation. There is a general tendency for high-rise building development in newly built residential areas independent of the income of the inhabitants. Generally, the differences between different income residents are not obvious in morphology but their services such as guidance, security and some facilities in the gated borders. Although there are guidelines and more facilities behind borders in high-income residential areas, the building morphology and distribution of facilities do not support people in their activity patterns for a livable and sustainable neighborhood in terms of social and ecological aspects. Due to the gaps in planning in terms of livability, this housing tendency requires a paradigm change. Accordingly, this can be provided by the logical distribution of facilities and public spaces. In this sense, it is very important to provide a sense of place and safe spaces in these environments following the concept of "smart density planning", which takes into account the acceptable density level and density distribution of the area (Gideon, 1967; Scarr, 1973).

Smart density planning can be a reflection of "smart city" approach to planning that is minimized to technological developments. While smart city is a broad concept and does not propose a specific morphology, "smart density planning" differs from smart city/planning with the definition of morphology based on smart and efficient density planning in terms of population and facilities/services. The proposed concept defines a morphology that takes into account the human scale and optimization of activity patterns in terms of smart planning giving priority to ecological and social sustainability, rather than the well-known definition of "smart cities". This morphology can be achieved with a low rise and high density, including well-interrelated micro-scaled spatial arrangements which support efficient activity patterns of neighborhoods opposite to the current tendencies of building developers in Turkey. This provides integration of spaces in order to connect them in a good meronymy, commercial viability, sustainable movement system, efficient and adequate green spaces, functional efficiency and flexibility which provide the ease of use and appropriate human scale.

The purpose of the study is to demonstrate the importance of "smart density planning" which follows the arguments of "smart planning" focusing on density distribution. A qualitative approach has been used in order to understand the dynamics specific to the space. It has been defined principles of qualified/sustainable and livable urban environments in the urban transformation areas focusing on an efficient morphology supporting activity patterns with socio-spatial quality. Since the principles have been discussed in the case of urban transformation areas in Turkey, there are also general housing development problems in the form of mass production, and these principles can lead the future housing development as a guideline. Moreover, the study also creates awareness by transferring "smart city" concepts under the shadow of technology to the urban design/planning with the priority of ecological and social sustainability of neighborhoods.

2 Conceptual Framework

2.1 Problems of Urban Transformations in Turkey

Since the pre-industrial agricultural cities turned into modern industrial cities, there has been a transformation with the increasing population and changing dynamics in lifestyles. In Turkey, the urban population increased up to 80% relative to the total population of the country between the years 1950 and 1960, and this increase reached its highest level between 1965 and 1970 (Osmay, 1998). As a result of this situation, the need for housing has increased incredibly and the response to this requirement has been a challenge, especially for big cities. The number of existing legal housing could not cover the demand in Turkey. Some people find their own self-organized solution as illegal squatter areas. In the 1980–2000s, mass housing construction was started by cooperatives and initiatives of the government, parallel to mass housing construction and the number of illegal buildings increased during this period. The urban transformation has been legalized with laws after the 2000s and housing supply was higher than housing demand during this period. However, new neighborhoods had not met the need for efficient accessibility of the services. After the 2000s, several actors have the right to the implementation of projects in parcels, and comprehensive/qualified planning and design has not been a real concern in urban transformation implementations (Koca, 2012).

For example in Ankara, after being the capital city of Turkey in 1923, there was a migration process from rural areas to the city; however, huge numbers of people could not afford to settle in close distance from their working areas such as factories. Then, they started to create their own houses on weakly controlled urban land, mostly near the decentralized factories (Uzun, 2005). From the 1960s the housing policy of Turkey developed in a way designed to build a high number of housing units with minimal investment (Tekeli, 2012). Reconstruction operations between 1955 and 1970 began to change the face of cities, especially in Istanbul. In the 1980s, with the awareness of the redeveloped squatter lands, authorities developed a model which is called "Urban Transformation Projects" in order to transform illegal houses into regular/legal houses. Even the aim is to provide settlements for the low-income group which was living in squatter settlements before the transformation. Within this time the target group has turned into

middle-high income groups and this situation causes a debate in Turkey (Ozdemirli, 2014). These projects mostly have been conducted by demolishing the existing settlement and rebuilding a new standardized model without a distinct character which also demolishes collective memory, social structure, cultural values, daily life routines of dwellers, neighborhood relations, existing urban tissue and patterns (Şenyel, 2006; Tekeli, 2010; Alkışer & Yürekli, 2011).

The general morphology of these areas is high-rise apartment buildings on a mass scale which also has expanded to the periphery in the form of satellite cities. The sprawl in Turkey with its high-density and high-rise expansion to the city edges differs in this sense from the sprawl in America and Europe. In metropolitan cities of Turkey, there are illegal houses at the periphery, however, with the urban transformation projects on the land, squatter settlements have been demolished and new high-rise mass housing residential areas have occurred. These newly built transformed areas caused a housing problem with regard to social and spatial qualities as a result of the focus on the satisfaction of quantity (Ataöv & Osmay, 2007; Tekeli, 2010; Türkün & Kurtuluş, 2005). The unconsidered distribution of density of these urban transformations, which were expected to respond to the problems of modern cities, went beyond their purpose of existence and resulted in a lack of livability with inefficient infrastructure, services, greenery and social inadequateness (Figs. 1, 2, 3, 4). The transformations in the form of high-rise buildings have turned into an arbitrary tendency, exceeding the necessities in terms of housing demand. After the 2000s, housing has

Fig. 1 Siluets from the Sentepe, Ankara, Turkey (from authors' archive)

Fig. 2 Apartment blocks in
Sentepe, Ankara, Turkey (from
authors' archive)

Fig. 3 A top view from urban
transformation project in Mühye,
Ankara (Ankara Büyükşehir
Belediyesi, 2014)

Fig. 4 Around transformation project in Mühye, Ankara (from authors' archive)

become a way of investment for some and a statute determiner for some others, instead of being only an accommodation possibility for people which gives people the opportunity to make a profit (Koca, 2012). The socio-spatial needs around the housing and the neighborhood structure in terms of livability and sustainability left their significance to land use compulsions and unearned income. Considering the problems of high-rise/density settlements from different perspectives, this tendency requires a shift that can lead to better implementations with strategies providing smart distribution of the functions in an adequate morphology and relationship which responds to socio-spatial requirements efficiently. This shift can ensure the sustainability and livability of these neighborhoods.

2.2　Sustainable and Livable Neighborhoods

The usual definition of sustainability means not to leave future generations a much worse environment than the present (Chapman & Gant, 2007). The meaning of the word sustainability is "to hold up" or "to support from below". Although the concept of sustainability was first mentioned for the first time in the Brundtland Report named "Our Common Future", its history is as old as human life. In the Brundtland Report it has been described as: "the development that meets the needs of the present without compromising the ability of future generations to meet their own needs" (Brundtland, 1987, p. 51). The term emerged after the

oil crisis in the 1970s in order to consider the balance between humanity and nature and reduce the consumption of natural resources. Although it began with economical motivation, the ecological and social aspects have integrated into the concept after a while and sustainability has been a fluid concept with broad definitions including social and spatial qualities. Cooper (1997) indicates four main principles of sustainability as "futurity, environment, equity, and participation".

Sustainability has been put forward recently as a special phenomenon, and the terms such as "energy efficiency" and "ecology" have been assumed as a distinct movement in the discipline of architecture and planning. Actually, sustainability should not be understood as a style but as an ideology that every architect/planner should take into consideration in the design process (Gyson, 2012). All people and all disciplines are responsible to protect nature and provide the balance between technology, nature and human. Koolhaas and Whiting (1999) indicates that architecture is the task to create a plausible relationship between the formal and the social. However, the concerns and urban/building design implementations today are away from the direction of providing healthy environments that increase life quality by decreasing the negative effects of construction and technology. Although architecture and urban design have been both crucial and central to social life, their importance has declined in recent years as a result of economic growth and progress in technology. The new chaos of city life converted people from being "part of the environment" to being "separate from the environment". This makes our living environments economically, ecologically and socially deprived of adequate qualities. The ideology and principles of sustainability have come into life as an opportunity for successful living environments which connect people and their needs, as well as humans and nature. The main idea of sustainability is to consider humans and nature in the center to minimize the undesirable aspects of the constructions and enhance the life of quality. According to Guy & Farmer (2000), there are six competing qualities of sustainability; ecological, smart, aesthetic, symbolic, comfort and community. Gyson (2012) summarizes the concept of sustainability in the field of architecture and urban planning as having special design qualities, being contemporarily technical and socially compatible. It is also crucial to sustain provided quality in the future.

Consequently, people are motivated to live today and in the future in sustainable environments because these places respond to various requirements of the residents of today and the future with their safe, inclusive, well-planned characteristics and contribute to the quality of life with its equal opportunities and good services (Odpm, 2003). Key principles of ensuring sustainable communities are promoting accessibility, walking, cycling, public transport, decreasing the need to travel with mixed-use development and promoting efficient land use by providing quality of life in safety and convenience, ensuring an attractive, well-maintained appearance with a distinct sense of place by promoting social integration, enhancing green infrastructure and playgrounds.

In the field of urbanism, some urban theories have discussed the livability and sustainability of neighborhoods without mentioning the term "sustainability". Some of these are the "Neighbourhood Unit" theory of Perry, New Urbanism, Smart Growth, "Garden City" concept of Howard. According to Perry, the social integration between the people is the distinction between good and bad working neighborhoods. Thus, the socio-spatial quality of the design is an important aspect that also has an influence on the life cycle of a neighborhood in the context of sustainability. With the growing population filling the gaps between villages and the city, Perry interpreted this expansion as a growing attenuation of community characteristics and in some emerging regions residents continued to relate to their neighbors, while in many of these areas of expansion the relationships cannot be sustained. There are different spatial needs required for different social groups who live in the same neighborhood. In this regard, Perry indicates that the different needs of different social groups have an important influence on neighborhood design (Perry, 1929). Perry's proposal is the idea of a self-sufficient neighborhood unit. There are specific architectural design requirements for the unit design that define a local community, have a sense of

belonging to the environment, have a sense of safety and trust-based relationships, have an environmentally sensitive land use, reduce urban car use and travel with its walkability. Moreover, these units create safe spaces for pedestrians and children. Perry has titled these requirements under the name of the neighborhood-unit principles and his proposal is the idea of a self-sufficient neighborhood unit in a holistic approach (Perry, 1929).

New Urbanism is also an urban design movement that encourages walking habits in residential environments and environmentally friendly habits which are crucial for creating neighborhood relationships and a sense of belonging. This has a huge effect on many aspects of subsequent urban design and land use strategies. Creating a sense of community and ecological practices are the main concepts of this movement (Katz et al., 1994). Architectural design perspective according to the context, the provision of social infrastructures such as sports facilities, libraries and community centers in the right distribution and density balanced with workplaces and residences are prominent considerations of New Urbanism.

Another approach for a livable and sustainable neighborhood is smart growth which aims to manage the sprawl (Song, 2005). The mixed land use, a compact building design, a strong sense of place, consideration of the need of society, improved transportation, cost-effective measures in development etc. are the main considerations of smart growth in order to provide an economic, environmental and social development (Susanti et al., 2016).

"Garden City" is the concept derived by Ebenezer Howard with the aim of intersecting the positive matters of urban and rural life. The main purpose of the approach is to prevent dense, unhealthy and limited-access neighborhoods. It expresses the composition of the settlements, recreative spaces, commerce and socio-cultural facilities in a low-density neighborhood. Green/rural areas around the neighborhood and face-to-face relations between inhabitants has prioritized with an approach that supports the recovery of urban living conditions with social purposes as the most important issue of planning (Ersoy, 2016).

All these concepts demonstrated various principles and considerations which show parallelism in substance in order to provide qualified and livable urban environments. Although human-centered approaches should be the main concern for architects and planners, neighborhoods have been transformed without consideration of the efficient distribution of services, activity areas, green spaces etc. because of the focus on unearned income and land use compulsions. The livable neighborhoods which also respond to the onto-logic requirements of the people with low-rise and acceptable density have been transformed into high-rise settlements with the urban transformation projects in order to provide more houses and to gain an income through this type of project.

The situation today in the urban transformation of Turkey's neighborhoods is not compatible with the idea of sustainability in terms of social and ecological aspects and livability. This has been the motivation for the development of "smart density planning" as a concept for the urban regeneration areas in Turkey. The authors propose that the contribution to livable and sustainable neighborhoods can be provided with this concept, which follows demonstrated principles in the context of the smart distribution of population and services in an efficient and interrelated morphology.

3 Materials and Methods

The method of the article is to elaborate a theoretical framework for smart density planning by reviewing literature and studies focusing on density, livability and sustainability. As the main manner, deriving guidelines from the previous movements/concepts and studies about planning in terms of livability and sustainability and how to reconsider/reflect them for a positive response to the urban transformation challenges in Turkey has been considered.

The paper first discusses the meaning, context, relation and importance of the terms. Moreover, a critical review of the smart city concept has been conducted in relation to the presented "smart density planning" in order to rethink urban transformations. Then the study composes a framework to evaluate and enhance the urban transformation implementations in Turkey in general and link desired guidelines to sustainable/smart planning. The indicators to assay how smart density planning can be achieved have been defined in order to present an analytical framework for the future evaluation of neighborhoods.

4 Smart Density Planning

The right way of controlling density is the fundamental approach to sustainable development. A concept of "smart density planning" has been introduced by the authors in order to ensure human-centered, livable and sustainable living environments. The emerging point of the concept has been based on a "smart city" vision from the perspective of urban planning/design on the scale of neighborhoods. The proposed "smart density planning" concept supports that the daily activity patterns in neighborhoods can be achieved with low-rise and high-density approaches and micro-scaled public spaces integrated with housing environments efficiently. The "density" has been emphasized in "smart density planning" which focuses on the logical distribution of facilities/services according to the density and in an efficient morphology that provides ease of use. Smart density

planning can be a subset of "smart city" approach in terms of planning, which is a more broad concept and does not propose a specific morphology. In addition to the smart city approach, "smart density planning" proposes a morphology based on smart and efficient density planning in terms of population and facilities/services. The study criticizes "smart city" based on technology and ignores smart planning approaches. For this reason it is crucial to comprehend and discuss the "smart city" concept in order to understand the gaps.

A smart city was first linked to technological developments depending mainly on economic perspective (Gibson et al., 1992). Similar to the destiny of the term "sustainability", the integration of social dimensions has become a crucial issue since the value of the livability of place and quality of life has been a discussion. Giffinger et al. (2007) emphasize that although the term is not used in a holistic way, it presents several aspects from IT solutions to the smartness of the inhabitants of the city. Even though there are a lot of definitions of smart cities related to technology, there are several studies that relate "smart city" with planning patterns in recent years (Caragliu et al. 2011; Berry & Glaeser, 2005; Ateş & Önder, 2019). A smart city can be classified according to its approach in three categories (Greco & Cresta, 2015):

- Defining a smart city as a technologically advanced city, highlighting the "hardware" approach (Cairney & Speaks, 2000; Washburn & Sindhu, 2010)
- Defining as a city that has the ability to manage the resources intelligently in order to contribute to the quality of life in a human-centered approach (Partridge, 2004; Berry & Glaeser, 2005)
- Defining a smart city as a holistic approach for the integration of technology and human and social capital (Kanter & Litow, 2009; Campbell, 2012).

Hollands (2008) points out the lack of studies which correlate smart city implementations with the most crucial characteristics of the city and its transformations. This causes a smart city based on a high-tech motivated entrepreneurial city. On the other hand, Cohen (2012) indicates the vision of a smart city, including several dimensions such as "Smart cities use ICT to become more intelligent and efficient in the use of available resources, with the effect of reducing costs and energy consumption and at the same time, improve the delivery of services and quality of life citizens, reducing the ecological footprint and developing innovative and sustainable economy".

Despite broad definitions of smart cities, it can be concluded that a smart city is a city that is sustainable, competitive and self-sufficient. This means efficient and adequate solutions as a response to the challenges of the modern cities in urban transformation should be improved in order to achieve real smartness from the perspective of urban planning regarding the smart city. This is possible with the design of a successful urban environment. It is a very crucial aspect of how planners can contribute to making these cities smarter in the context of livability and sustainability in complex and chaotic cities. One of the most important elements of creating a livable and perceived urban environment for smartness is the density distribution of spaces. Density is not mentioned here just as the population/number of people but as a logical distribution of social and public activities to the number-of people. The social, cultural and economic identity of the residents of the neighborhood has a relation to the morphology in which the distribution of this density has been designed (Susanti et al., 2016). This morphology defines the activity patterns of the residents. For livable neighborhoods, it is important to reach services efficiently and to create communication with other people and facilities. For this reason, building compact and walkable cities that allow smaller-scale movement both saves time and avoids investment and operational expenses. To achieve this goal, a morphology with compact and micro-scaled arrangements of the density allowing pedestrian movement is of particular importance. Smart density planning, with the design of special places where children can play as a part of urban life, establishes residential units with adequate and efficient public spaces, roads that can also be easily used by disadvantaged social groups such as the elderly, disabled and pregnant, and enables all segments of the society to improve the sense of belonging, integration and adaptation to the city. The indicators of smart density planning have been determined by rethinking discussions and principles in terms of livable and sustainable neighborhoods in light of the specific problems of urban transformations in Turkey. These problems have been reconsidered with the focus of an approach based on the smart distribution of density which can be achieved with a proper morphology supporting social relations between people in themselves and with their physical environment. These indicators have been described as follows.

4.1 Density and Proportion of Buildings

Density is a term that clarifies the number of people on the land vis-à-vis its size (Cambridge, 2020). The density of buildings and street accessibility are essential factors of the city which has a dense concentration of people (Jacobs, 1961; Ye et al., 2018). Perception of density can change at different places because of separate spatial features of lands. Bonnes et al. (1991) have mentioned that the relation between vacant lands and built areas, the width of streets, size and the height of buildings has an impact on the perception of density. The relationship that people establish with the neighborhood spaces is affected by the density of the neighborhood in terms of population, structuring and activity patterns. Planning the density level and density distribution of the area to form is very crucial for livable and sustainable neighborhoods (Gideon, 1967; Boggs, 1965; Scarr, 1973).

In high-density residential areas, problems arise in sharing urban areas and ensuring neighborhood privacy. Usually, public spaces and building entrances are shared by many people in these places. The fact that the neighborhood does not have sufficient/adequate green areas, parks, children's playgrounds, educational areas or their use is not planned according to their capacity weakens the social relationship that the neighborhood residents establish with that place. As the number of people using these spaces increases, the relation and unity of families with their ground decreases. In addition, it has been revealed in the studies that the feeling of belonging and adopting the common areas decreases when the number of people living in dense settlements increases (Gehl, 2011).

The morphology/proportion of the building blocks is also an important indicator that influences smart distribution of density for the livability and quality of life in the neighborhoods. According to Al-Kodmany (2011), like spacing, alignment and coherence, the height of the building affects the sense of place in a human-scale environment. Creating zones of influence is also determined by the morphology of the building blocks in the same density. Assigning some definitive areas for small groups of people is preferred to undefined empty flying spaces around tall buildings in order to create these zones (Newman, 1996). The advantage of having more spaces on the ground turns into a disadvantage due to the lack of sense of belonging which was observed in Pruitt-Igoe (Newman, 1996).

"Tall building" has signified density of people per base-land because of its various floor/story. According to Al Kodmany (2011), "tall building" is a relative term because of the characteristic of the city; for instance, a 20-story building can be named "tall" if it is located in Damascus, but it is not a "tall" building for Chicago. Similar to the definition of "tall", its outcomes—deficits, benefits—are controversial. Although tall buildings have been accepted as an efficient way of land use and it has been promoted by Mayors in many countries (Buchanan, 2008), it causes chaos and becomes a trigger to stress and it gives a feeling of workaholic/placeless people with many other accompanying disadvantages (Brown et al., 2009; Jacobs, 1961; Gehl, 2011). The people except on the first few floors of the building cannot get meaningful contact with the ground level in multi-story buildings (Gehl, 2011). Moreover, there are also some negative effects on the physical and psychological health of children (Alexander, 1977; Van Vliet, 1983). It is necessary to build "short blocks" which provides permeable and perceivable urban landscape on a human scale in order to ensure livable neighborhood and sustainable urban facilities (Jacobs, 1961). As mentioned in the "Introduction" and "2.1", urban transformations in Turkey have been implemented in a high-rise morphology in order to get unearned income from land use. The discussion about building proportions reveals the importance of a paradigm shift toward a better distribution of the density with low rise and high density for livable neighborhoods in Turkey.

4.2 Accessibility and Street Network

Accessibility can be described as the extinction of the difficulty in reaching the destination of the users and visitors, the ease of the use in their anticipated activities and facilities for their required purposes (Voordt & Wegen, 2005). Even in the sixteenth century, to provide accessibility by arranging a street network in order to form connections between the landmarks was the main objective while establishing a new city (Madanipour, 2007).

In smart density planning, accessibility of planned spaces is necessary in order to sustain a livable environment and successful development. Lynch (1981, cited in Carmona et al., 2010) has mentioned the term "access" as the reachability of one to services, places, resources, activities and other persons. Visual and physical accessibility is needed to become integrated with its surroundings (English Partnership, 2000). In addition to physical and visual accessibility, symbolic accessibility which means reachability of each group of people without discrimination should be considered (Carmona et al., 2010).

Streets—*as a site of interaction* (Vidler, 2002)—provide the access that the neighborhood unit uses in relation to the inside and outside of the neighborhood by creating a network system according to their size and capacity. Additionally, the design of streets creates a morphology passing through the periphery or the center of the neighborhood.

According to Ersoy (2015), how the roads should be planned in hierarchical order is decided by considering what purpose they will be used for. For instance, although a high level of the road should be planned in order to provide continuous and fast-moving car traffic, a low level of the road should be planned with the aim of ensuring accessibility between houses and social facilities.

The most efficient solution for an adequate traffic pattern is to promote public transportation and discourage car usage by improving public transport (Gehl, 2011, Jacobs, 1961). Bus stops should be designed by arranging roads with considering the safety of cars/vehicles and pedestrians. The location of stops should be planned in a position where all inhabitants of the neighborhood can reach easily.

Besides the car-oriented roads/streets and efficient public transportation, the provision of a good, safe and sufficient vehicular-pedestrian concept is a crucial indicator of the quality and adequateness of the circulation. The needs of pedestrians, cyclists, especially children and older people, or people with impaired mobility should be responded to by considering the measures efficiently. The network of pedestrians and vehicles must ensure convenience, safety and security for all intended users. Moreover, the hierarchy and division of these routes must be clear and adequate. A priority should be assigned to the pedestrians and the car-parking zones and similar services should be designed conveniently in the neighborhoods.

Farr (2008) indicates that the public places created with urban uses should be accessible to everyone living in the neighborhood. Therefore the walkability of mixed uses such as shopping opportunities and working areas from the housing environments has become very crucial (Gehl, 2011). Moreover, the interconnected and walkable street pattern is an important planning principle in urban transformation areas. In this pattern, it is emphasized that the side length of a building block should not exceed 180 m (Farr, 2008). Thus, the presence of small building blocks and frequently located intersections are important indicators to increase accessibility. Since most of the high-rise areas have a crowded population, they create a traffic load that influences walkability and transport in a negative way. In these areas there is a need for an improvement of an adequate network system which also requires an improvement in terms of walkability, street hierarchy and public transport. In the case that the traffic system has been improved in an infrastructure that can solve the problems, this development is in contradiction with ecological sustainability and dynamics of neighborhood ideas in Turkey.

4.3 Green Spaces and Ecological Considerations

Green areas are spaces that contribute to a better life in the neighborhood such as open spaces, gardens, squares and parks/playgrounds. They provide the intersection of nature and cities. Apart from its aesthetic aspects, green spaces have benefits for comfort, health and physiologic well-being of people by preserving pollution and absorbing the noise of the others in an urban land (Shaftoe, 2008; Ersoy, 2015). Additionally, green spaces encourage walkable areas for inhabitants and provide safe areas for pedestrians by separating pedestrians and vehicles in settlement areas.

The size of green spaces in a neighborhood should be determined in compliance with population, characteristics of settlement, topography, ecologic system and climate (Ersoy, 2015). Overall, the green system should be continuous, and in this way there will be a corridor that helps airflow.

In Turkey, there is a lack of consideration in the transformation areas as well as other new residential areas. In mass housing tendency, there are big building blocks surrounded by undefined areas. Even if there are defined green spaces, these are macro-scaled in just one place which does not allow a well integration for all of the building blocks. Smart density planning suggests that these areas should be micro-scaled and well integrated with ease of reachability with close contact to the houses.

As a result, smart density planning offers an efficient distribution of green spaces for inhabitants of a neighborhood in order to increase the ecological aspects of the neighborhood. Moreover, these micro-scaled spaces will be assigned to the houses adequately and increase the sense of belonging opposite to the floating high-rise building blocks. The design of the green areas with the low-rise morphological characteristics ensures interrelation with the housing and other services in accordance with the population, and neighborhoods can provide more livable and sustainable areas.

4.4 Services

Neighborhoods are like real organisms which people meet their daily needs. These service areas which also act as public places are busy and important areas that create a lot of circulation in the neighborhoods. The nature and position of these mixed-use environments are important to establish a safe territory which also has a positive effect on encouraging walkability and a sense of belonging in the neighborhoods (Jacobs, 1961; Gehl, 2011). Moreover, public space improvement and mixed-use development decrease social segregation and encourage social cohesion in the neighborhoods (Madanipour, 2007). Thus the coexistence of residential areas in the form of mixed-uses with shopping and commerce and workplace etc. will ensure people save time and energy as it allows them to meet their needs at a close distance.

The nature and position of mixed uses which establish a territory have been positively associated with security in many studies due to the fact that businesses create diversity by merging with residential areas and especially because of its feature of increasing the eyes on the street by encouraging the use of this diversity by pedestrians. Safety is one of the most important indicators in the urban transformation areas in the cities. In order to ensure safety in living environments, a new urban pattern has been created in the form of gated communities with security guidance, and the pattern of neighborhoods in Turkey has been sacrificed for this reason.

For an efficient neighborhood design, it is beneficial to allocate suitable residential areas with public spaces, education, health and shopping facilities to the places which are convenient to neighborhood centers in a location that can provide ease of transport within a reasonable walking distance in case of emergency. Moreover, the capacities of these areas should be considered in accordance with the density of the people living in the area for smart density planning.

Although most of the new urban transformation projects in Turkey includes commercial, recreational and sport facilities inside, these meet the needs of only their residents in the form of gated communities. This situation should be handled by considering all neighborhoods instead of focusing on a few blocks.

4.5 Flexibility

Inhabitants' needs can change over time because of the social, demographic and economic changes in their life and usable spaces become essential. The capability of buildings to change in physical structure and adaptation to spaces is expressed with the term flexibility. According to Schineder and Tilll (2005) flexibility is to give chance to inhabitants in accordance with their desires and decisions about the future usage of space. Buildings and surrounding urban spaces can gain flexibility at the design stage. This ensures a better performance of spatial organizations during different periods which is a very crucial condition for the provision of sustainable design.

Due to the changes in lifestyles, spatial conditions such as sizes and types require implications in terms of adaptation over time. It is observed that people prefer to adapt to their environment instead of changing their living environments because of the fluctuations in social and demographic circumstances (Hasgül & Özsoy, 2016; Schneider & Till, 2005; Habraken, 2019). For example, the situation of the need for additional spaces in the buildings results in the preference for detached houses, sometimes with the addition of a new floor to the top floor of apartments. The standardized and fixed housing sites built today in Turkey do not allow possible interventions needed within the time and this brings along the forced adaptation of the user to the environment instead of the preferred opposite version. As a result, this perspective ignores the adaptation to the needs and patterns of the inhabitants.

The ability of living environments to adapt to the time is a logical solution for different reasons to current problems of modern cities. The efficient use of space considering adaptations decreases space consumption, carbon footprint and damage to the environment. Moreover, there are a lot of economic benefits because as a result of the considerations of flexibility these living environments will sustain longer with accompanying cost savings (Cellucci & Di Sivo, 2015, Zairul & Geraedts, 2015; Schneider & Till, 2005). This contributes to the sustainability of the neighborhoods. Consequently, flexibility becomes one of the main elements that should not be underestimated in smart density planning. Rather than finding a solution to housing needs by building high dense and tall buildings, it will be a more sustainable method in the long term to build the houses flexibly and to offer solutions in this direction. This can be achieved with a shift from the strict classification of rational functionalism of modernism to a flexible and adaptable urban environment considering the social dynamics of everyday life. Instead of standardized approaches in housing and urban design, some decisions should be left to user preferences. Some architectural values such as types, patterns, themes and systems should be followed but the dominant approaches such as coding and assigning just playground equipment for children should be replaced with flexible solutions (Habraken, 2017; Alexander, 1977). This also allows adaptation and contributes to the architectural quality and identity of spaces.

Most of the urban transformation in Turkey has been focusing on designing houses for inhabitants who live under bad conditions in the neighborhood. Architects/institutions, which have responsibilities for planning, implement standardized designs in order to find quick solutions for the inhabitants' needs. However, in practice, these solutions do not meet future requirements. The characteristics, design and materials of buildings and their environments do not meet the future requirements and cannot provide sustainable solutions because of a lack of consideration in terms of flexibility. Flexibility ensures the adaptability of the spaces according to the changing requirements of the users that support sustainability ideas. The planning of the neighborhoods should foresee future adaptations. In this way, even if the users of units change, there is no need to demolish all of the areas for new users' needs. The adaptation can be managed with small interventions. The flexibility can be possible with detailed considerations in the planning phase taking into account possible additions and extractions in the long term.

4.6 Playgrounds

Clarence Perry has taken playgrounds and small parks as the elements of each neighborhood unit (Carmona et al., 2010). Playspaces are very crucial for mental freedom, which provides the deviation from the rules and for the development of intelligence (Lefaivre, 2007; Groos, 1973). Children's playgrounds also contribute to the liveliness of the environments and reinvention/communication of the children as well as their parents (Kalfaoglu Hatipoglu, 2016).

In Turkey, transformed/regenerated urban settlements contain playgrounds that enable children to spend their time; however, the planning of playgrounds has not responded according to the density of settlements or housing units and their quality is underestimated. Moreover, the design of these playgrounds should be considered adequately and should not be assumed just as standard plastic play equipment. All the public areas should provide the possibility for children to spend their time in a flexible and creative manner and these areas should be sufficient both qualitatively and quantitatively for the inhabitants to provide smart density planning. All the public areas should provide the possibility for children to spend their time in a flexible and creative manner and these areas should be sufficient both qualitatively and quantitatively for the inhabitants to provide smart density planning. Accordingly, these playgrounds should be designed in accordance with the density of settlements or housing units. Moreover, the design of these playgrounds should be considered adequately and should not be assumed just as standard plastic play equipment. Instead of putting standardized play equipment, some creative solutions specific to the neighborhood should be designed as several public spaces in the neighborhood. Sandpits and some graded topographies are some suggestions in order to create playgrounds without equipment. Moreover, micro-scaled public spaces integrated with housing environments efficiently provide visibility and easy access from houses instead of disconnected macro-scaled playgrounds.

4.7 Hierarchy/Transition of Public/Private Zones

Designing a hierarchy of private, semi-public or public space is crucial in order to create defined areas at the entrance of residential units (Newman, 1996). This is provided by real or symbolic barriers. These areas, which are defined as the hierarchy of transition zones, are necessary to create boundaries that define this hierarchy in the transition from public streets to private units—buildings, flats—to define these areas. These spaces, which are defined as private, semi-private, semi-public and public spaces, create perceptible transition zones for use, belonging and neighborly relations. Therefore, it is important to ensure a sense of belonging and safe spaces which are the main indicators for the sustainability and livability of a neighborhood (Lawson, 2010; Altman, 1975; Hall, 1966; Watson, 1970; Newman, 1996). Building types and heights are directly related to the formation of this hierarchy because these different types define the grading of these zones (Heng & Malone-Lee, 2009). The solution of a campus of the same density with different types affects the relationship of the residence with the street and the situation of creating a sense of belonging, privacy, and social relationship with different transitional zone occurrences. In high-rise buildings, the perception of private areas such as indoor shared spaces as public areas and the lack of transitional zones such as semi-public and semi-private areas have a negative influence on accustomed neighborhood patterns. Consequently, the consideration of adequate planning of these zones influences smart density planning.

In urban transformation projects in Turkey, it is observed that there is not an efficient transition zone due to the high-rise buildings and walled island residential areas with security measures. There are some differences in the space hierarchy. For example; some urban transformation projects which are transformed from squatter settlements, do not have semi-public/semi-private spaces anymore. In Fig. 5, it can be perceived that the circulation route of one building represents a street in terms of its inhabitants and the hierarchy of transition zone has been lost in these floating buildings.

The motivation for the concepts is the lack of social and ecological aspects which have an influence on livability in these projects regardless of the income levels of the inhabitants. Figure 5 presents the housing tendency which is also a characteristic of urban transformation areas. These high-rise mass housing blocks cause damage to the socio-spatial structure of the neighborhood which has been discussed in several parts of the study. The criteria for introducing "smart density planning" are the guidelines that have great importance when designing new neighborhoods by urban transformation projects. The discussed indicators lead to how to provide a balanced distribution of services and public spaces which ensure sustainable and livable environments. Additionally, these indicators are the guidelines for arranging physical environments in an ecological way and establishing adequate neighborhood relations. With interrelated morphology and smart distribution of population

and services of low-rise, high-density housing, "smart density planning" has the potential to provide livable sustainable neighborhoods.

5 Conclusion

Following the modernization processes, urbanization and then urban transformation processes have affected cities in a negative way in terms of sustainability and way of living in such concretized urban contexts. In order to minimize these negative effects, we need a rethinking of regeneration processes which may lead to an alternative future path for urban transformations in terms of urban practices, approaches and implementations. To achieve this goal, smart density planning has been introduced as a solution to the challenge of coincidental planning which ignores the smart distribution and socio-spatial quality of the living spaces.

The discussion of several concepts and movements in order to set up a framework of indicators for smart density planning revealed how important it is for livable urban transformation areas. Moreover, the "smart city" is discussed critically and reconsidered in terms of planning and sustainable neighborhoods. The definition of the indicators of smart density planning, which also act as a guideline, shows that density is not just correlated with population but other qualitative and quantitative parameters, such as logical distribution of functions, adequate vertical and horizontal morphology design, sensitive circulation concept etc.

In addition, the indicators present problems of high-rise buildings for neighborhood patterns and social relations and

Fig. 5 Mass housing project from Bursa (Alagöz, 2011)

promote more human-oriented and human-scaled living environments. As a result of the new morphology with high-rise development in these transformation areas in Turkey, the previous morphology has been changed dramatically. Moreover, the accustomed neighborhood structure, previous activity patterns, the communication of the inhabitants themselves and their interaction with the ground, and the safety of these spaces have been demolished. Thus the lack of consideration for smart density planning in these urban transformation areas has an influence on the socio-spatial structure, cultural values, and daily life routines of dwellers, neighborhood relations, urban tissue and existing green patterns. The discussions and evidences in the indicators reveal the gaps in considerations of these implementations. It is highlighted that most urban problems can be minimized by following the content of the indicators of the suggested concept which promotes solutions with directions. Moreover, the study evokes an awareness of the possibility and necessity for a shift toward planning that considers smart density planning for a human-centered livable environment. This is very important for the socio-spatial quality and sustainability of neighborhood structure in Turkey, which responds to the ontological needs of the inhabitants beyond physical requirements.

The authors' intention for their future studies is to implement the analytical framework on an urban transformation case in Turkey in order to highlight the problems and development potentials concretely.

Acknowledgements This research did not receive any specific grant from funding agencies in the public, commercial or not-for-profit sectors.

References

Alagöz, B. (2011). *TOKİ'nin Bursaya Tokadı: Doğanbey Kentsel Dönüşüm Projesi hakkında*. ARKİTERA. https://www.arkitera.com/gorus/tokinin-bursaya-tokadi/

Alexander, C. (1977). *A pattern language: towns, buildings, construction*. Oxford University Press. https://arl.human.cornell.edu/linked%20docs/Alexander_A_Pattern_Language.pdf

Alkışer, Y., & Yürekli, H. (2011). Türkiye'de "Devlet Konutu" nun dünü, bugünü, yarını. *İtüdergisi/a, 3*(1), 63–74.

Al-Kodmany, K. (2011). Placemaking with tall buildings. *Urban Design International, 16*, 252–269. https://doi.org/10.1057/udi.2011.13

Altman, I. (1975). *The environment and social behavior: Privacy, personal space, territory, and crowding*. Brooks/Cole.

Angerbauer, S. (2001). Zeitgemaeszer Sozialwohnbau [Master thesis, Technische Universität Graz, Institut für Städtebau und Umweltgestaltung, Graz].

Ankara Büyükşehir Belediyesi (2014). *Güneypark Kentsel Dönüşüm ve Gelişim Projesi*. T.C. Ankara Büyükşehir Belediyesi. https://www.ankara.bel.tr/genel-sekreter/genel-sekreter-yardimcisi-mustafa_kemal/emlak-ve-stimlak-dairesi-baskanligi/yeni-yerlesimler-sube-mudurlugu/hayat-sebla/guneypark?web=1

Ataöv, A., & Osmay, S. (2007). Türkiye'de kentsel dönüşüme yöntemsel bir yaklaşım. *Metu Jfa, 2*, 57–82.

Ateş, M., & Önder, D. E. (2019). 'Akıllı Şehir' kavramı ve dönüşen anlamı bağlamında eleştiriler. *Megaron, 14*(1), 41–50. https://doi.org/10.5505/MEGARON.2018.45087

Berry, C. R., & Glaeser, E. L. (2005). The divergence of human capital levels across cities. *Papers in Regional Science, 84*(3), 407–444. https://doi.org/10.1111/j.1435-5957.2005.00047.x

Boggs, S. L. (1965). Urban crime patterns. *American Sociological Review, 30*(6), 899–908. https://doi.org/10.2307/2090968

Bonnes, M., Bonaiuto, M., & Ercolani, A. P. (1991). Crowding and residential satisfaction in the urban environment: A contextual approach. *Environment and Behavior, 23*(5), 531–552. https://doi.org/10.1177/0013916591235001

Brown, L. J., Dixon, D., & Gillham, O. (2009). *Urban design for an urban century: Placemaking for people*. Wiley.

Brundtland, G. (Eds) (1987). *Our common future: the world commission on environment and development* (p. 51). Oxford University.

Buchanan, C. (2008). *The economic impact of high density development and tall buildings in central business districts*. British Property Federation.

Cairney, T., & Speak, G. (2000). *Developing a 'smart city': Understanding information technology capacity and establishing an agenda for change*. Centre for Regional Research and Innovation, University of Western Sydney.

Cambridge Online Dictionary. (2020). Retrieved from: https://dictionary.cambridge.org/tr/sözlük/ingilizce-türkçe/density

Campbell, T. (2012). Beyond smart city: How cities network, learn and innovate. *Earthscan, NY*. https://doi.org/10.4324/9780203137680

Campbell, K., & Cowan, R. (1999). Finding the tools for better design. *Planning, 2*, 16–17.

Caragliu, A., Del Bo, C., & Nijkamp, P. (2011). Smart cities in Europe. *Journal of Urban Technology, 18*(2), 65–82. https://doi.org/10.1080/10630732.2011.601117

Carmona, M., Tiesdell, S., Heath, T., & Oc, T. (2010). *Public places, urban spaces the dimensions of urban design*. Routledge.

Cellucci, C., & Di Sivo, M. (2015). The flexible housing: criteria and strategies for implementation of the flexibility. *Journal of Civil Engineering and Architecture, 9*(7), 845–852. https://doi.org/10.17265/1934-7359/2015.07.011

Chapman J., & Gant, N.(2007). *Designers, visionaries and other stories: A collection of sustainable design essays*. Earthscan.

Cohen, B. (2012). *The top 10 smart cities on the planet*. Fast Company. Retrieved from: www.fastcoexist.com/1679127/the-top-10-smart-cities-on-the-planet

Cooper, I. (1997). Environmental assessment methods for use at the building and city scales: constructing bridges and identifying common ground. In P. S. Brandon, P.L. Lombardi, & V. Bentivegna (Eds.), *Evaluation of the built environment for sustainability*. E&FN Spon.

Ersoy, M. (2015). *Kentsel planlamada standartlar*. Ninova Yayıncılık.

Ersoy, M. (2016). *Kentsel planlama kuramları*. İmge Kitapevi.

Farr, D. (2008). *Sustainable urbanism: Urban design with nature*. Willey.

Gehl, J. (2011). *Life between buildings: Using public space*. Island Press.

Gibson, D. V., Kozmetsky, G., & Smilor, R. W. (Eds.). (1992). *The technopolis phenomenon: Smart cities, fast systems, global networks*. Rowman & Littlefield.

Gideon, S. (1967). *Space, time and architecture: The growth of a new tradition*. Harvard University Press.

Giffinger R., Fertner C., Kramar H., Kalasek R., Pichler-Milanovic N., & Meijers E. (2007). *Smart city—ranking of European medium-sized cities*. Centre of Regional Science of Vienna.

Greco, I., & Cresta, A. (2015). A smart planning for smart city: the concept of smart city as an opportunity to re-think the planning models of the contemporary city. In *International conference on computational science and its applications* (pp. 563–576). Springer, Cham. https://doi.org/10.1007/978-3-319-21407-8_40

Groos, K. (1973). *Die spiele der menschen, 1899.* Hildesheim.

Guy, S., & Farmer, G.(2000). Contested constructions: The competing logics of green buildings and ethics. In W. Fox (Ed.), *Ethics and the built environment.* Routledge.

Gyson, B. (2012). Sustainable design, a statement. In H. Drexler & S. Khouli (Eds.), *Holistic housing: Concepts, design strategies and processes.* Detail Edition.

Habraken, J. (2017). Back to the future: The everyday built environment in a phase of transition. *Architectural Design, 87*(5), 18–23. https://doi.org/10.1002/ad.2211

Habraken, J. (2019). *Supports: An alternative to mass housing.* Routledge.

Hall, E. T. (1966). *The hidden dimension.* Doubleday.

Hasgül, E., & Özsoy, A. (2016). Konut Tasarımında Esnekliğin Farklı Konut Tipolojileri Üzerinden Tartışılması. *Tasarım+ kuram dergisi, 12*(22), 69–79. https://doi.org/10.23835/tasarimkuram.315699

Kalfaoglu Hatipoglu, Hatice. (2016). Improving turkish housing quality through holistic architecture: assessment framework, guidelines, lessons from Vienna. Vienna.

Heng, C. K., & Malone-Lee, L. C. (2009). Density and urban sustainability: an exploration of critical issues. In *Designing high-density cities* (pp. 73–84). Routledge.

Hollands, R. G. (2008). Will the real smart city please stand up? Intelligent, progressive or entrepreneurial? *City, 12*(3), 303–320. https://doi.org/10.1080/13604810802479126

Jacobs, J. (1961). *The death and life of great American cities.* Random House.

Kanter, R. M., & Litow, S. S. (2009). Informed and interconnected: A manifesto for smarter cities. Harvard Business School General Management Unit Working Paper, 09–141.

Katz, P., Scully, V. J., & Bressi, T. W. (1994). *The new urbanism: Toward an architecture of community* (Vol. 10). McGraw-Hill.

Koca, D., (2012). Türkiye'de 2000 sonrası toplu konut üretimine genel bir bakış. Türkiye Mimarlığı ve Eleştiri (pp. 43–52). TMMOB Mimarlar Odası Yayınları.

Koolhaas, R., & Whiting, S. (1999). Spot check: A conversation between Rem Koolhaas and Sarah Whiting. *Assemblage, 40,* 36–55. https://doi.org/10.2307/3171371

Lawson, B. (2010). The social and psychological issues of high-density city space. *Designing high-density cities for social and environmental sustainability,* pp. 285–292.

Lefaivre, L. (2007). *Ground-up city: Play as a design tool.* 010 Publishers.

Madanipour, A. (2007). *Designing the city of reason: Foundations and frameworks.* Routledge.

Newman, O. (1996). *Creating defensible space.* Diane Publishing.

Odpm(Office for Deputy Prime Minister). (2003). Sustainable communities: building for the future.

Osmay, S. (1998). 1923'ten Bugüne Kent Merkezlerinin Dönüşümü. *Yılda Değişen Kent Ve Mimarlık, 75,* 139–154.

Ozdemirli, Y. K. (2014). Alternative strategies for urban redevelopment: A case study in a squatter housing neighborhood of Ankara. *Cities, 38,* 37–46. https://doi.org/10.1016/j.cities.2013.12.008

Partridge, H. L. (2004). Developing a human perspective to the digital divide in the 'smart city'. In *Australian library and information association biennial conference*

Payne, G. K. (1977). *Urban housing in the Third World.* Routledge.

Perry, C. (1929). The neighbourhood unit, a scheme of arrangement for the family-life community. *Regional Plan Association of New York.*

Scarr, H. A. (1973). *The nature and patterning of residential and non-residential burglaries.* US

Schneider, T., & Till, J. (2005). Flexible housing: Opportunities and limits. *Architectural Research Quarterly, 9*(2), 157–166. https://doi.org/10.1017/S1359135505000199

Şenyel, A. (2006) Low rise housing development in Ankara [Master Thesis, METU, Turkey]. https://etd.lib.metu.edu.tr/upload/12607341/index.pdf

Sezer, M. (2009). Housing as a sustainable architecture in Turkey: A research on toki housing. [Doctoral dissertation, Master Thesis, METU, Turkey]. https://etd.lib.metu.edu.tr/upload/12610551/index.pdf

Shaftoe, H. (2008). *Convival urban spaces-creating effective public spaces.* Earthscan.

Song, Y. (2005). Smart growth and urban development pattern: A comparative study. *International Regional Science Review, 28*(2), 239–265. https://doi.org/10.1177/0160017604273854

Susanti, R., Soetomo, S., Buchori, I., & Brotosunaryo, P. M. (2016). Smart growth, smart city and density: In search of the appropriate indicator for residential density in Indonesia. *Procedia-Social and Behavioral Sciences, 227*(1), 194–201. https://doi.org/10.1016/j.sbspro.2016.06.062

Tekeli, İ. (2012). *Türkiye'de yaşamda ve yazında konutun öyküsü (1923–1980).* Tarih Vakfı Yurt Yayınları.

Tekeli, I. (2010). *Konut sorununu konut biçimleriyle düşünmek,* Ilhan Tekeli Toplu Eserleri 13, İstanbul, Tarih Vakfı Yurt Yayınlari.

Türkün, A., & Kurtuluş, H. (2005). *İstanbul'da Kentsel Ayrışma.* Bağlam Yayınları.

English Partnership The Housing Corporation. (2000). *Urban design compendium.* Homes and Communities Agency.

Uzun, N. (2005). Residential transformation of squatter settlements: Urban redevelopment projects in Ankara. *Journal of Housing and the Built Environment, 20,* 183–199. https://doi.org/10.1007/s10901-005-9002-9

Vidler, A. (2002). *A city transformed: Designing 'defensible space.'* The MIT Press.

Vliet, W. V. (1983). Families in apartment buildings: Sad storeys for children? *Environment and Behavior, 15*(2), 211–234. https://doi.org/10.1177/0013916583152005

Van der Voordt, D. J. M., & van Wegen, H. B. (2005). *Architecture in use. An introduction to the programming, design and evaluation of buildings.* Architectural Press of Elsevier.

Washburn, D., & Sindhu, U. (2010). *Helping CIOs understand "smart city" initiatives.* Forrester Research, Inc. 2010.

Watson, O. M. (1970). *Proxemic behavior: A cross-cultural study,* vol. 8. Mouton De Gruyter.

Ye, Y., Li, D., & Liu, X. (2018). How block density and typology affect urban vitality: An exploratory analysis in Shenzhen, China. *Urban Geography, 39*(4), 631–652. https://doi.org/10.1080/02723638.2017.1381536

Zairul, M. N., & Geraedts, R. (2015). New business model of flexible housing and circular economy. In *Proceedings of the future of open building conference.* ETH-Zürich. https://doi.org/10.3929/ethz-a-010581209

Strategies to Improve Urban Performance

Livable Streets Determinants in Egypt: A Study on Streets' Physical Attributes in New Urban Communities

Aly Elkhashab

Abstract

One of the most integral parts of urban landscape and transit infrastructure in cities is city streets. City streets/streets are widely known for their significance as a medium for urban transit and commuting and also act as part of the public realm, where people usually gather to fulfill their social needs and aspirations. Lately, urban planning disregarded the secondary role of streets as public spaces and as a catalyst for public life. This was mainly driven by rising pressure on urban planning to accommodate the increasing number of new vehicles going road daily, crowding the already congested streets. Recently, urban planning acknowledges the significant role of streets in maintaining the city's vibrant and lively atmosphere. Several recent studies shifted focus toward traffic management and began regarding it as a key determining factor of streets' livability. However, research addressing people's perception of physical qualities and attributes with regards to their impact on the livability of streets tends to separate. This study examines the understanding of people to physical attributes of Egypt's streetscapes and is built on reviewed literature from 1980 to 2020 for the most recognized and referenced urban space architecture. Questionnaires and observations were used to define the main factors affecting street livability in two chosen multifunctional streets. About 15 physical attributes of streets were recognized and evaluated from reviewed literature, questionnaires and site observation. Accordingly, the proposed study will explore how the provision of basic facilities such as paving, roads, maintenance, cleaning, parking space and traffic lights will affect streets' livability. The final outcome of this study can be utilized by practitioners and policymakers in Egypt, to provide a more holistic understanding of key factors affecting the livability of streets in new urban settlements.

Keywords

Livable streets • Streets' physical attributes • Determinants of livability • City streets

1 Introduction

Streets occupy 25–30% of all urban developed land areas (Jacobs, 1993); hence, they are an important part of the landscape and users' daily life (Bohl, 2002) and serve as places of public expression (Leinberger, 2008). It is safe to say that the experience/environment that a street creates has a direct relation to the quality of urban life by means of its form function and organization within the community. Thus creating better streets would ensure a pleasant urban experience/environment to be developed.

Architects, urban planners and designers have continuously contributed to the significance of streetscape's physical attributes in creating a vital/livable surrounding environment and promoting local amenities. Studies showed that streets are a fundamental element of the overall built environment and are considered to be crucial elements of the public realm (Jacobs, 1961). Although Jacobs and Appleyard (1987) developed the livability definition almost 50 years ago, it was implemented as a consequence of several research works in the last 20 years on various aspects of post-modern cities and criticized numerous urban space problems such as toxic, noisy, dirty, poor quality and unwelcoming environments (Hartanti, 2012).

To tackle these issues for improving and humanizing the open spaces in modern cities, the concept of livability as a vital objective for creating a good urban setting was defined by several experts. Livable space was conceptualized as:

A. Elkhashab (✉)
Department of Architecture, Faculty of Engineering, Arab Academy for Science Technology & Maritime Transport, Smart village, Cairo, Egypt
e-mail: alielkhashab22@gmail.com

© The Author(s), under exclusive license to Springer Nature Switzerland AG 2022
C. Piselli et al. (eds.), *Innovating Strategies and Solutions for Urban Performance and Regeneration*,
Advances in Science, Technology & Innovation, https://doi.org/10.1007/978-3-030-98187-7_8

- The ideal street has to play as a safe sanctuary, giving a healthy, green, and pleasant environment, become a neighborhood territory that engages the community, become a place for play and learning for the children and have a unique quality that becomes the place identity (Appleyard, 1981).
- The eight characteristics of a good street as illustrated by Jacobs (1995) refer to livable street, i.e., a place that affords for people to walk with some leisure, affords physical comfort, clear definition, eyes-catching quality, transparency between inside and outside, complementarities in building design, good maintenance and good quality of construction and design.
- Livable spaces are places that attract all strata and classes (rich and poor, educated and uneducated), suit all ages (children, youth and elders) and encourage various activities. The livable spaces are affordable to all people, easily accessible and connected to the surrounding neighborhoods. They are available to all people, irrespective of their racial background, age, or gender. They provide a forum for individuals and society to be democratic. They have gathering spaces and foster socialites. Livable spaces form an area's cultural identity and provide a place for local communities (Zalloom, 2017).
- In the lifeless spaces, people who live and work in a given area are left without a place to interact in an informal, pleasant environment, and the people who pass through lose the possibility to experience a unique sense of place (Places, 2017).

So, the concluded aspects gathered up could be: Livable space is considered to be a safe and healthy space, where everyone can live in relative comfort. Also, it is a place that encourages the community to engage freely and offers a well-managed environment (green, healthy and pleasant) to people that is well maintained. It should be affordable for all people (different classes, ages and ethnic origin) and encourage various activities. Also, it acts as a place for play and learning for children and has a unique quality that becomes the place identity.

Almost all of the cultural, spiritual, com-mutual, physical and recreational activities take place in commercial streets throughout the neighborhood (Mehta, 2014). Livability has been strongly associated with streets that tackle the above-mentioned problems and fulfill community needs and serves the public needs for recreational activities (Francis, 1991) and (Gehl, 2001). This research aims to define livable streets determinants in Egypt by analyzing streetscapes physical attributes in two multifunctional streets in new urban cities, known as El Mostakbal street and El Bostan street, both located in Sheikh Zayed City. Questionnaires

and structured observations will be used to examine society's perception and attitudes to streets' physical attributes and their effect on street livability which will be discussed later on.

2 Literature Review

Appleyard (1981) explained the concept of "livable streets" through his popular book "livable streets". This book introduced the approach of traffic calming in various cities worldwide toward a friendlier urban environment against a rapid increase in the number of vehicles in cities. Appleyard and Lintell (1972) addressed the harmful/negative traffic effects on decreasing inhabitants/citizens' quality of life, based on the "livable street" project in 1969. Traffic speed and traffic noise were the main factors with a great impact on the livability of the surrounding environment.

In 1982, Kaplan & Bush argued that a city-dweller appreciates "green" places more. In addition to efficiency of urban open spaces in providing a user friendly environment to get away from urban chaos, road congestion and overloading. Hartig et al. (1997) and Kaplan & Bush (1982) discovered evidence of relaxing and rehabilitative powers for human beings in natural scenes.

In 1985, urban open space was defined by Jackson as an urban form that brings people together through passive enjoyment. Also in the early 1980s, Lynch took the view that urban open space includes elements/attributes intended to engage groups of people and foster meetings. Urban open space has been distinguished from sidewalks by asserting that the first is a space on its own, instead of just an area to cross (Marcus & Francis, 1998).

J. Davis (2002) indicated that streets determine the character of a city. But often due to poor management and preservation, these streets let us down. Street clutter erodes an area's distinctive identity, importance and distinguishability. Such urban clutter is simply a product of the insufficient and sometimes distracting knowledge provided by the existing policy guidance to urban designers. These single-issue guides tend to take a detailed perspective of the urban streetscape thinking about whether a qualified engineer should be present to manage the overall image and integrate overall requirements.

In 2009, Collins and Shantz stated that city streets, public gardens and squares have always been crucial spaces for either community activities, economic or political ones. They act as places/hubs of interaction where societal customs may take place. Also, they express the norms of their society in many collective activities that may occur, either in the context of everyday life or in special/relatively frequent events, such as festivals and public events (Carr et al., 1992).

Nevertheless, streets are designed as places for traffic in the modern, post-industrial and contemporary conceptualization of space.

In (2005), characteristics of livable streets were examined by Dumbaugh and Gattis in correlation to street safety. Streets are defined as motorist thoroughfares; commercial streets also double as public gathering areas for residents and visitors. In urban streets people can shop, communicate, socialize and participate in different social and leisure activities that make urban living enjoyable. Findings revealed that the livability proponents support the installation of street trees, landscaping, attractive street lights and other roadside installations along the edge of the vehicles traveling routes, both to improve the esthetic attractiveness of a street and physically protect pedestrians from potentially hazardous oncoming traffic.

Sauter and Huettenmoser (2008) studied and evaluated five streets regarding the effects of the amount of traffic on the quantity and quality of street life in Basel, Switzerland. It was found that quieter streets are attaining a better community life. Park (2008) also concluded that calming traffic can have great effects on street walkability, as well as encouraging people to walk too.

Forsyth et al. (2008) found out that urban planning proposed that improving mobility in central neighborhood areas such as commercial streets could make urban lifestyles healthier, safer, cleaner and undoubtedly better and more efficient. Moreover, studies showed that ease of mobility (accessibility) improves streets' quality and livability (Jacobs, 1961), and also asserted that improving streets' physical attributes such as roads pavement, shelters (shading devices), lighting elements and aesthetic values facilitate walking and sports/outdoors activity, hence encouraging streets' walkability and establishing a more sustainable, healthy and livable environment (Rehan, 2013). Dumbaugh and Gattis (2005), Portella (2007) and J. Davis (2014) stated that commercial signage is classified as street clutter, and those physical details of streets affect the perception of users of community identity and sense of belonging. Layne (2009) has highlighted that social space of a street could be enhanced by adequate landscape design, and also facilitates generational engagement as well as promotes social interaction.

Mackett et al. (2008) created an app for assessing handicapped accessibility to the environment and investigated the details of the streetscape, including road crossing challenges and entrances to buildings. They noticed that modifications to such details have an effect on street use, improved access to handicapped facilities and encouraged street livability.

Hartanti, N. B. (2012) explained that originally streets are not designed as thoroughfares for the vehicle only. Streets, in fact, double function as public spaces. People usually get to use streets either to stroll/walk, shop, interact, or engage in a variety of social and recreational activities which make urban living enjoyable for users. So, the concept of street design has to be changed from creating more road bays struggling to catch up with the rapid growth in the number of vehicles. In an attempt for making space for people, whether on foot or by automobile, that is the livable street. Street livability is defined primarily by better integration of the interests and safety of pedestrians with the mobility of vehicles in conjunction with land use and activity. Gössling (2020) and Abdel-Aziz et al. (2020) highlighted the increasing availability of cars flooding on the city streets, as well as their detrimental effects on biodiversity and quality of life, leading to increased traffic congestion. In addition to psychological and social costs, it results in both time and cost losses. Also, an increase in emissions and rising energy consumption for automobiles had detrimental effects on air quality.

Tawil et al. (2014) studied El Medina street in the West Amman case, either the challenges it faces or the solutions needed to resolve the absolute occupation of traffic by converting it into a pedestrian-friendly street. As a consequence, strategies were introduced to reclaim traffic space as people's space, develop new street concepts in Jordan and divert congested routes into local ones. All of the previous are recommended concepts with more public spheres that can steer the development of streets in Amman.

According to Appleyard B. (2017), livable streets are about more than just providing a safe and pedestrian-friendly setting. Livable streets also provide an urban space/setting that encourages social interaction with the environment in order to promote physical and psychological development. Livable streets elements are as follows:

- A safe space.
- A good community.
- A healthy and livable environment.
- A friendly territory.
- A place for learn and play.
- A green space.
- A unique historic place.

Zhan D. et al. (2018) studied the understanding of residents' satisfaction with urban livability in China. They distributed their designed questionnaires conducted in 2015 in 40 major cities. Surveyed dimensions regarding urban livability were as follows: public facilities, natural environment and the sociocultural environment, urban security, environmental health and transportation. Their results revealed that all six dimensions of urban livability have significant and positive effects on overall user satisfaction, especially concerning the natural environment, transportation and environmental health are the greatest impacted. Other attributes that have an effect on overall satisfaction with urban livability are location, housing type, education, size of family and age, although their effects are far less than that of the dimensions of urban livability.

Combs S. and Pardo F. (2021) illustrated the significant shift in demand for safe walking, bicycling, and outdoor activities due to the COVID-19 epidemic. Globally, cities enacted a range of laws and initiatives aimed at addressing this shift such as turning out most of the road space for pedestrians, lowering speed restrictions and promoting bike services. Also, the study assessed the future developments of pedestrian and bicycle infrastructure design and implementation, and how transportation professions might evolve in response to lessons learned during and after the pandemic.

In their study, Moreno C. et al. (2021) focused on socioeconomic repercussions on cities during the COVID-19 pandemic leading to total/partial lockdown in most cities worldwide, in order to maintain decent levels of health. Constraints created by the epidemic have necessitated a radical rethink of the city, resulting in the re-emergence of the concept of "15-min City" that was stated before by the author in 2016. The concept complements the existing concepts of smart cities and the rhetoric of creating more humane urban fabrics, as well as creating safer, more resilient, sustainable and inclusive cities. The "15-min City" concept means that basic urban services must be given in close proximity to urban centers, without discriminating people based on their socioeconomic level or age. The study found that Shanghai city (case study) basic utilities are not well distributed along city areas/districts equally and orderly. As poor urban services distribution at peripheries in residential areas necessitated the need for automobiles to access all the available services/facilities, thus making walkability impossible to move around between urban amenities.

Arefi and Nasser (2021) stated that most urban designers realize the necessity of transportation infrastructure in cities, however, sometimes they overlook some of the repercussions. And specifically micro-scale ones: bus stops and sidewalk design impacts on placemaking, pedestrian perceptions of local safety, parking lots width and spacing, and traffic calming devices as viable design solutions. The study addresses a few of these seemingly insignificant, although being really crucial factors of the transportation infrastructure and street network that are sometimes disregarded in urban design practice. Also in their study they examined the relation between urban design, perceived safety, street livability and accessibility in particular.

NABIL T. et al. (2021) discussed the necessity of establishing spaces for social interaction, as well as how to change them into pedestrian-friendly and sustainable, taking into account environmental and social factors. According to their findings, most Iraqi cities lack sidewalks that promote better social interaction and provide individuals with a healthy and safe environment that improves their recreational levels and health. Availability of continual social and recreational activities along the day creates livable and sustainable streets. Also, pedestrian streets express the local identity of the region and could also create an economic investment opportunity for the city, through developing commercial and job opportunities.

Abdulmughni M. et al. (2021) in their study stated that enormous urban expansion had negative effects on the human dimension, as streets became more devoted to transportation rather than to pedestrians. In their research, they studied about two streets in Riyadh that were developed pedestrian-friendly. They investigated in their study the physical aspects of Riyadh streetscapes and defined their influences on the livability and quality of spaces. The study concluded that the length of the street, sort of commercial activity, crossing facilities, the width of the sidewalk and facilities for the disabled are all factors that contribute to the streets' livability. Also, the presence of street greenery, adequate lighting, shadings & canopies, and seating areas/benches all influence streets' livability.

The examined literature, however, assessed the livability of the streetscape and concentrated on a few numbers of physical attributes that affect streets' quality and livability, considering all other factors are the same. Notably, the study exposed that only a few experimental studies tackled the people's perception of certain physical attributes such as affecting the livability of the street. Table 1 indicates a comprehensive review of literature during the last 40 years among the most prominent and referenced urban space architecture researches about streets' physical attributes and the determinants of streets' livability. This research defines 15 physical attributes and makes an attempt to identify which attributes have the greatest impact on street livability.

3 Methodology

Many urban space scholars in the late twentieth century concluded that people's lives are affected by the physical attributes of the built environment. The physical environment may represent people's perceptions, emotions and behaviors that undermine their environmental values (Rapoport, 1982; Sanoff, 1991). The research methodology used in this study is built on Ahmad and Mahmoudi (2015). Similar to the initial study, it examined some physical attributes that contribute to street livability. However, in this research, the methodology was modified to suit the local context. The new research suggests similar trends to those mentioned in the original report and reviewed further literature, and introduces extra features and also additional attributes that were not previously tested as shown in Fig. 1.

Table 1 Livability physical attributes derived from literatures

Photos	References	Specs	Surveyed questions
 Paving	– Appleyard (1981) – Rubenstein (1992) – Dumbaugh and Gattis (2005) – Mackett et al. (2008) – Forsyth et al. (2008) – Amr (2015) – Abdelhafeez et al. (2010, 2013) – Mehta (2014) – J Davis (2014) – Helmy (2018) – Yassin (2019) – Ahmad and Mahmoudi (2015) – S. Combs and F. Pardo (2021) – Arefi and Nasser (2021)	– Wideness of pavement (Mackett et al., 2008; Elsawy et al., 2019; Mehta, 2014) – Size and scale of road and pavement (Ahmad & Mahmoudi, 2015) – Pavement quality (Ahmad & Mahmoudi, 2015) – Patterns of pavement especially in the children's area (Nassar, 2015)	– Pavement wideness? (Mackett et al., 2008; Elsawy et al., 2019; Mehta, 2014) – Size and scale of road and pavement? (Ahmad & Mahmoudi, 2015) – Pavement quality? (Ahmad & Mahmoudi, 2015) – Pavement patterns? (Nassar, 2015)
 Seating	– Rubenstein (1992) – Mackett et al. (2008) – Abdelhafeez et al. (2010) – Mehta (2014) – J Davis (2014) – Nassar (2015) – M. Abdulmughni et al. (2021)	– Availability of seats (Nassar, 2015; Mehta, 2014) – Sufficient seats (Amr, 2015) – Comfortableness (Amr, 2015) – Seating areas well distributed in order to improve various social activities (Nassar, 2015)	– Are street elements landscapes (street furniture) comfortable? If not please state the reason, are they enough and covering all spaces? If not where are the spaces that are lacking them? (Amr, 2015) – Do you consider the landscape street furniture (benches, seats, lighting fixtures, receptacles…) enough? If not, where? (Amr, 2015; Mehta, 2014)
 Shelter & canopy	– Rubenstein (1992) – Forsyth et al. (2008) – Amr (2015) – Abdelhafeez et al. (2010) – Mehta (2014) – J Davis (2014) – Nassar (2015) – M. Abdulmughni et al. (2021)	– Availability of Shaded areas (Nassar, 2015; Mehta, 2014) – Sufficient/Enough shelters/shadings (Amr, 2015; Nassar, 2013, 2015)	– Do you find sufficient shading on campus? Please state what provides shade (trees, pergolas, concrete shading devices…etc.) (Amr, 2015; Nassar, 2015)

(continued)

Table 1 (continued)

Photos	References	Specs	Surveyed questions
 Lightening	– Rubenstein (1992) – Forsyth et al. (2008) – Amr (2015) – Nassar (2015) – J. Davis (2014) – Helmy (2018) – M. Abdulmughni et al. (2021)	– Improve lighting as it increases the safety factor for space users (Nassar, 2015) – Well distribution of lighting items, increases safety level, reflects a clean image and welcoming urban space (Nassar, 2015; Elsawy et al., 2019)	– Sufficient no. of lighting elements? (Nassar, 2015) – Distribution of lighting elements? (Nassar, 2015) – Good lighting (Elsawy et al., 2019)
 Signs	– Rubenstein (1992) – Shalaby (2004) – Dumbaugh and Gattis (2005) – Abdelhafeez et al. (2010, 2013) – J. Davis (2014)	– Are directing signs legible (Amr, 2015) – Signage and landmarks illustrate the public health importance of physical activity (Nassar, 2015)	– Are directing signs legible (easy to read and easy to follow)? (Amr, 2015)
 Planting	– Appleyard (1981) – Rubenstein (1992) – Forsyth et al. (2008) – Shalaby (2004) – Amr (2015) – Abdelhafeez et al. (2010, 2013) – Nassar (2015) – Aulia (2016) – Appleyard B (2017) – M. Abdulmughni et al. (2021)	– Toward increasing space exposure and preventing splitting space into isolated dispersed parts, planting elements should be well designed (Nassar, 2015) – To improve space consistency, minimize high-surrounded planting. (Nassar, 2015) – Greenery condition (Nassar, 2015)	– Do you see that the numbers of different landscape elements (trees, plants, paths, plazas, furniture......) are enough? (Please answer for each element) If not where? (Amr, 2015) – E: Are there any sustainable vegetation on your campus? Why are they sustainable? (Amr, 2015) – Greeneries condition? (Nassar, 2015)
 Sculpture & fountain	– Rubenstein (1992) – Nassar (2013) – Abdelhafeez et al. (2013) – J. Davis (2014) – Ahmad and Mahmoudi (2015)	– Availability of water features (Abdelhafeez et al., 2013) – It's some kind of work of art and considered as centerpiece for people coming from several parts of the street to take pictures besides the fountain. Also water sound's really relaxing. So, the fountain is found to be of great quality in the street context. (Ahmad & Mahmoudi, 2015) – In most places people feel safe beside the lake (Nassar, 2013)	– Importance of water feature? (Nassar, 2013) – Do water elements improve health? (Ahmad & Mahmoudi, 2015)

Table 1 (continued)

Photos	References	Specs	Surveyed questions
Proportions of space	– Lynch (1981) – Forsyth et al. (2008) – Abdelhafeez et al. (2013) – Nassar (2015) – Aulia (2016)	– Human scale (Elsawy et al., 2019)	– Street to building ratio (Elsawy et al., 2019)
Harmony of architectural style	– Lynch (1981) – Rubenstein (1992) – Abdelhafeez et al. (2013) – Ahmad and Mahmoudi (2015) – Elsawy et al. (2019)	– Lack of harmony between various buildings and contrast adversely affect streets' visual integrity (Ahmad & Mahmoudi, 2015) – Harmony of architectural style (Elsawy et al., 2019)	– Harmony of architectural style? (Ahmad & Mahmoudi, 2015; Elsawy et al., 2019)
Facilities for disabled people	– Lynch (1981) – Rubenstein (1992) – Mackett et al. (2008) – Helmy (2018) – Ahmad and Mahmoudi (2015) – M. Abdulmughni et al. (2021)	– Ease of movement (Mackett et al., 2008) – Adequate facilities for the disabled (Ahmad & Mahmoudi, 2015) – Availability of wheelchair facilities (Elsawy et al., 2019)	– Availability and ease of movement of wheelchair? (Elsawy et al., 2019)
Parking	– Appleyard (1981) – Rubenstein (1992) – Shalaby (2004) – Amr (2015) – Tawil et al. (2014) – Aulia (2016) – Arefi and Nasser (2021)	– Availability of parking space (Amr, 2015; Tawil et al. 2014; Ahmad & Mahmoudi, 2015) – Distance of parking from the site (Amr, 2015)	– Do you have a parking problem on campus? If yes, why? (Amr, 2015) – Insufficient parking – Parking charges – Far parking spots – Unsafe parking – Other
Accessibility	– Lynch (1981) – Shalaby (2004) – Mackett et al. (2008) – Amr (2015) – Aulia (2016) – Helmy (2018) – Elsawy et al. (2019) – Yassin (2019) – Zhan D. et al. (2018) Arefi and Nasser (2021)	– Adequate access to public transportation (Ahmad & Mahmoudi, 2015) – Ease access to all facilities (Elsawy et al., 2019)	– Ease of access to public transportation? (Ahmad & Mahmoudi, 2015; Elsawy et al., 2019)

(continued)

Table 1 (continued)

Photos	References	Specs	Surveyed questions
Traffic management	– Appleyard (1981) – Shalaby (2004) – Dumbaugh and Gattis (2005) – Mackett et al. (2008) – Forsyth et al. (2008) – Tawil et al. (2014) – Ahmad & Mahmoudi (2015) – S. Combs and F. Pardo (2021) – Gössling (2020) – Abdel-Aziz et al. (2020) – Arefi and Nasser (2021)	– Quality traffic management (Ahmad & Mahmoudi, 2015)	– Rating quality of traffic management? (Ahmad & Mahmoudi, 2015)
Maintenance &cleaning	– Lynch (1981) – Appleyard (1981) – Rubenstein (1992) – Amr (2015) – J Davis (2014) – Ahmad and Mahmoudi (2015) – Nassar (2015) – Helmy (2018) – Elsawy et al. (2019)	– Quality of maintenance and cleaning of facades and streets (Nassar, 2015; Ahmad & Mahmoudi, 2015; Elsawy et al., 2019)	– Rate quality of maintenance and cleaning of the surveyed street? (Ahmad & Mahmoudi, 2015; Elsawy et al., 2019)
Street Clutter	– Dumbaugh and Gattis (2005) – Amr (2015) – J Davis (2014)	– Irregular distribution of signs (Ahmad & Mahmoudi, 2015)	– Distribution of street furniture? (Ahmad & Mahmoudi, 2015)

Fig. 1 Diagram showing the broken cycle between livability research and streets' planning in Egypt. *Source* Author

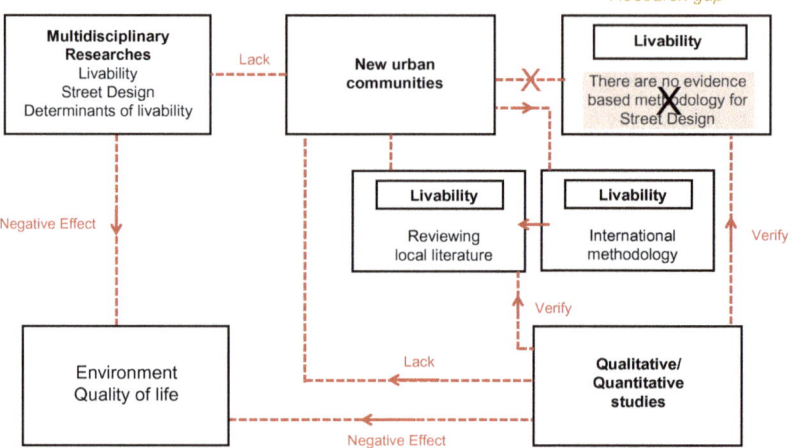

This study is based on reviewed literature from 1980 to 2020, providing a framework for studying physical attributes that has the greatest impact on streets' livability. Among the most well-known and referenced urban space architecture researches about streets' physical attributes and the determinants of streets' livability, there is a group of attributes consistently found to be investigated throughout the literature such as seating, traffic management and shelter. On the other hand, fewer studies reviewed attributes like the orientation of the street or street skyline. To provide this context, these attributes have been chosen among several frequently referred ones defined by various sources as the most frequently cited physical attributes.

An international case study for vibrant livable streets would be studied carefully, assessed and evaluated, and to find out which physical attributes determine their livability. Then, the concluded physical attributes from these case studies would be filtered. Finding out which physical attributes are missed in the Egyptian new urban cities context, not only this but also the ones that could be adapted and implemented. On the other hand, two local streets are chosen to be studied, analyzed and evaluated from new urban cities streets' in Egypt. The chosen streets to perform the study are allocated in El Sheikh Zayed city (one of the most well-known Egyptian new urban cities). These two streets are widely known for their high accessibility, vitality and vibrancy and are also economically and socially important for city life.

First, El Bostan street is one of the most trafficked streets in the city. The importance of this route/street is that it works as a central route serving numerous destinations that are parallel to the main axis of the Mehwar road and binds it to the city of El Sheik Zayed. It includes a high intensity of different activities from educational centers (e.g., Cairo University), commercial centers (Americana plaza, Arkan plaza, Capital business park, Tivoli dome, etc.), office areas (Arkan plaza, Capital business park, Juhayna headquarter, Edita headquarter, etc.) residential dwellings/compounds to different domestic uses. El Bostan street has four main intersections: Al Nozha street, Dorra Circle, El Safa street and El Amal street. It is noted that El Bostan street is one of the main anchors for local use; it is also highly attractive and accessible by passing vehicles that connect various neighborhoods/districts. El Bostan street is made of four sides; each direction consists of two: main and service roads. The main road consists of three lanes and service one consists of two lanes. It offers multiple uses for the local community; recreational, working places, shopping etc. The street offers multiple uses/services such as restaurants, bars, malls and other attractions that provide for the daily needs of the street inhabitants as well as those who visit the route.

Secondly, El Mostakbal street is one of the most notable streets in the city. The importance of this route/street is that it works as a central route serving numerous destinations. It includes a high intensity of different activities such as educational centers (schools, nurseries, etc.), commercial centers, residential dwellings/compounds and also religious ones. El Mostakbal street has four main intersections: Atef Sedky street, El Hekma street, El Safa street and El Amal street. Not only this but also El Mostakbal street is one of the main anchors for local use as it serves as a highly attractive and accessible by passing vehicles that connect various neighborhoods/districts. El Mostakbal street is a two-way street with four lanes on each way. It provides multiple uses for the surrounding community: recreational, working places, shopping etc. The street offers multiple uses/services such as restaurants, cafes, supermarkets and other attractions that provide for the daily needs of the street inhabitants as well as those who visit or even just walk the route. Throughout this context, to reduce the various deficits that occur along the street, it should be able to develop. Also, the pedestrian requirements should be given more attention. In order to attract users, some elements need to be developed.

The methods of data collection used in this study are intended to cover all facets of the thesis while considering the different viewpoints of associated parties. Both qualitative and quantitative data will be used in this study as shown in Fig. 2. Qualitative data will be collected by literature review, questionnaire and site systematic analysis (structured observation), while quantitative data will be retained as measurements and values that can help to explore the ones that promote street livability. Structured observations and questionnaires will be used to gather data on street users' preconception of the streets' physical attributes that encourage the livability of streets. Both selected multifunctional streets are located in El Sheikh Zayed city. Streets were selected based on the fact that future developments in Egyptian countries focus on developing new cities. These streets are highly accessible, crucial, economic and socially important streets in the selected city (El Sheikh Zayed city). To generalize the findings of the research according to (Yin, 2003), the need for findings to be repeated so that at least two (case studies/examples) to be tested. And secondly, the limitations of this study, which are time and budget are constraints for conducting this research.

Questionnaires discover the viewpoint of participants on the efficiency and livability of areas studied, and the impact of findings on street livability and define the most prominent ones that are determinants of street livability. At the analysis stage, observation outcomes would be utilized to interpret the outcome of the questionnaire/survey. A questionnaire will be designed to define livability determinants in streets; in order to assess the point of view of the users, a pilot study will also be performed. A developed questionnaire using physical attributes concluded from reviewed literature would be circulated over 15 days among 110 users from passersby

Methodology

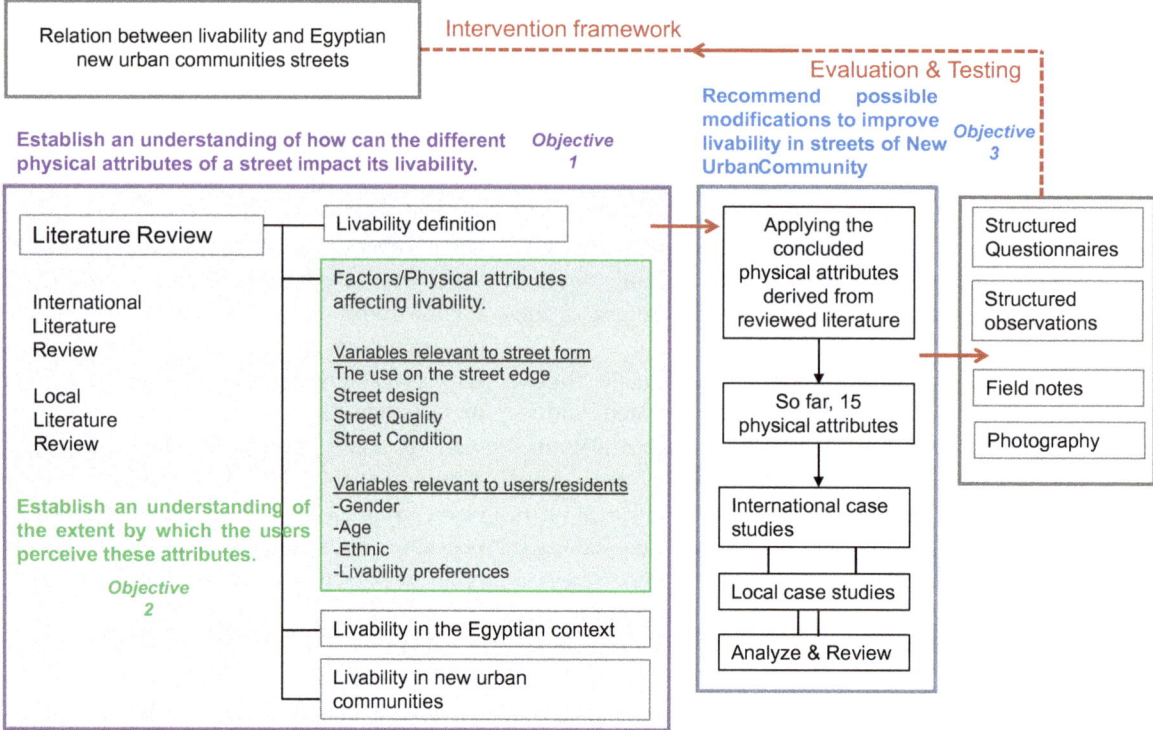

Fig. 2 Research framework illustrating the structure of study flow. *Source* Author

of studied areas who recognize those streets very well. Also, the physical attributes listed will be included in the questionnaire and questioned to participants if they affect the area's livability or not. And other attributes that might be confusing to participants, such as buildings' heights, street skyline and streets' orientation will be neglected. So, 15 physical attributes were listed as major contributing factors to the livability of streets, which are identified as pavement, seating, shelter (shading) and canopy, illumination/lighting, signage, planting, sculpture and fountain, space proportions, harmony between the various buildings' architectural style, handicapped facilities, parking, accessibility, traffic management, maintenance and cleaning, and street cluttering.

This research focuses on identifying the streets' main physical attributes that lead to livable streets in Egypt via providing empirical evidences. Also aiming to answer the following research questions:

– What are the physical attributes of streets that are acknowledged in literature to impact streets livability?
– To what extent are these attributes implemented in the streets of new urban communities?
– To what extent do the street users perceive the presence/absence of these attributes?

– To what extent are these attributes impacting livability in new urban communities?

Also, this research aims to:

– Establish an understanding of how can the different physical attributes of a street impact its livability.
– Establish an understanding of the extent to which the users perceive these attributes.
– Recommend possible modifications to improve livability in the streets of the New Urban Community.

Observations are among the most widely used post-occupancy analysis techniques used in many urban space researches (FRANCIS, 2003). The qualitative analysis involved quality, harmony and adequacy. Direct observation and an objective examination of the physical attributes of surveyed streets were pursued for visual evaluation by taking field notes and photos. To ensure liability of concluded attributes, areas chosen will be observed frequently at different times of the day all over the week and also at peak hours over three months. Data collection either by observation, field notes or photography for each of the attributes examined is registered and tabulated together and given a database for each attribute to be tested. Therefore quality, adequacy and harmony of each attribute were explained by analyzing these details, and findings from this part will be used later on to replicate the results of the questionnaire.

A survey would be designed and distributed, targeting to explore the perspective of users on found physical attributes and their impact on street livability. As discovered by several researches, streets' livability and efficiency cannot be determined without taking into consideration the understanding of people who typically inhabit the space (Nasar 1988). And to achieve this, a brief/simple definition of livability and livable streets to ensure that all participants understand well the meaning of these expressions will be added as an introduction to the survey. Then, users' attitude toward the listed subjects would be measured through the Likert scale. Five alternatives from "very poor to very good" with users' points of view will be included in the model and people who have no idea labeled "average". Bardo et al. (1982) recognized that the reliability of the scale decreased as the number of answer points surpassed two. Also, this was reinforced by the fact that the reliability of the Likert scale increased from 2 to 5 (Lissits & Green, 1975). Not only this but also a pilot study will be conducted to evaluate the questionnaire's validity for uncertainty and ease of comprehension. Once feedbacks are obtained and their recommendations were incorporated, the questionnaire will be completed and prepared for delivery.

Users of the two multifunctional surveyed streets, Al Mostakbal and Al Bostan streets, who are residents, workers or passers-by and know the area very well, are the targeted ones. A survey sample consists of 100 participants and this is the least amount for data analysis (Dooley, 2001). Residents, workers and passers-by of such multifunctional streets are usually too large and their sociodemographic characteristics are diverse. Furthermore, the population size is unknown as there is no knowledge about the average daily number of pedestrians on these streets. In this study, the questionnaire design was derived from questionnaires/surveys included in the reviewed literature. Participants will be chosen 18 years of age or above from both genders. A total of 110 participants will therefore be selected to make the study feasible and to fulfill this requirement. Participants will be randomly picked. Distributed questionnaires will be circulated throughout the day (working days and weekends) with an average of 10 min each.

4 Quantitative Review

From the 110 questionnaires provided, 86 (78.2%) questionnaires were completed and empirically verified. However, 12 (nearly 11%) questionnaires obtained were incorrectly completed and thus excluded. Figures 3, 4, 5 and 6 demonstrate the type of users, people's level of familiarity with selected streets, also how often do they visit them and the type of activities they perform in their leisure time. Table 2 demonstrates people's recommendations/answers to surveyed streets. The outcome of this survey reveals how the design, quality and condition of surveyed spaces affect the livability of them. About 68% of Al Bostan's participants agreed that the street is livable; on the other hand, 24% objected livability of the street. Almost 72% of the participants in Al Mostakbal agreed with the area's quality and livability, and only 18% of the participants disapproved to consider the space livable. Furthermore, 8 and 10% of participants were neutral and had no clear information about Al Bostan street and Al Mostakbal street quality and livability, respectively.

5 Review of Data

After evaluating the results, this research aims to explore the level of acceptance of users on found physical attributes that determines streets' livability and their impact on street livability. As noted before, this questionnaire will be introduced to five categories of users' responses as "very poor, poor, average, good and very good". So participants' behavior will be defined according to these categories in order to improve their understanding of responses. Moreover, the answers of participants who are ignorant of the livability of surveyed areas and the consistency of defined attributes were excluded. The quantitative data analyses are then used to measure the overall outcome and expose the effect of physical streets' characteristics on the promotion of the area's efficiency and livability. Quantitative analysis will be used for evaluating the results from questionnaires after sorting the categories. To examine the validity of the research, a reliability check is also carried out to test the reliability of variables, which will be measured using SPSS analysis as shown in Tables 3, 4 and 5.

6 Findings, Discussions and Visual Evaluation

Throughout the last 9 months, from December 2019 to August 2020, observational and questionnaire surveys were performed to provide a database for assessment of the area's livability as well as prominent physical attributes leading to livable/vital streets. First, there will be a quick illustration of an international case study. Then, an illustration of two surveyed streets in El Sheikh Zayed city is shown in Figs. 7, 8, 9, 10 and 11. Not only this but also some field notes accompanying the pictures were written throughout the assessment of the areas discussed below. The characteristics

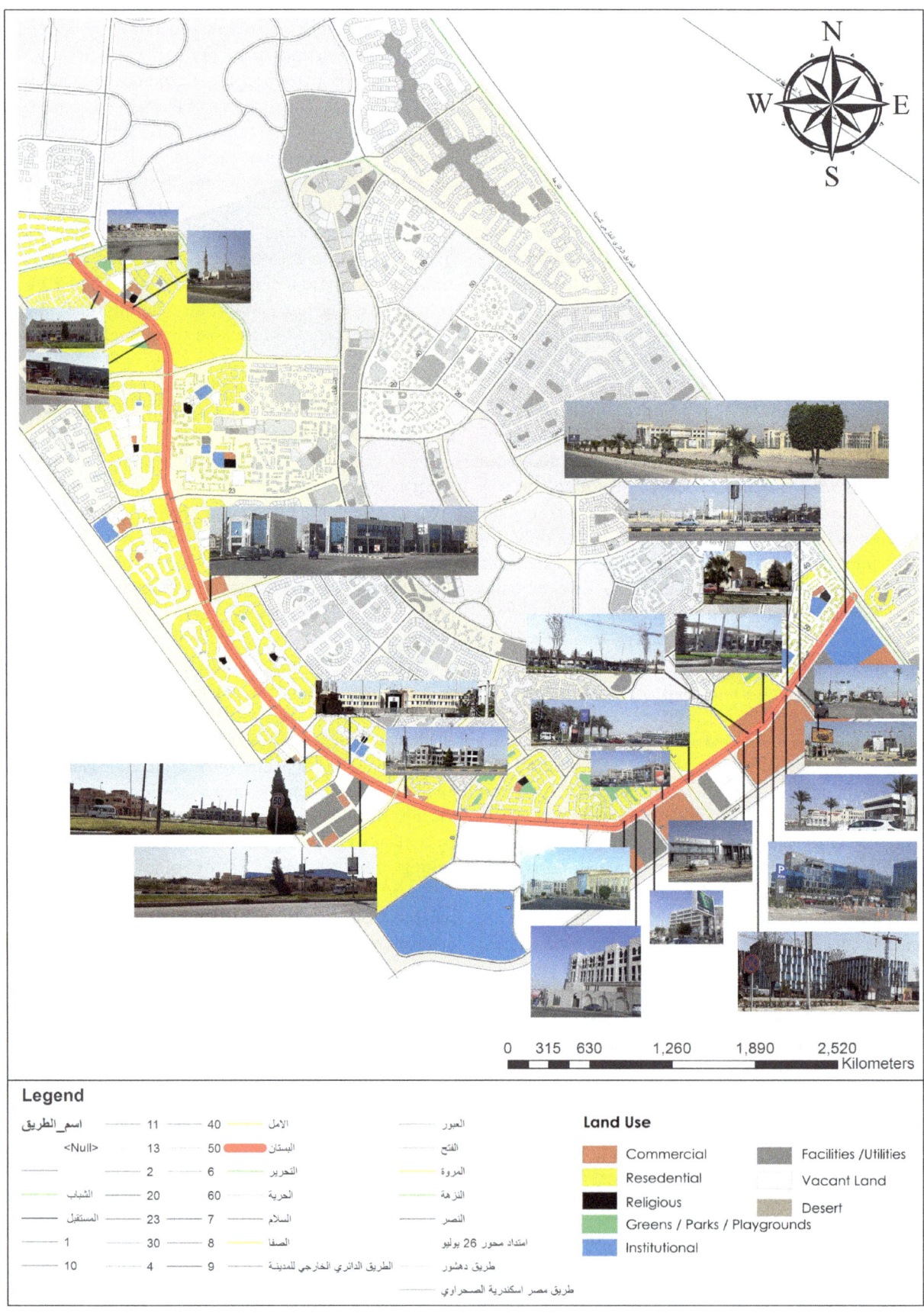

Fig. 3 Familiarity to the space

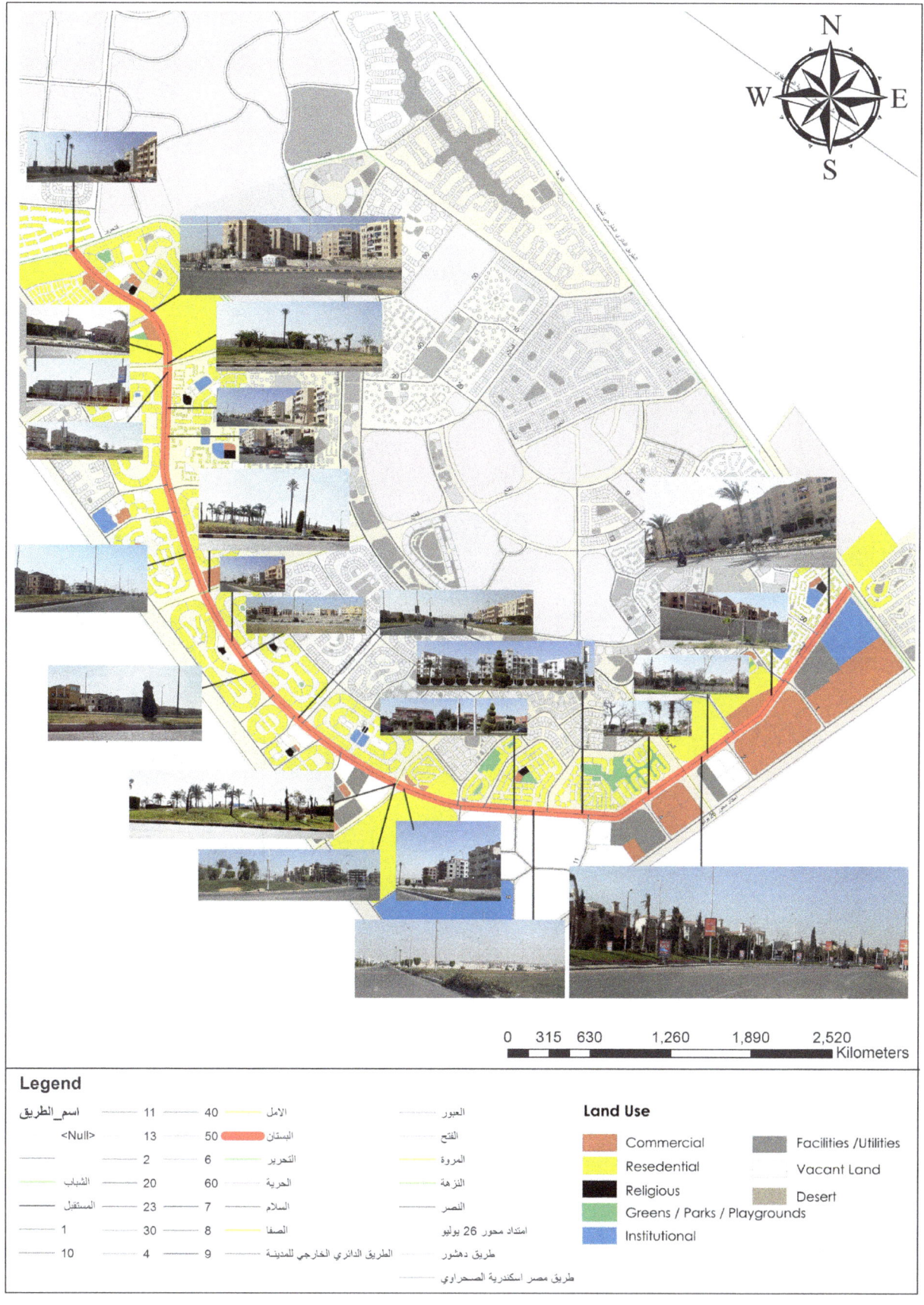

Fig. 4 The number of times for visiting the space

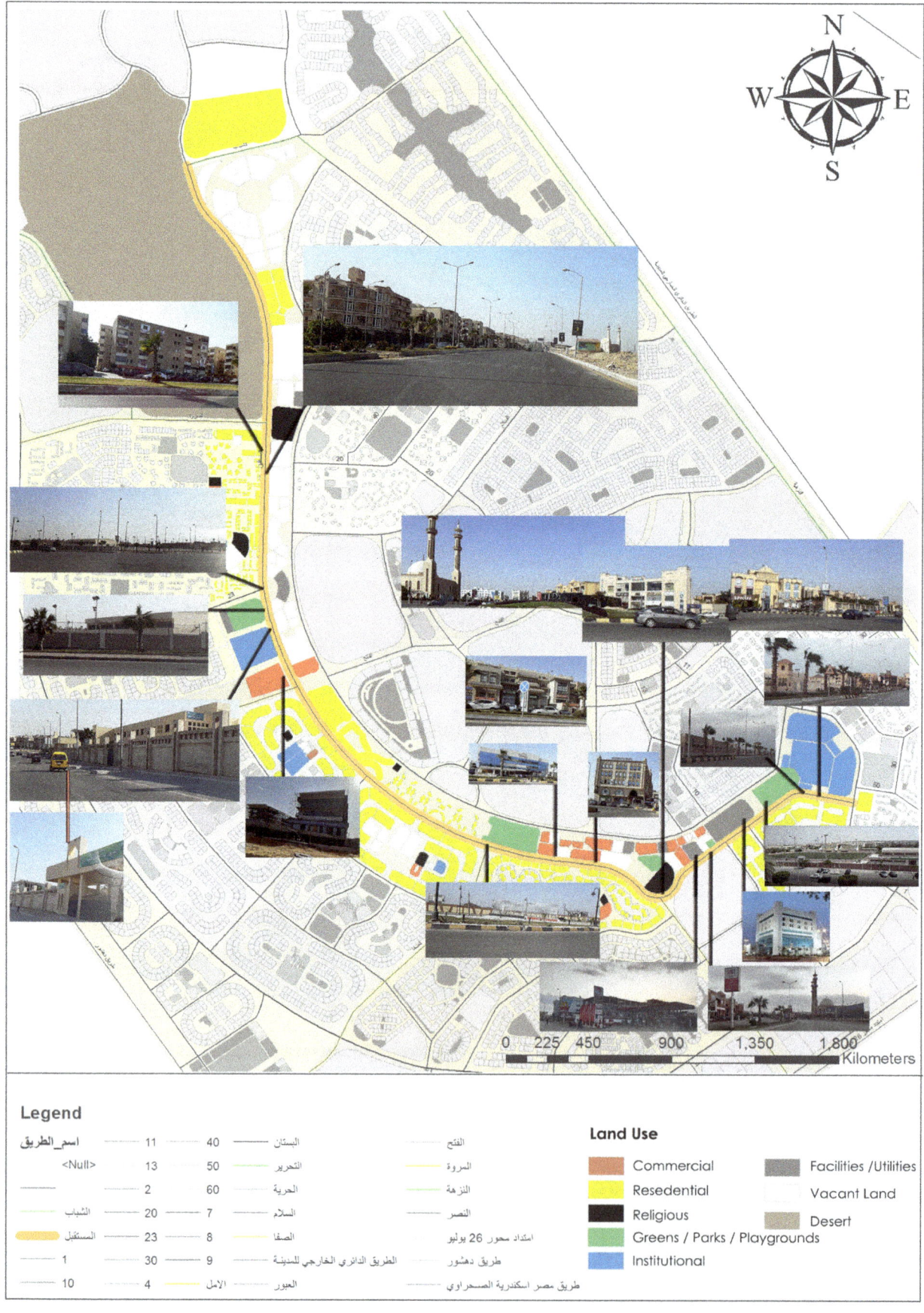

Fig. 5 Type of users using the space

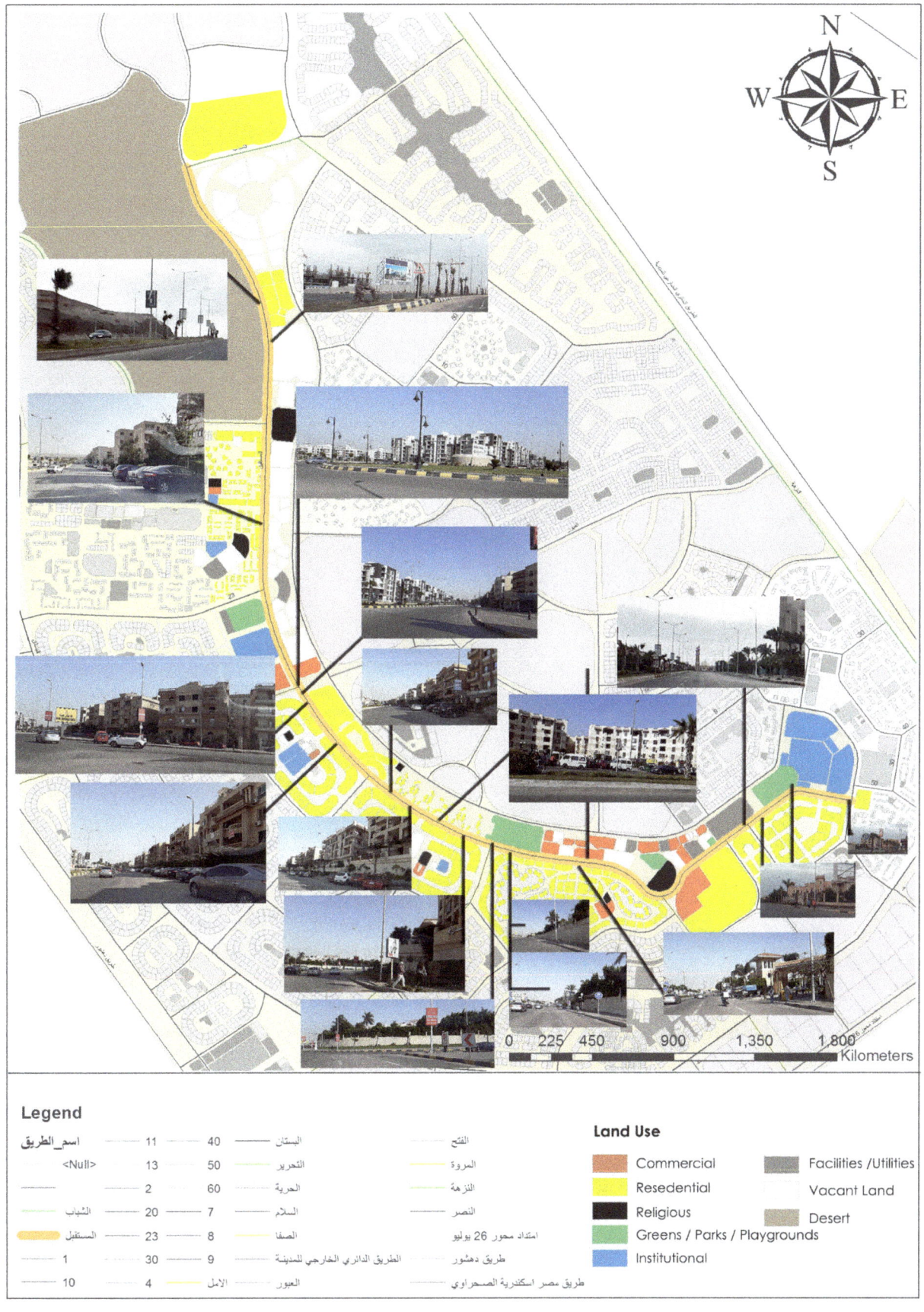

Fig. 6 Activities/hobbies that may take place in the space

Table 2 Participants most repeated answers/recommendations to survey questions

Question	Participant's answers/recommendations
Leisure activity suggested to be added to the space	– More wide pavements – Increase seats for social interaction – Provide more shaded areas – Improve night lighting – Improve greeneries/planting maintenance condition – Adding a water feature to the space – Traffic management between pedestrians and vehicles – Improves maintenance and cleaning of the space – Reduce/remove unnecessary furniture, signs, etc.
Suggested activities to be added to the space	– Walking/running track – Playing area for children – Outdoor space for physical exercise/activity
Users recommendations	– Enhancing the space exposure and preventing dividing it into isolated fragmented sections by redesigning planting elements – Improve lighting increases the safety factor for users of the space – Pedestrian paths surrounding the space should be redesigned to abide them from being used as car parking areas – Increase parking lots

Source Author

Table 3 Users' gender percentage of surveyed streets

Gender		El Mostakbal	El Bostan
Valid	Male	68.8	58.3
	Female	31.2	41.7
	Total	100.0	100.0

Source Author

Table 4 Average age of users of surveyed streets

Age	El Mostakbal	El Bostan
Mean	29.92	29.92
Median	28.50	28.50
Std. deviation	10.068	10.068
Range	30	30
Minimum	18	18
Maximum	48	48

Source Author

of physical attributes of the areas examined are exposed through photos and notes.

7 Quantitative Examination

Survey results reveal how these spaces are perceived as livable and quality ones by people. Almost 72% of El Mostakbal street respondents confirm the streets' livability, and 18% disagree. While in El Bostan street, 24% of respondents disagree, and 68% confirm the livability of the street. Yet, 10 and 8% of respondents were neutral, respectively, and their answers' revealed the unawareness of surveyed spaces. Research findings show that majority are not satisfied with surveyed streets as livable and quality ones, although El Mostakbal street is higher in quality.

8 Analysis Results

From observation and questionnaire results, many disabled facilities are applied and implemented on both surveyed streets, although most of them are not implemented/maintained correctly (e.g., implementation of small pillars in the middle of crossings on the middle island on both edges to prevent cars from using it as illegal U-turn). So, these pillars are considered an obstacle in front of wheel chairs' movement. Also, pedestrian ramps should be distributed along the pavement orderly (every 200 m). User-friendly pavement types are recommended to be used instead of the one implemented, to provide friendlier and more comfortable facilities to disabled users. All of these reinforce the findings of (Mackett et al., 2008) concerning

Table 5 Descriptive statistics for participants most repeated answers/recommendations to survey questions

Physical attributes	El Mostakbal Street				El Bostan Street			
	Min.	Max.	Mean	Std. deviation	Min.	Max.	Mean	Std. deviation
Pavement	−2.00	2.00	0.2500	1.13818	−2.00	1.00	−0.2500	1.13818
Seating areas	−1.00	1.00	−0.0833	0.79296	−1.00	1.00	−0.4167	0.66856
Shading and canopy	−1.00	2.00	0.4167	0.90034	−2.00	1.00	0.5000	0.79772
Lightening	−1.00	2.00	0.5833	0.79296	−1.00	1.00	0.0833	0.79296
Signs	−2.00	2.00	0.0833	1.16450	−2.00	2.00	0.4167	1.08362
Planting	0	2.00	1.0000	0.85280	0	2.00	1.0833	0.79296
Sculpture & fountain	−2.00	2.00	0.1667	1.19342	−2.00	2.00	0	1.20605
Harmony between architectural style of different buildings	−2.00	2.00	−0.1667	1.40346	−2.00	2.00	0.5000	1.16775
Proportions of space	−2.00	2.00	0.0833	1.50504	−2.00	2.00	0.2500	1.28806
Facilities for disabled people	−2.00	1.00	0.0000	0.95346	−2.00	1.00	−0.5000	0.90453
Parking space	−1.00	1.00	0.2500	0.75378	−1.00	1.00	−0.0833	0.79296
Traffic management	−1.00	1.00	0.0000	0.85280	−1.00	1.00	0.3333	0.77850
Accessibility	−1.00	2.00	0.6667	0.98473	−1.00	2.00	0.8333	0.83485
Maintenance & cleaning	−1.00	2.00	0.3333	0.98473	−2.00	1.00	−0.2500	0.96531
Street clutter	−2.00	1.00	−0.3333	0.98473	−2.00	1.00	−0.5833	0.99620

Source Author

the importance of basic street facilities for disabled people for ease of access and other needed services. Proper paving, planting, maintenance, cleaning, traffic management, adequate parking spaces and street clutter in respondents' point of view are the main attributes that improve the livability of surveyed streets. Also traffic calming effects have a significant effect on streets' livability, and this reinforces the findings of (Appleyard & Lintell, 1972) and (Appleyard, 1981) that defined traffic impacts on streets' livability. Most results are common in studied streets, revealing that proper paving, planting, maintenance, cleaning, traffic management, adequate parking spaces and street clutter are the main physical attributes of El Sheikh Zayed City to have quality and livable streets. These street improvements will encourage users to walk, reinforcing the findings of Forsyth et al. (2008), Sauter and Huettenmoser (2008) and Park (2008) that had a significant impact on encouraging walkability through traffic management. Also, studies provide the same results when examining physical attributes in different urban settings, as testing the relation between planting and livability of streets that reinforces the results of (Layne, 2009) and (Bosselmann et al., 1999) emphasizes landscape significance in urban settings. To sum up, most research results confirm the previously reviewed literature with few exemptions. Lack of design, for signage distribution, has a minimal impact on the quality and livability of streets. This might be a result of users' ignorance about the negative effects of this problem. A livable concept, in general, and

livable streets especially and the significance of the physical environment were identified 40 years ago. Despite the fact that most of our streets are still not livable and miss basic physical attributes (e.g., adequate/good/proper planting) that improves and beautifies space quality and promotes street livability.

9 Conclusion

This study examines how people understand streets' physical attributes effect on Egypt's streetscapes. Accordingly, this study explores the provision of basic facilities such as proportions of space, maintenance, cleaning, parking and traffic management that will affect street livability in new urban communities. The research aims to define determinants of streets' livability in Egypt by analyzing streetscapes' physical attributes in two multifunctional streets in new urban communities. Research results revealed that streets' physical attributes promote livability and quality of city streets. Furthermore, some suggestions are proposed for improving the quality and livability in areas examined and also improving the efficiency of determinants for streets' livability. So, 15 physical attributes were concluded that determine streets' livability: paving, seating, shelter and canopy, good lighting, legibility of signs, planting (good greeneries condition), availability of water features, the proportion of spaces, harmony of architectural style between buildings, handicapped

Fig. 7 A comparison between the 15 physical attributes that determines streets' livability between the two surveyed streets (*Source* Author)

facilities, parking, ease of accessibility, traffic management, good maintenance, cleaning and street clutter. Hence, most of these attributes already existed, so problems found by structured observations and questionnaires are related to a lack of proper management of streets' and enhancing attributes quality. It is recommended to increase the pavement's width, repair/renovate deteriorated paving, and maintain and clean the surveyed areas. Moreover, it is highly recommended to collect/remove rubbish more frequently at different times of the day. Also providing the surveyed spaces

Fig. 7 (continued)

Traffic Mangement		
Maintainance & Cleaning		
Street Clutter		

Fig. 7 (continued)

with shading devices (e.g., plants that provide shade) and seating areas is a must; removing unnecessary signs, advertisements and street furniture to eliminate/reduce street clutter is highly recommended; allocating water features for the site if possible would increase streets' livability and quality; and lowering the speed of cars, especially in most crowded zones of streets (e.g., by using Woonerf, street bumps, road markings and signs). Also, redesigning, repairing and renovating the affected parts/zones will improve the quality of the physical environment. Moreover, providing more parking spaces and separating pedestrians' paths from cars movement needs improvement. Furthermore, pedestrian crossings need to be improved to be safer for users (e.g., the use of crossing buttons/devices). In addition, the location of surveyed streets' from a new monorail that is now under construction will ease the accessibility to public transportation in the future and will encourage walkability in El Sheikh Zayed streets. This will probably raise sufficiency, quality and reliability of public transport and also would reduce streets' congestions significantly because of lowering in use of private cars. Besides, providing cheap/free parking is highly recommended, that will avoid the issue of double parking and parking of cars on both sides of streets, causing streets' congestions and leading to less safe streets. There are some unused/neglected parts in studied streets that are recommended to be used as parking spaces.

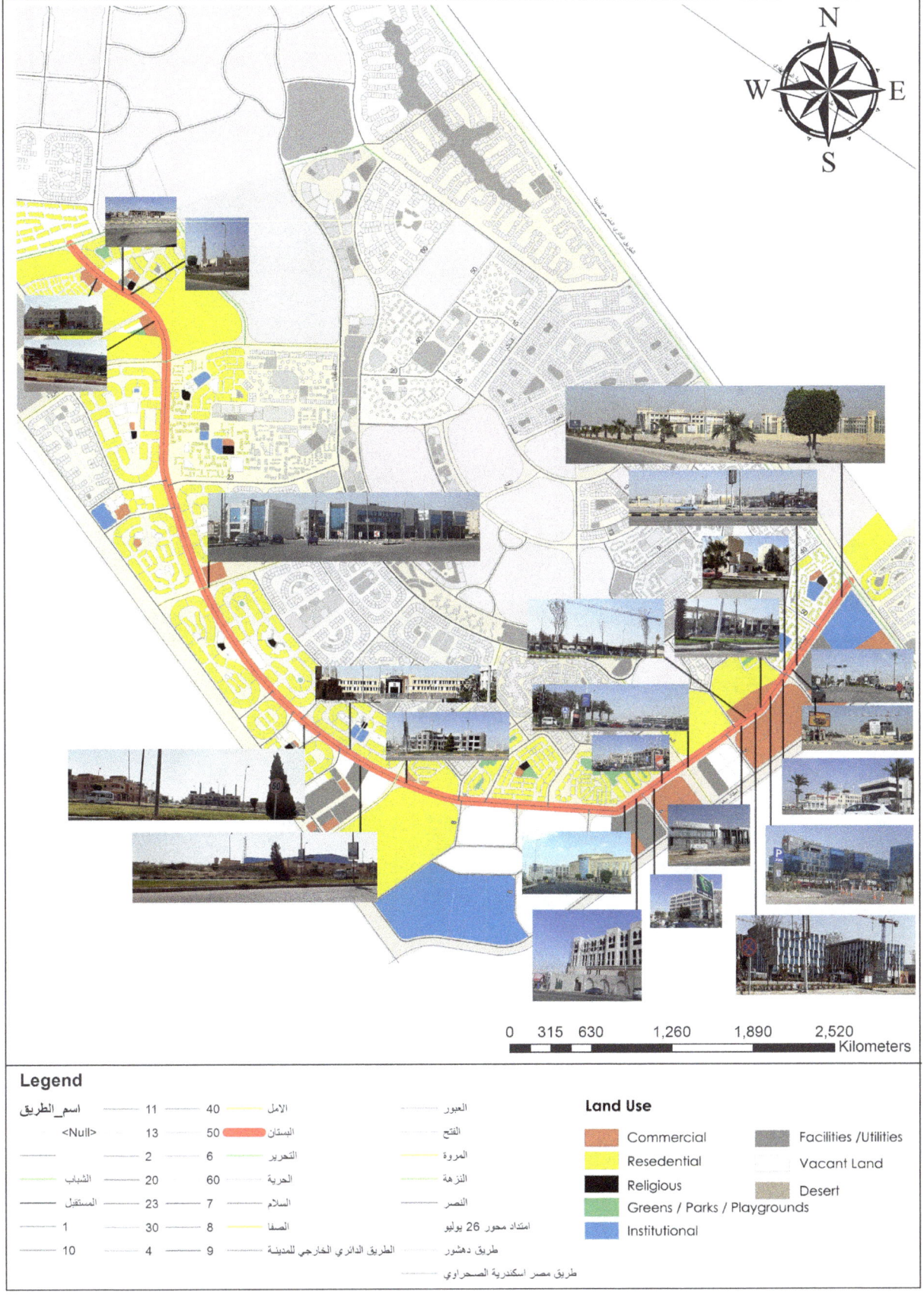

Fig. 8 Map showing commercial and services areas/zones/units located on both sides of studied street/area (El Bostan street) *Source* Author

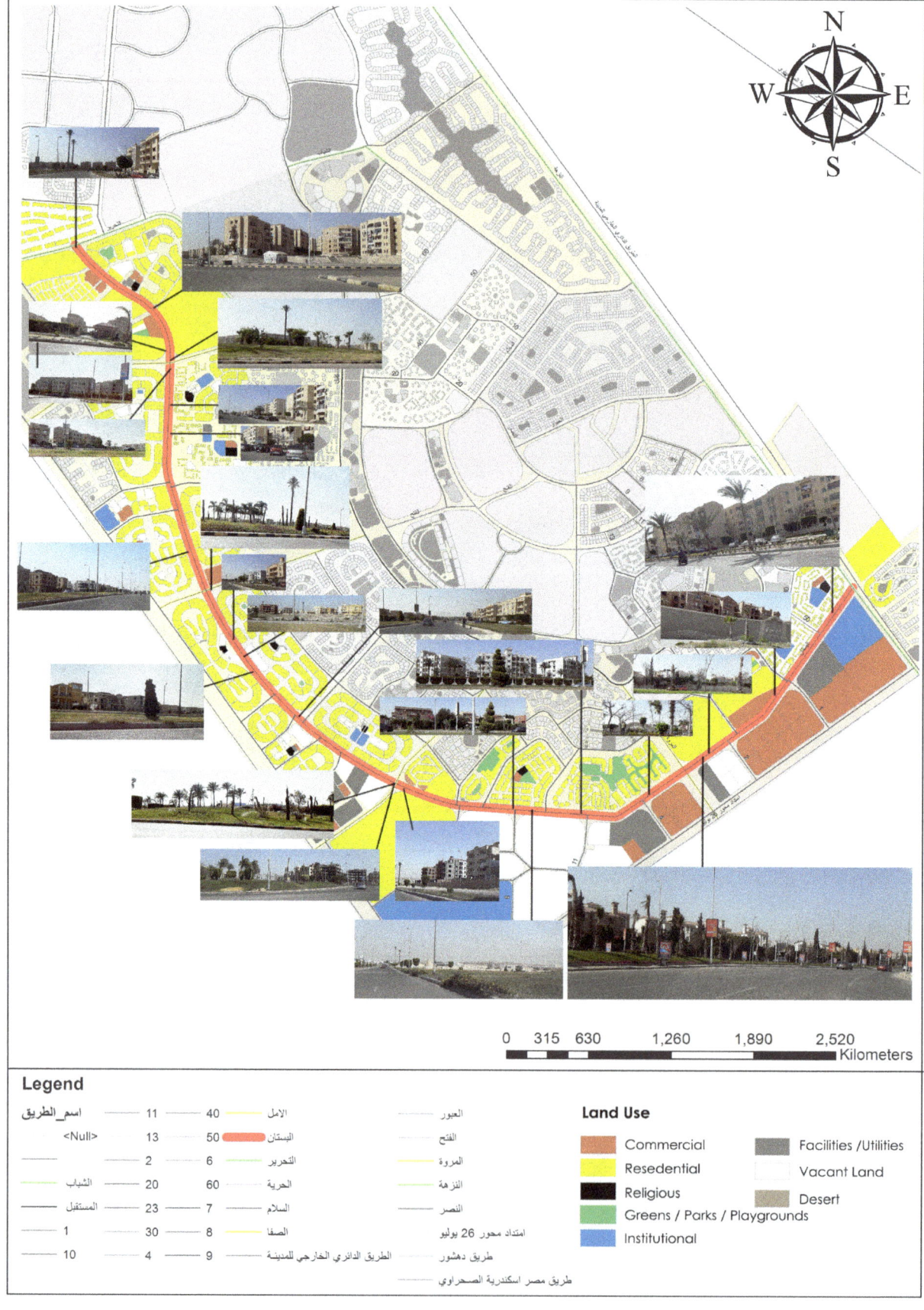

Fig. 9 Map showing residential areas/zones/units located on both sides of studied street/area (El Bostan street). *Source* Author

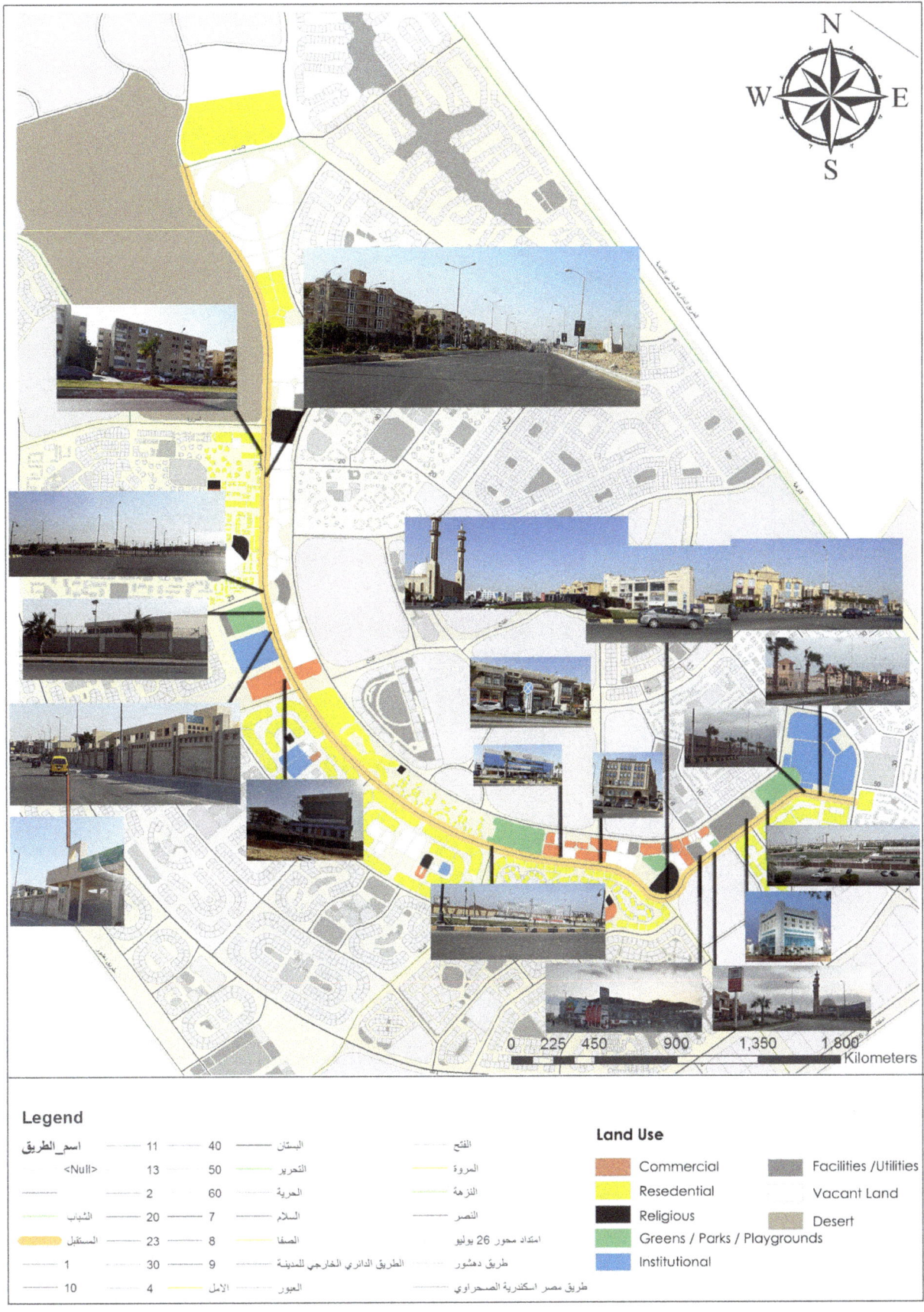

Fig. 10 Map showing commercial and services areas/zones/units located on both sides of studied street/area (El Mostakbal street). *Source* Author

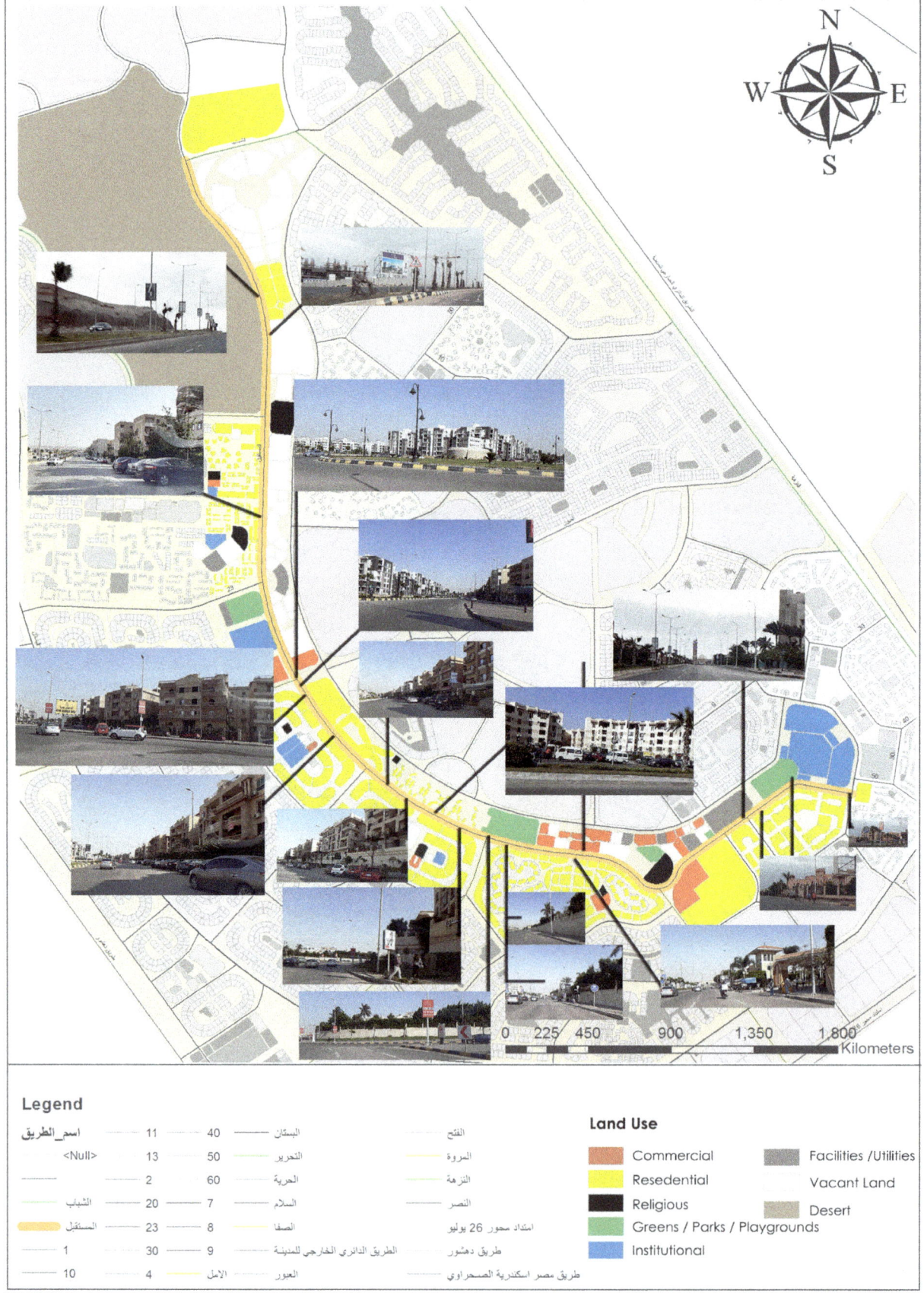

Fig. 11 Map showing residential areas/zones/units located on both sides of studied street/area (El Mostakbal street). *Source* Author

References

Abdel-Aziz A., Abdel-Salam H., El-Sayad Z. (2020). Reshaping the urban experience: Prospects for digital streetscape towards better livability in public spaces.

AbdelHafeez, M., Badran, E., & Nassar, U. (2010). *Principles to evaluate historic designed landscape of urban parks, case study of Al-Azhar Park.*

AbdelHafeez, M., Badran, E., & Nassar, U. (2013). *Experiential landscape as a tool to enhance behavioral response of users in urban parks, case study of Al-Azhar Park.*

Abdulmughni, M., Alzamil, S., & Alabed, M. (2021). The characteristics of livable streets: A study of physical aspects of two streets in Riyadh. *Journal of Urban Research, 39*(1), 43–58.

Ahmad, & Mahmoudi. (2015). *Determinants of livable streets in Malaysia: A study of physical attributes of two streets in Kuala Lumpur, Malaysia.*

Amr. (2015). *Sustainable landscape in university campus urban design.*

Appleyard, D. (1981). *Livable streets.* University of California Press.

Appleyard, B. (2017). The meaning of livable streets to schoolchildren: An image mapping study of the effects of traffic on children's cognitive development of spatial knowledge. *Journal of Transport & Health.*

Appleyard, D., & Lintell, M. (1972). The environmental quality of city streets: The residents' viewpoint. *Journal of the American Institute of Planners, 38*(2), 84–101. https://doi.org/10.1080/01944367208977410

Arefi, M., & Noha, N. (2021). Urban design. *Safety, Livability, & Accessibility.* https://doi.org/10.1057/s41289-021-00155-9

Aulia. (2016). *A framework for exploring livable community in residential environment. Case study.* Public Housing in Medan, Indonesia. www.sciencedirect.com

Bardo, J. W., Yeager, S. J., & Klingsporn, M. J. (1982). Preliminary assessment of format-specific central tendency and leniency error in summated rating scales. *Perceptual and Motor Skills, 54*(1), 227–234.

Bohl, C. C. (2002). *Place making: Developing town centers, main streets, and urban villages.* Urban Land Institute.

Bosselmann, P., Macdonald, E., & Kronemeyer, T. (1999). Livable streets revisited. *Journal of the American Planning Association, 65*(2), 168–180.

Carr, S., Francis, M., Rivlin, L., & Stone, A. (1992). *Public space.* Cambridge University Press.

Combs, T. S., & Pardo, C. F. (2021). Shifting streets COVID-19 mobility data: Findings from a global dataset and a research agenda for transport planning and policy. *Transportation Research Interdisciplinary Perspectives.* https://doi.org/10.1016/j.trip.2021.100322

Dooley, D. (2001). *Social research methods* (4th ed.). Prentice Hall.

Dumbaugh, E., & Gattis, J. (2005). Safe streets, livable streets. *Journal of the American Planning Association, 71*(3), 283–300.

Elsawy, A. A., Hany, M. A., & Saadallah, D. (2019). Assessing livability of residential streets—Case study: El-Attarin, Alexandria, Egypt. www.elsevier.com/locate/aej, www.sciencedirect.com

Forsyth, A., Hearst, M., Oakes, J. M., & Schmitz, K. H. (2008). Design and destinations: Factors influencing walking and total physical activity. *Urban Studies, 45*(9), 1973–1996.

Francis, M. (2003). Urban open space.

Francis, M. (1991). The making of democratic streets. In A. V. Moudon (Ed.) *Public streets for public use.* Columbia University Press.

Gehl, J. (2001). *Life between buildings: Using public space.* Danish Architectural Press.

Gössling, S. (2020). Why cities need to take road space from cars—And how this could be done. *Journal of Urban Design, 25*, 443–448.

Hartanti, N. B. (2012). Street as livable space in the urban settlement (p. 9). Trisakti University.

Hartig, T. A., Korpela, K., Evans, G. W., & Garling, T. (1997). A measure of restorative quality in environments. *Scandinavian Housing & Planning Research, 14*, 175–194.

Helmy. (2018). *Rethinking public space: Livability as a strategy for safe places.*

Jackson, J. B. (1985) Vernacular space. *Texas Architect, 35*(2).

Jacobs, J. (1961). *The death and life of great American Cities.* Random House Digital, Inc.

Jacobs, A. B. (1993). *Great streets.* MIT Press.

Jacobs, J. (1995). Uncanny Australia. *Ecumene, 2*(2), 171–183.

Jacobs, A., & Appleyard, D. (1987). Toward an urban design manifesto. *Journal of the American Planning Association, 53*(1), 112–120.

J Davis, C. (2002). *Proceedings of the Institution of Civil Engineers—Municipal Engineer* (Vol. Volume 151 Issue 3). https://doi.org/10.1680/muen.2002.151.3.231

J Davis, C. (2014). *Street design for all.* Public Realm Information and Advice Network (PRIAN). www.PublicRealm.info

Kaplan, R. M., & Bush, J. W. (1982). Health-related quality of life measurement for evaluation research and policy analysis. *Health Psychology, 1*(1), 61.

Layne, M. R. (2009). *Supporting intergenerational interaction: Affordance of urban public space.* Ph.D. Thesis, Graduate Faculty of North Carolina State University.

Leinberger, C. B. (2008). *The option of urbanism: Investing in a new American dream.* Island Press.

Lissitz, R. W., & Green, S. B. (1975). Effect of the number of scale points on reliability: A Monte Carlo approach. *Journal of Applied Psychology, 60*(1), 10.

Lynch, K. (1981). A theory of good city form. In *Chapter Two—Mapping method: Physical & spatial characteristic of environment.* MIT Press.

Mackett, R. L., Achuthan, K., & Titheridge, H. (2008). AMELIA: Making streets more accessible for people with mobility difficulties. *Urban Design International, 13*(2), 80–89.

Marcus, C. C., & Francis, C. (Eds.). (1998). *People places: Design guideline for urban open spaces.* John Wiley & Sons.

Mehta, V. (2014). Evaluating public space. *Journal of Urban Design, 19*(1), 53–88.

Moreno, C., Zaheer, A., Didier, C., Catherine, G., & Pratlong, F. (2021). Introducing the "15-Minute City": Sustainability, resilience and place identity in future post-pandemic cities. *Smart Cities.* https://www.mdpi.com/journal/smartcities.

Nabil, T., Samaan, M., & Nabil, M. (2021). Role of pedestrian streets in improving urban environment and livability in the city: Al-Tabou street – Baqubah city -Diyala- Iraq.

Nasar, J. L. (1988). *Environmental aesthetics: Theory, research and applications.* University Press.

Nassar, U. A. (2015). *Urban space design to enhance physical activities and motivate healthy social behavior in Cairo, Egypt* (p. 11). Istanbul, Turkey.

Park, S. (2008). Defining, measuring, and evaluating path walkability, and testing its impacts on transit users' mode choice and walking distance to the station. Ph.D. Thesis, University of California.

Places in the Making: MIT Report Highlights the "Virtuous Cycle of Place making." Retrieved May 16, 2017, from https://dusp.mit.edu/sites/dusp.mit.edu/files/attachments/project/mitdusp-places-in-the-making.pdf

Portella, A. A. (2007). Evaluating commercial signs in historic streetscapes: The effects of the control of advertising and signage on user's sense of environmental quality. PhD Thesis, Oxford Brookes University.

Rapoport, A. (1982). *The meaning of the built environment: A nonverbal communication approach*. Sage Publications.

Rehan, R. M. (2013). Sustainable streetscape as an effective tool in sustainable urban design. *Housing and Building National Research Center Journal, 9*(2), 173–186.

Rubenstein, H. M. (1992). *Pedestrian malls, streetscapes, and urban spaces*. John Wiley & Sons, Inc.

Sauter, D., & Huettenmoser, M. (2008). Liveable streets and social inclusion. *Urban Design International, 13*(2), 67–79.

Shalaby. (2004). *Sustainable urban landscapes in neighbourhoods*.

Shantz, A., & Latham, G. P. (2009). An exploratory field experiment of the effect of subconscious and conscious goals on employee performance. *Organizational Behavior and Human Decision Processes, 109*(1), 9–17.

Tawil, M., Reicher, C., Ramadan, K., & Aljafari, M. (2014). Towards more pedestrian friendly streets in Jordan: The case of Al Medina Street in Amman. *Canadian Center of Science and Education, 7* (1913–9071), 16. https://doi.org/10.5539/jsd.v7n2p144

Yassin. (2019). *Livable city: An approach to pedestrianization through tactical urbanism*. www.elsevier.com/locate/aej, www.sciencedirect.com

Yin, R. K. (2003). *Case study research design and methods* (3rd ed.). Sage publication.

Zalloom, B. (2017). Creating livable public spaces. In *The European Conference on Sustainability, Energy & the Environment 2017 Official Conference Proceedings,* Zarqa University, Jordan.

Zhan, D., Kwan, M. P., Zhang, W., Fan, J., Yu, J., & Dang, Y. (2018). Assessment and determinants of satisfaction with urban livability in China. *Cities, 79,* 92–101.

Using Cool Coating for Pavements, Asphalt, Façades and Building Roofs in the Urban Environment to Reduce the Summer Urban Heat Effect in Giza Square, Egypt

Ahmed Abdel Moneim Al Qattan

Abstract

Sunrays fall daily on the facades, roofs, pavements and asphalts, causing an increase in surface temperatures and Summer Urban Heat, especially during summer time and in countries of hot climates. This results in an increase in the rate of energy consumption as a result of the presence of reflective materials, such as Cool Coating for asphalt, facades, and roofs, which is manufactured and applied to buildings and on sidewalks to reduce the temperature emissions within the surrounding urban environment. This paper measures the success of using Cool Coating for pavements and asphalt as well as Cool Coating for building facades and roofs in Giza, Egypt. It also draws up a comparison between them to identify which is better, in terms of performance, in reducing air temperatures and their Physiological Equivalent Temperatures (PET), as an indicator of a Human physiological response to the urban heat-island effect in Giza, Egypt. A study is conducted using the ENVI-met computer simulation software, to mimic the microclimate in the urban environment and select the best results. The study concluded that the use of Cool coating for building facades and external surfaces, in Giza Square, exhibited excellent performance in reducing air temperature and PET value, optimizing climatic performance of existing buildings and urban areas, and helping to reduce energy use and impact of the summer urban heat.

Keywords

Cool coating • ENVI-met • Urban heat effect • Urban heat island • Physiological equivalent temperature • Urban environment

A. A. M. Al Qattan (✉)
Faculty of Engineering, Department of Architecture, Al-Azhar University, Cairo, Egypt
e-mail: dr_qattan@azhar.edu.eg

1 Introduction

The predicted increase in temperatures of urban areas around the world is triggered by the increased absorption of solar radiation by surfaces, such as asphalt and sidewalks, and buildings facades and roofs. The resulting Urban Heat Island (UHI) phenomenon is a major contributor to climate change in the built environment and poses a major threat to human health. It has encouraged researchers to allocate ways to reduce the absorbed radiation. Current Research and Literature has presented data on the extent and characteristics of the Urban Head Island Phenomenon in European and American Countries, all of which stated that the consumption of cooling energy could double due to an increase in surrounding temperatures in affected areas (Hassid et al., 2000; Santamouris et al., 2001). At the same time, the environmental quality of extremely hot areas continues to deteriorate, leading to an increase in pollution, which then affects the city's ecological footprint, creating a dangerous cycle (Santamouris et al., 2007).

As a result of technological innovations and their application (Santamouris et al., 2007), advanced materials for the urban environment that are capable of extinguishing, dissipating, and reversing heat and solar radiation on surfaces, such as facades of buildings, pavements and asphalt, have now become available. The world is now introduced to the uses of 'Cool materials' in the urban built environment, which have highly reflective surfaces, a high emission factor and deliver a high solar reflectance (Santamouris et al., 2011). These materials can be used in corridors, roads and other urban areas as well as on rooftops. According to various recent studies, they can contribute significantly to reducing surface temperatures (Doulos et al., 2004; Zinzi, 2010).

Urban areas in cities are subject to tough thermal conditions due to an increase in population. The concentration of heat in urban areas differs from that of surrounding rural ones, thus increasing energy consumption to achieve summer cooling. According to the UN predictions that the urban population will increase by 80% by 2050—from 3.3 billion people in 2007 to 6.4 billion people in 2020—(Zinzi, 2010), the main challenge for architects and urban planners, in the twenty-first century, will be improving the performance of urban areas by reducing the impact of summer urban heat (Jha et al., 2013). The goal of this paper is to propose an improvement of heat conditions and human comfort in urban areas in existing cities, especially public squares that are crowded with walkers. This is done by testing cool materials, both, on building roofs and facades or on asphalt floors and pavements and is best achieved through the use of the ENVI-met climate simulation program to support architects and city planners in implementing decisions that improve thermal performance and reduce the impacts of summer urban heat.

2   The Summer Urban Heat Effect

The urban summer heat effect indicates that a city's climate differs from, in terms of Temperature and Humidity, that of other surroundings during the summer season. This difference is due to the effect of urban environmental components on the city's weather in a degree that differs from the impacts on the surrounding environment (Nikolopoulou & Lykoudis, 2007) (Fig. 1). Therefore, the higher the effect of the summer urban heat on urban areas within cities, the higher the energy consumption and electricity needed for cooling.

3   The Albedo

Albedo is measured on a scale of 0–1, where 0 corresponds to surfaces or black bodies that absorb all incident radiation absorbing all light striking it and 1 means that a surface reflects all incoming energy. In other words, a 1 on the albedo scale means 100 reflection, whereas a 0 translates to no reflection (Givoni, 1998) (Fig. 2).

4   Physiological Equivalent Temperature (PET)

The physiological equivalent temperature (PET) was put forward by a German research group, headed by Peter Hoppe. According to Chirag Deb and Ramachandraiah Alur, it "is already included in the new VDI guidelines (German guidelines for urban and regional planners) for assessing the thermal component of microclimate" (Deb & Ramachandraiah, 2010). They define the PET as the air temperature at which, in a standard indoor environment (without wind and solar radiation), the heat resources of the human body are balanced with the same core and surface temperature as under the complex outdoor conditions assessed. The advantages of using PET are as follows (Deb & Ramachandraiah, 2010):

- It is a universal index and is irrespective of clothing (clo values) and metabolic activity (met values).
- It has a thermos physiological background and gives the real effect of the sensation of climate on human beings.
- It is measured in °C and so can be easily related to the collective experience.
- It does not rely on subjective measures.
- It is useful in both hot and colder climates (Knez et al., 2009).

5   Cool Coating for Façades and Roofs & Cool Coating for Pavements and Asphalt

Surfaces, such as Roofs or façades of any existing building, transfer heat into the building due to sun exposure, which, as a result, increases the Temperature within and around the building. Floor coverings, specifically pavements and asphalt streets, also transfer heat into building interiors and affect the climate around them. Several studies and research in the field of Environment and Climate Change resulted in recommendations on the use of cool coating to transform any roof or façade of an existing building into a cold surface.

29°C 85°F
RURAL FARMLAND

31–32°C 88–89°F
COMMERCIAL

33°C 92°F
DOWNTOWN URBAN

30–31°C 86–88°F
SUBURBAN RESIDENTIAL

30°C 86°F
PARKS

Fig. 1 The effect of the summer urban temperatures on urban areas and on surrounding rural areas (Nikolopoulou & Lykoudis, 2007)

Fig. 2 Albedo measured on a scale of 0–1(10)

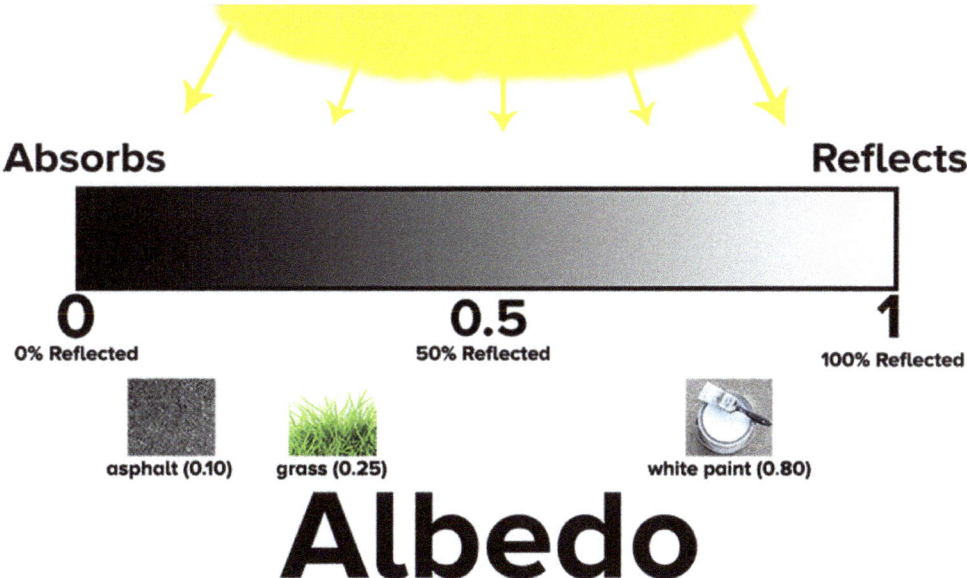

Fig. 3 The use of cool coating on surfaces such as roofs or facades of existing buildings as well as on flooring surfaces such as pavements and asphalt streets (Santamouris et al., 2012; Tukiran et al., 2016)

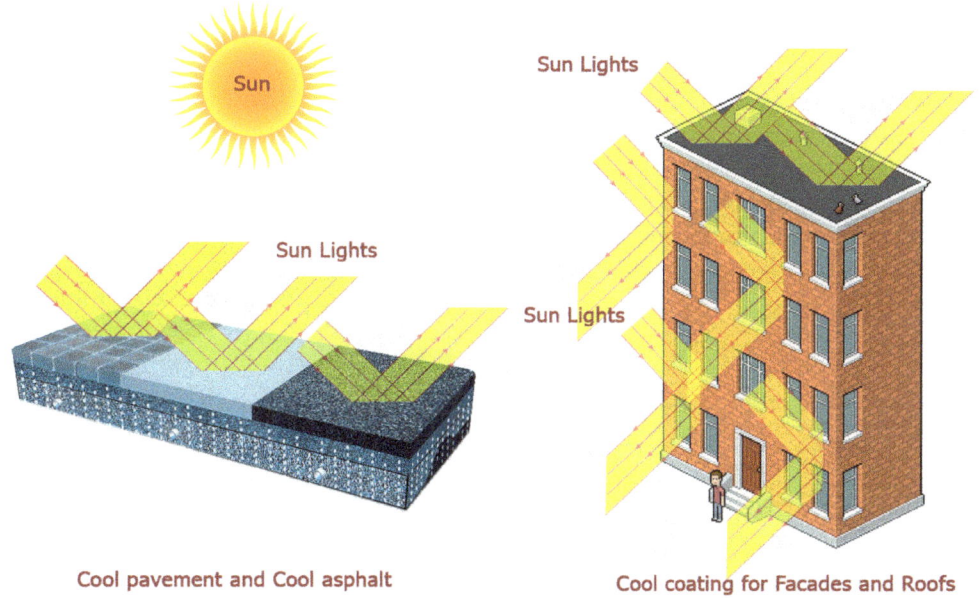

This also applied to pavements and asphalt streets where working sunlight would be reflected to decrease the surface temperature, essentially reducing the Temperature within the building itself along with that of the surrounding urban environment (Santamouris et al., 2012; Tukiran et al., 2016) (Fig. 3).

In order to reduce the effects of summer urban heat, there exist many types of cool coatings that offer the highest performance of sun reflection, according to Albedo and the Solar Reflection Index (SRI) and their physical properties, with respect to climate, roof quality, color, and more (Wan et al., 2012). Moreover, Table 1 illustrates the types of Cool Coatings and their respective properties and abilities of reflecting sunlight (Deb & Ramachandraiah, 2010) and reducing roof and facade temperatures.

6 The Methodology

This research paper aims to analyze the use of cool roof coatings on façades and roofs and pavements and asphalt, as well as identify the better of the two, in terms of solar radiation absorption during summer peak times in Egypt, particularly around the Giza Square. It aims to reduce the

Table 1 Showing the types of cool coating, their properties and ability to reflect sunlight according to albedo and the solar reflection index (SRI) and its physical properties in relation to climate, roof quality, colors, etc. (Deb & Ramachandraiah, 2010)

Example SRI values for generic roof materials	Solar reflectance	Infrared emittance	Temp rise (°)	SRI
Gray EPDM	0.23	0.87	38	21
Gray Asphalt Shingle	0.22	0.91	37	22
Unpainted Cement Tile	0.25	0.90	36	25
White Granular Surface Bitumen	0.26	0.92	35	28
Red Clay Tile	0.33	0.90	32	36
Light Gravel on Built-up Roof	0.34	0.90	32	37
Aluminum	0.61	0.25	27	56
White Coated Gravel on Built-up Roofing	0.65	0.90	16	79
White Coating on Metal Roof	0.67	0.85	16	82
White EPDM	0.69	0.87	14	84
White Cement Tile	0.73	0.90	12	90
White Coating-1 coat 8 mils	0.80	0.91	8	100
PVC White	0.83	0.92	6	104
White Coating-1 coat 20 mils	0.85	0.91	5	107

impacts of summer heat by testing cold materials, both on the surface and the facades of buildings or on asphalt floors and pavements. Their thermal performance is tested by using the ENVI-met computer program, for climate simulation, in order to support architects and city planners in making decisions and in their efforts to improve thermal performance and reduce physiological equivalent temperatures (PET) as shown in Fig. 4.

7 Case Study Framework

To prove the effectiveness of cold surfaces and translate that into practical application, whether on the roofs and facades of buildings or on the floors of pavements and asphalt, the simulation modeled the conditions in Giza Square in Cairo, Egypt, on June 23, 2018, at the peak of daylight-saving time. The steps taken for this specific case study are as follows:

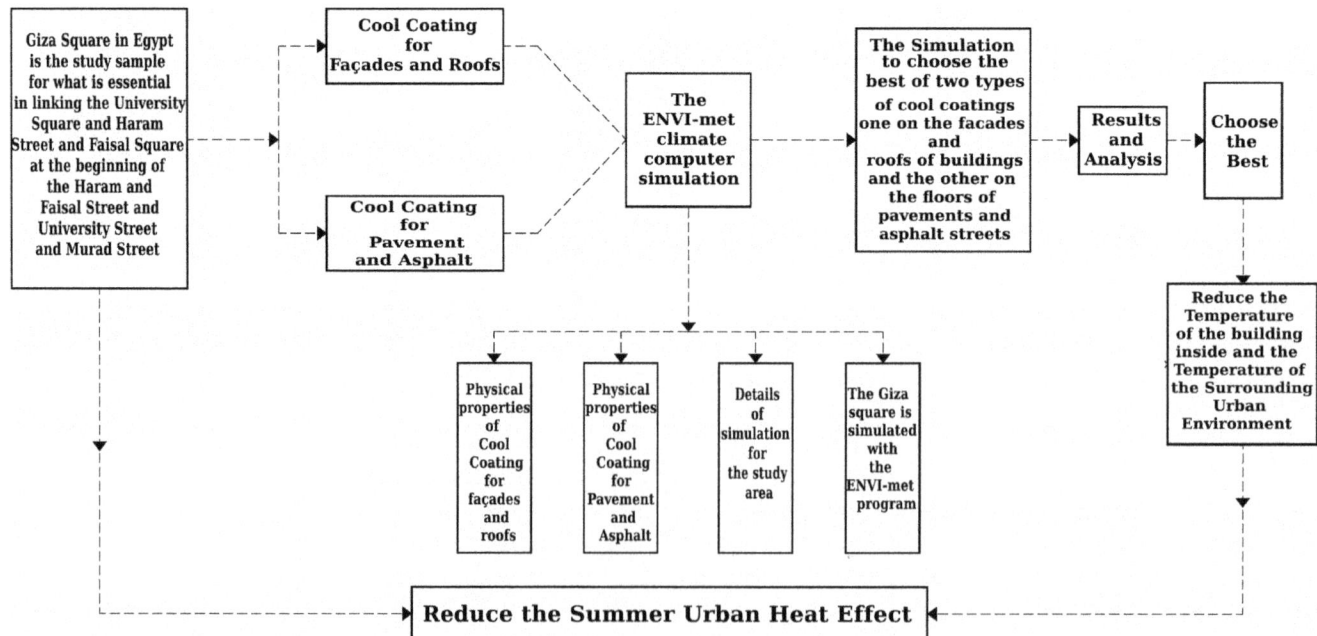

Fig. 4 The use of Envi-met climate computer simulation to choose the best of the two cold paints, (1) facades and roofs of buildings or (2) pavement floors and asphalt streets, to reduce the absorption of solar flares and then reduce the summer urban heat effect

1. Insertion of the Giza square drawings in the ENVI-met simulation program inclusive of the number of buildings, their areas, and the number of floors.
2. Insertion of additional details for the simulation, of the investigated area, in the ENVI-met program, such as starting time, wind speed, wind direction, temperature, relative humidity in 1.5 m, relative humidity in 1.5 m, building interior temperature, etc.
3. Input Physical properties of Cool Coating for façades and roofs in the ENVI-met program including default thickness, absorption, reflection, emissivity, thermal conductivity, and density.
4. Input Physical properties of pavements and asphalt in the ENVI-met program, such as default thickness, absorption, reflection, emissivity, thermal conductivity, and density.

To confirm the results, three simulations were performed, running for more than four hours for each of the three cases presented. The results were extracted and analyzed in a table to lay out the best results for reducing air temperatures and the Physiological Equivalent Temperature (PET) value through the use of cool Coating on facades and exteriors of the surrounding buildings, as well as on pavements and asphalt floors, in Giza Square, Egypt,

8 The Experiment

A. Climate Performance Simulation of the Urban Environment

The ENVI-met software can simulate climates in urban environments to evaluate the effects of climate change, vegetation, architecture, and materials. Furthermore, the software facilitates the creation of sustainable living conditions in a continuously changing environment. With ENVI-met's interactive tools, we can dive into any aspect of the complex microclimate and examine how the performance of different designs. The modeling software is applicable worldwide, from the Tropics to the Polar Regions. With more than 3,000 independent studies, ENVI-met is the most evaluated microclimate model available, proving its capabilities through the accurate simulations carried out of the outdoor microclimate for any place around the world.

B. Case study

The Giza Square in Egypt has been chosen for this study due to its significance in linking the University Square with Al

Haram Street, Faisal Street, University Street and Murad Street. Among the most important features of the area are the Giza Bridge and the Giza Tunnel established in 2008. The area houses the headquarters of the Faculty of Veterinary Medicine of Cairo University, the Giza Complex, a bus station and a service car park. Pedestrians and vehicle drivers face tremendous difficulties in bypassing this vital and busy area as shown in Fig. 5. In other words, the square was selected for the case study and for investigation mainly due to its high density. The day selected for simulation was June 23, 2018, which accompanied the necessary details about the area, as shown in Table 2.

The study sample selected is in the form of a square with a length of 180 m, a 180 m width and a surface area of 34,400 square meters. The sample contains a group of buildings, as shown in Fig. 5, that include the Egyptian Telecomm Company building, the Giza Mosque, the services building, the Nile Hospital, and five residential buildings around the area across from a taxi stand/car park offering public transportation.

C. Cool Coating for façades and roofs (Höppe, 1999)

There exist many cool coating materials used on building roofs and facades. However, the cool coating material with a 0.05 thermal conductivity, an albedo index of 0.85, was chosen as shown in Tables 1 and 3.

D. Cool Coating for pavement and asphalt (Wang et al., 2015)

There are also many cool coating materials available for pavement floors and asphalt streets that reduce the surface temperature. Cool coating materials with an Albedo value of 0.40 and characteristics are shown in Table 4.

Table 5 shows the properties of cold coatings for the three cases: the Base Case, the Cool Coating for façades and roofs case, and the Cool Coating for pavement and asphalt case (Kinouchi et al., 2003; Tsoka et al., 2019; Wang et al., 2015), all of which will be inserted into the ENVI-met program.

The Giza square is simulated in the ENVI-met program as shown in Fig. 6, which reflects the physical properties of cool coatings of the three cases as entered into the program, to prepare the study sample for the simulation process, as follows: the Base Case, the Cool Coating case for façades and roofs, the Cool Coating case for pavements and asphalt, building heights and finishing materials.

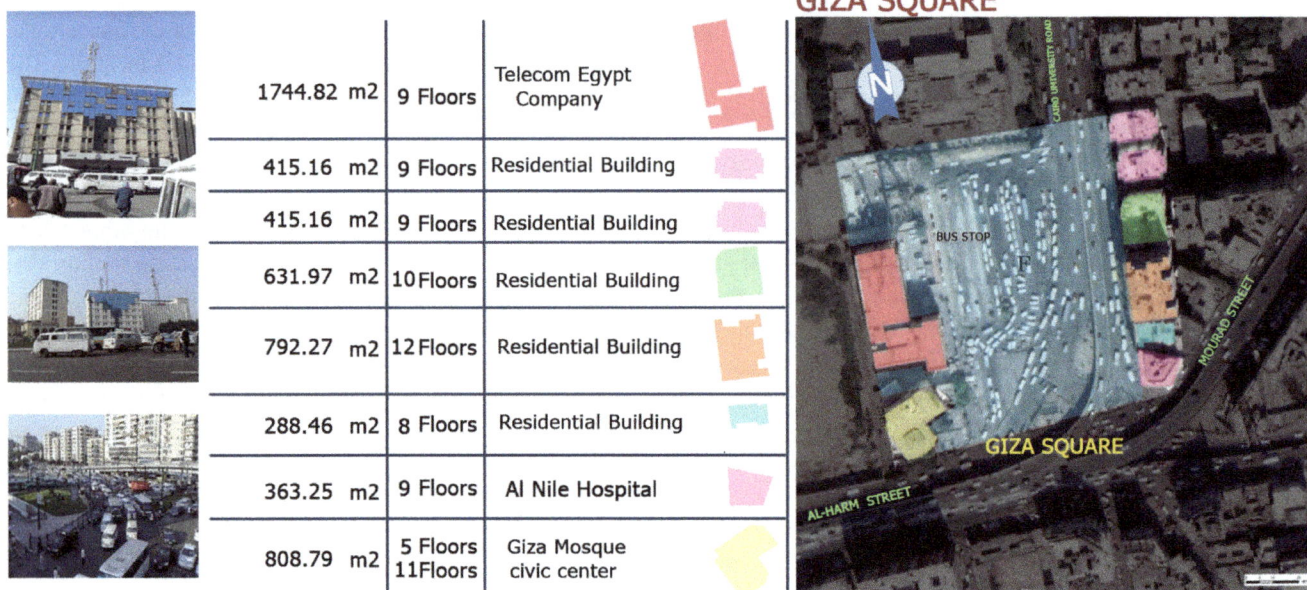

Fig. 5 The Giza Square in Egypt is the study sample because of its importance in linking the University Square, Haram Street and Faisal street. The table highlights the data inserted into the simulation program lsuch as number of buildings, their areas and the number of floors of each building

Table 2 Simulation details for the study area

23 June 2018	
Starting time	13:00
Wind speed	2 m/s
Wind direction	Northwest
Temperature	30 °C
Relative humidity in 1.5 m	55%
Building interior temperature	28 °C

Table 3 Physical properties of cool coating for façades and roofs used in the simulation program (Höppe, 1999)

Building material	Thermal conductivity (W/m K)	Thickness (mm)	Conduction resistance (m^2 K/W)	Density (kg/m^3)	Specific heat (J/kg K)
Air[b]	0.03	–	–	1.23	1008
Concrete[b]	0.65	100[a]	0.15	2450	840
Cool coating[a]	0.05[a]	0.5[a]	0.01[a]	1053[a]	[c]
Glazing[b]	0.70	3.5[a]	0.01	–	–
Plastered concrete block[b]	0.10	125[a]	0.12	800	920
Plaster[b]	0.25	10[a]	0.04	850	1000
Wood[b]	0.15	50[a]	0.33	608	1630

[a] Thermophysical properties are obtained from the manufacturers
[b] Thermophysical properties are taken from the Energy Plus material dataset (2013)
[c] Thermal mass, mC, is O J/K due to negligible mass of coating

Table 4 Physical properties of Cool Coating for pavement floors and asphalt (Wang et al., 2015)

		Surface Albedo	Thickness (m)	Volumetric He at capacity (kj/m³ k)	He at capacity per unit[a] (kj/m³ k)
Asphalt road	Asphalt	0.2	0.250	2251	623
	Loam soil	–	0.050	1212	
Cool pavement	Concrete	0.4	0.035	2083	398
	Sand	–	0.015	1463	
	Loam soil	–	0.250	1212	

[a]Per unit: 1 m × 1 m × 0.3 m (length × width × thickness)

Table 5 The physical Properties of cool coatings for the three base cases (Kinouchi et al., 2003; Tsoka et al., 2019; Wang et al., 2015)

Cool coating for pavement and asphalt	Cool coating for façades and roofs	Base case	Properties of materials
0.035	0.05	0.25	Default thickness
0.5	0.70	0.20	Absorption
0.4	0.85	0.12	Reflection
0.90	0.90	0.90	Emissivity
0.50	0.05	0.75	Thermal conductivity
792	1035	2243	Density

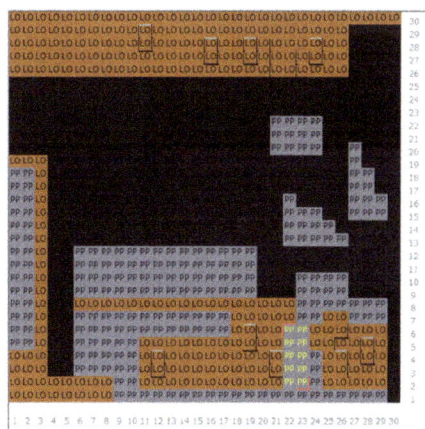

Fig. 6 The study sample—Giza Square in Egypt after simulation on the ENVI-met program and setting building heights and finishing materials

9 Results and Analysis

After conducting the simulation process for the Envi-met program, the maps shown for Air Temperature and results of the Physiological equivalent temperature (PET) were unloaded as an urban sector with height of 1.5 m and a general location. Thus, the results shown in the following figures were reached:

A. **Air Temperature sector in Giza Square in Egypt on June 23, 2018, at 1:00 PM**. as shown in Fig. 7

A-A. **General location of air temperatures in Giza Square in Egypt on June 23, 2018 at 1:00 PM**. as shown in Fig. 8

B. **An Urban sector for Physiological equivalent temperature (PET) in Giza Square in Egypt on June 23, 2018 at 1:00 PM**. as shown in Fig. 9

B-A. **Public site for a Physiological Equivalent Temperature (PET) in Giza Square in Egypt on June 23, 2018 at 1:00 PM**. as shown in Fig. 10

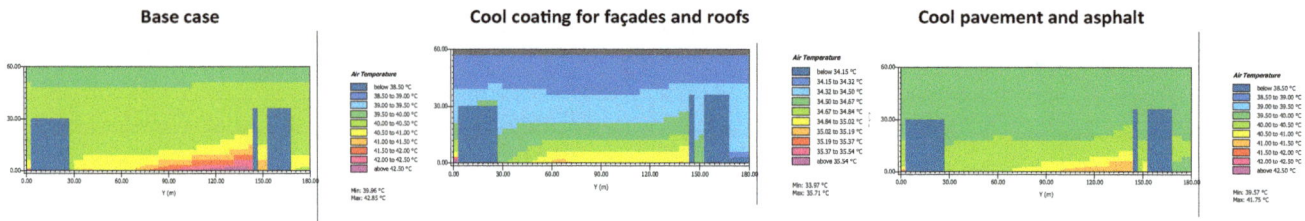

Fig. 7 As for the urban sector: air temperatures in the base case ranges between 39.96 and 42.85 °C. In cool coating for facades and roofs, the air temperature ranges between 33.97 and 35.71 °C. Finally, in cool pavements and asphalt, the air temperature ranges between 39.57 and 41.75 °C

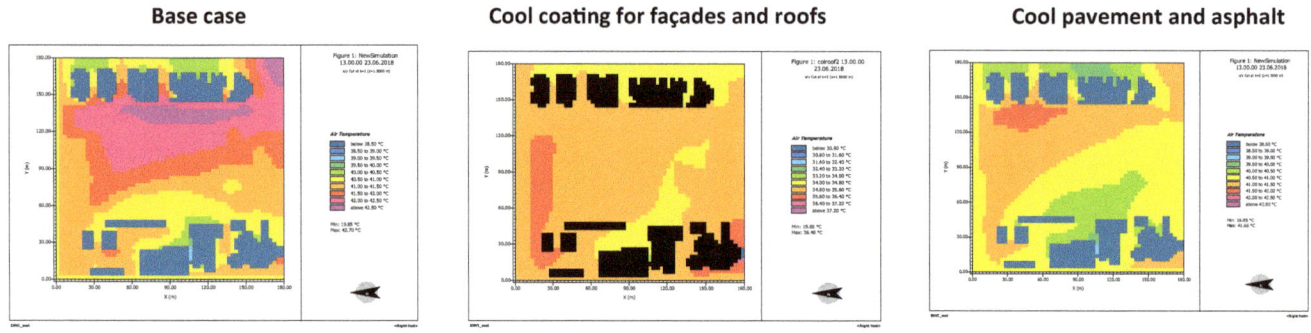

Fig. 8 A general location for the air temperatures in the base case where they range between 19.85 and 36.48 °C. As for the cool pavement and asphalt, the air temperature ranges between 19.85 and 41.66 °C

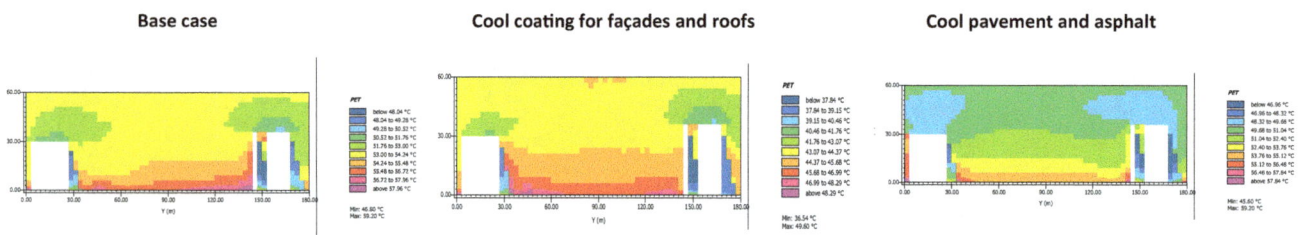

Fig. 9 An urban sector of the Physiological Equivalent Temperature (PET) in the base case of the study sample states that an individual's sense ranges between 46.80 and 59.20 °C while, in the cool coating used for facades and roofs, the physiological temperature of an individual's sense ranges between 36.54 and 49.60 °C. Finally, the cool pavement and asphalt have a physiological temperature of the individual's sense ranges between 45.60 and 59.20 °C

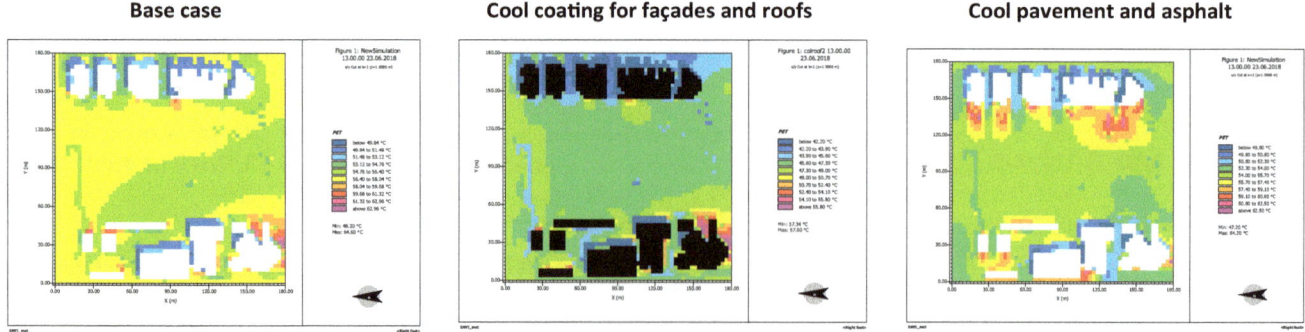

Fig. 10 A general location for Physiological equivalent temperature (PET) in the base case, where the physiological temperature of the individual's sense ranges between 48.20 and 64.60 °C, while in cool coating for facades and roofs, where the physiological temperature of the individual's sense ranges between 37.34 and 57.00 °C. Finally, the cool pavement and asphalt where the physiological temperature of the individual's sense ranges between 47.20 and 64.20 °C

Fig. 11 a The layout of air temperature in Giza square, Egypt, on June 23, 2018, at one in the afternoon. **b** Section of air temperature in Giza square, Egypt, on June 23, 2018, at one in the afternoon. **c** The layout of the Physiological equivalent temperature (PET) in Giza Square, Egypt on June 23, 2018, at 1:00 pm. **d** Section of Physiological equivalent temperature (PET) section in Giza Square, Egypt on June 23, 2018 at 1 pm

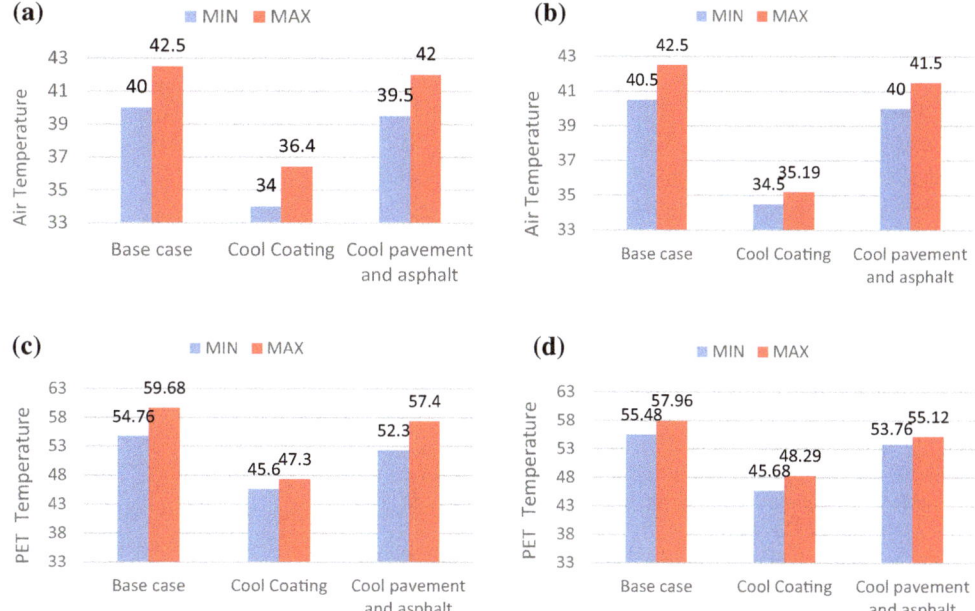

In counting the illustrative maps of the three cases: (1) Cool pavement and (2) asphalt Base case, (3) Cool coating for façades of Air temperature and Physiological equivalent temperature (PET). After performing a simulation for each case separately, we see that the best results of PETs are those found on the facades and roofs when compared to coating cold material on asphalt and pavement floors in Giza Square, Egypt. Nonetheless, developers and Urban Planners can decide whether to use these cold materials for coating in Urban Areas on asphalt floors and sidewalks or on roofs and facades of buildings.

According to the results (Fig. 11), it is recommended to use cold materials coating on the roofs and facades of buildings than coating them on asphalt floors and sidewalks in the event that there are fields with asphalt floor surfaces and large sidewalks with multiple buildings around them with the same climate specified in the field study.

10 Conclusion

1. The resulting air temperatures in the general location for the base case range between 40 and 42.5 °C. Meanwhile, the use of Cool Coating resulted in less than 6 °C, with a record of 34–36.4 °C. In the case of Cool pavement and asphalt use, the resulting air temperature was less than that of the base case by half a degree where a range of 39.5–42 °C was recorded.

2. The resulting air temperatures in the Base case section ranges between 40.5 and 42.5 °C, while in Cool Coating, it is less than 6 °C at the minimum, whereas the maximum is less by 7 °C with a range of 34.5–35.19 °C. The base case, where cool coating was used for pavements and asphalt, a range of 40–41.5 °C was recorded along with half a degree at the minimum and a maximum of one degree.

3. The Physiological equivalent temperature (PET) resulting from the general location for the Base case ranged between 54.76 and 59.68 °C, while with Cool Coating it resulted in 9° less as a minimum and by 12° as a maximum, recording a range between 14.6 and 47.3 °C. However, for Pavements and asphalt, the PET resulted in half a degree lower as a minimum and by one degree as a maximum resulting a recording between 40 and 41.5 °C.

4. The resulting Physiological equivalent temperature (PET) in the general location for the Base case ranged between 55.48 and 57.96 °C. In cool Coating, it resulted in less than 10 °C with a score of 45.68–48.29 °C. In the case of Cool pavements and asphalt, PET came out as less than that of the Base case by 2 °C and recorded a range between 53.76 and 55.12 °C.

We conclude from the results of the simulation for the three cases is that the Cool Coating case comes first in reducing air temperature by 6–7 °C. According to the Physiological equivalent temperature (PET), the air temperature decreases by 10° for the Base case while the Cool pavement and asphalt case comes second as the temperature is reduced by one to two degrees from the Base case.

The objective of the research was to reach the best results in reducing the air temperature and the Physiological equivalent temperature (PET) in Giza square, Egypt, by using Cool Coating on the facades and roofs of the buildings. The square is considered one of the most iconic

Egyptian centers, used by the spectators and vehicles and can achieve lower temperatures than that of the base case to help reduce energy consumption and mitigate the effects of the Urban Heat Island (UHI). In conclusion, the analysis of the results indicates that the use of cool Coating for the facades and exterior surfaces of the surrounding buildings and for the floors of pavements and asphalt proved most effective in reducing the air temperature and Physiological Equivalent Temperature (PET) in Giza Square, Egypt.

References

Deb, C., & Ramachandraiah, A. (2010). The significance of physiological equivalent temperature (PET) in outdoor thermal comfort studies. *International Journal of Engineering Science and Technology, 2*(7), 2825–2828.

Doulos, L., Santamouris, M., & Livada, I. (2004). Passive cooling of outdoor urban spaces. The role of materials. *Solar Energy, 77*(2), 231–249.

Givoni, B. (1998). *Climate considerations in building and urban design.* Wiley.

Hassid, S., Santamouris, M. N. A. N. C., Papanikolaou, N., Linardi, A., Klitsikas, N., Georgakis, C., & Assimakopoulos, D. N. (2000). The effect of the Athens heat island on air conditioning load. *Energy and Buildings, 32*(2), 131–141.

Höppe, P. (1999). The physiological equivalent temperature—A universal index for the biometeorological assessment of the thermal environment. *International Journal of Biometeorology, 43*(2), 71–75.

Jha, A. K., Miner, T. W., & Stanton-Geddes, Z. (Eds.). (2013). *Building urban resilience: Principles, tools, and practice.* World Bank Publications.

Kinouchi, T., Yoshinaka, T., Fukae, N., & Kanda, M. (2003). 4.7 Development of cool pavement with dark colored high albedo coating. *Target, 50*(40), 40.

Knez, I., Thorsson, S., Eliasson, I., & Lindberg, F. (2009). Psychological mechanisms in outdoor place and weather assessment: Towards a conceptual model. *International Journal of Biometeorology, 53*(1), 101–111.

Nikolopoulou, M., & Lykoudis, S. (2007). Use of outdoor spaces and microclimate in a Mediterranean urban area. *Building and Environment, 42*(10), 3691–3707.

Santamouris, M., Papanikolaou, N., Livada, I., Koronakis, I., Georgakis, C., Argiriou, A., & Assimakopoulos, D. N. (2001). On the impact of urban climate on the energy consumption of buildings. *Solar Energy, 70*(3), 201–216.

Santamouris, M., Paraponiaris, K., & Mihalakakou, G. (2007a). Estimating the ecological footprint of the heat island effect over Athens, Greece. *Climatic Change, 80*(3), 265–276.

Santamouris, M., Pavlou, K., Synnefa, A., Niachou, K., & Kolokotsa, D. (2007b). Recent progress on passive cooling techniques: Advanced technological developments to improve survivability levels in low-income households. *Energy and Buildings, 39*(7), 859–866.

Santamouris, M., Synnefa, A., & Karlessi, T. (2011). Using advanced cool materials in the urban built environment to mitigate heat islands and improve thermal comfort conditions. *Solar Energy, 85*(12), 3085–3102.

Santamouris, M., Gaitani, N., Spanou, A., Saliari, M., Giannopoulou, K., Vasilakopoulou, K., & Kardomateas, T. (2012). Using cool paving materials to improve microclimate of urban areas—Design realization and results of the flisvos project. *Building and Environment, 53*, 128–136.

Tsoka, S., Tsikaloudaki, K., & Theodosiou, T. (2019). Coupling a building energy simulation tool with a microclimate model to assess the impact of cool pavements on the building's energy performance application in a dense residential area. *Sustainability, 11*(9), 2519.

Tukiran, J. M., Ariffin, J., & Ghani, A. N. A. (2016). Comparison on colored coating for asphalt and concrete pavement based on thermal performance and cooling effect. *Jurnal Teknologi, 78*(5).

Wan, W. C., Hien, W. N., Ping, T. P., & Aloysius, A. Z. W. (2012). A study on the effectiveness of heat mitigating pavement coatings in Singapore. *Journal of Heat Island Institute International, 7*(2).

Wang, Y., Berardi, U., & Akbari, H. (2015). Urban Heat Island effect in the city of Toronto: An analysis of the outdoor thermal comfort CCTC.

Zingre, K. T. (2014). *Building energy savings using high-albedo-high-emittance (cool) roof materials.* Doctoral dissertation, Ph.D. thesis, Nanyang Technological University, Singapore.

Zinzi, M. (2010). Cool materials and cool roofs: Potentialities in Mediterranean buildings. *Advances in Building Energy Research, 4*(1), 201–266.

Towards Sustainability in Resettlement Plans, the Necessary Conditions and Their Interplay: The Case of Mumbai

Kanchan Sen Sharma

Abstract

Researchers have developed models, metaphors, and analytical tools over the years to evaluate and analyze policies. But the precise judgement of effectiveness with definite units of analysis is still a challenge. This study addresses the very definition of sustainability in involuntary resettlement in development and infrastructure projects in Indian context and attempts to establish definite terms for sustainability of a resettlement plan (RP). There is an absence of well-defined units and analytical methods for impacts of resettlement on the affected families. As per international institutions of investment and finance like World Bank (WB) and Asian Development Bank (ADB), this is the primary reason behind implementation discrepancies. The study makes use of expert opinion surveys and pair-wise comparison matrices to come to a few conclusions.

Keywords

Land acquisition • Rehabilitation • Sustainable resettlement • Period of sustainability • Compensation

1 Introduction

The basic definition of sustainability is the capacity to be maintained at a certain level. The level, rate, or rank of the minimum requirement of maintenance may be subjective. The sustainability discussed in this article is the capacity of people displaced/resettled by a development/infrastructure project in Mumbai to live a decent quality of life for a certain amount of time. This article attempts to establish some groundwork for evaluating the sustainability of a resettlement plan (RP).

Ideally, the resettlement process in which the project affected families (PAF) are not driven to impoverishment and the compensation is based on principles of equity and equivalence is called sustainable resettlement (OECD-FAO, 2009). But the issue arises with the vagueness of these terms and the lack of definition and standards regarding the sustainability of RPs across the nation. The policyscape (Mettler, 2016) of Land Acquisition, rehabilitation and resettlement (LARR) in India has been an interesting and fertile ground for legal and political contestation over the years. The major causes of conflict as per researchers have been the lack of sensitivity towards Project affected population (PAP), insufficient compensation, misuse of the term public purpose, top-down planning approach, procedural delays, land being left unused after acquisition, and people left unrecognized as entitled PAP (Wahi, 2017). It is quite evident that LARR policies have often been used as a tool to catalyze urban development. LARR tends to create critical imbalance of power, and it has been found that legal reforms of the act are a necessity but not the sufficient to reduce this imbalance. The series of amendments this act has been through displays an increasing number of court case litigations with each amendment. Despite of the legal reforms of the act, there is a constant dissatisfaction among the affected people. Irrespective of what the Act directs the authorities, the supply of compensation and resettlement usually falls short. In Indian megacities, there is a constant attempt to regularize the irregular. Unplanned settlements are considered as urban sprawl resulting from neglect and overpopulation. Studies conducted in Shivaji Nagar of Mumbai revealed that the local governments and municipal corporations are more interested in putting labels of 'legal' and 'illegal' on Mumbai's population rather than proceeding with a 'provide for all' approach (Björkman, 2014) which results in extreme neglect, illegalisation and degradation of the people affected by these convulsions of policies and acts. When it comes to Mumbai, the Mumbai Metropolitan Region (MMR) which was planned with inspirations from several global cities like London, New York, Tokyo, etc. (Sassen, 2001) resulted into chaos and

K. S. Sharma (✉)
CUSE, Indian Institute of Technology, Bombay, Powai, Mumbai, 400076, Maharashtra, India
e-mail: kanchansensharma@gmail.com

© The Author(s), under exclusive license to Springer Nature Switzerland AG 2022
C. Piselli et al. (eds.), *Innovating Strategies and Solutions for Urban Performance and Regeneration*,
Advances in Science, Technology & Innovation, https://doi.org/10.1007/978-3-030-98187-7_10

confusion in the urban development of the city's outskirts (Phadke, 2013). This again brings us to the conclusion that the cost of hasty enthusiasm for development is paid by the PAP through dislocation and loss of livelihood. This article will highlight the problem areas of India's resettlement polices with few implementation examples from Mumbai and then try to make a few statements regarding sustainability standards through both literature and opinion surveys.

2   LARR in India and the Litigations

It is important to understand the policy structure of the Land Acquisition Act that stood stubbornly for so long. The authoritarian outlook of the government towards the citizens since the colonial period took many years to noticeably change. Land Acquisition Act (LAA) was initially practiced to abolish the system of *Zamindari* in the country which was basically the practice of stocking up of land by the rich. The purpose was to fairly allocate the land to the poor through a just and transparent procedure. Post-independence, the Act tuned into the government's primary tool to implement pro-development resolutions (Ramanathan, 2011). The process of acquiring land as per this Act was a linear process. With the basic assumption that government has the power of Eminent Domain over all the privately-owned plots, this Act laid out a conventional path of acquisition. The public notice in the official gazette was the first step of acquisition. It had mainly these purposes—Informing the public about intentions of the government and cautioning them to not invest further in such land, making it legal for the land to be inspected by an authorized person even without the owner's permission and warning the general public to not show any interest in such properties (Sinha & Singh, 2016). As soon as the landowners are aware of the acquisition it is natural for them to raise objections. The Act provided them a chance to raise their objections through a written document addressed to the collector, which was considered as a substantial right. But the time span provided to do so was merely 30 days. After hearing these objections, the collectors would submit a report to the government which contained all their recommendations and the records of the proceedings. On the basis of this report the government would decide whether to go ahead with the acquisition or not. After and if the decision was taken positively towards acquisition, a declaration would be issued by the government which consisted of all the details of the acquisition. These details included the precise area measurements, the public purpose, and conclusive statements. This would be followed by a notice issued to the interested parties (landowners and other) followed by an enquiry and announcement of the award and

compensation. The compensation would have to be paid within two years post which the government would acquire the land. The compensation was calculated on the basis of market value. An interest of 12% per annum was to be paid if the compensation wasn't provided even after the acquisition. A solatium of 15% (later increased to 30%) was to be provided as well.

Initially there were no rights for claims for original landowners after the acquisition was complete. However in the later Amendment of 2007, there was a proposal that directed companies to give back some portion of the profits to the original owners in the form of shares and debentures *(Land Acquisition Amendment, 2007)*. Then in 2009 amendment the compensation amount had to include solatium, and several charges including administrative and acquisition charges of the resettlement property *(Land Acquisition Amendment, 2009)*. Monetary aid here refers to the aid provided for resettling and re-establishing lives in a new place and home. In the 1894 model, the compensation was only of monetary nature and the total solatium was the only aid that was provided *(Land Acquisition Act, 1894)*. As per the 1894 Act, the only provisions for rehabilitation and resettlement were 5% reservation for the PAP families in government departments. There was no provision for the people who lost irrigable land and they were given compensation only as per the measurement of their plots. The Land Acquisition Amendment, 2011 provided PAP with the option to choose between mandatory employment in projects, a one-time payment of 5 Lakh rupees or annual compensation of rupees 2000 per month per family for the duration of 20 years. This was the first significant step towards livelihood restoration. The next Amendment stated that there should be compulsory employment provided to at least one member of an affected family of farm laborers *(Land Acquisition Amendment, 2013)*. Farmers were to be provided equivalent area of cultivable wasteland in case an irrigated multi cropped land is acquired for some exceptional case (LARR, 2013, c. 3). The 2011 Amendment Bill stated that if the acquired land wasn't used and sold to a third party then 20% of the profits shall be shared with the original owners of the land (LARR, 2011, c. 9). The Amendment also announced a one-time allowance of 50,000 rupees as resettlement aid. Further, it stated that it was mandatory to provide the PAP land for a house as per the Indira Awas Yojana (IAY) in rural areas or a constructed house of at least 50 square meters plinth area in urban areas. The 2013 Amendment increased the profit sharing on sale percentage from 20 to 40% *(Land Acquisition, Rehabilitation and Resettlement Act, 2013, c. 13)*. A deep analysis of Supreme Court cases involving land acquisition in India was done by several scholars from the Centre of Policy Research ranging from 1940 to 2016. The review of these cases uncovered the most common and recurring issues over so many years and the reasons behind them. As per the review by Wahi (2017), the largest numbers of acquisitions were for planned development (16%), housing (9%), industry (8%), and infrastructure (6%).

2.1 Mumbai and LARR

The main Acts which govern the Acquisition process in Mumbai are the Sections 125–129 of the Maharashtra Regional & Town Planning Act, 1966 and amended in 2017. This Act also commands the authorities to meet the terms of the LARR Act 2013 and its substantiating provisions. Newspapers have often reported that the Brihanmumbai Municipal Corporation (BMC) has lost several pieces of land on which they had intended to build social infrastructure facilities (hospitals, schools, recreation grounds) to the plot owners in various court cases. According to BMC's statements in multiple newspapers, the compliance with LARR 2013 makes it difficult for the authorities to acquire lands for these social and urban infrastructures and the Municipal commissioner Ajoy Mehta had urged the State government to amend the Land Acquisition policy in such a way that acquisition becomes easier (Pinto, 2018). As per the Maharashtra, regional and town planning act, the land can be acquired for public purposes stated in the plans through one of the options—by paying an amount agreed upon by both the parties, by paying an Amount in compliance with the Right to Fair Compensation and Transparency in the Land Acquisition, Rehabilitation and Resettlement Act (*RFCTLARR*) 2013, Floor Space Index (FSI) OR Transferable Development Rights (TDR) provided free of cost or by making an application to the State government for acquisition and complying with Act. As per the Maharashtra Regional and Town Planning Act, for urgent purposes like defense, national security, natural calamities etc., the land can be acquired after providing a 15 days' notice, Provided that compensation for standing crops is provided, damages caused by sudden dispossession are compensated *(Right to Fair Compensation and Transparency in the Land Acquisition, Rehabilitation and Resettlement Act 2013, 28)*, Interest and other compensation finalized by the local development authority, and the advance payment of maximum 2/3 of the total amount be made. It is evident that the whole process derives its backbone from the Land Acquisition Amendment Bills and the authorities consider this as a reason for complexity.

3 Mumbai Projects

3.1 Mumbai Urban Transport Project (MUTP) and Its Resettlement

Mumbai Urban Transport Project (MUTP) was the widening of two highways, namely Jogeshwari Vikroli Link road (JVLR) and Santa Cruz—Chembur Link Road (SCLR). In the project's 2000 report, the Government of Maharashtra, Housing and Special Assistance Department Mantralaya provide the details of their R&R Plan which had to be designed for about 23,000 PAF (1,20,000 PAP). Since the project is of early 2000s the resettlement plan consisted of basic provisions of either monetary compensation or the choice of moving into a 25 square meter developed plot on a Greenfield site, or a 20.91 square meter tenement for squatters under slum redevelopment scheme (MUTP, 1997). The meagre difference between the incentives for the landowners and squatters in the plan was concerning. For the resettling families, the basic facilities of water supply, toilets, tap water pathways and the community facilities of dispensary, primary school, playground shop etc. was to be be provided on site (MUTP, 1997). There were some basic incentives for livelihood related issues in the plan, like seed capital loans, training facilities in association with NGOs, but the focus was on preserving the current livelihood patterns by minimum displacement and transportation allowances (MUTP, 1997). In her article Renu Modi writes about the drawbacks of this scheme which is appreciated on international levels for providing free settlements to the people the project displaced. Renu Modi says that the biggest flaw in the plan was the similar treatment towards squatters and encroachers. It was decided to provide a free residential tenement of 225 square feet irrespective of legal title ownership of their commercial or residential property. This led to several protests by people who had much bigger plots and well-constructed houses and business plots acquired by the authorities in exchange of a small housing unit. These cases were then dealt with case by case (Modi, 2013). The loss of prime locations of their enterprises and businesses again hit the PAP hard and the need for a stronger livelihood restoration plan was felt. The whole point of resettlement is not just about getting the site cleared off the people living on it and relocating them elsewhere. It is also about development not being at the cost of some families. New houses at the resettlement colonies should also have sizes proportional to the lost homes and the recognition must be more categorical and systematic.

3.2 Navi Mumbai Airport Acquisition

The interesting acquisition process of Navi Mumbai Airport Influence Notified Area (NAINA) and their policy package drew attention and criticism from the experts and the media. In this section, the proposed schemes in documents and the contrast with the real implementation scenario are outlined. In their working paper, Matthews et al. (2018) describe NAINA's alternate acquisition mechanism of transferable development rights (TDR) and allowing landowners to aggregate and surrender 40–50% of their land parcels to the authorities as innovative and efficient. By 2018 January, The City and Industrial Development Corporation of Maharashtra (CIDCO) had received 341 building permit requests. Apparently, the landowners had complied with the scheme and the genuine rewards of a portion of land and additional TDR seemed assuring to them. There are detailed flattering mentions of NAINA acquisition in several documents and news articles. Although the lack of proper R&R package for landless and indirectly dependent families was pointed out as well (Matthews et al., 2018). The Modified draft report of the project also consisted of details mostly about the reservation criteria, TDR, investment and tenders (MDDP CIDCO, 2017) but not enough details about the humbler landowners and their resettlement policy package. Media reports on the contrary show several occurrences of struggle, protests and resistance when it comes to NAINA's acquisition of smaller private land plots in over ten villages. An elaborate article in The International Alliance of Inhabitants addresses the timeline of PAP reaction to the project which evicts about 3500 village families to some resettlement sites. Unnecessary acquisition, unfair compensation, and the fact that the resettlement sites had not even been ready by the eviction date were major causes behind farmers' hearing boycotts in 2010 and violent protests in 2017 ("Navi Mumbai Airport," 2018). In January 2018, Maharashtra Cabinet had passed the proposal of providing 22.5% of acquired land to PAP as compensation for the 671 hectares of private land the authorities had required (Rawal, 2018). CIDCO had devised a resettlement policy package for the displaced families and 3 alternative sites. The resettlement plan claimed to provide land for land compensation, shifting allowance, minimum agricultural wage for certain period, civic amenities at new site, choice of plot location, and several other well intended incentives. The studies conducted by Rajak and Roy in 2016 over 100 PAP reveal the following perceptions they had. Despite of being availed homestead plots with basic civic amenities on site, it was felt by the people that they were incompetent to construct their own houses from scratch and that they should've been provided with construction aid and assistance. Apart from the shortage, the lack of financial management and structure in the compensation reward, and the loss of agricultural and fishing livelihood were also pointed out by the PAP who felt disruption in accessing their previous sources of income (Rajak & Roy, 2016). Thus, it is evident that even in this project, which a very recent one, the gaps of plans and implementations are quite visible and relevant.

3.3 Mumbai-Nagpur Expressway Acquisition

This mega infrastructure project, also known as the Samruddhi Mahamarg Project is India's first and longest Green Expressway which is supposed to be about 701 km long and needed a total of approximately 8311 hectares of land. This project has a government website which has land related information and the amounts of land acquired by voluntary and involuntary methods. This project has practically no resettlement program or social impact assessment and the acquisition and voluntary settlements have all been solely made on the basis of handsome compensations (mahasamruddhimahamarg.com). The State has recently given a free pass to Maharashtra government from Social Impact Assessment (SIA) and R&R incentives through tactful application of acts (Maharashtra Highways Act, c. 15). As per the official website of the project (mahasamruddhimahamarg.com) the farmers who lost agricultural lands were provided with as much as 3–4 times their asset value, and when they compared it to their farming paltry income, they wilfully sold their land. A hefty amount of compensation always allures the affected population and diverts their decisions and reduces their initial resistance to the acquisition but to what extent the unsupervised transfer of wealth supports their lives ahead is a vital concern. A piece of land taken away from farmers does not just take away their assets but also their source of livelihood and sense of belonging to a place. This again raises the concern regarding the sustainability of the monetary compensations provided. Several renowned researchers in the field of R&R have often advised against solely monetary compensation. Monetary compensation is said to be problematic as the PAP are mostly not used to handling a large sum of money suddenly bestowed upon them (Cernea, 2003). Several other issues regarding this type of acquisition include theft, risk of assault, fraud regarding disbursement and substantial delays in the provision of the money (Perera et al., 2013; Tagliarino, 2017).

3.4 Mumbai-Ahmedabad Bullet Train Acquisition

The ambitious Mumbai Ahmedabad bullet train project is an excellent recent example of the government's radical ideas of pushing development agendas or the façade of it as rapidly as possible. The R&R policy package of this project, unlike several other recent projects in Mumbai, is detailed and well organized on documents. The Joint feasibility report of the main funding company Japan International Cooperating Agency (JICA) and the Ministry of Railways (MOR) narrates the whole policy framework of LARR which rests on the following acts majorly—Railways Act, 1989 and Railways (Amendment) Act (RAA), 2008, National Rehabilitation and Resettlement Policy, 2007, The World Bank OP-4.12 on Involuntary Resettlement, JICA Guidelines for Environmental and Social Consideration. The report also provides the vivid categories into which the PAP would be classified and grouped for better aid as per—income status, religion, education, age, monthly expenditure, and interestingly also on the basis of travel distance to primary source of income (RAP Final Report, 2015). The three major Acts which form the framework are different and overlapping in several areas but basic claims of trying alternatives of involuntary displacement, PAP participation and having mechanisms to restore lost livelihoods were acknowledged formally in the project feasibility report. One of the major reasons these acts, reports and packages seem elaborately conscious of the people the project affects is their crafty wordplay. It is mentioned that the displaced people who also lose their source of primary livelihood must be 'sufficiently compensated and provided support for alternate means of sustainable livelihoods by the project proponents' (RAP Final Report, 2015). Here, the compulsion stress is put on providing 'support' through the word 'must' which is a non-specific statement in our opinion. Instead, if the statement was—It's mandatory to provide such people with an alternative source of income by the authorities, it would have been a comparatively rigid regulation as the term 'support' can mean anything ranging from lending some cash to providing a few months of training. When it comes to analyzing the affected and lost structures, the household survey by the authorities has revealed several features in detail like —structure pattern, construction material, roofing and wall type and a strong mechanism for the uplifting of vulnerable (socially, economically and other) groups has been stated (RAP Final Report, 2015). The real situation although, as captured by media outlets, again doesn't seem as impressive as the public would want it to be. Mumbai Mirror has published a series of reports from 2017 to 2019 covering the protests by tribal farmers and residents losing their homes to the project. The 508 km long bullet train corridor, apart from affecting forest lands and mangroves, is overrunning 828 commercial and residential structures in Vasai-Virar in exchange of very meager compensation (Ganapatye, 2019). The affected people from Vasai-Virar are also being forced to relocate further in Kalyan with the compensation insufficient to even purchase a new flat which approximately costs about 30 lakhs in the area (Naik, 2019). The woes of the

tribal farmers of Palghar district came out violently in 2017 after the foundation stone was laid by the PM and their loss of irrigated land and being given no say in the project proceedings was raised by tribal organizations like Bhoomi Sena (Rajemahadik, 2017). Resistance and protests are natural for such large-scale projects as people find it difficult to come to terms with such a huge change, but incompetence in terms of R&R implementation is also a reality that cannot be denied.

3.5 Mumbai Metro Rail Project

Metro Rail Projects in India have become the staple ingredient of development felt necessary by the government in every moderate to mega city. Several roads are dug up and Metro project under construction since a long time in several Indian cities. Recently there has been huge uproar over Mumbai Metro's massive tree felling and environmentally damaging approach. The 2012 Social Impact Assessment report of Mumbai Metro Rail Corporation (MMRC) refers Land Acquisition Act 1894 as the basis of the initial proceedings. Although, several other Acts have been considered for the process which include—the Mumbai Metropolitan Region Development Authority (MMRDA) Act 1974, the National Rehabilitation & Resettlement Policy 2007, JICA regulations and the R&R Policies of Mumbai Urban Transport Project (MUTP). The social due diligence report by ADB and MMRDA has declarations that the metro rail project being an urban project, has no purpose with acquiring agricultural lands and if some crop losers do exist as PAP they must not be legal titleholder practitioners. The report clearly states that legal title or lease holders are eligible to a specific set of compensation, livelihood restoration and resettlement incentives. For a titleholder the livelihood restoration measures included one-time aid of 5,00,000 INR, or 2000 INR per month for 20 years or skill development training for a family member, and for legal commercial unit losers, an alternate site, transportation arrangement and even three years' travel expenses for increased distance are to be provided (ADB, 2018). Apart from this there are special provisions for scheduled caste (SC), scheduled tribe (ST) and vulnerable households. The loss of permanent source of income would be compensated by provision of one year's income equivalent and aid to search for alternative jobs (ADB, 2018). These provisions are quite practical in nature but they mostly focus on the PAP possessing papers and documents of their ownership and lease while it has been observed over time and various cases that such mega projects affect people in varying direct and indirect ways. Ashwini Bhide, the Managing Director of MMRC interacted with the media and explained how the 350 resident PAP (out of total 650) in Girgaon and Kalbadevi would be resettled in a single building of a superior quality and more carpet area than their original homes and the slum encroacher would be sent to MHADA resettlements in Chakala (Chacko, 2016). The Land Acquisition process is based on the Maharashtra Regional Town Planning (MRTP) Act. Pagdhare, 2018 mentions how Metro line 3 directly and indirectly affects 2622 people out of which 1837 are residents while others lose their commercial assets. These people thrive on their commercial network of shops, enterprises and small businesses whose fabric would be deeply affected by the project (Pagdhare, 2018). Apart from this, the Kokna Tribal community's members of the Aarey Colony were forced to live in slum rehabilitation colonies and about 302 households were demolished. The government officials while speaking to the media have usually been seen presenting oversimplified statistics and scenarios. The government reports and data also seem to be contradictory to their reassuring statements to the newspapers and reporters.

4 Methodology

Since this is an experimental study, a crucial part of the main objective is to develop a methodology for checking sustainability of resettlement and rehabilitation plans. The methodology applied is based on expert opinion and case studies of the above mentioned Mumbai projects. The method of extracting information through expert opinion surveys helps the researcher to establish some validity and reliability in the study (Bogner et al., 2009) The methodology consists of the following main steps—Identification of the weak links and areas of shortage in the biggest acquisition projects of Mumbai and their resettlement and rehabilitation plans. This is followed by surveying experts regarding the most important and basic resettlement site facilities which must definitely be provided in a RP. The reflection of responses on the idea of sustainability and preparation of the next set of questions, establishing an equation of sustainability and time period of survival as 'period of sustainability' and the demonstration of the idea through basic calculation are the subsequent steps. Even for selecting the experts for the iterative testing, a few points had to be kept in mind—they must be aware and updated regarding the latest policies, political environment and other details of the scenario should possess a combination of specialization, professional and competence, must be open to innovative implementation of their knowledge, and must be able to see the creativity and vision of the researcher (Iriste & Katane, 2018).

4.1 Survey and Defining Sustainability

A qualitative iterative testing was done through expert interviews of 25 respondents, out of which most (67%) were urban scientists and researchers and the rest were urban practitioners (15%) and government and urban organization officials (18%). The various recent Mumbai projects and their rehabilitation plans highlighted and pointed towards the lack of clear understanding of what sustainability in terms of involuntary resettlement is. There are several things which contribute towards the overall sustainability in a resettlement scenario but here in this article, the focus is to enlist few basic constituents and understand their contributions. This survey was done based on four major classifications of contributing elements which shape a family's experience and duration of stay in a resettlement colony. They are—the basic physical infrastructure, the access to livelihood, social and community facilities and the resettlement buildings' comfort factors. To understand their interplay, inputs, and the researchers' basic idea of period of sustainability (POS) this survey provided an underpinning. This survey was done in two levels, the initial a small five question one followed by another more detailed one. The intention was to know the duration a person can withstand being under a certain state of mind, ranging from being very unhappy and stressed to being content and satisfied with their living conditions. The responses were taken as 'number of years', so that the next level of questions could be framed accordingly. Once the respondents provided their desired ranges of time, being in various states of mind, the next level was based on the duration a person can survive at a resettlement site with family under the given conditions of physical infrastructure. The focus was on physical infrastructure for this detailed questionnaire as it is the most important factor for sustenance at a site as per the initial feasibility and literature studies. However, the importance of other contributing elements was also assessed by asking the respondents to fill a pair-wise comparison matrix of all the four categories.

5 Responses and Results

Most of the respondents were from well-to-do backgrounds but had understanding of urban scenarios of involuntary resettlement. The first level questions were based on the following five states of mind one can be in while living at a particular place—(1) Very unhappy and desperate to shift elsewhere (2) Dissatisfied and looking for alternatives (3) Living normally but looking for upgrade and (4) Living comfortably. For each state of mind (SOM), the respondents mentioned duration (in years) one was capable of living in that SOM. For example, a very unhappy and desperate person would be able to stay that way for a few months up to about maximum 5 years. Similarly, for all the 5 SOM, respondents entered relevant duration. We tried to form few ranges of time based on their responses and came up with— (0–5), (5–10), (10–20), (20–30), and (30–50). The responses were scattered across these ranges for the 5 SOM. The next set of questions was about the range of years respondents would be able to survive under given facilities. Electricity is an important part of physical infrastructure but since power supply is not seen as a huge issue and is mostly provided for most planned and unplanned settlements, hence we left it out of our survey. The four major classifications of physical infrastructure for the survey—water supply, sanitation, drainage and solid waste management. Four major types of water supply connections observed in various resettlement sites of Mumbai are provided and the respondents provided majority of their votes to a particular range for each facility (Table 1). The sites which have hand pumps have a major part of their residents' day time spent in collecting water for daily function. Tube wells have similar impact on people's lives and hence living with both hand pumps and tube wells is challenging and have both been scored with minimum time range. Respondents felt that it's not possible to peacefully live with such facilities for more than 5 years. The best condition of water supply at a resettlement site would be a 24 h one with personal connection to each house

and accordingly, it has been voted with the maximum range of 30–50 years.

Table 2 depicts results of similar questions regarding the sanitation facilities at the resettlement site. Community or common toilets can be seen in several resettlement sites and mostly in slum resettlement sites. Living with such facilities is difficult although the less fortunate are forced to continue staying that way for years. Here the intention is to know the POS a normal person considers for such type of sanitation conditions who is not under the extreme pressure of homelessness. The person must also be aware and educated enough to express opinions. Having a single toilet and bath in the household is still a dream for so many displaced and slum dwellers in Mumbai. Although it cannot be denied that a family of 5 or 6 sharing a single toilet bath is inconvenient and hence the respondents allotted it a time range of 10– 20 years after which one would surely wish to upgrade for comfort i.e. multiple toilets in their houses. Mumbai being a flood prone city has drainage as one of the primary concerns for settlers. Low lying areas have had terrible consequences during Mumbai's prolonged monsoons, which is one of the reasons why clusters of encroachers have settled on hill tops. Drainage being one of the topics discussed in the survey, the respondents' scores made it clear that they do not consider areas with open kuchcha and pucca drains fit for living past 5 years. Even covered drains were considered sustainable for up to 20 years. Underground and pucca drains are essential for living over generations in peace (Table 3).

Table 1 Water supply types and respective POS (*Source* Primary Suvey)

	Type of water supply connection	0–5	5–10	10–20	20–30	30–50
W1	Hand pumps (manual collection)	**100%**	–	–	–	–
W2	Tube wells (limited supply)	**90.4%**	9.6%	–	–	–
W3	Common taps (full time supply)	3%	8%	**62%**	27%	–
W4	Personal connection (limited supply)	–	2%	10%	**76%**	12%
W5	Personal connection (full time supply)	–	–	16%	25%	**59%**

Table 2 Type of sanitation services and respective POS (*Source* Primary Survey)

	Type of supply	0–5	5–10	10–20	20–30	30–50
S1	Community toilets and bath (shared among 8–10 households)	**100%**	–	–	–	–
S2	Common sharing toilets and baths (shared among 2–3 households)	**80.9%**	19.1%	–	–	–
S3	Personal attached single toilet and bath	4%	8%	**62%**	19%	7%
S4	Personal multiple toilet and bath facilities per house	–	–	6%	13.1%	**80.9%**

Table 3 Type of drainage facility and respective POS (*Source* Primary Survey)

	Type of supply	0–5	5–10	10–20	20–30	30–50
D1	Open Kuchcha drains	**100%**	–	–	–	–
D2	Open Pucca drains	**71.4%**	15%	13.6%	–	–
D3	Covered drains	–	–	**62%**	22%	16%
D4	Underground drains	–	–	4%	16%	**80%**

Fig. 1 Maintenance issues at Umarkhadi Resettlement Colony, Mumbai

Fig. 2 Waste disposal and management issues at resettlement colonies, Mumbai

Solid waste management is considered here as an impactful constituent towards sustainable living because of apparent reasons. Heaps of disposed garbage and waste around the city are very visible urban issues and have serious negative consequences to the residents' health (Figs. 1 and 2). Also, while left with handling their own waste themselves, people tend to either dump them nearby or burn them, both of which degrade the air quality. Table 4 displays the results and the respondents felt that even a colony or building wise waste collection system would be enough to live peacefully for a decade or two. Door to door collection is the best and most civilized procedure. For this article and the results through the pilot expert survey, we consider all the physical infrastructure classifications as equally important and assign equal weight to them. Hence all the different type of facilities listed above can form numerous combinations and each combination would result into different POS.

Table 4 Type of solid waste disposal system and respective POS (*Source* Primary Survey)

	Type of solid waste management	0–5	5–10	10–20	20–30	30–50
Sw1	Door to door collection	**100%**	–	–	–	–
Sw2	Building wise collection	6%	**72%**	18%	4%	–
Sw3	Colony wise collection	–	–	12%	**76%**	12%
Sw4	Self (landfill dumping site)	–	–	4%	8%	**76%**

Table 5 Possible combinations of periods of sustainability

W1, W2, W3, W4, W5	=320 combinations
S1, S2, S3, S4	
D1, D2, D3, D4	
Sw1, Sw2, Sw3, Sw4	

Every site would have separate combinations of these facilities and hence a unique POS. There are a total of 320 possible combinations out of the parameters considered as given in Table 5.

5.1 Pair Wise Comparison

The above assessment consisted of only physical infrastructure attributes. The intent although initially was to know the importance of several other attributes in comparison to physical infrastructure. The other attributes here refer to— social and community infrastructure (SI), Basic Livelihood Access (LA) and Building Comfort Factors (BCF). The respondents were asked to fill up pair wise comparison matrix between these three attributes with each other and compared to physical Infra (PI). The scoring of the pair-wise comparison was done and the results are in Table 6. The first comparison between Physical Infrastructure and Social Infrastructure was bound to incline towards PI as people are generally more concerned about basic connections about water, electricity and sanitation in their homes over community facilities, health and education in the housing community. Moreover these facilities can be availed by travelling to nearby neighborhoods as well. Although, the inclination wasn't as big as it was expected to be, as the weight of PI was 0.32 as compared to 0.16 of SI. This meant that the respondents gave Physical Infra double the importance as Social Infra. The responses indicated the weight of

Access to basic livelihood and PI as 0.39 and 0.32 respectively. This means people felt livelihood access is even more important than physical infrastructure amenities in the houses. This indicated the importance of proper livelihood restoration and maintenance in a resettlement plan, something which is still missing from a lot of our recent project plans. Building comfort factors like lighting, ventilation, thermal comfort were given the weight of 0.13 compared PI's 0.32 i.e. People give PI almost 2.5 times more importance than these comfort factors for sustainability. One reason for such scoring might be the fact that these factors are realized a while after one starts living in the new house rather than being obvious since the beginning.

6 The Best and the Worst Resettlement Scenarios

Through the survey the weight of other factors apart from Physical Infrastructure on the POS was known. But for the inclusion of their impacts on the final POS, we need more data regarding the condition of those facilities on site. Also a systematic method is needed to capture their impacts similar to but slightly different to what was done for PI. In the given data, the best scenario would be a site with—Personal water connection, multiple personal toilet and bath, underground drainage, and door to door waste collection. By adding the individual POS of these facilities and averaging them a total POS of 40 years is obtained. Meanwhile, the worst-case scenario being-manual hand pump for water, community toilets, open kuchcha drains, and self-disposal of waste would fetch us the total POS of just 2.5 years. There are a total of 320 possible scenarios with POSs ranging from a minimum of 2.5 years to a maximum of 40 years, but only the first and last four combination results are presented in Table 7.

Table 6 Cumulative matrix (*Source* Primary Survey)

	PI	SI	LA	BCF	Sum	
PI	0.01	0.46	0.53	0.31	1.3	**0.32**
SI	0.15	0.004	0.24	0.27	0.66	**0.16**
LA	0.22	1.57	0.003	0.32	2.11	**0.39**
BCF	0.22	0.19	0.12	0.0003	0.53	**0.13**
Sum	1	1	1	1		

Table 7 Values of first and last 4 combinations

First 4 combinations		Last 4 combinations	
W1 S1 D1 SW1	**2.5**	W5 S4 D4 SW1	30.625
W1 S1 D1 SW2	3.75	W5 S4 D4 SW2	31.875
W1 S1 D1 SW3	8.125	W5 S4 D4 SW3	36.25
W1 S1 D1 SW4	11.875	W5 S4 D4 SW4	**40**

7 Conclusion

The survey and analysis done in this article are primarily done concerning basic infrastructure amenities. There are several other factors like social wellbeing, community links, inter community networks of people which get deeply affected when they are relocated elsewhere. The details and in-depth analysis of those factors is not done in this study. There are certain losses that are irreversible and irreparable as development projects are prioritized. But the target here is to not push the PAP towards poverty and homelessness post displacement. The fact that till date several mega projects in the country are actually taking place without a resettlement plan and just monetary compensation (irrespective of its amount) is a major concern. The several basic aspects discussed in the article are still found missing in major RPs of today. Categorization and proper identification of a PAP is important in resettlement. A landowner cannot be treated and compensated the same way as the policies would to an encroacher or squatter. The difference between Slum Rehabilitation and Involuntary Resettlement due to development projects must be clearer despite basic living standards being provided in any case. The POS is affected by numerous aspects of the policy package. The study and analysis highlight the importance of access to livelihood and availability of basic physical infrastructure. A well-maintained balance of basic facilities and infrastructure is the key to peaceful resettlement and increasing the POS of the people at any site. Monetary compensation and its fair calculation and disbursement have been topics of discussions in resettlement policy research. We would just like to nod our heads in agreement to the fact that monetary compensation is an important component of RP and its proportionality to lost assets and structures is crucial. Another important finding of our study is that POS is a necessary but not sufficient condition to judge the quality and efficiency of a RP. State of Mind (SOM) is also a significant factor which indicates if the POS of the PAP at the site was spent very happily, satisfied, dissatisfied or extremely unhappily. The relation between the POS and SOM is probably directly proportional as our survey suggests that the better the SOM, the longer people wish to stay at a place (higher POS). The exact equation between these two is something that will be addressed in future studies. The abundance of incentives being provided in a RP is often mistaken for its efficiency. A clear gap between great number of incentives and a long term sustainable RP has been observed. This gap can be filled when we are capable of calculating the POS of a given plan and enumerate the difference between its claim and future outcome.

References

(2015). (rep.). *Joint feasibility study for Mumbai-Ahmedabad high speed railway corridor* (Vol. 1, pp. 1–122).

Asian Development Bank. Social Due Diligence Report. (2018). https://www.adb.org/sites/default/files/project-documents/49469/49469-007-sddr-en.pdf

Björkman, L. (2014). Becoming a slum: From municipal colony to illegal settlement in liberalization-era Mumbai. *International Journal of Urban and Regional Research, 38*(1), 36–59. https://doi.org/10.1111/1468-2427.12041

Bogner, A., Littig, B., & Menz, W. (2009). *Interviewing experts.* Springer.

Cernea, M. M. (2003). For a new economics of resettlement: A sociological critique of the compensation principle. *International Social Science Journal, 55*(175), 37–45. https://doi.org/10.1111/1468-2451.5501004

Chacko, P. A. (2016). CNT and SPT Amendment Bill 2016. *Indian Currents, 28*(49). Retrieved August 21, 2017, from http://www.indiancurrents.org/cnt-and-spt-amendment-bill2016-1444.php

Ganapatye, S. (2019, June 2). *Mumbai-Ahmedabad bullet train project: State to 'forcefully' acquire land, process to complete in two months.* Mumbai Mirror. https://mumbaimirror.indiatimes.com/mumbai/other/push-comes-to-shove-state-to-forcefully-acquire-land/articleshow/69616162.cms

Government of India. (2009, February 25). *Land Acquisition Amendment Bill, 2009.* https://www.prsindia.org/sites/default/files/bill_files/Land_97-C_of_2007.pdf

Government of Maharashtra. (2016). (rep.). *Pre Feasibility report for environmental impact assessment for Nagpur Mumbai Expressway* (pp. 1–25).

Government of Maharashtra, Mumbai Metro Rail Corporation Ltd. (2012). *Social impact assessment for Mumbai Metro Rail Line 3* (Vol. 1 Final Report).

Government of Maharashtra, Special Planning Authority Navi Mumbai Airport Influence Notified Area. (2017). *Modified draft development plan.*

Government of Maharashtra. (1997). (rep.). *Resettlement and rehabilitation policy for Mumbai Urban Transport Project (MUTP)* (pp. 1–16).

Involuntary Resettlement Portfolio Review Phase II: Resettlement Implementation. (2014, June 16). *Social Development Department, World Bank.* https://pubdocs.worldbank.org/en/96781425483120443/involuntary-resettlement-portfolio-review-phase2.pdf

Iriste, S., & Katane, I. (2018). Expertise as a research method in education.In *Rural environment. Education. Personality. (REEP): Proceedings of the 11th International Scientific Conference.* https://doi.org/10.22616/reep.2018.008

Lok Sabha, 77 The Land Acquisition, Rehabilitation and Resettlement Bill 1–60 (2011).

Lok Sabha. (2021, May 9). The right to fair compensation and transparency in land acquisition, rehabilitation and resettlement (Second Amendment) Bill, 2015. PRS Legislative Research. https://prsindia.org/files/bills_acts/bills_parliament/LARR_(2nd_A)_Bill,_2015_1.pdf

Matthews, R., Pai, M., Sebastian, T., & Chakraborty, S. (2018). State-led alternative mechanisms to acquire, plan, and service land for urbanisation in India. *World Resources Institute India, Working Paper* (pp. 1–68).

Mettler, S. (2016). The policyscape and the challenges of contemporary politics to policy maintenance. *Perspectives on Politics, 14*(2), 369–390. https://doi.org/10.1017/s1537592716000074

Modi, R. (n.d.). Reconstructing livelihoods? Observations from resettlement sites in Gujarat. *Beyond relocation: The imperative of sustainable resettlement beyond relocation: The imperative of sustainable resettlement* (pp. 154–182). https://doi.org/10.4135/9788132108238.n8

Naik, Y. (2019, February 26). 828 structures to go to make way for bullet train. *Mumbai Mirror*, 2.

OECD-FAO agricultural outlook 2009. (2009). *OECD-FAO Agricultural Outlook.* https://doi.org/10.1787/agr_outlook-2009-en

Pagdhare, D. (2018). A study on social and economical condition of project affected persons in Mumbai: special reference to metro line III. *Journal of Emerging Technologies and Innovative Research, 5*(10), 164–170. https://www.jetir.org/view?paper=JETIRF006026

Perera, C., Zaslavsky, A., Christen, P., & Georgakopoulos, D. (2013). Sensing as a service model for smart cities supported by Internet of things. *Transactions on Emerging Telecommunications Technologies, 25*(1), 81–93. https://doi.org/10.1002/ett.2704

Phadke, A. (2013). Mumbai metropolitan region: Impact of recent urban change on the Peri-urban areas of Mumbai. *Urban Studies, 51*(11), 2466–2483. https://doi.org/10.1177/0042098013493483

Pinto, R. (2018, November 10). *BMC chief: Amend land acquisition laws | Mumbai news—Times of India.* The Times of India. https://timesofindia.indiatimes.com/city/mumbai/bmc-chief-amend-land-acquisition-laws/articleshow/66562785.cms

Rajak, R., & Roy, A. K. (2016). 'Development project, land acquisition and resettlement in Maharashtra: A study of Navi Mumbai International Airport Project'. https://uaps2015.princeton.edu/papers/151200

Rajemahadik, V. (2017, September 15). *Thousands of Palghar tribals protest bullet train project; hold Black flags.* Mumbai Mirror. https://mumbaimirror.indiatimes.com/mumbai/other/thousands-of-palghar-tribals-protest-bullet-train-project-hold-black-flags/articleshow/60532076.cms

Ramanathan, U. (2011, November 5). *Review of Land Acquisition, Eminent Domain and the 2011 Bill. Economic and Political Weekly.* https://www.epw.in/journal/2011/44-45/commentary/land-acquisition-eminent-domain-and-2011-bill.html

Rawal, S. (2018, January 17). *Review of Compensation proposal gets nod, Navi Mumbai infra a boost. Hindustan Times.* https://www.hindustantimes.com/mumbai-news/compensation-proposal-gets-nod-navi-mumbai-infra-a-boost/story-CRSoM7VKq4elchhoCj4ZfI.html

Sassen, S., & Robert S. Lynd Professor of Sociology Saskia Sassen. (2001). *The global city: New York, London, Tokyo.* Princeton University Press.

Sinha, K., & Singh, N. (2016). Land acquisition in India: History and present scenario. *Journal of Legal Studies and Research, 2*(4), 21–37. http://www.jlsr.thelawbrigade.com/

Tagliarino, N. (2017). The status of national legal frameworks for valuing compensation for expropriated land: An analysis of whether national laws in 50 countries/regions across Asia, Africa, and Latin America comply with international standards on compensation valuation. *Land, 6*(2), 37. https://doi.org/10.3390/land6020037

The expressway. (n.d.). Maharashtra Samruddhi Mahamarg—Maharashtra Samruddhi Mahamarg. https://www.mahasamruddhimahamarg.com/the-expressway/

The Land Acquisition Act, 1894. (n.d.). Ministry of Road Transport & Highways, Government of India. https://morth.nic.in/sites/default/files/THE_LAND_ACQUISITION_ACT.pdf

The Land Acquisition Amendment Bill 2007. (n.d.). PRSIndia. https://prsindia.org/files/bills_acts/bills_parliament/1197003952_Land_20Acq.pdf

The Right to Fair Compensation and Transparency in Land Acquisition, Rehabilitation and Resettlement Act, 2013. (n.d.). Legislative Department | Ministry of Law and Justice | GoI. https://legislative.gov.in/sites/default/files/A2013-30.pdf

Wahi, N. (2017). Land acquisition in India: A review of Supreme Court Cases from 1950 to 2016. *SSRN.* https://ssrn.com/abstract=3378958

Urban Morphology, Environmental Performance and Energy Use: Holistic Transformation of Porto di Mare as Eco-District Via IMM

Carlo Andrea Biraghi, Tommaso Mauri, Maria Coraly Mazzucchelli, Elena Sala, Massimo Tadi, and Gabriele Masera

Abstract

The impact of urban development on climate change is now evident, threatening the living conditions of a growing number of people in developed and developing countries. On the one hand, there is the need to limit soil consumption and urbanisation rates, whereas, on the other hand, the presence of slums and brownfields within formal urban contexts compromises the living conditions in the neighbouring areas. Given the complexity of the topic, a holistic, multi-scale and multi-disciplinary approach is required. This article presents a case study of integrated actions applied at the neighbourhood scale, using the Integrated Modification Methodology (IMM) conceived at Politecnico di Milano, which refers to the built environment as a complex adaptive system (CAS). The IMM is an iterative multi-stage and multi-layer process applied to urban CASs to improve their metabolism and environmental performance. The IMM approach to sustainability is aligned to the UN Sustainable Development Goals (SDGs) 2030 as a systemic methodological interpretation of the SDG11, which suggests local-based actions firmly linked with targets and indicators. Inside IMM, this role is played by the Design Ordering Principles (DOPs), a system of integrated actions and evaluation measures for simultaneous improvements in environmental performance, social inclusion and urban metabolism. The study case is the area of Porto di Mare, located on the south-eastern border of the city of Milan, bounded by important infrastructures, between the city centre and the rural belt surrounding the Milanese metropolitan area. The proposal of an Eco-District for the Porto di Mare area is an opportunity to demonstrate the potential implications of a sustainable regeneration process on territorial low-carbon energy planning strategies, with a significant impact on a larger part of the city. This article presents the process of local optimisation of the Eco-District masterplan by acting on morphological and typological parameters to simulate alternative design scenarios and evaluate their performance using the visual programming interface of a BIM software. The main performance aspects that are considered are thermal loads of buildings, outdoor comfort and energy use intensity. This challenge raised some relevant research questions: is there a plausible amount of the Gross Floor Area (GFA) value which can be considered coherent from an energy and environmental point of view? Which could be the connection between this parameter, the morphological volume and the energy performance at a district scale? How can the Eco-District be evaluated from a sustainable point of view? The presented strategy for the exploration of alternatives, based on existing energy modelling tools, gave an answer to most of the open questions presented here, showing a replicable approach for similar problems and contexts.

Keywords

Urban regeneration • Eco-district • Parametric design • Sustainability

C. A. Biraghi (✉)
Department of Civil and Environmental Engineering, Politecnico di Milano, Milan, Italy
e-mail: carloandrea.biraghi@polimi.it

T. Mauri · M. C. Mazzucchelli · E. Sala
Building Engineering—Architecture, Politecnico di Milano, Milan, Italy

M. Tadi · G. Masera
Department of Architecture, Built Environment and Construction Engineering, Politecnico di Milano, Milan, Italy

1 Introduction

The impacts of urban development on climate change are evident. More than 70% of the environmental impact of people is determined by where they live, how they commute and what they eat (ANCE, 2013). This raises issues that threaten the living conditions of a growing number of people in both developed and developing countries. If, on one the hand, there is the need of limiting soil consumption and urbanisation rates, on the other hand, the presence of brownfields within formal urban contexts compromises the living conditions of the neighbouring areas.

Milan is rapidly gaining importance in the international panorama, attracting people and investments. This process leads to important urban transformations, which started before the Expo 2015 with the completion of the Porta Nuova and City Life areas, continuing also in other important areas through a series of competitions such as the Farini Railway scale one and the two editions of the Reinventing Cities C40, involving 11 sites in 2 years. In the next 10 years, an amount of nearly 10 billion of euro is expected to be invested into the redevelopment of the few remaining available areas (Dezza, 2019). The question, apparently, is therefore not if Urban Transformation Areas (ATU) will be developed or not, but only how and when this process will happen. In this context, designers are asked to minimise the impact of these inevitable transformations not only at the site level, but also trying to implement projects integrated with the surrounding urban subsystems. Given the complexity of the topic, a holistic, multi-scale and multi-disciplinary approach is required.

This article presents a case study of integrated actions applied at the neighbourhood scale and based on a rigorous methodology named "Integrated Modification Methodology" (IMM), which refers to the built environment as a complex adaptive system (CAS) (Tadi et al., 2020). The case presented here is the area of Porto di Mare (ATU 15, PGT[1] Milan 2030), located on the south-eastern border of the city of Milan, bounded by important infrastructures, between the city centre and the rural belt surrounding the Milanese metropolitan area. This area was included into the Europan 12 competition and has recently become property of the municipality after years of abusive settlements and abandonment.

To achieve these goals, an integrated proposal for an Eco-District has been elaborated in accordance with both public and private partners (Milano Depur S.p.a.–Depuratore Milano Nosedo; REsilienceLAB; Gruppo Ecologisti Est Milano–GREEM; Associazione Piccole e Medie Industrie di Milano–API Milano; Wageningen University and Research–WUR). The Eco-District systemic and bottom-up approach tries to act on people, making sustainable behaviour much easier than elsewhere and transforming users into more sensible agents (EcoDistricts, 2014).

---

[1] Piano di Governo del Territorio (PGT), urban planning tool of Milan Municipality.

2 The Integrated Modification Methodology (IMM) and Its Application to the Porto di Mare Area

The IMM is a phasing process comprehending an open set of scientific techniques for morphologically analysing the built environment in a multi-scale approach and estimating its performance for the actual state or alternative design scenarios. It is a nonlinear phasing process aiming at delivering a systemic understanding of any existing urban context, formulating the modification set-ups with the aim of improving its performance and examining the modification strategies to transform that system. The IMM methodology has been developed at the Department of Architecture, Built Environment and Construction Engineering (DABC) of Politecnico di Milano and considers cities as complex adaptive systems (CAS) consisting of different interacting subsystems: environmental (energy, water, food, transport, health, biodiversity), economic, social and cultural ones. It analyses the correlation between urban structure and environmental performance through a set of scientific and rigorous procedures, among which a wide range of indicators aligned with the 17 UN Sustainable Development Goals (SDGs) for 2030.

Two main factors distinguish this research from similar studies:

- the holistic approach through which the built environment is regarded as a multidimensional synthesis of various systems;
- the intention of visualising mathematical-based predictions of transformation reactions to possible intervention scenarios.

The IMM represents a support tool for transformation and management decisions, flexible enough to conceptually expand with respect to the multi-finality of the built environment and its various environmental and socio-economic subsystems. It is based on a nonlinear phasing process involving the following structure:

- Phase I. Diagnostic: Analysis and Synthesis;
- Phase II. Assessment and Formulation;
- Phase III. Intervention and Modification;
- Phase IV. Retrofit and Local Optimisation.

In the first phase, urban components (Volume, Void, Network, Type of use) are first investigated individually, and subsequently, the emergence of their synergetic interaction, called Key Categories (KCs), can be considered. In the IMM, KCs are a new organisation emerging from the elementary parts, although not as a simple additive result of their properties. By modifying the components or the way in which they interact, different organisations emerge with the possibility of either improving or worsening the current system condition. In the IMM, the Key Categories are as follows: porosity, permeability, diversity, proximity, accessibility, interface and effectiveness. The diagnostic of the system is composed of a horizontal (among components) and vertical (across components, among KCs) investigation, showing the weakest elements that are mostly responsible for the current system performance. To estimate the performance of the built environment, the IMM offers an open list of nearly 150 indicators, organised in 12 families, corresponding to the 12 IMM Design Ordering Principles (DOPs) and clustered on the basis of the SDG targets (United Nations, 2015), with the aim of approaching the performance of the built environment from different angles. Figure 1 shows a flowchart describing IMM elements and their interaction.

The IMM indicators' list includes some of the SDG11 indicators (1.1–2.1–6.1–7.1), although it is broader, as the smaller application scale (city vs. country) allows to have a wider set of measurable aspects. The second phase of the IMM process, called "Formulation", anticipates the design phase and is highly connected with it. It moves from the outputs of the Diagnostic Phase, identifying the transformation catalysts thanks to the KCs and their associated indicators. In this second phase, the DOPs play a significant role, represented by a set of integrated tools composed of 12 descriptive guidelines to orient the designer in the arrangement of the CAS structure.

The application of these principles intends to modify the structure of the CAS and, consequently, its performance. Hence, within the IMM methodology and its specific phasing process, a direct link between the SDG targets and DOP actions has been established (Fig. 2).

In the third phase, different transformation design scenarios oriented to local modification (neighbourhoods/local nodes) can be elaborated, with the aim of achieving a positive transformation of the whole urban system. As a consequence of this phase, a new structure of the system, with enhanced performance, will emerge. Subsequently, the new CAS resulting from the transformation phase is retrofitted in the last phase, where it is evaluated and compared with the old one using the same procedure and indicators previously applied in the investigation phase (Phase I); finally, it is locally optimised to reach the optimum modification plan.

The case study of the Porto di Mare area, an important and underdeveloped zone in the south-east of Milano, characterised by an unexpressed potential, is here presented as an example of an innovative Eco-District characterised by a full integration of the infrastructural systems (mobility, building, green–blue infrastructure, physical flows of energy, water and waste) in the design in order to reduce its carbon footprint. In the next section, the Eco-District masterplan, developed engaging local stakeholders, is presented and its performance is compared with the state of the art and with that of an alternative transformation scenario.

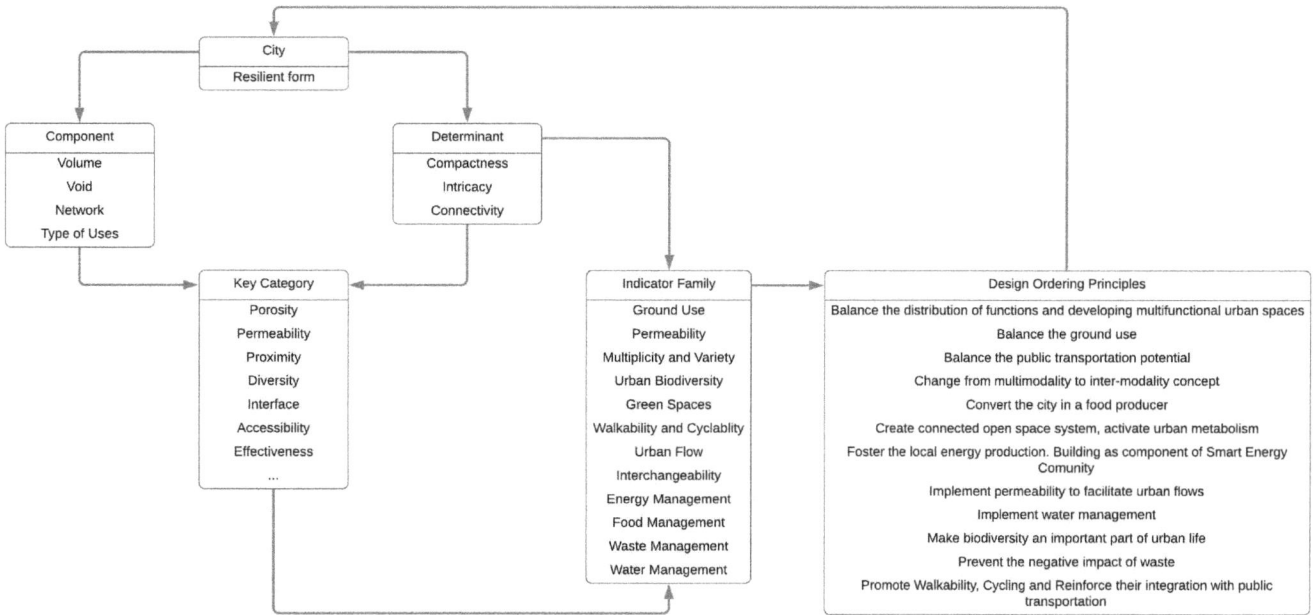

Fig. 1 Flowchart describing the IMM elements and their interaction (Biraghi, 2019)

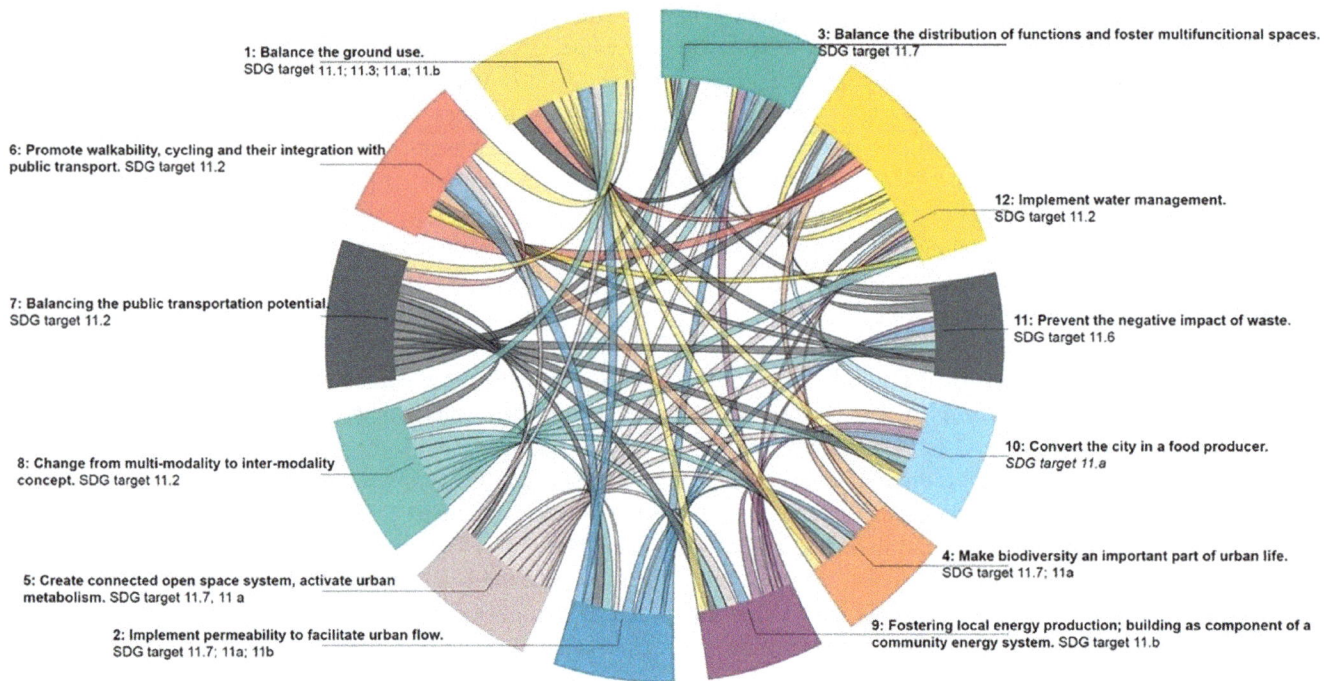

Fig. 2 Diagram showing the interdependency of IMM DOPs and the related SDG11 targets

3 A Masterplan for a New Eco-District at Porto di Mare in Milan

The first two phases, analysis and synthesis and assessment and formulation, widely discussed in Tadi et al. (2019), selected the horizontal (Void component) and vertical (Interface KC) catalysts and provided a site-specific DOP ranking with some design suggestions. Promoting walkability and cycling and reinforcing their integration with public transportation was identified as the priority principle to drive the transformation. A masterplan was then developed to give architectural consistency to all the hints and advices coming from the diagnostic activity. The design process was carried out through two international workshops hosted in the Lecco Campus of Politecnico di Milano in collaboration with the University of Cincinnati, focusing on the energy-related aspects, and the Wageningen University and Research (WUR), integrating the food system into the landscape and urban design. The result was not only an arrangement of volumes but an integrated proposal with attention on qualitative aspects of the open space, modifications to the existing public transportation network and type of uses indications.

Before going into the explanation of the masterplan, reported in Fig. 3, it should be mentioned that all built volumes have been established on the portion of soil, property of the municipality, already consumed and

currently occupied by informal illegal settlements whose demolition is already foreseen.

Starting from the northern corner of the area, the first modification involved the existing metro stop. A simple exit facing a highly congested street has been integrated into an interchange hub where it is possible to switch to both different individual and public transportation modes, encouraging soft mobility solutions. The street network has been modified to host a large pedestrian area in correspondence with the hub, characterised by safe paved ways surrounded by green cool islands. In this area, another key function of the district, the local food market, has been established because of its high level of accessibility given by the proximity to both roads and open landscape. This intervention connected an existing garden located in the bottom part of a large block with a dozen of social housing buildings to the public park "Gino Cassinis", fading into the countryside towards south. From that area, a north–south green spine delimitates the buildable area on the left from the surrounding natural one on the right. This vertical urban axis is balanced by the presence of a horizontal natural corridor connecting the countryside with the agricultural area located in the top-left corner beside Cascina San Giacomo and the Nocetum association offices. A system of finer grain green public and semi-public spaces permeates the residential blocks, offering a diversified outdoor experience. The natural landscape, result of the incomplete projects for the creation of a Milanese harbour developed in the twentieth century, is

Fig. 3 Porto di Mare masterplan
(Tadi et al., 2019)

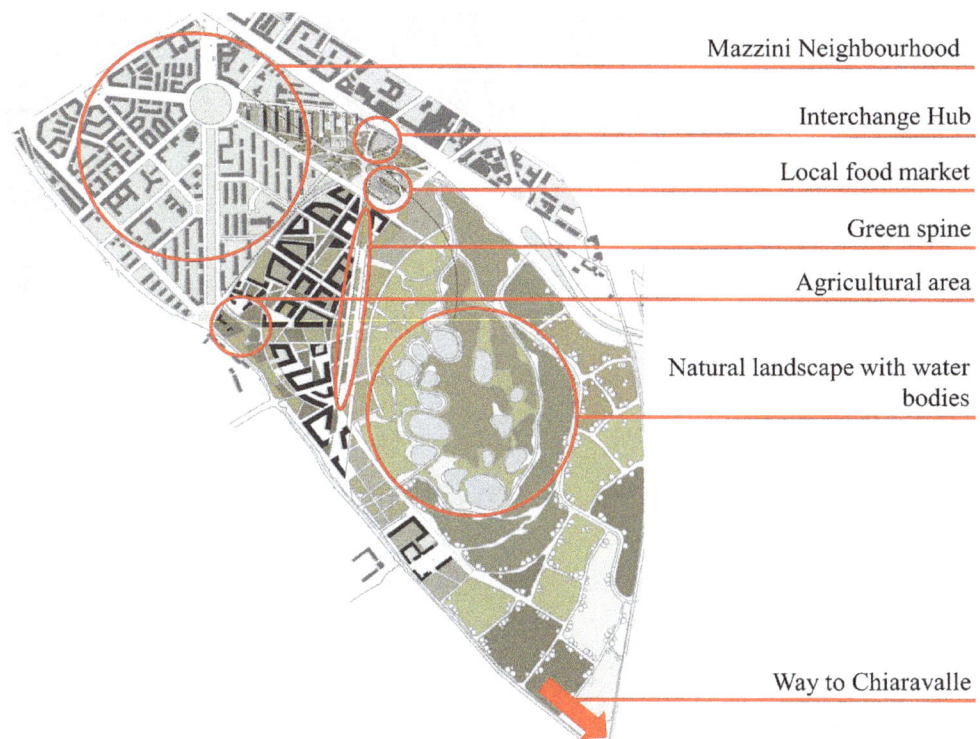

Mazzini Neighbourhood

Interchange Hub

Local food market

Green spine

Agricultural area

Natural landscape with water
bodies

Way to Chiaravalle

preserved and integrated with water bodies for climate mitigation and recreational fishing activity. The presence of water is somehow natural in this area, which is the lowest of the Milanese context in terms of height above sea level and where the aquifer is almost superficial.

To support the decision of a whole car-free district, a local electric bus line crosses the area, stopping along the spine, and into the existing street network, forming a loop that serves both the new and the existing inhabitants. In addition, the existing cyclable network coming from south and connecting to Chiaravalle and La Valle Dei Monaci is extended inside the district and over, reaching the three main city train stations (Centrale, Garibaldi and Rogoredo), intercepting and reconnecting existing network segments. Buildings along the main axis of the district are mix-use, with the presence of commercial and community spaces at the ground floor also to balance the lack of services of the nearby Mazzini neighbourhood, characterised by a zoning approach with few blocks containing all the public buildings surrounded by linear popular residential buildings fenced together. The typology of the court has been used to increase the sense of enclosure, totally missing, to limit the amount of residual space, and as a typology suitable for hosting social housing, representing more than 30% of the residential offer. The same set of maps used in the investigation phase to represent the state of the art (Tadi et al., 2019) is used here to show the Eco-District transformation scenario, drawing a profound vision transformation impact on the CAS. Key

category maps[2] (porosity, proximity, accessibility, diversity, effectiveness and interface), representing a functional symbiosis between the four component layers, are reported in Fig. 4.

To provide a better assessment of the impact on environmental performance, an alternative transformation scenario has been tested against the Eco-District 1 and the starting condition. This project has been developed by students of the course Architectural Design Studio 2 of the Politecnico di Milano, held by Professor Michele Caja, under the supervision of Arch. Carlo Andrea Biraghi and Sotirios Zaroulas in the academic year 2017/2018 (Biraghi, 2019). Each group was in charge to design a set of mix-use buildings in a predefined grid of blocks with maximum heights allowed. For doing that, a subset of 25 IMM Performance Indicators, covering 10 out of the 12 families,[3] has been used to obtain a preliminary understanding of the transformation impact. The full list of indicators is reported in Table 1, with each of them indicating the IMM family, the progressive number inside the IMM indicator list, the name and the acronym, the value for the three different system arrangements and the ranking position in respect to the other

---

[2] Permeability Key Category map has not been included as, at the time of the diagnostic of the state of the art (Tadi et al., 2019), its representation was not yet implemented.
[3] Waste management family has not been included because for a lack of data. Energy management family has been treated separately in more detail in Sect. 4.

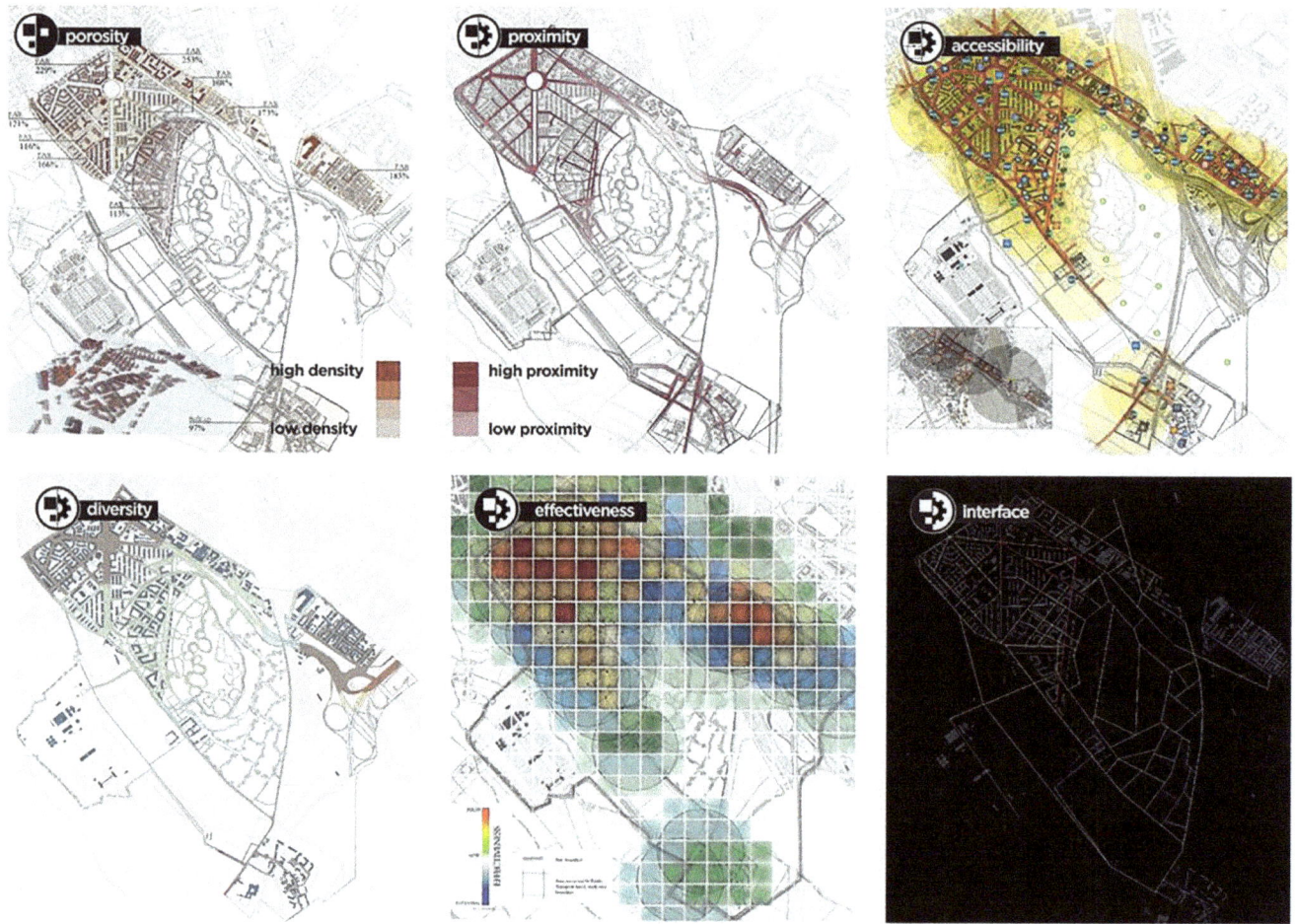

Fig. 4 Key category maps of the Eco-District proposal: porosity, proximity, accessibility, diversity, effectiveness and interface

Table 1 Performance assessment of the three compared scenarios using a subset of IMM indicators

Indicator family	Nr	Indicator	Acr	State of the art		Eco-District		Alternative scenario	
				Value	Rank	Value	Rank	Value	Rank
1 Ground use	1	Volume density	VD	2.42	45	4.19	26	3.77	29
	2	Building density	BD	417	43	401	43	419	43
	3	Population density	PD	8926	34	9405	32	9332	32
2 Permeability	5	Street cover ratio	SCR	0.18	46	0.24	31	0.22	36
	8	Block density	BLD	0.24	38	0.32	32	0.27	36
3 Multiplicity and variety	11	Population activities ratio	PAcR	13.65	33	12.7	33	12.85	32
	13	Job housing ratio	JHR	0.27	64	0.29	61	0.28	63
4 Biodiversity	17	Land use share	Lush	0.89	3	0.94	1	0.94	1
5 Green spaces	26	Green coverage ratio total	GCRt	0.35	35	0.32	39	0.41	27
	28	Green coverage ratio urban	GCRu	0.14	52	0.22	26	0.16	42
	29	Tree density	TD	2960	19	3176	16	3158	17
6a Cyclability	31	Bike lane density	BikeD	433	54	2381	5	1948	12
	32	Bike lane average length	BikeAl	63	47	521	1	485	1

(continued)

Table 1 (continued)

Indicator family	Nr	Indicator	Acr	State of the art		Eco-District		Alternative scenario	
				Value	Rank	Value	Rank	Value	Rank
6b Walkability	41	Node density	ND	104	48	113	43	107	48
	45	Accesses per block	AxBLP	1.88	59	1.86	60	2.15	53
	48	Ground floor activities	GFAc	0.49	41	0.58	32	0.59	32
7 Flows	50	Public transportation accessibility	PTA	0.93	65	1.26	58	0.96	63
	51	Road per capita	LlPR	1.32	67	1.7	46	1.3	68
	42	Dead-end nodes ratio	NDER	0.15	45	0.13	54	0.13	54
8 Interchangeability	67	Transportation mode share	Modesh	0.67	33	0.88	15	0.67	33
	68	Metro line share	MMsh	0.2	30	0.2	30	0.2	30
	69	Stop density	StopD	15.7	58	17.03	55	15.94	58
	70	Line density	LineD	5.5	73	5.8	71	5.5	73
10 Food management	78	Green coverage ratio agricultural	GCRa	0.22	22	0.1	22	0.25	18
12 Water management	86	Water area ratio	WAR	0	40	0.08	2	0	47
			Perf. index	41		14		21	

87 Nuclei of Local Identity (NIL), a neighbourhood-like partition existing in the city of Milan. A 3D view of the three layouts (State of the art, Eco-District and Alternative scenario) and a plan of NIL 35 Lodi-Corvetto, including the Porto di Mare transformation area, are shown in Fig. 5. An overall simplified performance index was used here to obtain a synthetic performance value, calculated giving the same weight to all indicators and using the rank as a shared unit of measure. In this sense, a higher overall performance ranking among the NIL means a more sustainable design proposal.

This analysis showed, as expected, an improvement in respect to the state of the art for both transformation scenarios, with better results for the Eco-District proposal mainly because of the higher attention to the network component both in terms of public transportation integration and relationship with the countryside, managed with a smoother transition. Having consolidated this result, a finer grain optimisation process was carried out, focusing on thermal loads of buildings, outdoor comfort and energy use intensity.

4 Local Optimisation Via the Multi-layered Modelling Process

Once the overall arrangement of the masterplan was defined based on the process described above, a local optimisation process was started to assess the impact of smaller scale design decisions, regarding typological aspects, such as the roof shape, but also micro-modifications to the overall design layout (Li et al., 2016) and even technological choices. In this case study, the investigation specifically regarded some energy-related aspects that are used to evaluate the achievement of the following design principles:

- roads should be sunny during winter and shaded in summer (Li et al., 2016);
- south facades should be exposed to sun in winter and mid-seasons and shaded in summer (Johansson, 2006);
- roofing should avoid receiving shade to maximise electricity production from PV panels (Compagnon, 2004).

The indicators that will be used to assess the effects of the design choices belong to the Energy Management Family and are therefore part of the wider set of indicators of the IMM methodology:

- energy use intensity (EUI);
- solar radiation falling on the south-oriented surfaces;
- potential production of renewable energy from roof-mounted photovoltaic panels;
- solar radiation falling on the ground.

The local optimisation process will focus on the variation of these aspects at the building scale:

- building height;
- floor plan depth;
- roof slope;

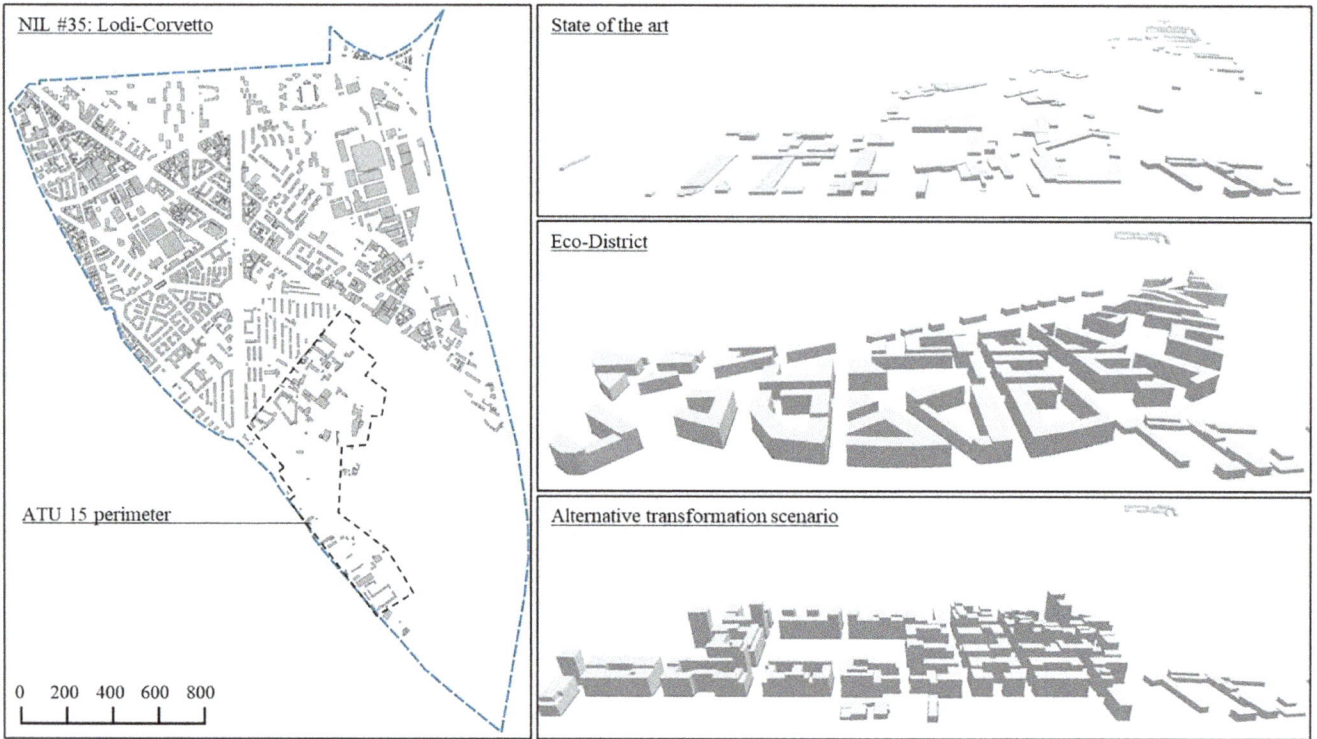

Fig. 5 NIL #35 plan with ATU 15 perimeter highlighted and 3D aerial view from Via Fabio Massimo of the three compared scenarios for ATU 15: state of the art; Eco-District; Alternative transformation scenario (Biraghi, 2019)

– average U-value of the building envelope (perimeter walls, windows and roofs);
– window-to-wall ratio (WWR).

To control the variation of these parameters and obtain the related values for the indicators, a multi-layered modelling process was implemented, exploiting the dynamic synergies between a proprietary BIM software (Autodesk Revit) and its visual programming plug-in Dynamo, an open-source computational software which extends the BIM domain with the data and the logic environment of a graphical algorithm editor. Dynamo was crucial to automate repetitive tasks through the creation of scripts that allowed to handle multiple combinations of parameters in a manageable time. The modifications to the geometry of buildings were managed via an algorithm inside Dynamo developed by the authors, and energy analysis was managed through Green Building Studio, a flexible cloud-based service by Autodesk, running building performance simulations to optimise energy efficiency from the early design process (Mousiadis & Mengana, 2016). During this relatively early stage of design, energy analysis was performed using a combination of Revit conceptual masses and building elements.

It must be highlighted that the intervention area was of particular interest for an additional reason related to institutional planning procedures. From one edition to the other of the PGT, the admissible GFA to be located increased from 127,719 m^2 in 2015 to 819,761 m^2 in 2020. This raised the question of which could be an acceptable amount of GFA in this extremely large range, considering the overall goals of energy efficiency and comfort. A preliminary analysis of the resulting building masses led the authors to cap the allowable GFA to 260,000 m^2, in the hypothesis of a relatively consistent height of the buildings across the site (i.e. no towers or very tall blocks) (Fig. 6).

To deal with a more manageable amount of data, the masterplan was divided into three sectors (1 pink, 2 yellow, 3 light blue) by the main south-east/north-west axis, going from the countryside to the city. Each building was then classified into five parametric typological families identified by a letter (C: "C" shape open court; CC: closed court; L: "L" shape; P: parallelogram; T: triangle; PE: customised shape). Combined, this information forms an alphanumeric code (i.e. C_1_A) uniquely identifying a specific building (Fig. 7). Moreover, height limits are defined in relationship to the surrounding context to assure a coherent urban transition from city to countryside, as indicated in PGT 2015.

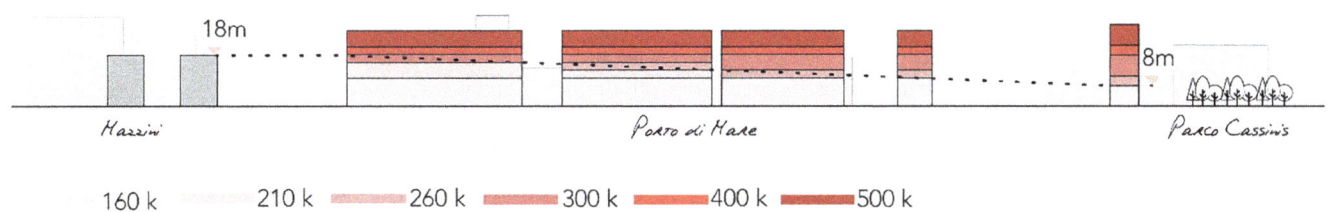

Fig. 6 Cross section for determining the maximum admittable height to provide a smooth urban transition from the existing city to the surrounding green areas

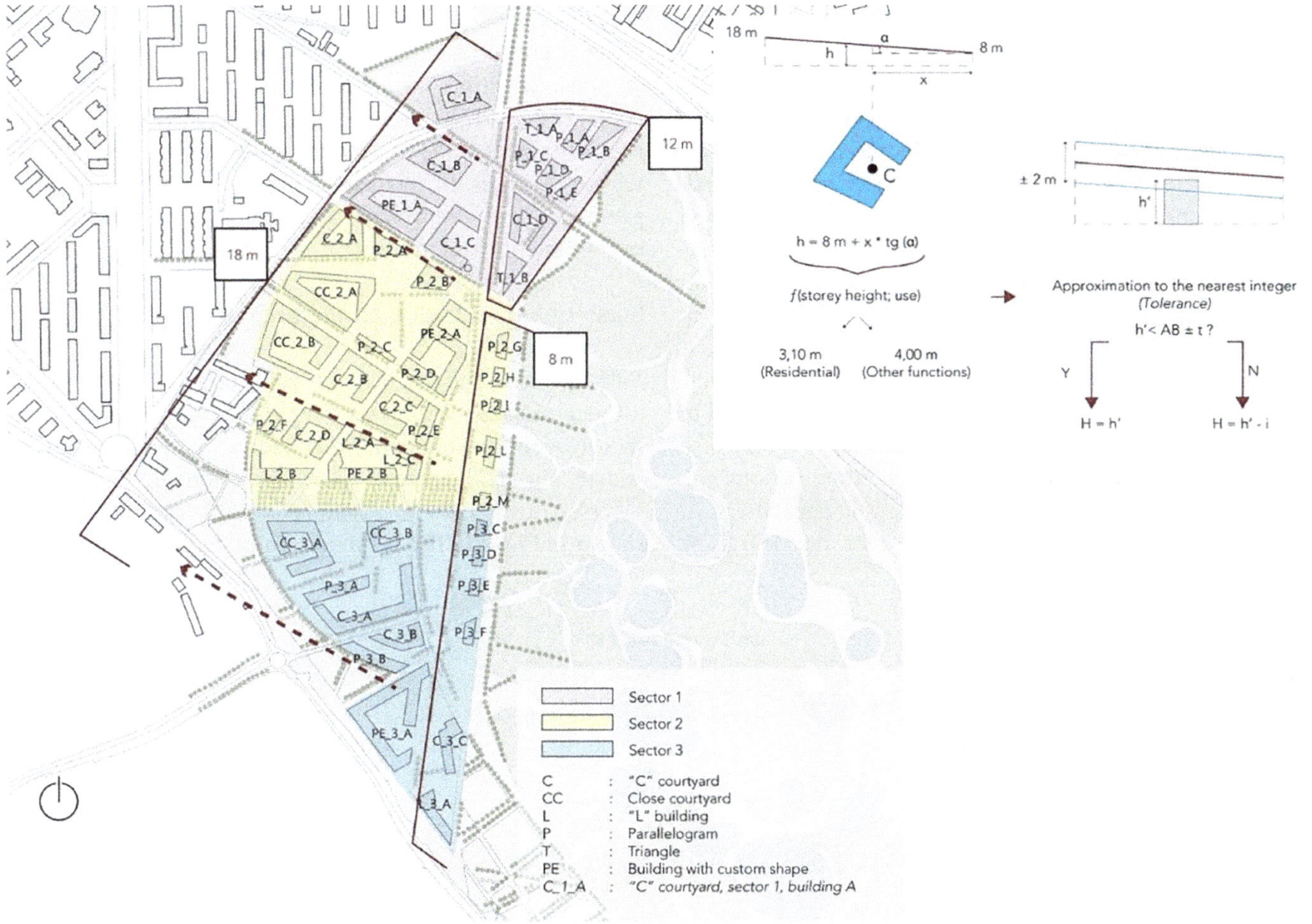

Fig. 7 Masterplan subdivision and building typological classification for modelling purposes

The height of the tallest buildings on Via Fabio Massimo (18 m) was used as the upper limit, whereas at the lower end, two-storey agricultural buildings (approx. 8 m) were used as reference (Mauri et al., 2018). Volume modelling followed three steps:

– creation of an inclined plane corresponding to the 8–18-m-high line;
– mathematical evaluation of building height (as a function of storey height and use);
– verification of the height value with the tolerance range.

Considering the aspects mentioned above, the multi-layered modelling therefore presents 54 scenarios resulting from all the possible combinations of the following variables:

– three GFA configurations: 160 k m^2–210 k m^2–260 k m^2;
– two U-value scenarios for building envelope components: "Law" (mandatory values) and "Top" (improved);
– three cases with different window-to-wall ratios (WWR): 15% (baseline), 30% and 50%;

– three footprints: the one defined by the masterplan (1), fixed at 12 m (2) or increased by 10% (3);
– two roof geometry types: constant height or decreasing heights (computed according to the previously illustrated principle).

Evaluation of energy use intensity (EUI)

The starting case for this investigation is named "Baseline, 1" and corresponds to the masterplan presenting the following characteristics: a GFA of 160 k m^2, a "Law" U-value, constant building height and no footprint modification in comparison to the original proposal.

To perform whole-building energy simulations, Revit exploits the built-in plug-in Energy Analysis. This add-in provides an estimate of the expected energy use (fuel and electricity) based on the building's geometry, climate, type, envelope properties and active systems (HVAC and lighting). Through this analysis, the 54 possible cases were analysed, and among them, only 7 showed better EUI performances than the reference case "Baseline, 1", as shown in Fig. 8. For each case of GFA, the best geometric configuration was Type 3, corresponding to a footprint increased by 10%, meaning a more compact arrangement of volumes.

The results show that the differences in the various scenarios refer mainly to the cooling and heating parameters. As an example, in Fig. 9, the outcomes for the 160 k m^2

configuration with the U-values required under the current law are shown according to the different window-to-wall ratios: since Type 3 always performs better, it was selected for further analyses.

Solar radiation on south-facing surfaces

Analysis of solar radiation falling on surfaces facing south shows that it decreases as the total GFA increases, as this is translated into taller buildings and a more mutual shadow cast by the volumes (Fig. 10). Furthermore, Type 3 showed a lower average annual solar radiation than the other configurations, resulting in a lower requirement of energy for cooling (Fig. 9); its more compact arrangement of volumes compensates for the reduced access to solar radiation in winter. The results obtained confirm the hypothesis that geometry and compactness are significantly linked to the overall energy requirement.

Solar radiation falling on the ground

Subsequently, the correlation between GFA and annual average radiation falling on the ground was examined as an indicator of potential comfort conditions in the open spaces, concluding that, for the selected masterplan, there is an inversely nonlinear correlation between the two variables, as shown in Fig. 11. The results given by the simulation of the

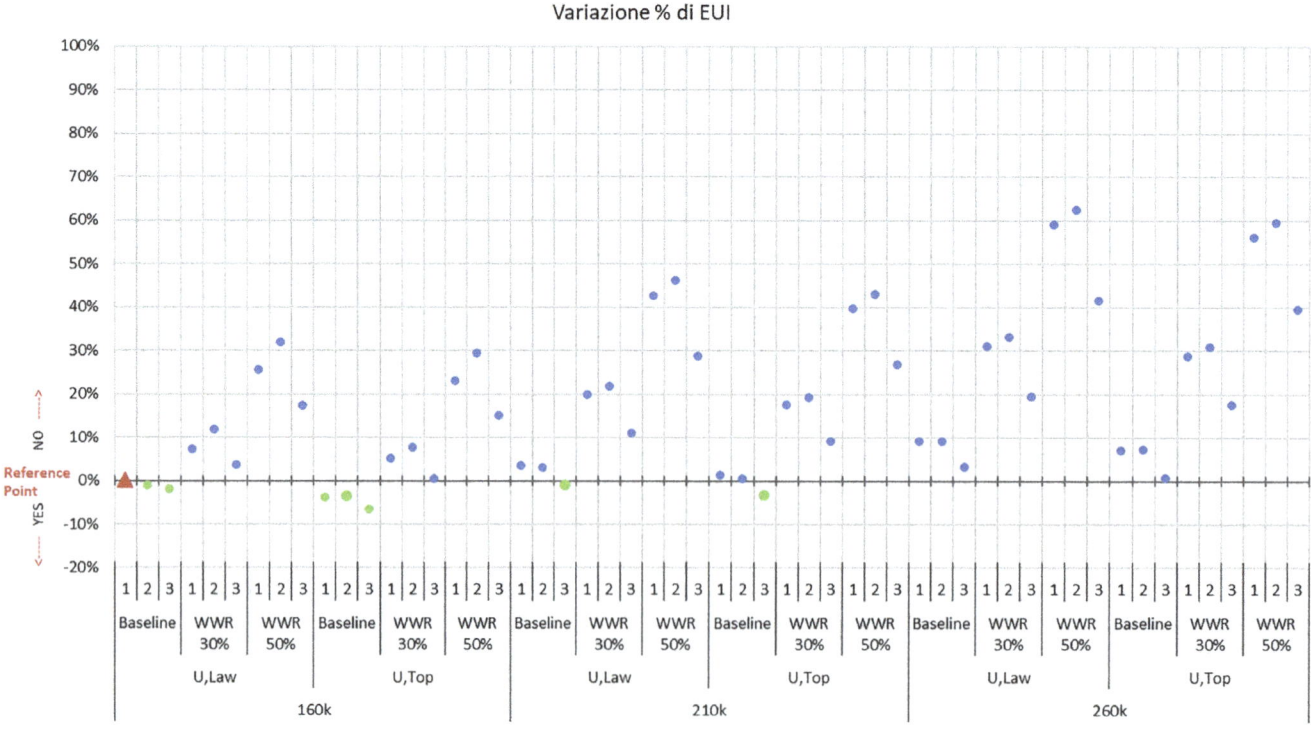

Fig. 8 Comparison of the 54 scenarios resulting from the combination of the modelling parameters

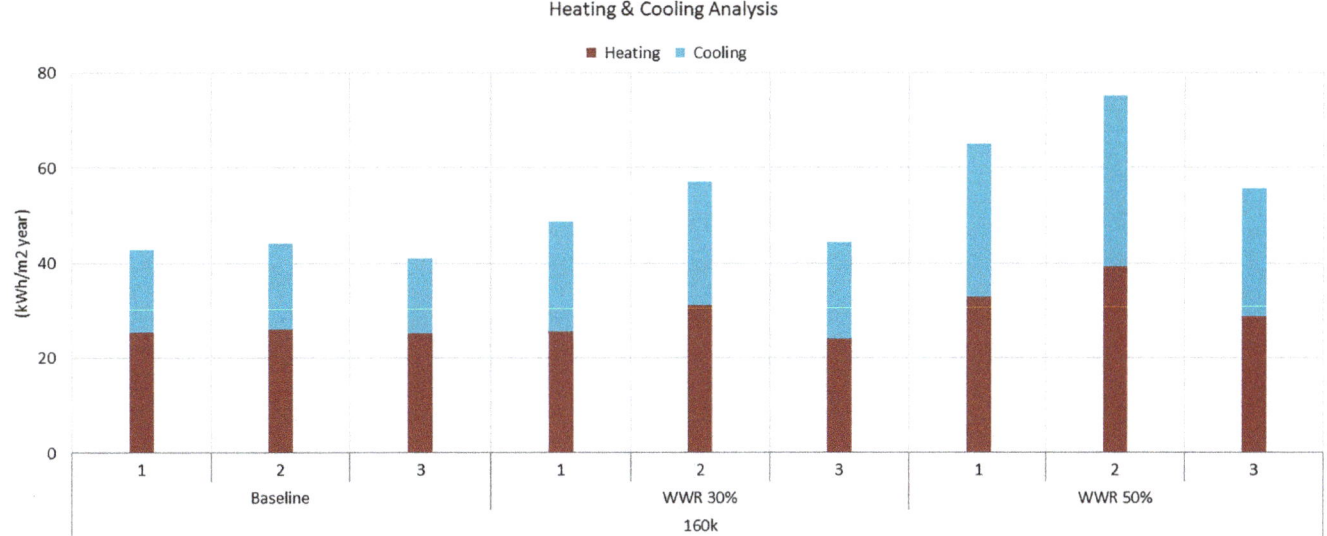

Fig. 9 Heating and cooling analysis for the different configurations of the 160 k m² scenario

Fig. 10 Graph showing the solar radiation on south-oriented surfaces: total and variation per m² compared to the reference case "Baseline, 1"

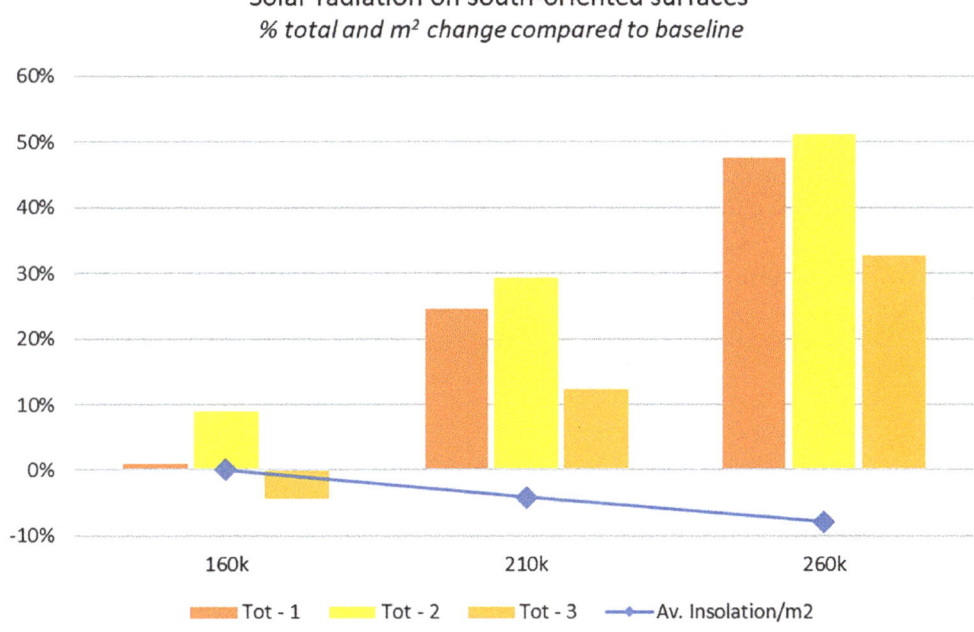

different scenarios were projected through polynomial trendlines, chosen for R2 values very close to 1 (indicating the maximum reliability).

Roof geometry

Once the analyses on solar radiation incidence were concluded, it was possible to investigate the correlation between the roof type of the buildings and their implication on the potential energy production by photovoltaic panels, designed to offset at least part of the energy consumption calculated above.

Three different roof geometry configurations were simulated: single pitch (10° inclination across the whole site), south-inclined (10° north–south inclination through the barycentre of each building to increase solar exposure) and flat. This led to a total of 66 different scenarios. By analysing the output given on the heating and cooling needs (EUI), it was clear that the roof morphology has a considerable impact on the energy requirement, since the inclined solutions showed an average energy saving of about 5% compared to the flat roof scenarios and a 10% higher electricity production by photovoltaic panels.

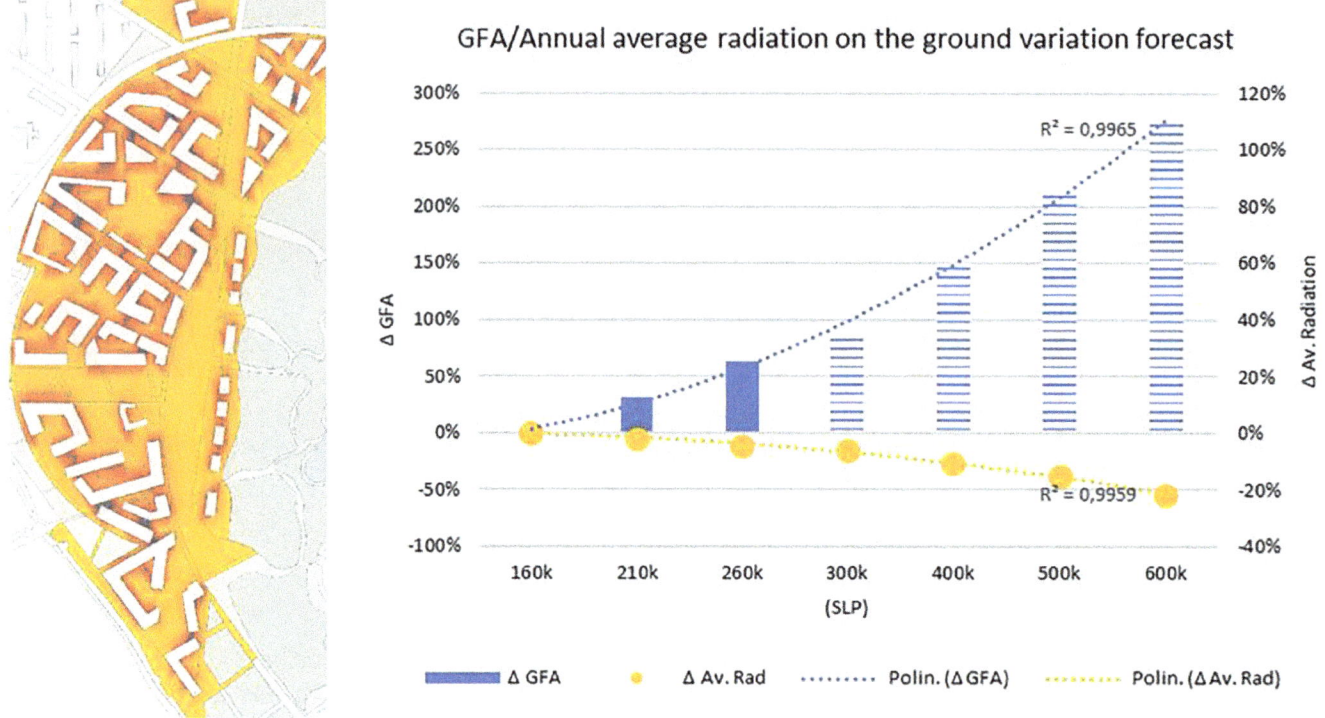

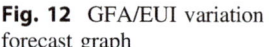

Fig. 11 Solar radiation on the ground and variation forecast

Based on these results, it is possible to understand the evolution of the trend of the overall EUI (including both energy required and energy produced by photovoltaic panels) as a function of the variation of the GFA (Fig. 12).

The forecast shows a widening gap between the two projections as the GFA increases. As the geometrical parameters (height and footprint) remain constant, a non-linear correlation between GFA and EUI is visible due to the growing density of the district, leading to more shade and a smaller impact of the energy produced by photovoltaic panels since the roof surfaces remain the same, while the floor area to be heated and cooled increases.

Fig. 12 GFA/EUI variation forecast graph

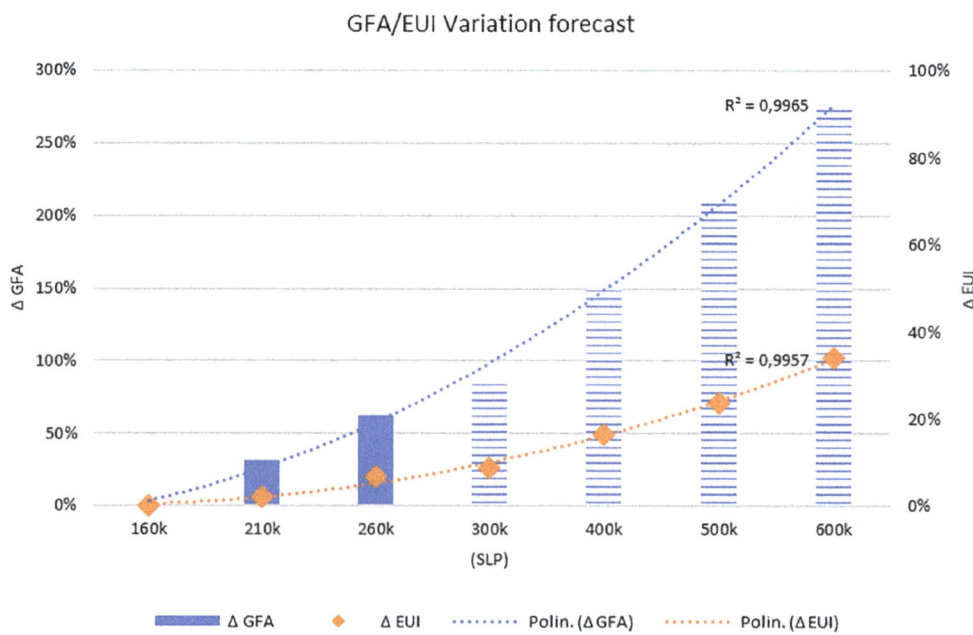

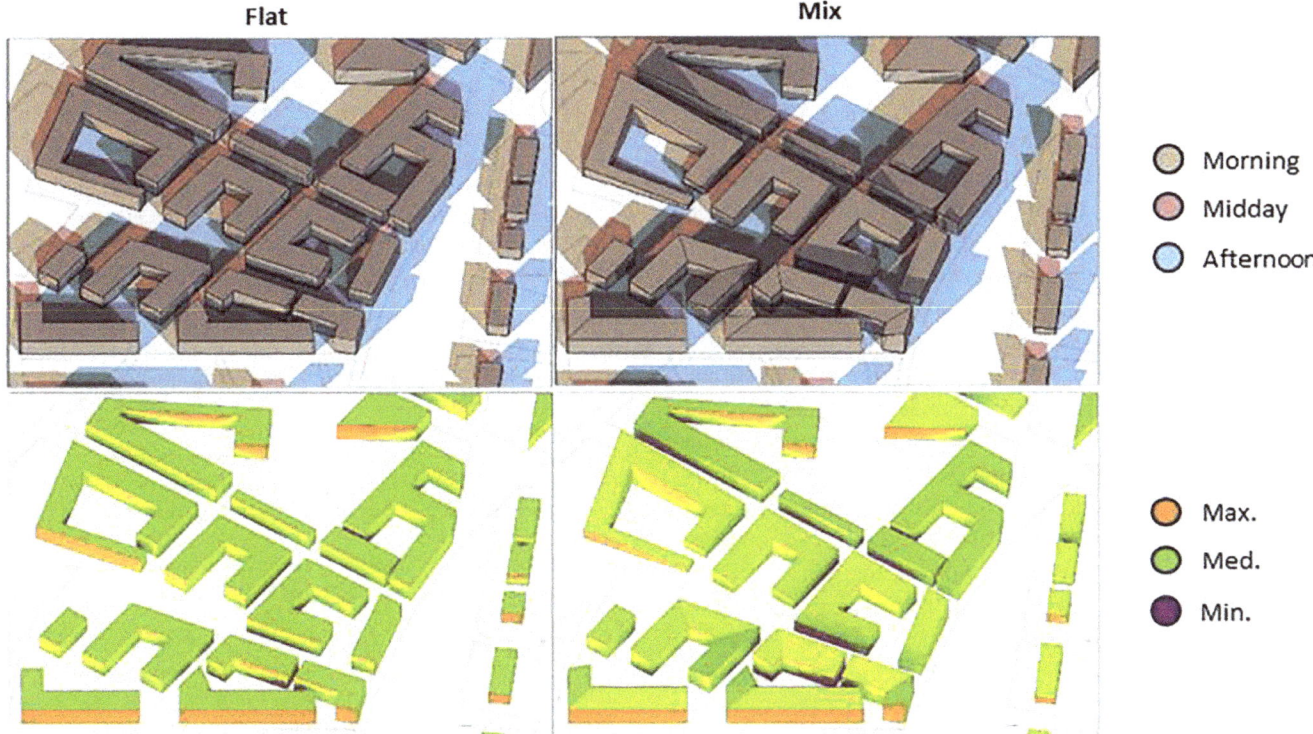

Fig. 13 Shadow (up) and solar radiation (bottom) analysis in sector 2. Comparison between flat and mixed roofing

Introducing variations

The process described above provides an overall picture of how the selected indicators react to design solutions applied to the whole district without any differentiation. To achieve more diversity in the volumes and the architectural expression of the new city, it was imagined that the actual design will be a combination of the configurations with the better performances.

The masterplan was therefore divided into six sectors, where different building types were mixed and shadow analyses were performed to assess the achievement of the goals stated at the beginning of this paragraph. For this study, only a qualitative analysis was carried out considering shadow cast on the ground and on the neighbouring buildings. The best configuration determined the typology of roofing for each specific sector; the new mixed district composition was analysed to verify the above criteria and to obtain details about energy use, solar radiation and shadows. This method was applied for each satisfactory GFA configuration in winter, summer and mid-seasons, evaluating the shadows in the morning, at mid-day and in the afternoon.

Finally, to most effectively evaluate the interactions among buildings, projected shadows in Sectors 2 and 3 were more precisely analysed by further subdividing the buildings into subgroups and evaluating the incident average radiation per square metre on facades and roofs. The analysis highlighted that with increasing building height, the impacts of the roof geometry decrease. Figure 13 shows the 160 k m^2 configuration as an example: it is possible to detect some differences between the flat and the mixed roofing, which can provide a better production by photovoltaic panels and a more efficient energy profile.

5 Conclusions

We used a case study transformation area to demonstrate the advantages of the application of the IMM methodology and its integrated and holistic approach for the redevelopment of an area. The application to an area in a large city of a developed country confirms the flexibility of the methodology, previously successfully applied also in informal settlements (Masera et al., 2020). While Tadi et al. (2019) focused on the Diagnostic and Formulation phases, here, the attention is on the Modification, Retrofitting and Local Optimisation phases. The overall assessment of a district performance via IMM methodology is effective and able to grasp differences between alternative transformation scenarios even in a partial application of its indicators. Local Optimisation demonstrated how a consistent improvement in the energy performance of a district can be obtained with minor morphological modifications.

This showed how a simplified volumetric masterplan can be sufficient to measure performance aspects related to urban dynamics and the interaction with the surroundings, whereas energy aspects benefit from a finer grain investigation. Parametric design tools were used not for the automatic generation of shapes (form-finding), but rather to adjust a given layout on the basis of minor typological building modifications (height, floor plan depth, roof, WWR).

The presented modelling strategy provides a possible answer to some open questions, showing a possible replicable approach for similar contexts. Given the urban features of the Porto di Mare context and the existing regulations, three scenarios of GFA were considered acceptable and coherent from both the energy use and the environmental point of view: 160 k, 210 k and 260 k m^2. Another interesting result is the benefit for the energy consumption provided by an increase of 10% of the actual building footprint (i.e. deeper floor plans) for each GFA scenario.

Given the lack of obstructions, in this case study, it is suggested to adopt inclined roofs towards the south to maximise the incidence of solar radiation and therefore increase the potential for energy production by photovoltaic panels. An accurate local evaluation of roof types, adopting mixed solutions at the smaller local scale, can provide advantages by avoiding self-shadowing and improving the outdoor comfort for users.

From a building technology point of view, the thermal transmittance values provided by the law are considered acceptable, but a better insulation, especially for the opaque components, is strongly suggested given the evident advantages of the improved scenarios. Finally, it is suggested to not exceed a value of 15% of window-to-wall ratio.

The potential given by a dynamic control of design parameters through specific tools can be exploited even while working on building masses at an intermediate design stage, providing precious insights for the designers.

A better investigation of social and economic aspects, albeit not neglected, could ultimately provide a complete and effective tool for decision-makers to evaluate the current level of urban system performance and the impacts of future transformations. In addition, new data sources such as satellite images, sensors and citizen science data will be explored and integrated in the performance assessment to grant a more reliable and up-to-date picture of the behaviour of the urban system.

References

ANCE. (2013). L'esperianza degli Ecoquartieri per ispirare la strutturazione di misure e progetti della nuova asse urbana dei Programmi Operativi Regionali 2014/2020 (Fondi europei FESR ed FSE). *Scheda Di Sintesi*.

Biraghi, C. A. (2019). *Multi-scale modelling approach for urban optimization: compactness environmental implications* [Politecnico di Milano]. http://hdl.handle.net/10589/150884

Compagnon, R. (2004). Solar and daylight availability in the urban fabric. *Energy and Buildings, 36*(4), 321–328. https://doi.org/10.1016/J.ENBUILD.2004.01.009

Dezza, P. (2019). Milano, in arrivo 10 miliardi: così si trasformerà la città entro il 2030. *Il Sole 24 Ore-Online.* https://www.ilsole24ore.com/art/casa/2019-05-06/milano-arrivo-10-miliardi-cosi-si-trasformera-citta–entro-2030-115745.shtml?uuid=ABPzORtB

EcoDistricts. (2014). *EcoDistricts Protocol Executive Summary.* June, 19. http://ecodistricts.org/wp-content/uploads/2013/03/EcoDistricts_Protocol_Executive_Summary_ISSUE_6.242.pdf

Johansson, E. (2006). Influence of urban geometry on outdoor thermal comfort in a hot dry climate: A study in Fez, Morocco. *Building and Environment, 41*(10), 1326–1338. https://doi.org/10.1016/J.BUILDENV.2005.05.022

Li, X., Li, W., Middel, A., Harlan, S. L., Brazel, A. J., & Turner, B. L. (2016). Remote sensing of the surface urban heat island and land architecture in Phoenix, Arizona: Combined effects of land composition and configuration and cadastral–demographic–economic factors. *Remote Sensing of Environment, 174*, 233–243. https://doi.org/10.1016/J.RSE.2015.12.022

Li, Z., Quan, S. J., & Yang, P. P. J. (2016). Energy performance simulation for planning a low carbon neighborhood urban district: A case study in the city of Macau. *Habitat International, 53*(53), 206–214. https://doi.org/10.1016/j.habitatint.2015.11.010

Masera, G., Tadi, M., Biraghi, C., & Zadeh, H. M. (2020). Polimipararocinha: Environmental performances and social inclusion—a project for the favela rocinha. In *Research for development.* https://doi.org/10.1007/978-3-030-33256-3_12

Mauri, T., Mazzucchelli, M. C., & Sala, E. (2018). *Eco-quartiere Porto di Mare : Modellazione e analisi multi-scalare alla scala urbana. Progettazione integrata di un Food district e Hub per i trasporti* [Politecnico di Milano]. http://hdl.handle.net/10589/142146

Mousiadis, T., & Mengana, S. (2016). *Parametric BIM: Energy Performance Analysis Using Dynamo for Revit* [KTH]. www.kth.se

Tadi, M., Biraghi, C. A., & Zadeh, M. H. M. (2019). Urban low carbon energy transition. The new Porto di Mare Eco-district in Milan based on IMM methodology. *Urbanistica, 160*.

Tadi, M., Zadeh, M. H., & Biraghi, C. A. (2020). The integrated modification methodology. In *Environmental performance and social inclusion in informal settlements A Favela project based on the IMM integrated modification methodology* (Research f, pp. 15–37). Springer.

United Nations. (2015). *Transforming our world: The 2030 agenda for sustainable development.* https://sdgs.un.org/sites/default/files/publications/21252030 Agenda for Sustainable Development web.pdf

Can Skopje Museums Regenerate the Social and Urban Sustainability, Through "Social Friday Activity"?

Arbresha Ibrahimi

Abstract

The Republic of North Macedonia is a multi-ethnic state, but the lack of treatment in legal, political, social, and cultural aspects has engendered a problem of socio-cultural identity. The political factor, especially in a state with a democracy (dictatorship imposed) like the Republic of North Macedonia, social institutions especially museums are exploited to creating a national identity that results in the state's assimilation of multi-ethnic identity. This misuse of museums destroyed the primary purpose of these institutions by converting them into institutions where transparencies, trust, truth, and non-discrimination are taboo. The revival of museums means a revival of institutional, social, and urban sustainability, where every city of the Republic of North Macedonia and, ultimately, the capital of the state—Skopje—deserve their resurrection. In this research, I focus on the possibilities of revitalizing museums through social activity, called Social Friday activity. The social, cultural, historical, art and crafts, distinctive transformation, and occupation potential of the museums enable the variety of activities that can be realized in institutional staff-community-other institutions collaboration. The goal of Social Friday Activity is to help the local community, the flexibility of choosing the activity and method of intervention enables us to be closer to the problems of community. Categorizing and prioritizing problems, based on opportunities and means of interventions, opens a way to glocalization. Glocalization in context is a linear process of local and global development and interaction in every sphere. Raising awareness of "the back to the local community" creates sociality and urban sustainability, catalyzed by museums.

Keywords

Museums • Regenerators • Social and urban sustainability • Social Friday Activity

1 Introduction

The Republic of North Macedonia is a multi-ethnic, multi-cultural, and multi-religious state. The cultural mosaic is assimilated by law, by state structures for a principal purpose, and the formation of a nation state (Бачева et al., 2015). The national pathology for the benefits of "authenticity belongs" results in historical, cultural and demographic assimilation, selective economic investment, individual, social and institutional discrimination, etc. The erroneous methodology of perceiving multiculturalism is a source of despair within the state and a source of many problems abroad, especially on the issue of the name "Macedonia" and "Macedonians" with the Greek state, and on the cultural issue with the Bulgarian state.

To which nationality the state belongs to (Albanians or Macedonians)?, who is autochthonous in this state?, who has arrived in this state?, which nationality, in which period migrated to this country?, who are Albanians, are they the ancestors of the Illyrians?, who are Macedonians, are they Slavic Macedonians or a tribe of Bulgarians?, are the present-day Macedonians hiding behind the antiquities of Macedonian history?, do Albanians belong only to the Muslim faith?, were the Albanians Christian, Catholic, Orthodox, or atheist?, are only Eastern Orthodox Macedonians?, is the cross a symbol of "authenticity", symbolizing only the Macedonians in this country?, etc.

The internal struggle of inter-political and institutional states to create answers of the above questions with "clear-cut" history timeline, resulted in a loss of population confidence in the state institutions, reflecting by not using institutional services up to the migration of the population

A. Ibrahimi (✉)
Faculty of Architecture, University of Ljubljana, Zoisova cesta 12, 1000 Ljubljana, Slovenia
e-mail: ai2525@student.uni-lj.si

© The Author(s), under exclusive license to Springer Nature Switzerland AG 2022
C. Piselli et al. (eds.), *Innovating Strategies and Solutions for Urban Performance and Regeneration*,
Advances in Science, Technology & Innovation, https://doi.org/10.1007/978-3-030-98187-7_12

outside the Republic of North Macedonia. The change of the social identity of the territory of Macedonia has been done and continues to be done through the state, cultural, and educational institutions, by drafting, presenting, and documenting the "new reality" of the national identity of this territory. Due to the essential task of museums, research, presentation, and education of visitors, these institutes are a prey to "national pathology" by abusing the primary ethics of museums (Бачева et al., 2015).

I base these research on the history of the formation of museums, and the analysis of the capacity of museums as catalysts for communication through the identity that it carries at the institutional, social, and regional or global level, testifying to the key role of museums in the state. As a misuse of the state institutions of the Republic of North Macedonia, including cultural institutions, especially museums, for the disintegration of the multi-cultural society in the country, I take as a sample the museums of Skopje. While Skopje museums serve a multi-cultural and discriminated society, I focused on eradicating this abusive "institutional culture" of cultural wealth (movable and immovable), theoretically experimenting with the benefits of applying SFA as a regular strategy in implementation in the museums of Skopje, and the reflection of the museums in the creation of social and urban sustainability of Skopje.

2 Museums in the Republic of North Macedonia

From the Neolithic period until statehood in 1991 (Fig. 1), Skopje was under the influence of various occupiers (Gazevic & Maudus, 1974, p. 653). Across different periods, these groups' social, cultural, political, economic, and strategic development has created nests, turning the city into a museum of architectural layers.

Abuse of museum ethics started as a state ideological strategy to assimilate the social multiculturalism of the Republic of North Macedonia, which has its origins in the beginnings of museum development. From a retrospective point of view, the initial data starts from the Middle Ages, usually collections with religious provenance, due to the institutionalization of the churches. In 1907, there were some attempts for the institutionalization of museum collections or museums. The first museum in Skopje was opened in 1907, the "Commercial—Industrial Museum", which was closed in 1908. In 1913, Serbian occupiers decided to establish a "National Museum in Skopje", which was opened in 1914. The task of this museum was to collect, arrange, and study material objects of historical, archaeological, ethnology, and of the nature of "free areas". According to the Zaharinka Baceva, Branislava Mihajlova, and Krste Bogoeski:

> The Serbian occupying authorities began to form various societies and institutions with the task of proving that Macedonia historically belongs to Serbia and that the Macedonian people are Serbian people. Through these institutions, the assimilation's steep of the Serbian bourgeoisie should have been realized (Бачева et al., 2015, p. 7).

As a result of this strategy, the "Museum of Southern Serbia" was formed in 1924 as a historical and archaeological museum with a lapidary in Kursumlian. To the museum ware added the departments of ethnographic—anthropogeographic in 1925, a zoological department in 1926 and a theological—petro-glyphic department in 1927. After 7 years, in 1934, a decision was made to establish a church museum, "St. Mina", when the damaged church was to be adapted. In this museum, various artifacts and bibliographic materials were kept, such as frescoes, books, icons, clerical garments, and others. Until the end of its existence, it was not open to visitors. Under the slavery of the Bulgarians, the National Museum opened on May 25, 1942, with a capital in a commercial-industrial chamber and partly in Kurshumlian. The Zoological Department of the Museum of Southern Serbia was abandoned in 1926, but it was reused in 1941 in another location (Oce Nikolov Street, Skopje), which was demolished in 1963. The museum had mineral petro-glyph, paleontological, botanical, and zoological divisions. In 1944, the objects from the National Museum and the Zoological Museum were placed under the supervision of the National Bank of Skopje, and a large part of these objects was transferred to Sofia. One year after the Second World War, in 1945, the National Museum in Skopje started functioning again with three departments: Ethnographic, Archaeological, and Middle Ages. On May 1, 1964, the museum was

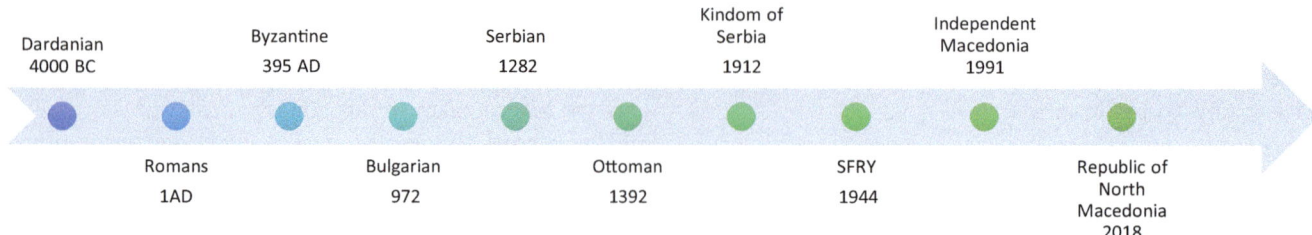

Fig. 1 Skopje through historical period and empires

officially opened to the audience. In order to collect materials from the flora and fauna of the today's geographic territory of the Republic of North Macedonia, departments were established for hydro-biological, botanical and paleontological, and ontological within the Natural Science Museum in Skopje.

The Ministry of Culture of the Government of the SFRY, in 1949, decided to reorganize the museums, specializing in those institutions. For this reason, from the National Museum in Skopje, are distinguished The Ethnological Museum, the Archaeological Museum, and the Art Gallery. The National Museum of Skopje remains with the status of a general museum of Macedonia. In 1951, the "Historical Museum of Macedonia" was formed, after the earthquake in 1964 the "Museum of Contemporary Art" was established. After the declaration of independence of the Republic of Macedonia in 1991, the "Mother Teresa Memorial House" in Skopje opens in 2009. The last attempt at creating an identity for Skopje is the aforementioned project "Skopje 2014" (Anon., 2018), which has been seen as a means of "uniting" the identity of the center by developing 28 buildings and 6 multi-story garages, while remodeling more than 6 facades, 34 monuments, 5 squares, 1 panoramic wheel, 2 underground garages, 4 bridges, 39 sculptures, 1 port, and 2 fountains. This project has turned the city's Brutalism architecture into "postmodern classicism" (Vasilievski, 2012), while simultaneously disrespecting history, urban legislature, architecture, and human rights law. As a part of this project was built the "Museum of the Macedonian Struggle" and the "Archaeological Museum of [the Republic of North] Macedonia". The elaboration and presentation of artifacts typical of the historical, archaeological, and natural heritage of Macedonia is a strategic task of the respective faculties only from the State University "St. Cyril and Methodius"; which is another strategy of assimilating the heritage of this territory under the conduct of the composition "University Museums" (Бачева et al., 2015). Until 2019, there are 11 museums in Skopje with their own facilities, which are (Fig. 2) (1) Natural History Museum of Macedonia, (2) Museum of the Republic of North Macedonia, (3) Museum of the Macedonian National Theatre, (4) Museum of the Macedonian Struggle, (5) Memorial House of Mother Teresa, (6) Museum of the City of Skopje, (7) Museum of Contemporary Art—Skopje, (8) National Gallery of the Republic of North Macedonia, (9) Memorial center for the Holocaust of the Jews from Macedonia, (10) Archaeological Museum of the Republic of North Macedonia, and (11) The Museum of the National Bank of the Republic of North Macedonia.

3 The Museum as a Catalyst for Sustainable Regeneration

According to International Council of Museums (ICOM) in 1974 "a museum is a non-profit-making, permanent institution in the service of society and of its development, open to the public, which acquires, conserves, researches, communicates and exhibits the tangible and intangible heritage of humanity and its environment for the purposes of education, study and enjoyment" (Sandahl, 2018). In a more general summary, the museum should combine social (Misiura, 2005, pp. 40–78), natural, and built environments. The union of these environments is done on the basis of the principle of purpose, functioning, physical, and metaphysical parameters as well as on the basis of development perspective. Each component of social, natural, and built environments communicates with themselves (between the fundamental constituent elements) and in relation to each other, due to their metaphorical and idiosyncratic nature of the circumstance use, the nature of relationships between them of their context, which may be incorporate or neglected (Rapoport, 1990, p. 17). This communication complexity of the environments united the museum can be analyzed in semiotic, symbols, and nonverbal communication models of meanings, in the level of element, system, or process (Table 1), which guide people's behavior, electing of feelings, or influence thoughts.

Museum as harmonization of social, natural, and built environment generates through the empowerment of forces, which enable the opportunity of personalization through taking possession, completing it and changing it. This design framework involves the transformation (from personalization to socialization) as a way of enrich and experiencing the different levels of meanings from different groups of identity. The different arrangement of the different elements embodies the diversity and complexity of identity levels, of a different viewpoint from different individuals or perceptual groups. Identity diversity has the capacity to generate cultural, economic, natural, and social sustainability, if development ethics are strictly adhered to, through elaboration of limits, needs (Lamberta et al., 2014, p. 2), and justice. Sustainability as dynamic process of museums, through their mission, must be an active and attractive part of the community by adding value to the heritage and social memory (Pop & Borza, 2016, p. 3). The activation of museums embodies the balance of cultural, natural, social, and economic environment, through the diagnosis and treatment of environmental imbalances. Based on the political functioning of the museum (Fig. 3), the museum staff compiles daily

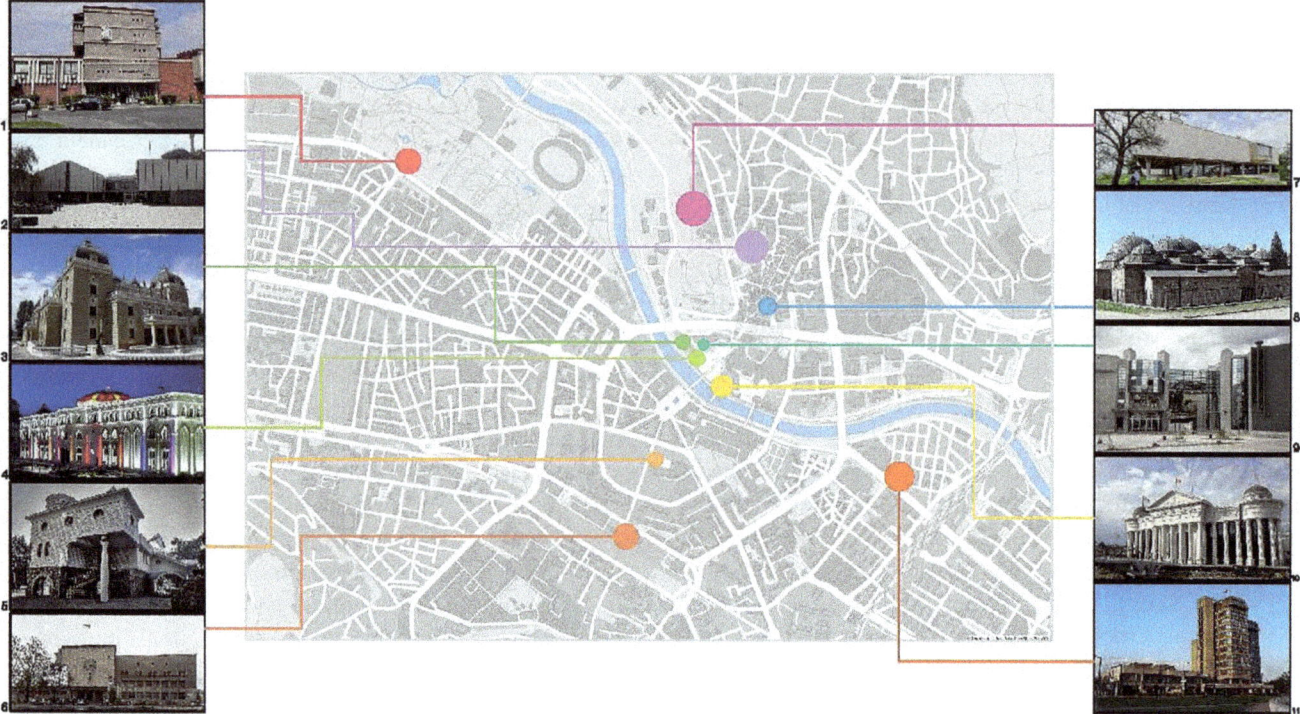

Fig. 2 Location of museums in Skopje, (1) Natural History Museum of Macedonia, (2) Museum of the Republic of North Macedonia, (3) Museum of the Macedonian National Theatre, (4) Museum of the Macedonian Struggle, (5) Memorial House of Mother Teresa, (6) Museum of the City of Skopje, (7) Museum of Contemporary Art —Skopje, (8) National Gallery of the Republic of North Macedonia, (9) Memorial center for the Holocaust of the Jews from Macedonia, (10) Archaeological Museum of the Republic of North Macedonia and (11) The Museum of the National Bank of the Republic of North Macedonia

practices as well as long-term plans to justify its existence and survival, and based on various factors, such as the museum's mission, collection field, staff philosophy and perception, funding, etc. utilizes all the elements that guarantee and strengthen cultural sustainability in a specific environment such as the museum, so that the museum can guarantee and generate social, urban, and economic sustainability (Lamberta et al., 2014, p. 6).

4 The Social Friday Activity Generates Skopje Museomorphosis

Inadequate functioning, first as an institution, then as an architectural building, and as a social and urban symbol, the re-conceptualization, re-legitimization, and regeneration of the museum in the Republic of North Macedonia are necessary. No matter how physically prosperous the intervention, the trust in these institutes cannot be restored. Restoration of trust in the museum should start from the constituent staff structure (according to Fig. 3) of the museums to the active involvement of the society to prove the transparency, the protection of originality at all costs, and its verification. Ethic (honesty, loyalty, and fairness),

professionalism, and cooperation creatively generate the institution. This generator can be activated via Social Friday Activity (SFA). SFA is an action taken by a group of people, organized during paid working hours, on Fridays, each 100 working days of the institute/company/organization, where employees are free to decide whether they want to participate in the event (Admin, 2019). The "100-day plan" provides estimates of responsibility for managers in partnership with employees, to create a personalized plan aligned with organizational goals, strategies, and measures. Stands for a quality and bearing responsibility for environments, BREON engineering, sprays visions not only for consumers and employers but also for society from 2017, by creating and implementing SFA (socialfriday.org, 2018). Mr. Fikret Zendeli, the COE of BREON Engineering and the Founder of Social Friday Activity (SFA), inspired by the fact that Friday afternoons are unproductive hours of the workweek, which are verified by the analysis made by the main Director of Maketagent, Mr. Thomas Schwabl (Schwabl, 2019), he uses this fact to create the activity in question for a sustainable society and healthy economy. By creating meeting points where institutions and society meet to support local social organizations and by regularly repeating the SFA, it enables the development of social awareness in corporate

Table 1 Environmental communication united in the museum (Rapoport, 1990), author contribution

Environmental communication united in the museum				
Ways of study		Structure		
Semiotic model based on linguistic		Element	The sign vehicle	What act as a sign
			The designation	To what the sign refers
			The interpreting	The effect on the interpreter by virtue of which a thing is a sign
		System	Syntactic	The relationship of the sign to the sign within the system of the sing; the study of the structure of the system
			Semantics	The relation of the sign to things signified, that is, how sings carry meanings, the property of the elements
		Process	Pragmatic	The relation of signs to the behavioral responses of the people, that is, their effect of those who intercept them as part of their total behavior; this then deals with the reference of the signs and system to a reality external of the system meaning
Relying the study of **Symbols**		Element	Traditionally	Any object, act, event, quality, or relation which series as a vehicle for a conception and also as any object in experience upon which man has impressed meaning
		System	Modern	
		Process	Contemporary	
Nonverbal communication model—that come from anthropology, psychology, and ethnology		Element	Analogy of metaphor	Environment present a form of nonverbal, because provide cues for behavior
			Directly	The relationship is very direct a real environment both communicate meanings directly and also aid other forms of meaning, interaction, communication, and co-acting
		System	Rules	Similar to language
		Process	Pragmatic	Relationship between particular nonverbal case and the situation, the ongoing behavior

culture and the establishment of self-sustainability (Admin, 2019). The flexibility of SFA organization is adapted based on the capacity of the institution/company/organization, the passion and enthusiasm of the participating employer, responds to or solves local needs, whether social, built, or natural environment. Simplicity, free join, playfulness, and creativeness of SFA generate strong roots of connectivity between individuals or interacting groups, absorbing and sharing vital positive vibes through doing good. Activities can be carried out by inviting business partners, friends, or people from community, with or without the physical support of the SFA team, but documented and promoted on the social media and website of Social Friday. SFA as a meeting cell and a bridge between institutions and the communities

metamorphoses social justice. From this point of view, the basic pillars of the SFA are as follows:

1. Equity—It is based on equality of results and not on the equality of methods, giving more to the needy (Worthington, 2006).
2. Diversity—It is differences in age, disability status, religion, ethnicity/race, sexual orientation, socio-economic status, indigenous background, national origin, and gander.
3. Supportive environments—It is both the physical and the social aspects of our surroundings. It encompasses where people live, their local community, their home, where they work and play. It also embraces the framework

Fig. 3 Shamrock organizational chart. This type of organization places emphasis on teams performing the needed tasks, but it is designed to bring some stability and structure to the teams and provides a flexible home based for staff members (Genoways & Ireland, 2003, p. 47)

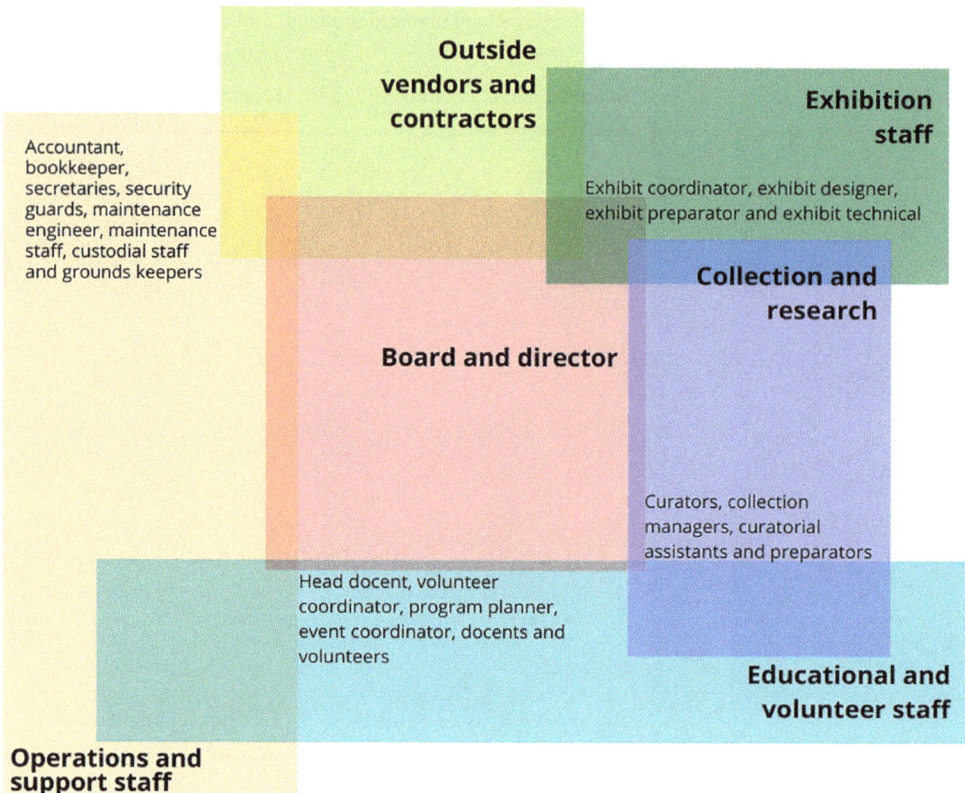

which determines access to resources for living, and opportunities for empowerment (World Health Oragnisation, 1991).

SFA through the interaction of equity, diversity, and supportive environment creates sustainability in micro- (within the composition of museum staff), meso- (within the community served by the museum), and macro- (within collaboration of different institutions at local, regional, and global levels) levels in Skopje. Micro-sustainability requires respect for staff diversity, whether in terms of individual identity, which may be in the context of age, disability status, religion, race, ethnic, sexual orientation, socio-economic status, indigenous background, and national origin, or identity collective where diversities are encountered based on professionalism, experience, culture, or political affiliation. Staff diversity often creates small groups of communication at the sector level, engagements and responsibilities, working hours, location, etc., where ignorance is formed between individual or staff groups, prejudices and stereotypes which result in more non-effective corporation. Regular application of SFA enables more frequent communication of the staff, regardless of the diversity they possess among themselves, by highlighting the human values in the individual, because of the focus on solving social problems, and thus reducing the prejudices and stereotypes created among the museum staff. Continuity of

communication contributes to effective discussion, staff well-being, and the formation of institutional sustainability unity. The organization of SFA by the Skopje museums' staff and the involvement of the community served by the museum can create a social sustainability at the meso-level. As a result of the individual and social diversity of the Skopje community, it often happens that museums do not meet the community requirements, in terms of facility, institutional, legal, administrative, educational, public program, informative, etc., and results in the loss of social trust in the Skopje museum institutions. Distrust in museums has consequences for functional destruction (economic, cultural, and social) due to the reduction in the number of visitors, and the same has an effect on the mortality of the building and the creation of urban metastasis. To enhance the vitality of the building, the SFA plays a key role in the inclusive continuity of the community, the creation of new contacts and ideas as a result of discussions on the solution of museum's institutional or supportive environmental problems. The strategic repetition of SFA creates the possibility of measuring, processing, and evaluating museum programs. The social and institutional sustainability created by Skopje museums through the application of SFA can also progress in the creation of urban sustainability, if in this activity, in addition to museum staff and the community, the participation of various institutions is integrated. Institutional diversity in aims, organization, functioning, and systematization of the purpose, as well as diversity in economic, political, technological, social, service, and legal status, can cooperate with museums and the community, when through SFA is created the possibility for inter-institutional communication, opportunities to complete information, data and legal procedures on certain issues, for the purpose of natural, built, cultural, economic, and social sustainability of Skopje (Table 2).

The coexistence of SFA under the principle of equity, diversity, and supportive environments creates a vital dimension of physical, social, spiritual, cultural, economic, and political renewals, coordinated at local, regional, state, and global levels, using the capacity to achieve an essential sustainable solution, converting the museum into a semiotic, symbolic, and nonverbal model of communication (Table 1).

The synthesis of museum institutions and SFA can constitute the social order and form the sense of Skopje sustainable social and urban identity.

5 Conclusion

The reasoning for the existence of museums in the Republic of North Macedonia should be based on the restoration of truth, transparency, and trust; to return the value held by these institutions as "socio-cultural buildings". The inclusion of research, collections, exhibitions, education, and enjoyment enables the merging or resolution of intercultural, intersocial, or interethnic problems that the state possesses inside and outside its borders. Due to state political

Table 2 Urban and social regeneration system based on sustainability pillars of museums (Pop & Borza, 2016, p. 3; Lamberta, et al., 2014, p. 5) and SFAs, author creation

Sustainability					
Museum	Urban planning and regeneration	**Natural and built**	**Supportive environments**	**Equity**	**Social Friday Activity (SFA)**
	Landscape planning				
	Active use of recycling				
	Green technologies				
	Eco buildings/energy efficiency				
	Environmental education				
	Eco events/exhibitions				
	Heritage preservation	**Cultural**			
	Cultural skills and knowledge				
	Memory/identity				
	New audience/inclusion				
	Cultural diversity/intercultural dialogue				
	Creativity and innovation			**Diversity**	
	Artistic vitality				
	Fundraising	**Economic**			
	Cultural tourism				
	Cultural employment				
	Economic revitalization of local community				
	Well-being	**Social**			
	Sense of place				
	Social responsibility				
	Active citizenship/participation				
	Engagement				
Society					

mismanagement, both in institutional, legal, and strategic terms; socio-political conflicts also appear in the functioning of museums - between museum staff, museums, as well as between museums, and competent state institutes. The consequences of the malfunctioning of museums are reflected in its underdevelopment, culture, socio-economic, and environment. From this point of view, the Social Friday Activity (SFA) is one of the most appropriate solutions for museums in the Republic of North Macedonia, because of the potential of this activity to the museum's awareness for supportive environment responsibility. The realization of each SFA implements the qualities of place for people, connectivity, safe public realm, legibility, diversity, ability to evolve local character, and identity. Museum community, formed from "visitors, stakeholders in museums which are the people who build and work in them, their boards of trust, donors and benefactors, the scholars and academics who use them for research, the producers, societies, and cultures whose creations are preserved there, and those persons, past and present, who represent" (Hein, 2000, p. 38) different levels of identities can regenerate the Skopje social and urban sustainability, through SFA, verified by Table 2.

Involvement of SFA in the development strategy of museums can develop a new methodological approach for the perspective of museums as generators of social and urban sustainability through phenomenological (Merleau-Ponty, 2013) principles and the re-evaluation of "the social and cultural mosaic of the city" (Badcock, 2002, p. 22), as a fundamental value of the city of Skopje and the state.

Acknowledgements This research paper is elaborated as a reflection on the author's Ph.D. progress entitled: "Museums as generators of identity in Skopje", at the University of Ljubljana-Faculty of Architecture, Ljubljana, Slovenia, supervised by Prof. Dr. Tadeja Zupančič.

References

Admin. (2019). *Mahnitëse: Fikret Zendeli nderohet për drejtësi sociale në US-Award* [online]. Retrieved April 10, 2020, from https://shtegu.com/archives/13342.

Anon. (2018). *Скопје 2014 под луна* [online]. Retrieved 2018, from http://skopje2014.prizma.birn.eu.com/.

Badcock, B. (2002). *Making sense of cities—A geographical survey.* Taylor & Francis Ltd.

Gazevic, N., & Maudus, S. (1974). Ratna privredna- Spahije. *Vojna Enciklopedija, 8*, 653.

Genoways, H. H., & Ireland, L. M. (2003). *Museum administration: An introduction.* AltaMira Press.

Hein, H. (2000). *The museum in transition: A philosophical perspective.* Smithsonian Books.

Lamberta, T. S., Boukasb, N., & Yerali, M. C., (2014). Museums and cultural sustainability: Stakeholders, forces, and cultural policies. *International Journal of Cultural Policy*, 1–22.

Merleau-Ponty, M. (2013). *Phenomenology of perception.* Taylor & Francis Ltd.

Misiura, S. (2005). *Heritage marketing.* Taylor & Francis Ltd.

Pop, I. L., & Borza, A. (2016). Factors influencing museum sustainability and indicators for museum sustainability measurement. *Sustainability, 8*(1), 101.

Rapoport, A. (1990). *The meaning of the built environment: A nonverbal communication approach.* University of Arizona Press.

Sandahl, J. (2018). *The museum definition as the backbone of ICOM.* s. l.: ICOM—International Council of Museums.

Schwabl, T. (2019). *Job productivity and social engagement.* Retrieved March 8, 2020, from https://socialfriday.org/wp-content/uploads/2019/02/Presentation_Job-productivity-and-social-engagement_February_2019.pdf

socialfriday.org. (2018). *Fikret Zendeli on the impact "Social Friday" has on his company.* Retrieved August 22, 2020, from https://socialfriday.org/fikret-zendeli-social-friday/

Vasilievski, V. (2012). *Skici od urbanizam i arhitektura* (2nd ed.). Grafotrejd dooel Skopje.

World Health Oragnisation. (1991). *Sundsvall statement on supportive environments for health* (pp. 1–14). Sundsvall, Sweden, World Health Oragnisation.

Worthington, S. (2006). *Equity.* Oxford University Press.

Бачева, З. А., Михајлова, Б., & Богоевски, К. (2015). *Музеите во Република Македонија.* Скопје: МНК-ИКОМ-Скопје.

Elementary Waste Insulation Panels in Hot Arid Regions

Farres Yasser

Abstract

Majority of housing projects in hot arid regions have very poor insulation. Strategies against overheating in the summer and overcooling in the winter significantly increase the heat island effect due to poor building materials and low heat capacities. The research investigates whether locally available waste materials could be used to build reliable and replicable exterior retrofit insulation panels. The research focuses on the panels' low thermal conductivity capabilities, cost-effectiveness, ease of construction, its buildability and acceptability by the locals and its capacity to fulfil a circular economy model. The methodology consists of a preliminary experiment where five chambers, including a control chamber, are built to test four types of insulation alongside a control room. The chambers are designed to resemble majority of buildings in hot arid regions and are tested for a week during the coldest and hottest periods of the year, five times a day. The results show that some materials are much more functional than others with the Styrofoam and Composite panels (cardboard along with Styrofoam) performing much better than polypropylene plastic and cardboard panels. The research suggests further development to these panels using a wider range of waste materials, or material subcategories, alongside incorporating new techniques, which would simplify the manufacturing process even further while keeping it cost-effective and would help direct the initiative toward practical realization. Furthermore, a deeper level of integration through larger groups of local laborers' experience-based opinions would make the application of the retrofits onto the facades easier, while also providing agency to the end users and locals.

Keywords

Thermal comfort • Local residents • Community engagement • Circular economy • Waste • Reuse • Insulation • Urban heat island effect • Thermal mass • Sustainability • Energy efficiency

1 Introduction and Literature

Seeing the extreme heat experienced in hot arid regions, buildings require thick external walls to reduce heat and delay its transferring from the exterior. Nonetheless, poor and cheap materials, like single red brick masonry brick walls (of 12 cm thickness), are frequently used to reduce labor and construction costs. This negatively impacts the urban environment, and the buildings' thermal comfort, thermal mass, HVAC costs, and energy consumption needed to adjust the poor thermal performance levels. The utilization of poorly performing materials such as single-layered bricks is usually due to populations residing in mid- to low-income settlements. These residents often build their own settlements due to lack of affordable formal housing settlements (AUC Centre for Sustainable Development, 2014).

A research by Bek et al. (2018) shows the many factors in urban environments which cause an increase in the heat island effect leading to a 1–4 °C increase in temperature. These factors are summarized into the following urban design and planning characteristics:

- Site geometry, which affects the flow of heat and air into and out of the buildings (Chun & Guldmann, 2014).
- Population density which amplifies the heat exhaust from the inside of the structures (Gardiner & Heller, 2008).
- Gas effusions which wane air transparency, decreasing ventilation and causing nocturnal re-radiation.
- Lack of vegetation due to built-up footprints causing less heat loss via transpiration and evaporation.

F. Yasser (✉)
School of Architecture, Design and the Built Environment, Nottingham Trent University, Nottingham, UK
e-mail: Farresyasser@gmail.com

- Poor building materials, which have low heat capacities, densities, and poor conductivity coefficients, leading to low thermal mass, solar radiation absorption, and less reflectivity, causing massive heat build-up (Fernández et al., 2015).

It is obvious that poor building materials, or single-layered red bricks, are the main amplifier of low thermal comfort levels thus increasing the use of HVAC systems and hence increasing the heat island effect by amplifying many of the aforementioned factors. Being largely applied on an urban scale, the single bricks substitute plantation in its reflective effects, and absorb and release more heat than they can insulate, due to the increasing population and their poor insulation properties.

It is therefore this paper's aim to tackle the possibility of retrofitting buildings using locally available and abundant waste to help mitigate the resulting heat island effect by providing high thermal mass exterior walls. The research inspects the cost-effectiveness and practicality of locally manufacturing the retrofit waste insulation panels, the acceptance of local laborers and end users of the panels' aesthetics and construction process, and, most importantly, the monitoring of the panels' performance during different seasons. This attempt helps to analyze the extent to which waste can act as a positive constituent in the circular model of insulation materials.

To better represent the impact of poor building materials on the urban heat island effect, Fig. 1 by Bek et al. (2018) represents the thermal imagery in the years 1990, 1998, and 2015 in two neighbouring settlements. An informal settlement, Dar El Salam (top satellite map), and a formal settlement, Maadi (bottom satellite map). A graph at the top of Fig. 1 with start and end mark between the two areas explains the drastic difference in temperature ranging from 1 to 4 °C between both areas. This is predominantly due to the use of poorly insulating building materials in the informal settlement area as opposed to the thick outer walls, well-finished and insulated in the formally planned area.

The research for retrofit insulation alternatives is needed because the only current insulation materials available in hot arid regions are expanded polystyrene (EPS) and X-foam. Since X-foam is ultimately too expensive, and therefore not widely used, expanded polystyrene is being nominated for wider endorsement in many hot arid countries including Egypt (H.A.B.N.R., 2008). Although less costly, expanded polystyrene remains unaffordable for the wider population. Meanwhile, expanded polystyrene is among a group of popular conventional insulation materials which emit detrimental toxic chemicals like ignition retardants, antioxidant additives, chlorofluorocarbons, and benzene. These are the

by-products and cost of having polystyrene as a structure's insulation material (Akovali, 2012).

Waste materials alone can uptake the role of a permanent substitute to raw materials extraction (RME). Consequentially, the use of a low cost or free materials like waste can be advocated for since their use is seminal to providing affordable and unhazardous insulating materials in the construction industry. This is fundamental for both the reduction of emission releases, through long-term buildings' operation, and for the mitigation of energy consumption in the preliminary stages of the buildings' lifecycle, particularly during construction. It is reported that upon combining resources extraction for construction and construction practices, they together are responsible for over 40% of both global emission releases and energy consumption (Lechtenböhmer & Schüring, 2011).

Furthermore, the use of waste materials for insulation is affordable, practical, and efficient, which brings us closer to the already passed zero-energy building goals of 2020. Pacheco-Torgal (2014) reinforced the importance of generally using waste as a primary building block for insulation materials. He states that the only way to meet the zero-energy or close to zero-energy building goals of 2020 is by incorporating thermal insulators which prove to be eco-efficient. This term is not only depictive of their low thermal conductivity output, but also of their construction technique and the emissions they disperse or capture during material collection, transportation, manufacturing, and post-installation.

An attempt to measure the thermal comfort performance of conventional walls in Hot Arid regions was performed by Mahdy and Nikolopoulou (2013). Although thorough and detailed onsite readings are usually absent, a detailed simulation model allowed for a clearer understanding of how single brick masonry walls affect the indoor temperature of apartments, as well as the amount of energy consumption used by HVAC appliances to offset the heat gains during the summer and the cooling gains during the winter. The research set out to test four external wall types as indicated in Table 1. "Half brick", mentioned in Table 1, refers to an Egyptian construction term for single masonry brick layered walls.

Mahdy and Nikolopoulou (2013) state that half red brick masonry walls are commonly used throughout Cairo's housing projects. This is because most of these settlements are built on low budget, requiring little investment in almost all building materials, while investing in 25 cm-thick red brick walls would double the costs of material acquisition and the labor costs. As indicated in Table 1, the full red brick external wall with an additional 2 cm expanded polystyrene thermal insulation, is recommended by the Egyptian

Fig. 1 The urban heat island effect between a well insulated and poorly insulated settlement. The informal settlement on the top uses single masonry brickwalls and constricted urban geometry, whereas the bottom formal settlement incorporates insulated thicker walls, leading to extreme differences in the heat island effects (Adapted from Bek et al., 2018)

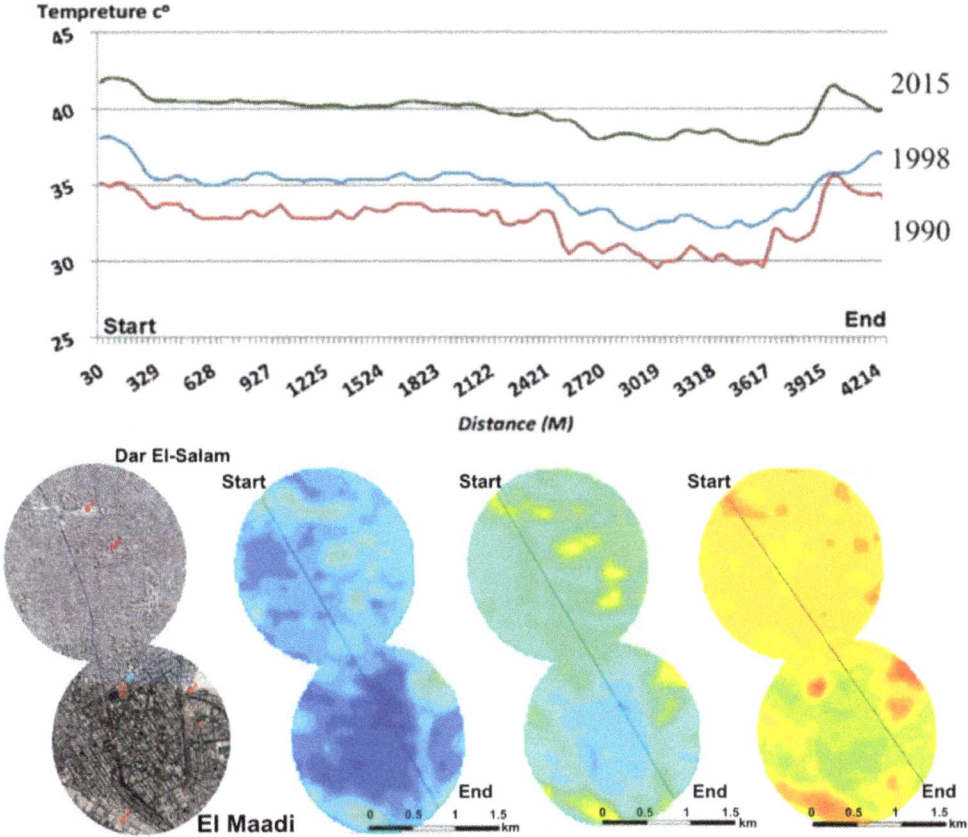

Table 1 External wall types tested in simulation software for comparison. The most commonly used materials for external wall insulation, the top two, are the poorest in terms of thermal conductivity performance (Mahdy & Nikolopoulou, 2016)

External walls	Thickness (cm)	U-value (W/m²k)	Description
Half red-brick wall	12	2.519	Commonly used
Full red-brick wall	25	1.898	Commonly used
Limestone bearing-wall	50	1.228	Egyptian vernacular architecture
Full red-brick wall plus additional 2 cm of expanded polystyrene thermal insulation layer	27	0.897	EREC minimum recommendation

Residential Energy Code, EREC (H.A.B.N.R., 2008). This solution is however highly impractical and unfeasible since the addition of 2 cm expanded polystyrene thermal insulation requires far more advanced levels of labor, which is subject to financial constrain, not to mention the purchase of both the expanded polystyrene and double the number of red bricks. Figure 2 shows the performance of each wall material composition mentioned in Table 1.

The simulation results performed on the software suite Design Builder prove that single masonry brick walls have the highest consumption of energy, with the expanded polystyrene and 25 cm-thick masonry walls using 31% less energy throughout the year. The simulation results also show that the latter saves dramatically on costs to maintain the required internal thermal comfort levels. Although the recommended insulation by the EREC proves to be very effective, it is unclear why residents would use EPS and 25 cm-thick walls or 50 cm limestone walls (vernacular architecture-based walls) when they provide slightly less operational costs and energy consumption levels but are more expensive than the 25 cm walls in terms of the materials themselves, their availability, and the labor that comes with applying them. Double brick walls require large sums of labor costs to be enacted as well, but vernacular bricks require even larger construction investments, let alone their transportation from distant locations to the greater Cairo region being a burden on its own. This is why neither of the three solutions is being used by the majority of Egypt's

Fig. 2 Performance of the most widely used wall materials in hot arid countries. Monthly energy consumption (kWh) of each external wall type (top). Yearly energy costs, in EGP, of each external wall type (bottom) (Mahdy & Nikolopoulou, 2016)

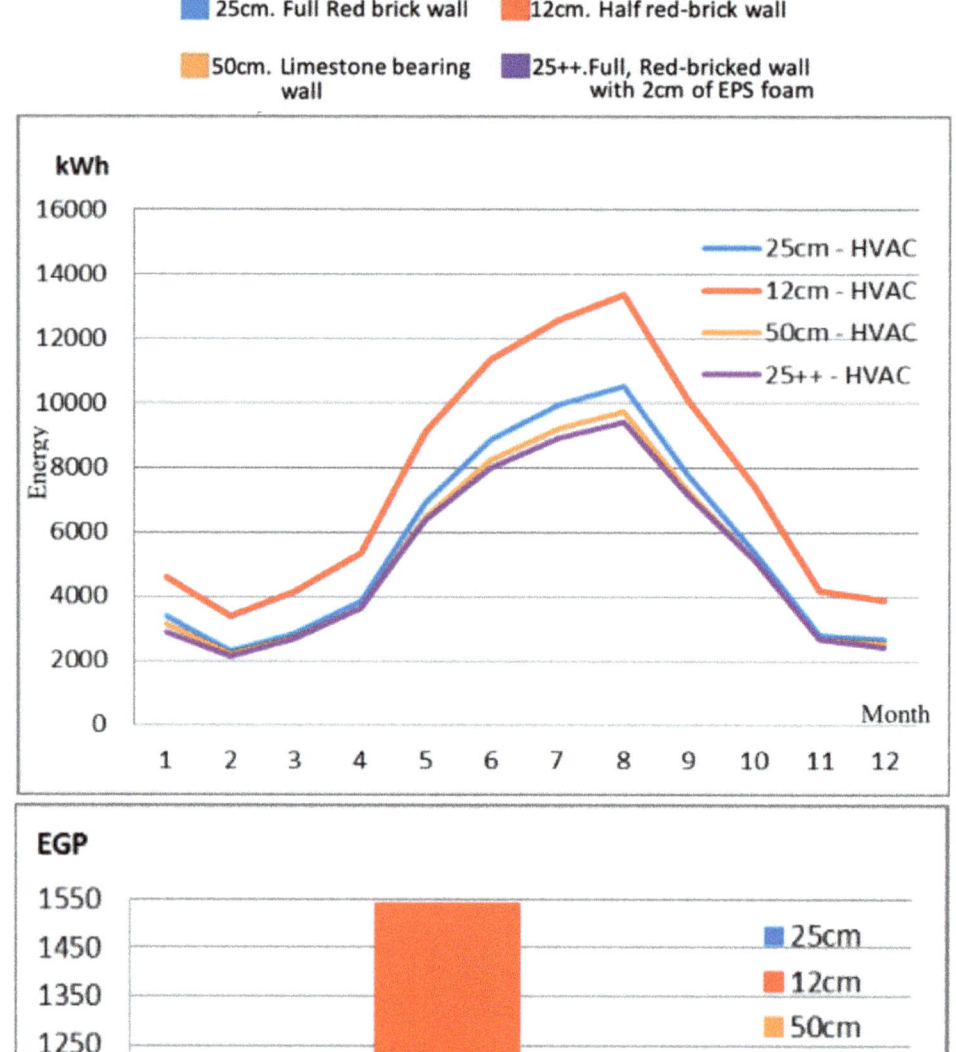

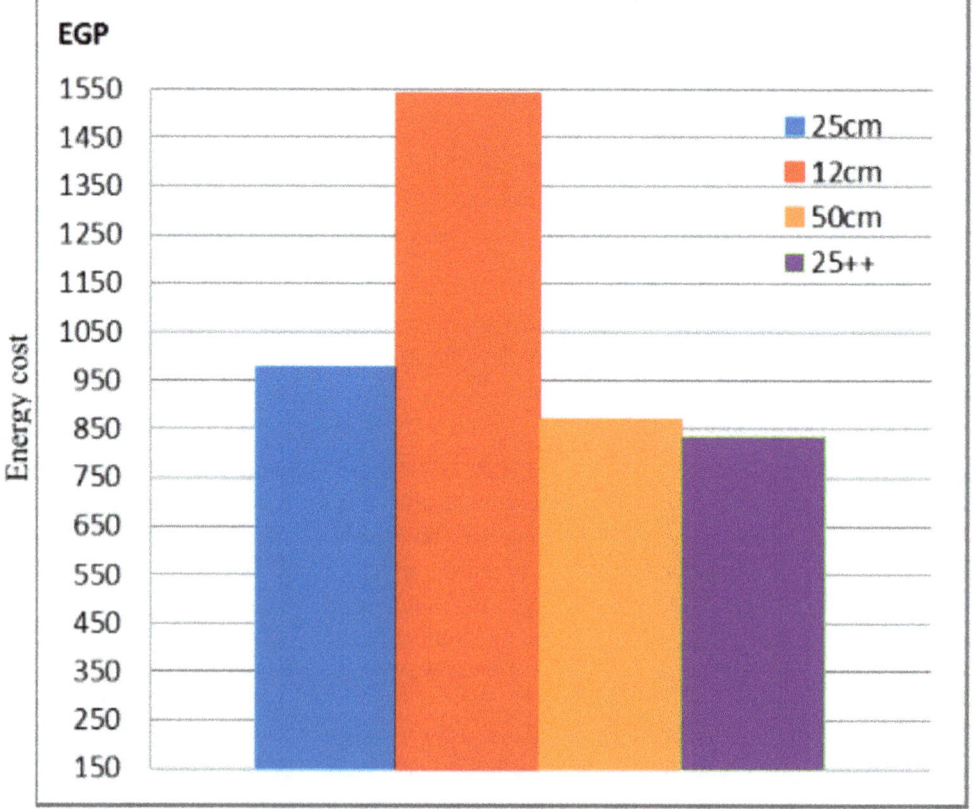

population, especially Cairo's residents. Meanwhile, a solution that can provide similar or stronger outcomes is yet to present itself. A research experiment was therefore designed to entertain whether waste materials can provide an amalgam of cost-effectiveness, low-to-non-existent manufacturing environmental effects, and low thermal conductivity through retrofit insulation panels.

1.1 Possible Sources for Waste in Cairo as a Case Study

The nature of waste in Cairo leads to limited choices of practicable waste for use, which are either plastic waste (polypropylene), cardboard, or Styrofoam. This is due to Egypt having 44% of inorganic waste predominantly divided into unspecified waste at 15%, plastic at 13%, paper and cardboard at 10%, glass at 4%, and metal at 2%. The inorganic waste along with the organic waste sources, which make up the remaining 56%, is only 20% of Egypt's reused waste. The remaining 80% is discarded. This data is only representative specifically of Great Cairo region, Alexandria, the Red Sea Region, Upper Egypt, and the Delta, with the rest of Egypt's 43 municipalities being greatly misrepresented or not present at all in the available charts (SWEEP, 2014). Figure 3 further illustrates the ratios of waste types in Egypt in general and its distribution among the main municipalities.

Since there is no strict legislation or framework around how much, how necessary, in what method and toward which outlets should waste be reused or recycled, this research acted as a trial

to test the practicality of using waste as insulation in a hot and arid city like Cairo, Egypt. A main concern for the experiment was observing whether the residents of Cairo's poorly insulated housing blocks will regard the realization of the panels' construction from waste as an attractive solution to overheating and overcooling. This is because they needed a visible and short-term incentive to encourage them to invest time, resources, and management techniques to reuse and recycle their generated waste without much transportation or allocation measures. The main residents, who judged the insulation panels proposals at hand, were the local labor men who live in the selected area. This was a crucial aspect for future progress. As emphasized by Ibrahim and Mohamed (2016), community involvement from inception is seminal for increasing both the quantity and efficiency levels of waste management in Cairo. This is then succeeded by incentives and tax waivers for recycling activities, alongside simple, and dedicated programs for specific kinds of waste. Meanwhile, the empowerment of low-income communities for recycling comes last.

The socio-economic question remains true as to whether low- and mid-income residents in hot arid regions will invest time and effort to separate the organic and inorganic waste to help build the insulation panels. Amid the constant struggle of Cairo's inhabitants, a resident from the low-income El-Zawya El-Hamara District stated the difficulty he finds in regarding waste management as anything but a luxurious activity when he and many others have to struggle to meet their basic wages, which are lower than what they need to stay afloat, and to fight rapid and high inflation as well as unemployment with no social service support (Sohair Milik, 2010).

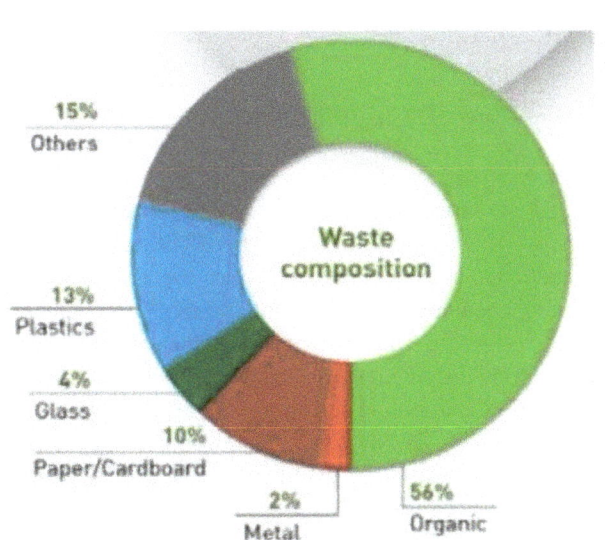
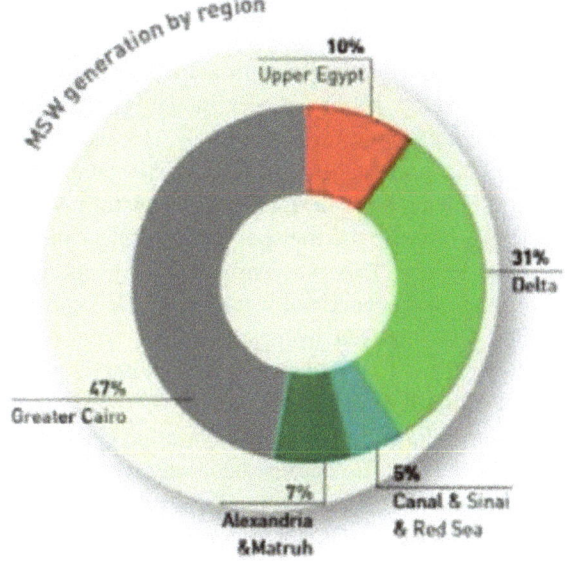

Fig. 3 Proportions of waste types in Egypt and their locations. Municipal solid waste types in Egypt (left), and by region in Egypt (right) (SWEEP, 2014)

1.2 Insulation Through Waste Within a Circular Economy

The potential existence of insulation panels from waste in hot arid countries could potentially offer a change in the circular economy of these countries. Reusing waste to create products will hinder the waste of diminishing natural capital, increase economic benefits from manufacturing with existing materials, and save costs of purchasing materials from other regions that could reach a potential savings cost of $1 trillion within the next decade (Esposito et al., 2017). In the mid of current economic stagnation in the Gulf and North Africa, a circular economy within the construction industry would provide almost 20 jobs per each 10,000 tons of recycled waste. This is contrary to the single job per 10,000 tons in the current linear economy of materials (Esposito et al., 2017).

Large-scale circular economy models could prevail over the existing linear models in hot arid regions with many megacities, which have large amounts of waste, pollution, and extreme levels of the heat island effect. In turn, these negative outcomes act as catalysts for the circular model to thrive when combined with the potential of large amount of waste per given area. For instance, in Dubai an exterior cladding tile "Nabasco® 8010" was made using local bio-based materials from various districts such as grass, reeds, recycled textiles, drinking water treatment waste, sanitary paper, and bio-based polyester resin. The strong versatile tile was used to reduce the interior temperature of an experimental apartment by up to 12 °C. This was possible due to the dissipation of the exterior temperature in the tiles' design which had triangular like molds with cavities that allowed airflow and also shading (VIRTUe, 2018, 2019) (see Fig. 4). This sort of movement toward a well-designed, all recycled, and waste-based insulation panel, which simultaneously works as an exterior facade tiling system, demonstrates the utility of the circular model in the construction industry within hot and arid regions. The tiles did not just consider the four stages of the circular economy cycle—collection, recycling, design, and retransformation of the materials—but also the transportation and the functional use of the materials. This was ensured through designing the tiles to be container-friendly, ensuring they can be easily installed and removed from any surface type and also by making them installable onto building corners (Pujadas-Gispert et al., 2020).

The inception of a waste insulation tile can take place at the micro-level of the circular economy which is usually in direct relationship with single consumers, local small and medium businesses, companies, or organizations. This is an advantage since the meso- and macro-levels, which operate at the levels of buildings, industries, and governments, are difficult to consolidate in many hot and arid regions due to the lack of cultural, financial, governmental, and technical themes which allow for proper circular economy development (Ferronato et al., 2019) (see Fig. 5). Specifically, many hot and arid regions suffer from the absence of a clear vision and legislation supporting circular economy, governments' investment into research, innovation and long-term investment, and a scarceness of qualified professionals in the field (Al Hosni et al., 2020).

1.3 Site Description

The experiment was conducted on a rooftop of a well-insulated building in Giza, Cairo. The building's rooftop was used to ensure that the chambers to be built had each of their four sides in parallel with the exact north, south, east, and west orientations. The chambers also had constant exposure to direct sunlight and wind to exclude external factors such as shade from neighboring buildings or disproportionate shading that could be a result of the chambers' elevations not being aligned to the true north, south, east, and west orientations.

As accessing the site represents a main concern, this building was particularly chosen as it allowed for examination of the chambers at any time during the possible monitoring periods, negotiated with the residents, which were during the hottest and coldest weeks of the year between January 25–31 and July 2–8. It was not possible to conduct the experiment on a university campus for national security reasons. This would have helped with constant onsite monitoring providing a much more accurate understanding of the chambers' performance overnight.

2 Methodology

2.1 Design of the Chambers

The chambers were 1.2 m × 1.2 m × 1.2 m. They were constructed of single masonry brick walls with a thermal conductivity of 0.45 W/mK (Al-Sibahy & Edwards, 2017), using traditional cement mortar between the bricks with a conductivity coefficient of up to 0.2 W/mK. The chambers were made in such a way to ensure that they were airtight. This was carried out by applying the mortar to the exterior of the walls multiple times and spreading it thoroughly along the bonds and repeating the process after the mortar dried. The chambers were built on a 15 cm base of polystyrene foam to prevent heat from the roof infiltrating into the chambers. Each chamber had a roof of 1.5 cm plywood topped with a 5 cm layer of expanded polystyrene foam. Each chamber also had

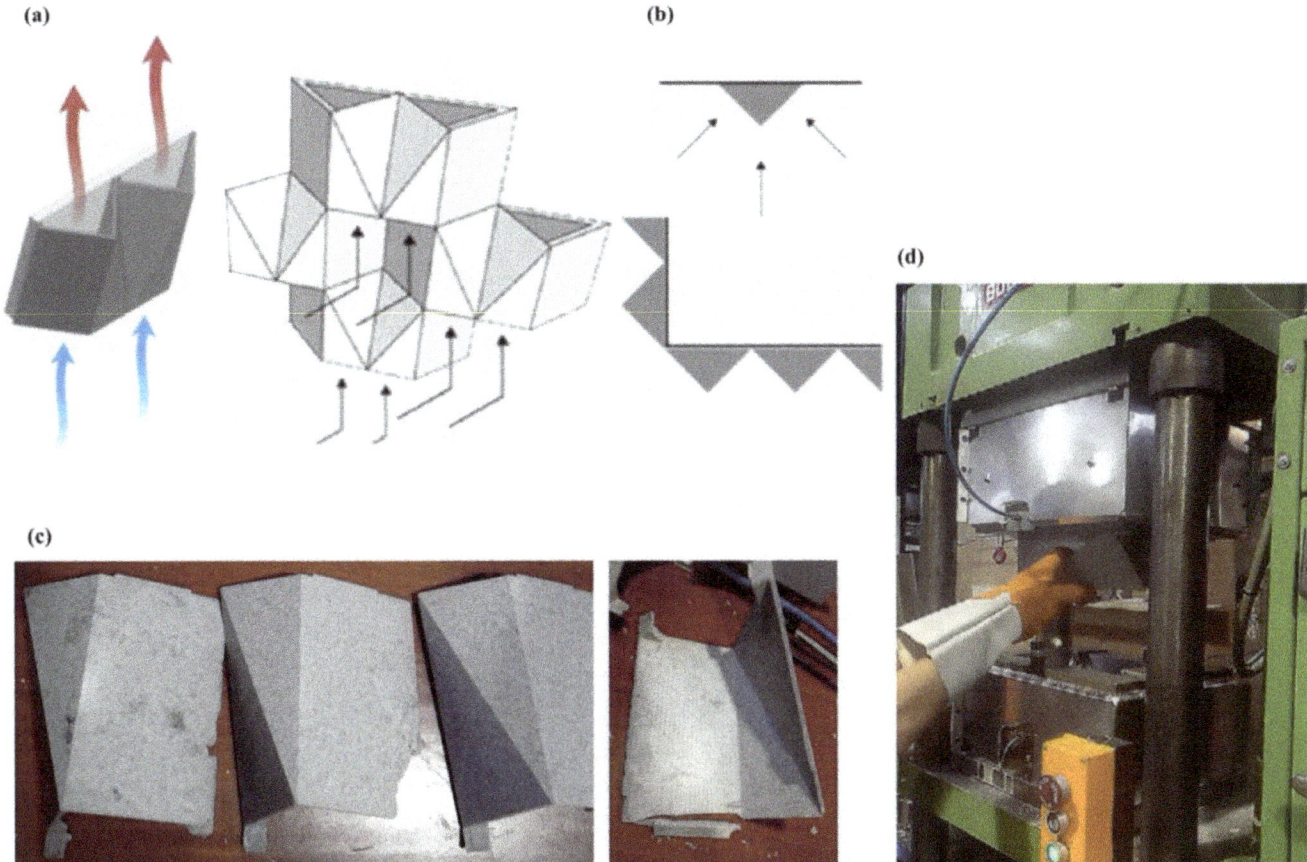

Fig. 4 Exterior cladding tiles made from bio-based waste materials, showing the ease of production within a potential circular economy in Hot and Arid regions. **a** Air-flows the tile cavities dispersing the air's temperature. **b** Smooth finish at apartment corners. **c** End product from bio-based waste materials. **d** Easy manufacturing process using mold pressuring system

two punctures on its sides to allow the thermometer probes through for readings when needed at the 25 cm height mark. The punctures were tucked with cork caps after each reading was taken to ensure air tightness (see Fig. 6).

After building the chambers, each one was appropriated to test a different type of insulation panel. The five chambers were divided as follows: one control chamber (without any further layers to its exterior) which reflected the traditional residential building exterior facades, a second which had panels of 4 cm cardboard from egg trays, a third chamber with 4 cm of Styrofoam, a fourth chamber with 4 cm of polypropylene plastic plate tops, and a fifth with a composite of 3 cm cardboard egg rays, filled with cardboard paper maché, against 1 cm of Styrofoam boards. All chambers except for the control chamber had an additional 2 cm of Eco mortar, made of sand, dirt, Adibond 65 (epoxy), and cement, placed on top to test if the panels could allow for further plastering and various finishes since protection from external environmental factors and application of further finishes for aesthetics are crucial for a successful retrofit solution.

2.2 Equipment Employed

The equipment used to monitor and record the readings from the chambers was a High-Precision Digital Thermometer with Bluetooth and a smartphone to transmit the readings precisely via Bluetooth without margins of error or fluctuation oversights. The thermometer was provided by Thermco products, model number ACCD790P Dual Probe-Pt100/Thermocouple/Rh High Precision. The thermometer's range of readings was 200–850 °C with an accuracy of ±0.015 °C. The probes were inserted into the opposite holes in each chamber to capture the readings using the third-party app provided by Thermco products and an average of the two probes' readings was then calculated.

2.3 Building Process with Locals

Upon inception, an open call was announced locally between the labor men to analyze the available waste materials that can be used as well as the shape, form, and

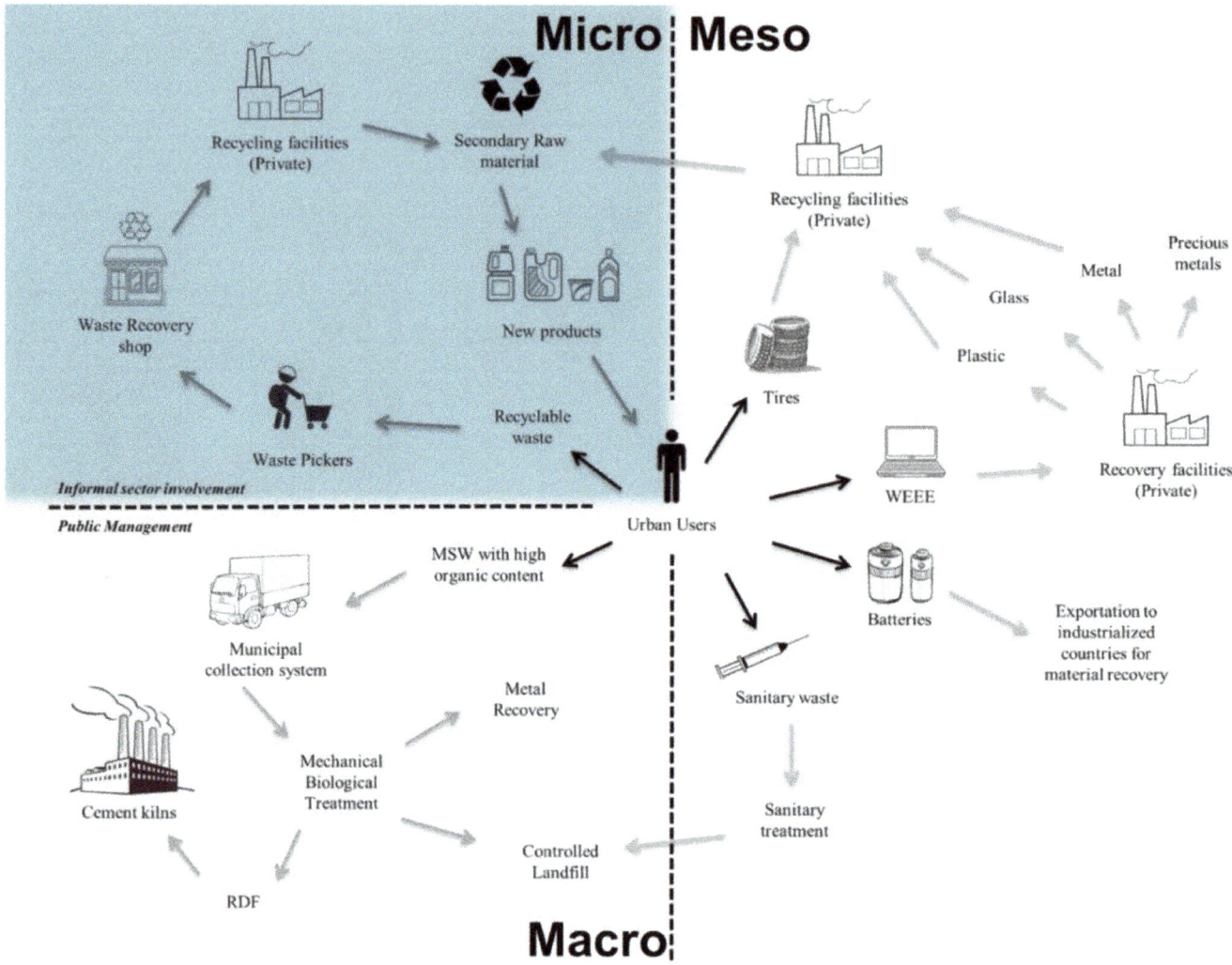

Fig. 5 The highlighted section shows the circular economy model utilized in this research paper. The overall figure represents the micro, meso and micro levels of the circular economy (Adapted from Ferronato et al., 2019)

installation method of the panels. This was crucial as the experiment's location, where the panels were tested, wasn't a neighbourhood that would ultimately use these panels if proven successful. The buildings at the experiment's location were formally planned, built with better means, and have moderately to well-insulated exterior walls. However, the labor men live in low-income settlements nearby and they would ultimately be the end users. Their approval of the panels was thus integral to the probability of their actualization. The five labor men involved in the experiment were all multipurpose repairmen with varying levels of experience in different trades with two of them being painters, one builder, one recycler, and one generic appliances repairer.

After careful research into the Giza area's available waste products in the form of plastics, foams, papers, or cardboards, a consensus was reached that the most usable forms of waste which are recyclable and widely available are cardboard boxes, cardboard egg trays, Styrofoam meal

plates, and disposable polypropylene plastic meal pots and their flat covers. Each phase in these materials' reuse process was examined to evaluate how productive or unproductive it would be to use. These phases incorporated waste identification, handling, separation, collection, transport, processing, transformation, and/or disposal. These materials were selected primarily because they are the most collected waste types by waste collection teams employed by small contractors at Ezbet El Hagana area in old Cairo. This is a result of their undemanding nature for reuse and reselling when soaked into paper maché or shredded into powder form substance. Their abundance also provides large quantities for resale, making them a targeted commodity.

The first best material suitable to apply was egg tray cardboard as its repeated hollow indentions allows for further padding of cardboard maché for a denser insulation material and also for its set square size of 30 × 30 × 4 cm. The second material was Styrofoam, with thermal

Fig. 6 A typical section of the constructed chambers for testing. The section shows the different layers existing in each of the 5 chambers built for experimentation and monitoring

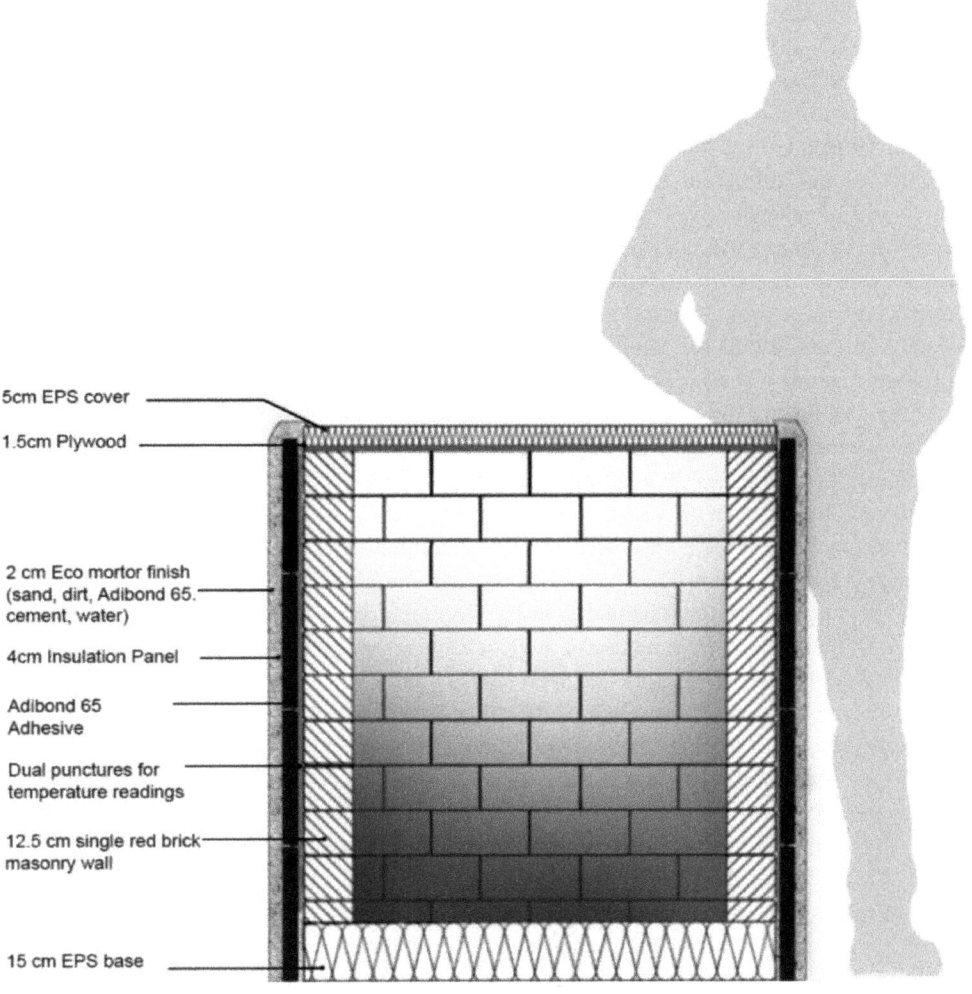

5cm EPS cover

1.5cm Plywood

2 cm Eco mortor finish (sand, dirt, Adibond 65. cement, water)

4cm Insulation Panel

Adibond 65 Adhesive

Dual punctures for temperature readings

12.5 cm single red brick masonry wall

15 cm EPS base

conductivity of 0.06 (Albert et al., 2013), also chosen for its set packaging size, which comes in plates of 25 × 18 cm after removing the elevated foam sides, but more importantly because it is a cacogenic polymer which does not biocompose nor can it be recycled or reused globally. It thus affects the environment on a molecular scale once its polymers are released through leaching and physically once it fragments (Farrelly & Shaw, 2017). Although successful efforts have recently been established in Montreal, Canada toward reusing Styrofoam (Polystyvert, 2018), these measures weren't implemented globally yet. The third and last material used was polypropylene plastic pot covers since they were easy to compress together. They have a thermal conductivity of 0.27 W/mK (Patti & Acierno, 2019). They are also abundantly available and do not require much remolding through compression machines and other recycling machinery that their plastic pot counterparts do in order to bring them to a form which can be used for insulation.

After selecting the materials, the next step was to construct the panels. The five labor men shared views on which

bonding agents to use for applying the panels onto the walls, the method of adhesion of the materials, cost-efficiency, material availability, and the methodology of construction for future replicability by lesser skilled or numbered labor men.

The utilized local adhesive was Adibond 65, a versatile adhesive with a wide range of applications according to the labor men's experience with it in various other projects, aside of its product specifications brochure. The adhesive is a latex dispersion admixture based on styrene butadiene rubber from the CMB company, and it was used to retain the panels onto the wall. The Adibond was then mixed with water, sand, dirt, and cement to create a semi-eco-friendly adhesive to use as a finish layer after the panels were applied with a respective ratio of 0.5: 2: 2: 1: 1. The mixture was tested externally after white glue was initially tested as the mixture's adhesive. While white glue is widely available and cost-efficient, it proved to be a poor adhesive at this occasion so Adibond was thus chosen although it is more expensive. The conventional use of hydrated lime was substituted with dirt since in Cairo the widely available materials throughout all housing blocks are sand and dirt, but lime comes from farther municipalities causing unsustainable transportation emissions and costs.

The polypropylene pot cover panels were made by stacking 10 plastic tops together forming 4 cm of insulation overall. A streak of adhesive was used on each layer, then 5 ml worth of adhesive was applied to the wall before the plastic tops were pushed against the wall. This was repeated until the tops filled the entire wall. Similarly, the Styrofoam plates were rid of their elevated sides and stacked together (10 of them) and glued the same way the plastic tops were until they reached a 4 cm thickness.

Afterward, the egg trays were flattened using a traditional wooden wagon nearby, relieving the need for a car or truck to compress the trays into the required 4 cm thickness. Two trays were compressed with a single layer of Styrofoam and glued to the bottom of the compressed trays. The inclinations of the egg trays were then filled with cardboard maché after they were stuck onto the chambers just before the Eco mortar was applied avoiding the use of any adhesive in order for the maché to adhere onto the compressed egg trays via the Eco mortar's weight. After each of the other three panel types was applied to its dedicated chamber, the Eco mortar was applied filing in the gaps and also creating a 2 cm layer of Eco mortar. The chambers were then left for around 3 days to ensure that the cardboard within the panels dried as well as the mortar itself (see Fig. 7a–e).

Upon completing the experiment, a trial was made to test for the possible elimination of an adhesive to administer the panels onto the walls through reusing waste plastic corrugated sheets to act as mullions that would hold the insulation panels in place. Some of the materials were shifted up or down the panel arrangement. This was only a trial; no further testing was conducted to compare the performance of each material when applied in such a manner (Fig. 7f).

2.4 Monitoring Process

The chambers were monitored five times a day for a week during the hottest summer days, July 2 to July 8, and the coolest winter days, January 25 to January 31. The times at which the readings were taken were 7 a.m., 10 a.m., 1 p.m., 4 p.m., and 7 p.m. Monitoring the chambers was not possible during other time frames due to lack of accessibility. The two probes in the high accuracy thermometer were used to take readings from both punctures on all five chambers. An average was then calculated ensuring the results better represent the in-chamber environment.

3 Results

A detailed graph (Fig. 8) was produced to create a visual presentation that conveys the real-time output of each insulation type at each time stamp on each day. To simplify the readings into a more encapsulated graph for comparison, Fig. 9a and b was developed with line graphs representing each insulation type's average temperature outputs in three time frames. The first was the average week readings in general, the second was the morning readings (7 a.m. and 10 a.m.), and the third was the average afternoon readings (1 p.m., 4 p.m., 7 p.m.). Figure 10a and b was the third and last explanatory line graphs aimed to help compare the average readings of each insulation type at each hour mark throughout the week.

The experiment's results provide nuanced results of how the panels performed during different seasons and various times of the day which showed varying degrees of similarity and contrast in different seasons and times (Fig. 8). When monitoring the average performance of each material five times a day over a 7-day period as depicted in Figs. 9a and 10a, the results showed much contrast to when the average weekly, morning, or afternoon readings were analyzed in Figs. 9b and 10b. This is due to the performance values of each insulation type fluctuating either up or down the temperature scale between readings, and thus coinciding at certain points on the graph, which would either represent them as warmer or cooler, in general, when, in fact, this is just an instance in time which does not reflect the actual difference in performance throughout the day or season. The results outlined below will discuss the differences between both types of figures, in order to provide a clearer and deeper understanding of how each insulation type behaves. The results of each panel's chamber for each season are organized by order of the best performing insulation type.

3.1 Summer Results

When monitoring the summer performance of the panels in Fig. 9a, the composite panels proved to be the most suitable panels providing the coolest temperature outputs, with readings of 33.4 °C and 37.9 °C at the 7 a.m., 10 a.m. averages, and at the 1 p.m., 4 p.m., and 7 p.m. averages making them perform much better than the control chamber, which came in last with a reading of 38.3 °C and 39.3 °C, respectively. However, when looking at Fig. 9b, it's obvious that the composite panels were superior only up until 11 a.m. and from 6 p.m. onward. Meanwhile, the other three panels proved to be better insulators throughout the day from 11

Fig. 7 The images represent the control experiment chamber (no insulation) and the 4 types of waste insulation developed for monitoring. **a** Control experiment chamber; **b** Styrofoam insulation; **c** Egg tray cardboard insulation; **d** Polypropylene insulation, and lastly. **e** a composite of compressed egg trays filled with cardboard mache and a bottom Styrofoam layer. **f** One further trial and error process that experimented with reusing waste plastic corrugated sheets as mullions for the panels to substitute the expensive Adibond 65 adhesive

a.m. till 6 p.m. This suggests that the composite panels may be the most suitable for insulation from late evening till next day pre-noon but not specifically for mid-day hours.

It is important to note that the average week temperature in Fig. 9a showed the composite panels' chamber to be slightly warmer than the polypropylene panels', making the composite panels the second best panels on an average weekly readings basis. This again is due to the composite panels' chamber having 5 recorded hours of cooler temperatures than the polypropylene panels', while the polypropylene had 7 total recorded cooler hours, which do not account for the actual hours outside the monitored time frames.

After 11 a.m., the second and third best panels appeared to be the Styrofoam and cardboard panels. Since Fig. 9b shows that the pattern is interchangeable with cardboard being cooler than the Styrofoam at the start by around 0.2 °C, increasing to a 0.7 °C difference up until 3 p.m., then both graph lines overlap and exchange functions making the Styrofoam cooler than the cardboard by 0.5 °C till the end of the day and into the evening. This concludes that the Styrofoam and cardboard panels provide cooler

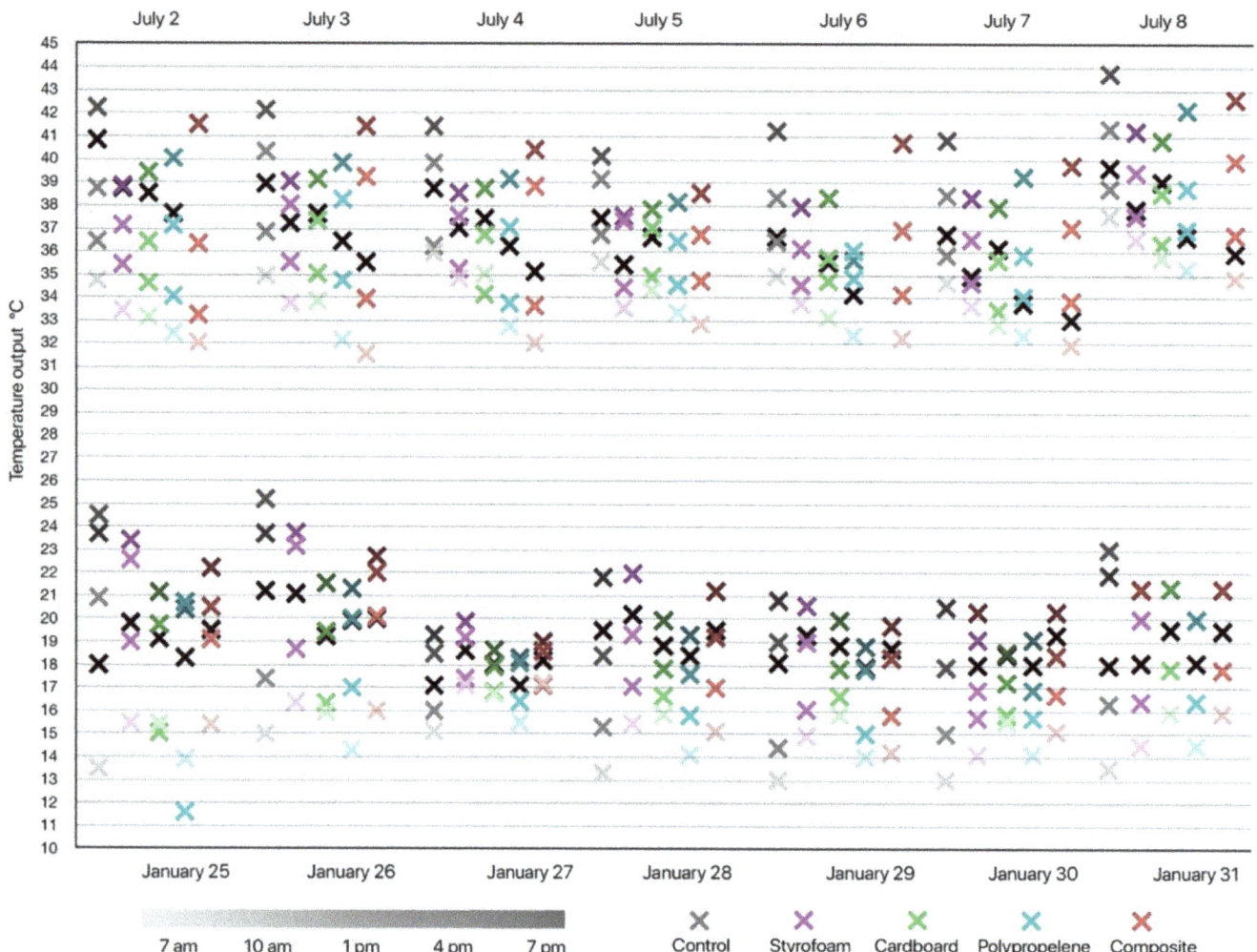

Fig. 8 A Scattered graph with detailed readings for each of the 5 chambers. The graph provides individual readings for each specific chamber insulation type on each of the 7 monitored days in both the hottest and coldest times of the year

temperatures than the composite panels from 11 a.m. to 6 p. m. by up to 2 °C, but they fall behind the composite panels after 6 p.m. with warmer temperatures that range between 2.2 °C and 1.5 °C, respectively, for each material.

Furthermore, in Fig. 9a, values of 34.3 °C and 33.8 °C for the cardboard and Styrofoam panels' chambers, respectively, during the second averages along with the same value of 37.6 °C at the third set of averages from 1 to 4 p.m., deemed them both the second best insulation types for that specific time period and put the composite panels third at just 37.9 °C.

The polypropylene panels proved to be the fourth best insulation for cooling during the summer. Though seemingly the second most cooling material it changed drastically during the day. From 7 a.m. till noon, polypropylene had an output of 33.8 °C at the 7 a.m. and 10 a.m. averages, and the

best insulation panel type at the weekly average at 1 p.m., 4 p.m., and 7 p.m. averages with a value of 35.9 °C and 37.2 °C preceding the composite panels by a cooler temperature of 0.2 °C and 0.7 °C, respectively. This was further demonstrated in Fig. 9b, when the polypropylene had an inverse effect to the composite panels effect throughout the day, making it warmer than the composite panels' effect by 1 °C till 11 a.m. then cooling down by an average of 1 °C to 1.6 °C at the peak of 4 p.m., only to return to its warmer temperature at 6 p.m., making the composite panels more suitable for the evening as an insulation panel. This could prove the polypropylene insulation as more suitable for buildings with more activity during the noon and afternoon hours.

The control chamber showed the highest temperature readings regardless of the figure observed. The values of the

Fig. 9 **a** Summer average results for whole week; for the 7 a.m. and 10 a.m. readings; and 1 p.m., 4 p.m. and 7 p.m. readings. **b** Summer average readings of each insulation type at each hour mark throughout the selected week

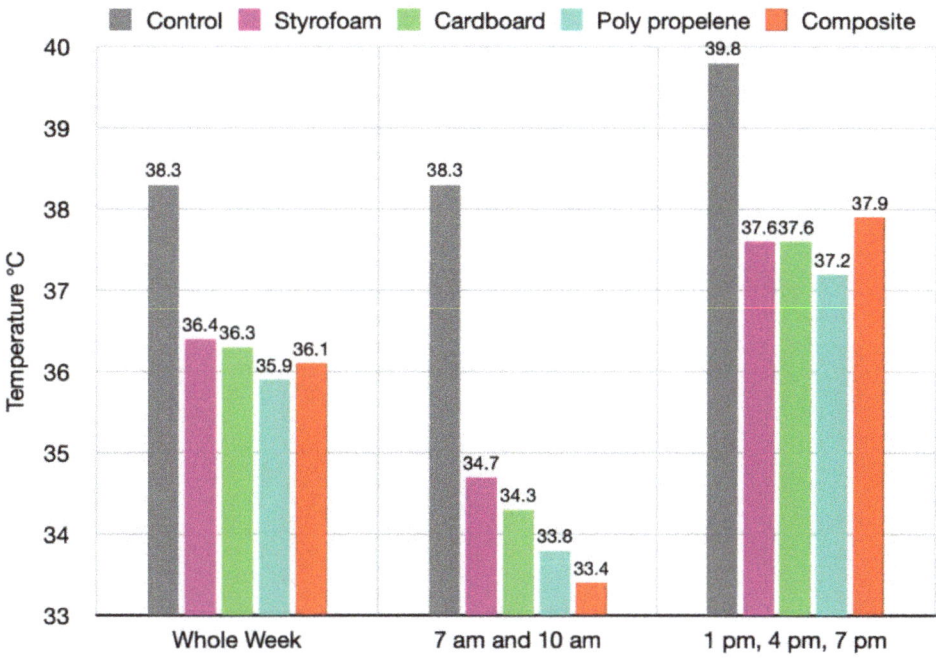

Figure 9a. Summer average results for the whole week; for the 7am and 10am readings; and 1pm, 4pm and 7pm readings

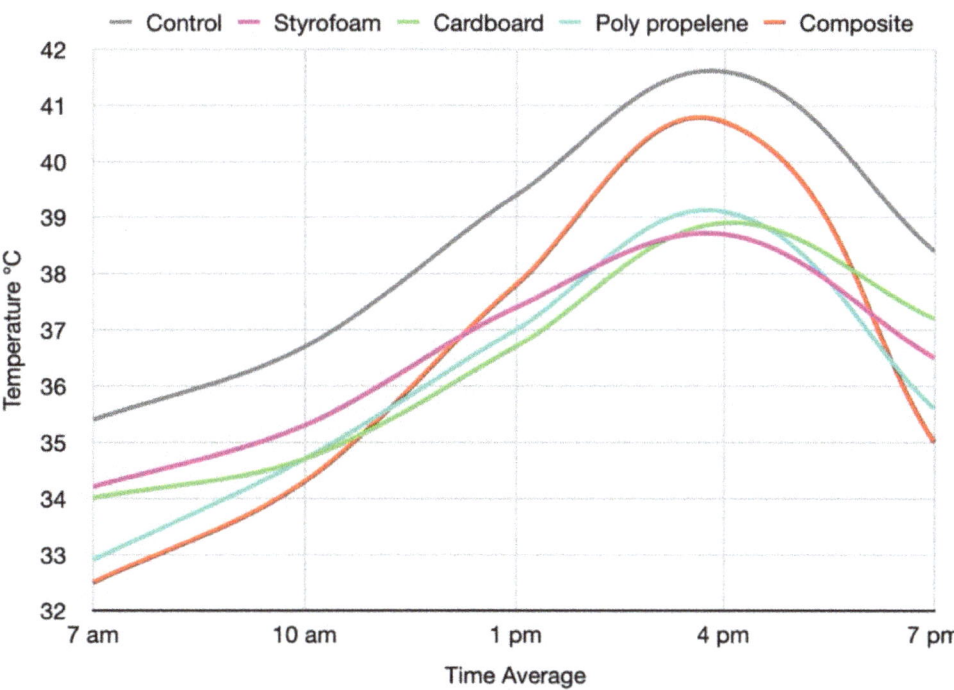

control chamber in Fig. 9a are higher than the fourth most suitable insulation panel by 2 °C, 3.6 °C, and 2.2 °C, respectively. In Fig. 9b, however, the control chamber was warmer than the fourth warmest insulation panel reaching a maximum and minimum difference in temperature of 1.2 °C and 0.9 °C, respectively.

3.2 Winter Results

According to Fig. 10a, the composite panels proved to be the best insulator with its primary average being indistinguishable from the Styrofoam panels at 18.6 °C. The composite panels then went on to provide warmer temperatures than the

Fig. 10 **a** Winter average results for the whole week; for the 7 a.m. and 10 a.m. readings; and 1 p.m., 4 p.m. and 7 p.m. readings. **b** Winter average readings of each insulation type at each hour mark throughout the selected week

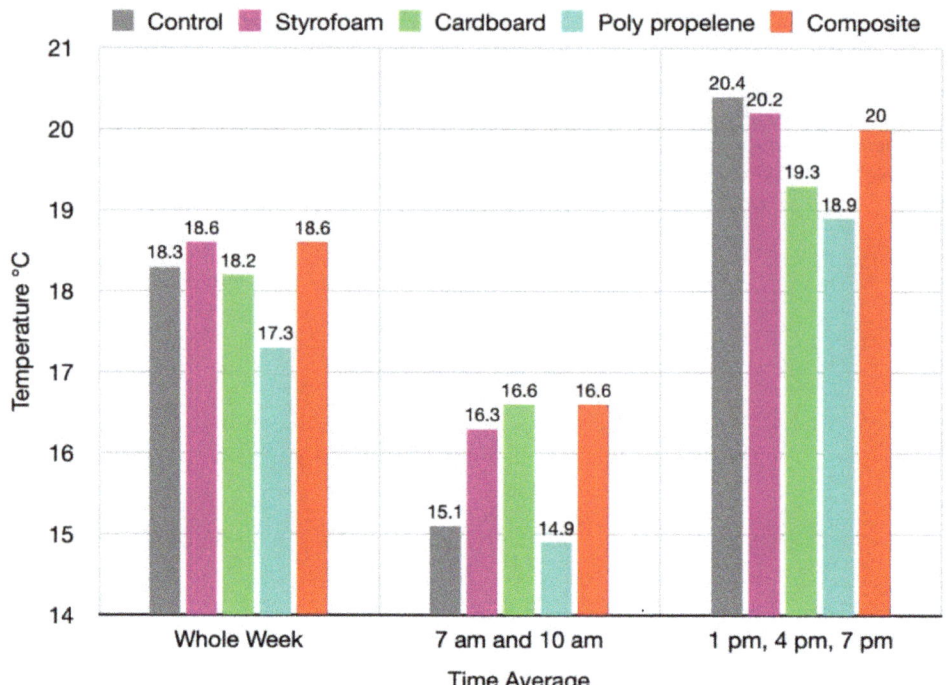

Figure 10a. Winter average results for the whole week; for the 7am and 10am readings; and 1pm, 4pm and 7pm readings

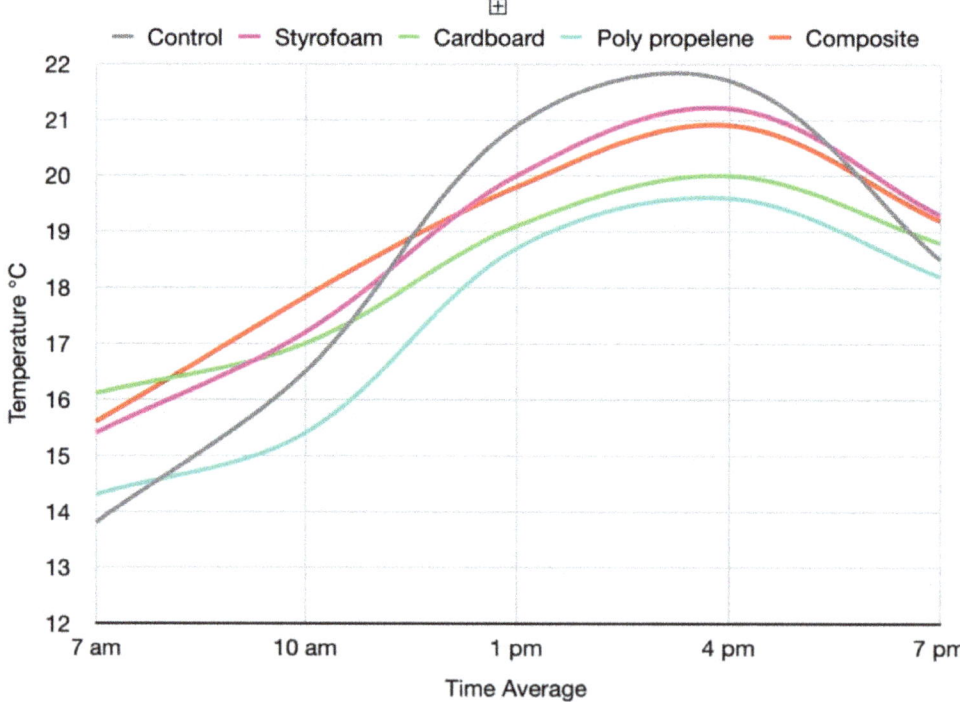

Styrofoam panels at the second average by a margin of 0.3 °C, Styrofoam then supersedes the composite panels by 0.2 °C from the afternoon onward.

Figure 10b shows the composite panels outperforming the Styrofoam panels till 12 p.m. by an average warmer margin of 0.6 °C, after which the Styrofoam panels performed better by an average margin of 0.2 °C till the end of

the day. This reveals once again that the composite panels are better insulators than the Styrofoam panels up until 12 p. m., but the Styrofoam panels are slightly better insulators from 12 p.m. until the end of the day.

However, the cardboard panels performed similar to the composite panels only in the secondary average values in Fig. 10a at 16.6 °C. But, in general, the cardboard panels gave lower temperature readings than both the composite and Styrofoam panels with a difference of 0.8–1 °C. In Fig. 10b, the cardboard panels' chamber proved warmer than both the composite and Styrofoam panels by 0.5 °C at the 7 a.m. reading at 16.1 °C, as opposed to 15.4 and 15.6 °C, only to cool down again from 8.30 a.m. to 9 a.m. changing its rank in performance to the third most suitable insulation panel for the winter reaching a maximum of 20 °C. It retained its position on the graph till the last reading of the day with an average margin difference of 0.9 °C.

The control chamber came forth in performance, ahead of the polypropylene insulated chamber, since it provided cooler temperatures than the composite, cardboard, and Styrofoam panels throughout the graph. Though it showed slightly warmer levels than the cardboard panels at the weekly average, and during the 1 p.m., 4 p.m., and 7 p.m. average in the graph, these results do not conflict with Fig. 10b. While it would seem that the control chamber provided better heating levels than the rest of the materials between 11:20 a.m. and 5:30 p.m., this is because the control chamber may have gained substantial amounts of heat much quicker than the Styrofoam, composite, and cardboard panels during the heat of noon giving it an advantage of around 1 °C. However, the other panels were able to sustain this heat and preserve it through the evening till the next day, making them significantly warmer by a range of 2 °C in the morning and 1 °C in the evening. These values may have increased during the night.

The polypropylene panels provided especially low-temperature readings in all the average sets of Fig. 10a by a range of 1–1.8 °C, reaching low temperatures of 17.3 °C, 14.9 °C, and 18.9 °C, respectively. In Fig. 10b, the polypropylene still showed very low temperature levels with a significantly lower difference in temperature than the Styrofoam panels. The polypropylene and Styrofoam showed a 1 °C average gap at 7 a.m. with the readings being 14.3 °C and 15.4 °C, and a 1.8 °C at 10 a.m. with readings 15.4 °C and 17.2 °C. This difference remained until the last reading on the day where the readings changed to 18.2 °C and 19.3 °C, respectively.

4 Conclusion

The insulation panels trials proved successful in being feasible from a cost and labor perspectives by means of adopting a micro-circular economy model. They also provided satisfactory levels of thermal comfort through decreased temperature levels.

The materials were collected from local suppliers, kiosks, and takeaway kitchen shops, which had Styrofoam, cardboard, and egg tray waste that they needed to discard of shortly. The methods of building the panels required little to no machinery whatsoever, and fossil fuel-free methods were used to collect, compress, and install the panels. The local laborers, who also represented the end users, accepted this initiative after their involvement in the selection, manufacturing, installation as well as the observation process of the panels' performance throughout both seasons. The Eco mortar proved useful in that it provided both safety for the panels and a reliable finish that allowed for various types of future finishes. This in itself was a milestone for the laborers as they saw the panels as fragile at first and inaesthetic regardless of how visually unpleasing the bare brick facades were, which the residents grew accustomed to.

The most effective insulation type during the winter from 7 a.m. till 11 a.m. is the composite panel, and the Styrofoam during the afternoon from 11 a.m. till 6 p.m. before its temperature results rise higher than that of the composite panels by 7 p.m. The composite panel was cooler by 3 °C than the control room and by almost 2 °C than the next warmest material, Styrofoam. The Styrofoam was cooler by an average of 1.7 °C than the composite material during the day before becoming 1.5 °C warmer by 7 p.m. Polypropylene and cardboard panels seem to provide the third and fourth coolest temperatures during the morning from 7 a.m. to noon. However, after 6 p.m., the composite panels followed by the polypropylene, Styrofoam, and cardboard panels, respectively, presented the coolest temperatures by differences of around 1 °C. The control chamber proved extremely deficient exceeding the other materials in temperature readings with a range of 1–6 °C. The 2 °C cooler temperature difference that Styrofoam provides during the day may be compensated for through fenestration openings and natural ventilation. This is because the 2 °C degree of cooler indoor temperatures the composite panels provide during throughout the night into noon is crucial for improved thermal comfort levels, less use of HVAC systems thus reducing the overall urban heat island effect. It would seem that if a building is used primarily for day operations such as workshops, cuisines, warehouses, etc. Styrofoam followed by cardboard panels would provide the best insulation throughout the day, decreasing the overall heat gains for the building operators during working hours.

Similarly, during the winter, the composite panel provided the warmest temperatures from 7 a.m. till noon outperforming Styrofoam by almost 1 °C. Styrofoam then provided warmer temperatures of around 0.2 °C from noon till the evening. The cardboard seems to provide the warmest temperatures at night as indicated by its performance from 6 p.m. to 7 a.m. but ultimately unknown due to the limited access to the chambers during the experiment. However, in the afternoon, it performed poorly throughout the monitored hours, providing even cooler temperatures than both the Styrofoam and composite panels. Though the control chamber provided warmer temperatures than both the

Styrofoam and composite panels by day with a 1 °C difference, it fluctuated to extreme low temperatures overnight until noon, making it an unstable solution. The polypropylene chamber provided the least effective temperature changes by it being the coolest chamber during the day and night.

The experiment as a whole proved successful as it reinforces the literature on how waste can and should take the place of conventional non-eco-friendly insulation materials globally, but also regionally in hot arid countries. The varying levels of high thermal mass and possible HVAC reduction that the insulation panels provided for each chamber demonstrated how installing these panels in the future on a large and mass scale could improve thermal comfort levels as well as decrease the energy consumption for cooling and heating thus decreasing the urban heat island effects in hot arid regions. This is true regardless of the other factors which the insulation panels cannot address such as lack of greens, overpopulation, pollution, and site geometry.

The study reveals the possibility of using unsophisticated, low-cost, and locally available methods to build insulation panels from widely available sources. Many studies were conducted using various molding methods which require the widespread availability of expensive fuel as well as costly and unpopular machinery that provide molding, pressing molding, hot-pressing molding, injection molding, etc. Both fuel and advanced machinery are rare commodities among the majority of locals in many hot arid regions, resulting in research papers of this nature stopping at the experimental stages. Few to no research papers have approached the subject from a practical and available methodology to the locals. In the meantime, this research provides an opening to communities in hot arid regions and those who work with them to refurbish existing buildings for better thermal comfort levels. It makes possible the idea of refurbishing 70% of housing in some of the hot arid countries mentioned in the paper through cost-efficient and simple methodological means.

Though this research paper was designed under a low budget, therefore the utilization of hot plate method and the impossibility of round the clock access to the chambers, it provides a strong starting point that is extremely economic and that any group of locals can adopt with or without the presence of an engineer or an expert. With further refining and development, this research can provide much more detailed results of the materials' performance within specific urban situations and times of the year.

Limitations

Due to limited resources, the lack of advanced monitoring materials, like a set of five heat flow meters to monitor each chamber, prevented higher levels of detail in understanding the nature and exact time frames in which the insulation panels decreased or increased in temperature. Further experimentation is needed to provide more reliable thermal conductivity coefficients. Testing the panels on rooms with different orientations and geographical locations through environmental simulation software would bring about this level of reliability.

Moreover, the chambers were not designed based on ISO standard guidelines, which may have provided more detailed or clearer results. This was again due to a lack of resources for the experiment.

Valuable information as to how the panels perform throughout the night wasn't available due to a lack of access to the chambers from 7 p.m. until 7 a.m. Lastly, the experiment was not performed in a mid- or low-income site, which hindered the interaction of a wider audience of less skilled residents. Had there been a wider spectrum of interaction, a better understanding of the panels' potential and acceptability would have been observed and recorded.

Implications

If this experiment were to be replicated, an automated day round surveillance of the chambers using heat flow meters should be utilized to provide accurate and frequent readings. This would incorporate spring and fall readings, which were absent in this experiment, allowing for a more holistic approach toward the panels.

The panels' construction process could have benefitted from more laborers sharing their views and expertise. This may provide easier methods of construction which do not include using the expensive adhesive for the panels' construction, though it was used in small quantities making the process cost-efficient. The extenuation of the adhesive altogether would make the panels even more cost-efficient and sustainable. It is also advised to use the reused plastic corrugated sheet mullions mentioned earlier to test the performance levels of the panels without the use of the expensive adhesive and compare them against the adhesive-based application method.

The mechanical properties of the panels need further testing via simplified versions of laboratory-based tests for the methodology to stay as simple as possible for the lay person to conduct. Moreover, testing using methods such as water splashes, mild-to-moderate battering, as well as natural means of exposure including a year-long exposure to the sun, rain, and natural weather are advised.

References

Akovali, G. (2012). 3—Plastic materials: chlorinated polyethylene (CPE), chlorinated polyvinylchloride (CPVC), chlorosulfonated polyethylene (CSPE) and polychloroprene rubber (CR). In F. Pacheco-Torgal, S. Jalali, & A. Fucic (Eds.), *Toxicity of building materials* (pp. 54–75). Woodhead Publishing. https://doi.org/10.1533/9780857096357.54

Al-Sibahy, A., & Edwards, R. (2017). Characterization of the clay masonry units and construction technique at the ancient city of Nippur. *Engineering Structures, 147*, 517–529. https://doi.org/10.1016/j.engstruct.2017.06.017

Al Hosni, I. S., Amoudi, O., & Callaghan, N. (2020). An exploratory study on challenges of circular economy in the built environment in Oman. *Proceedings of Institution of Civil Engineers: Management, Procurement and Law, 173*(3), 104–113. https://doi.org/10.1680/jmapl.19.00034

Albert, R., Juraj, S., Štefan, B., Henrich, L., Vladimír, I., & Andrej, P. (2013). Reuse of old corrugated cardboard in constructional and thermal insulating boards. *Wood Research, 58*(3), 505–510.

Bek, M. A., Azmy, N., & Elkafrawy, S. (2018). The effect of unplanned growth of urban areas on heat island phenomena. *Ain Shams Engineering Journal, 9*(4), 3169–3177. https://doi.org/10.1016/j.asej.2017.11.001

Chun, B., & Guldmann, J.-M. (2014). Spatial statistical analysis and simulation of the urban heat island in high-density central cities. *Landscape and Urban Planning, 125*, 76–88. https://doi.org/10.1016/j.landurbplan.2014.01.016

Esposito, M., Tse, T., & Soufani, K. (2017). Is the circular economy a new fast-expanding market? *Thunderbird International Business Review, 59*(1), 9–14. https://doi.org/10.1002/tie.21764

Farrelly, T. A., & Shaw, I. C. (2017). Polystyrene as hazardous household waste. *Household Hazardous Waste Management, April*. https://doi.org/10.5772/65865

Fernández, F. J., Alvarez-Vázquez, L. J., García-Chan, N., Martínez, A., & Vázquez-Méndez, M. E. (2015). Optimal location of green zones in metropolitan areas to control the urban heat island. *Journal of Computational and Applied Mathematics, 289*, 412–425. https://doi.org/10.1016/j.cam.2014.10.023

Ferronato, N., Rada, E. C., Gorritty Portillo, M. A., Cioca, L. I., Ragazzi, M., & Torretta, V. (2019). Introduction of the circular economy within developing regions: A comparative analysis of advantages and opportunities for waste valorization. *Journal of Environmental Management, 230*(September 2018), 366–378. https://doi.org/10.1016/j.jenvman.2018.09.095

Gardiner, K., & Heller, A. (2008). *Green design at any size, urban heat island effect & Tod.*

Ibrahim, M. I. M., & Mohamed, N. A. E. M. (2016). Towards sustainable management of solid waste in Egypt. *Procedia Environmental Sciences, 34*, 336–347. https://doi.org/10.1016/j.proenv.2016.04.030

Lechtenböhmer, S., & Schüring, A. (2011). The potential for large-scale savings from insulating residential buildings in the EU. *Energy Efficiency, 4*(2), 257–270.

Mahdy, M. M., & Nikolopoulou, M. (2013). From construction to operation: Achieving indoor thermal comfort via altering external walls specifications in Egypt. *Advanced Materials Research, 689* (September 2016), 250–253. https://doi.org/10.4028/www.scientific.net/AMR.689.250

Pacheco-Torgal, F. (2014). Eco-efficient construction and building materials research under the EU Framework Programme Horizon 2020. *Construction and Building Materials, 51*, 151–162. https://doi.org/10.1016/J.CONBUILDMAT.2013.10.058

Patti, A., & Acierno, D. (2019). Thermal conductivity of polypropylene-based materials. *Polypropylene—Polymerization and Characterization of Mechanical and Thermal Properties*. https://doi.org/10.5772/intechopen.84477

Polystyvert. (2018). Polystyvert unveils the world's first polystyrene dissolution recycling plant. *Resource Recycling News, June*. https://resource-recycling.com/plastics/wp-content/uploads/sites/4/2018/08/2018-08-20-EN-Press-Release-plant-inauguration-FINAL.pdf

Pujadas-Gispert, E., Alsailani, M., van Dijk (Koen), K. C. A., Rozema (Annine), A. D. K., ten Hoope (Puck), J. P., Korevaar (Carmen), C. C., & Moonen (Faas), S. P. G. (2020). Design, construction, and thermal performance evaluation of an innovative bio-based ventilated façade. *Frontiers of Architectural Research, 9*(3), 681–696. https://doi.org/10.1016/j.foar.2020.02.003

Sohair Milik. (2010). *Assessment of solid waste management in Egypt during the last decade in light of the partnership between the Egyptian government and the private sector.*

SWEEP. (2014). *Country report on the solid waste management in EGYPT 2013*. April, 1–83. http://www.sweep-net.org/country/egypt

VIRTUe. (2018). *Solar Decathlon Middle East 2018*. http://www.solardecathlonme.com

VIRTUe. (2019). *Eindhoven University of Technology*. https://teamvirtue.nl/

Comparative Review of Different Rating Systems Approach and Responses to Pandemic Situations

Waleed Salaheldin Hussien

Abstract

This study is going to spot the light on the shortcomings of the current sustainability and green buildings rating systems in terms of protection against pandemics especially COVID-19 pandemic. Six of the globally widely used, most developed, well-known Green buildings rating systems were carefully picked, studied and compared in scientific comparison according to certain criteria The aim is to show the common aspects of sustainable built environment and the credits/criteria which are aligned with COVID-19 protection precautions and measures. The reliable source of COVID-19 protection precautions is the official publications by the world health organisation on their Website after the crisis started. It is to be noted that only precautions with impact on architecture and urban levels were used by the researcher. The study/comparison showed gaps in the selected systems in global pandemic situations.The response of the different systems around the globe was studied and presented by the researcher. It is not intended to provide any criticism to the studied systems rather than providing an overview on different approaches of responses of different systems in a variety of geographical areas. Finally, the paper shows the recommended approaches in overcoming such gaps and shortcomings which were concluded after the study. It is expected that this study would provide practitioners in the field of architecture, Enginnering, Construction, green buildings industry and public health professionals some guidelines to help them overcome shortcomings in the current green standards and rating systems.

Keywords

Architecture • Sustainability • Pandemic • Pandemic architecture • Health and safety

1 Introduction

In 2020, the whole world was under brutal and relentless attack of new and mutated type of respiratory virus of SARS family, which is currently known as COVID-19, the spread of the virus was rapid that the world health organization declared the state of pandemic emergency, the effect of the pandemic reached almost everywhere on planet earth, several industries were severely affected and the impact is to be evaluated in post pandemic stage, but the initial readings predict that it will be catastrophic.

The sustainability experts and professionals around the globe must carry their responsibility and to be in the front line in the army fighting the current pandemic and the future potential ones. What can be done in this current historical turning point can save millions of lives in the near and far future. The war against pandemics will not end quickly and it is believed that it is time now to take further step in this war and not to stay in defensive mode but to take a proactive approach as "the best defence is a good offence".

The world Health Organization issued recommendations to all countries to fight the current pandemic COVID-19. These recommendations were in several formats including brochures, bulletins, reports and awareness campaigns (World Health Orgnisation). With hundreds of recommended precautions there are main protection measures that are reiterated by the organization and all public health professionals. These precautions include (and not limited to).

- Promotion of hygiene practices including regular hand washing and use of sanitizers. Hand hygiene stations availability.

W. S. Hussien (✉)
International Sustainability Academy, Hamburg, Germany
e-mail: Salaheldin@isa-fellow.com

- Promote respiratory etiquette by all people at the workplace and the use of personal protective equipment.
- Development of regulations and policies that mandates the use of face shields/masks and other protective tools.
- Social distancing and avoidance of unnecessary gatherings and reduction of people density inside buildings.
- Minimize the need for physical meetings or events and promotion of virtual conferences and meetings.
- Adjustment of working hours to avoid people gathering (working from home).
- Travel restrictions, to and from high risk countries.
- Space and surfaces cleaning and disinfection.
- Efficient and effective waste management especially hazardous waste.
- Responsible procurement of sanitizers, disinfectants and all similar products.
- Avoid the excessive use of sprayers or disinfectants in indoor and outdoor areas. And restrict smoking in indoor areas.
- Promote awareness campaigns and provide regular trainings and guidance.

A	Meets the proposed criteria Fully.
B	Meets the proposed criteria Partially.
C	Does not meet any Criteria.
Blank	Information not available.
N/a	Not applicable.

2 Structure

This paper consists of the following sections:

Overview of rating systems, Review Criteria, Analysis and elaborations, Results and Finding and Conclusions.

3 Overview of Rating Systems

Green building rating tools—also known as certification tools—are the tools which used to evaluate the performance of buildings in terms of sustainibility and provide recognition and publicity to it. The said rating tools, often voluntary, provide rewards to the buildings which has significant greener performance. Green Building Councils, which are members of the WorldGBC global network, develop and administer many of the world's ratings tools. By 2016, more than one billion square meters of green building space (an area as ten times as the size of the french capital) had been certified globally through member Green Building Councils (World Green Building Council).

Table 1 Studied ratinng systems and countries (Politia and Antoninib, 2016)

Rating system	Country
Leed	US/International
GSAS	Qatar
Estidama	UAE
Bream	UK/International
Green star	Australia
Greenglobes	Canada/US

Nowadays, there are hundreds of green buildings and sustainability ratings systems, standards in the global market. These tools were made to help guide, demonstrate, and document efforts to deliver green and sustainable, high-performance buildings. It is widely believed that more than six hundred 600 green product certifications around the globe with approximately one hundred in use in the united states and the numbers continue to grow.

The vital role of sustainability rating systems is currently in protecting the built environment against pandemics is being examined nowadays. In the following section (Table 1), several prominent rating systems are going to be studied and reviewed against WHO COVID-19 recommendations. The rating systems are namely LEED, GSAS, ESTIDAMA, BREAM, GREENSTAR, GREENGLOBES. These picked tools are currently the most well-known, most influential and technically developed green rating tools available (Fowler and Rauch, 2006).

4 Review Criteria

There are different approaches of evaulating the sustainability rating systems. During the literature review of this paper, the author reviewed different approaches of evaluation and concluded that there are four different sets of criteria; RSMEANS, B.K. Nguyen/H. Altan, E. Bernardi/S. Carlucci/C. Cornaro/R. Bohne and H.M. Karmany.

The first approach, which is called RSMEANS (Table 2), cited as follows "*there are four main principles that should be taken into consideration when evaluating a building rating or certification system:*

- *Science-based—the ability to reproduce the results,decisions by other stakeholders.*
- *Transparent—evaluation and award process is transparent and examinable.*
- *Objective—Conflict free, no corruption certification entity.*
- *Progressive—Tools are crafting a postive impact on the market and the industry.*"

Table 2 RSMEANS criteria

RSMEANS approach	
Science-based	*The ability to reproduce the results,decisions by other stakeholders*
Transparent	*Evaluation and award process is transparent and examinable*
Objective	*Conflict free, no corruption certification entity*
Progressive	*Tools are crafting a postive impact on the market and the industry*

The second approach was introduced by Professor and Scholar of School of Architecture at the University of Sheffield Dr. Binh K. Nguyen and Prof. Hasim Altan in 2011. The proposed criteria for assessment can be summerised as follws in Table 3.

The third approach was introduced by Researchers and Professors in the Faculty of Engineering of Norwegian University of Science and Technology and the Department of Enterprise Engineering in the University of Rome; Elena Bernardi, Salvatore Carlucci, Cristina Cornaro and Rolf Andre Bohne. Published on the web.

Table 3 Nguyen/Altan approach and criteria/sub-criteria

Binh K. Nguyen/Hasim Altan		Remarks
Popularity & influence	Well known Importance Numer of countries adopting the system Number of projects Involved Versatility	Versatility: the usability of the system as comparison basis or benchmark
Availability	1-Availability to the system itself: Accessibility System's format Information available for public System cost Certfication fees 2-Vailability of references: Vailability of on-line information vailability of non on-line information availability of case studies avilability of user's interaction System's openess	
Methodology	Methodology summary Weighting Rating levels Standardization Quantitative criteria Qualitative criteria Whole life cycle assessment Complexity Efficiency of assessment method	
Applicability	1-Stages of Building lifecycle incfluenced: Pre-design, design, construction, operation and post occupancy Demolition, second life 2-Technical Contents:	
Data collection process	Data gatherer Data collection method Documentation Measurability Convenience	Measurability: the ability to use tangible and measureable methods
Accuracy & verification	1-accuracy of data processing stage 2-Accuracy of data outputting stage 3-Verification: Assessor qualifications Level of details to check Third party Results aknowledgement	

(continued)

Table 3 (continued)

Binh K. Nguyen/Hasim Altan		Remarks
User-friendliness	1-Ease of use: 2-Product support Availability of responsive assisstance FAQ & Record inquiries Training & courses available Instructions or helps	
Development	Systems maturity Systems stability Update Development approach Future development	
Results presentation	Presentation method Clarity Comparability Result usability	

Table 4 E. Bernardi/S. Carlucci/C. Cornaro/R. Bohne approach and criteria

E. Bernardi/S. Carlucci/C. Cornaro/R. Bohne	
Focus	An inclusive focus on buildings
Scientific interest	Cited in at least 20 scientific researches and papers in reputable journals
Widespread adoption	More than 500 Certfied projects
A consolidated development state	Exceeds five years of service

The criteria can be summerized as in Table 4.

The fourth and last approach was presented by Heballah Mostafa Karmany in her research and study about green buildings rating systems for Egypt (Karmany, 2016). The approach can be summerized and presented as in Table 5.

From the above discussion, comparison and analysis study of different approaches was conducted. Using the analytical approach to merge all approaches in one approach which combines the most significant criteria and rearrange the criteria in a new set of evaluation criteria. The analysis is presented hereunder using colour mapping (Table 6). See Appendix 2 for more details.

After the analysis, the combined criteria can be categorized into 5 main groups (Table 7).

- Science base/Technical content
- Transparency/Accessibility
- Objectivity/Focus/Proactivity
- Progression/Development/Adabpatability/Diversity
- Miscellaneous asepcts (number of projects,assessment method,weighting..etc.).

Table 5 E. Hebaalla Mostafa Karmany evaluation criteria summerised

H.M. Karmany approach criteria
Technical content & sustainability aspect metrics
System accessibility
Assessment cost
Local context (regional Priority)
Assessment and weighting methodology
Registration & assessment process
Maturity
Life cycle approach
Validity
Adaptability

Table 6 Evaluation categories

Category
Science Base/Technical content
Transparency/Accessibility
Objectivity/focus/proactivity
Progression/development/Adaptability
Miscellaneous aspects (number of projects, assessment method, weighting).

Table 7 Evaluation criteria summary and points

Number	Aspect	Score
1	Science base/Technical content (Pandemic precautions inclusion)	20 Points
2	Transparency/Accessibility	20 points
3	Objectivity/Focus/Proactivity/Responsive actions	20 points
4	Progression/Development/Adaptability/Diversity	20 points
5	Miscellaneous aspects (number of projects, assessment method, weighting)	20 points
Grand total		100

Since the focus of this paper is pandemic protection measures and its inclusion in green buildings rating systems, the researcher focused his study on the criteria that relate to such topic. The above criteria adress five main aspects of the rating system response to the pandemic situation. It can be elaborated as follows:

- Protection measures included in technical contenet of the system.
- How accessible/transparent is the system in pandemic situations.
- The reaction to situation was responsive/proactive and focused.
- Is the system developing over time to absorb global challenges (especially what relates to biological threats).
- General overview of system capacity, setup and assessment methodology.

It has to be noted that the only reliable source currently for pandemic protection measures is the published WHO recommendations which are available on its website.

Table 8 shows the detailed points system for each category and subcategory.

The above criteria are going to be implemented on all studied rating systems. The following section provides more details.

5 Analysis and Elaborations

The following will be the detailed anaylsis of each rating system against the proposed criteria. The below (Table 9) is the key of sysmbols used in the anaylsis to reflect compliance or non compliance.

LEED (Leadership in Energy and Environmental Design) is the most widely used green building rating system in the world. Available for virtually all building types. It was checked versus the summarized criteria.

The studied version is LEED v4.1. Out of selected 12 WHO recommendations, 5 were found in line with LEED documentation and 1 credit contradicts with social distancing (Reduced parking footprint-option 3 car share). See Appendix 1 for detailed scores.

While the COVID-19 response is remarkably impressive due to issuance of several publications and pilot credits. The system is aligned with many other sustainability tools like WELL. However, health and safety certification are not mandatory as part of LEED certification.

BREEAM is the world's leading sustainability assessment method for master planning projects, infrastructure and buildings. It recognizes and reflects the value in higher performing assets across the built environment lifecycle, from new construction to in-use and refurbishment. The studied version is 2016 version as V.6 is yet to be available.

Table 8 Review criteria with scores (detailed)

Number	Aspect			Score
1	Science base/Technical content (Pandemic precautions inclusion)			20 pt
	1.A	Alignment with WHO recommendations		10
		1.A.1	Number of Credits/Criteria aligned	
		1.A.2	Accuracy and effectiveness of application	
		1.A.3	Applicability to certified or existing buildings	
	1.B	Credits/Criteria contradicting with WHO recommendations or Not covered		10
		1.B.1	Number of Credits/Criteria not aligned	
		1.B.2	Importance (Mandatory/optional)	
		1.B.3	Number of Credits/Criteria not covered	
		1.B.4	Importance (Mandatory/optional)	
2	Transparency/Accessibility			20 pt
		1.C.1	Easy to access/Material available	
		1.C.2	Cost of system/Certification fee	
		1.C.3	Availability of user interaction/support	
		1.C.4	Availability of case studies	
Total				40
3	Objectivity/Focus/Proactivity/Responsive actions			20 pt
	2.A	Action taken		10
	2.B	Number of Credits/Criteria added		5
	2.C	Addendums or chapters published		5
Total				20
4	Progression/Development/Adaptability/Diversity			20 pt
	3.A	Synchronization with other tools		10
	3.B	Cultural & Geographical zone adaptation		5
	3.C	different building types coverage		5
Total				20
5	Miscellaneous aspects (number of projects, assessment method, weighting)			20 pt
	4.A	Number of countries using the system		5
	4.B	Number of projects		5
		4.B.1	Registered	
		4.B.2	Certified	
	4C	Different building types of coverage		10
		4.C.1	Prescriptive based	
		4.C.2	Performance based	
		4.C.3	Solutions based	
Total				20
Grand total				100

Table 9 Analysis key

Key	Explainayion
A	Meets the proposed criteria Fully
B	Meets the proposed criteria Partially
C	Does not meet any Criteria
(Blank)	Intformation not available
N/A	Not Applicable

Out of selected WHO recommendations; 6 were in line with BREEAM documentation (Cerdit Tra05 Travel plan is in addition to 5 credits similar to LEED). None were found in contradiction with WHO recommendations (Carpooling is not an option of credit. It is generally recommended unlike LEED).

No additional credits or courses were available. The system is aligned with several tools of sustainability (e.g. CARES). The system has low adaptability to different regions despite the wide spread of certifications and projects. See Appendix 1 for detailed scores.

Global Sustainability Assessment System (GSAS) is the first performance-based system in the Middle East and North Africa (MENA) region, developed for rating green buildings and infrastructures. The studied version is GSAS V4.

Out of selected 12 WHO recommendations, 5 were found in line with GSAS documentation. no additional criteria added in response to COVID-9. However, some awareness campaigns were conducted (for example testing of local sanitizers).

The system is aligned with other local tools (like Gulf green mark and Qatar carbon trust). It covers narrow range of countries and no medium for cultural adaptation. The system combines performance and evidence-based methodologies. See Appendix 1 for detailed scores.

ESTIDAMA (PBRS), The aim of the Pearl Building Rating System (PBRS) is to promote the development of sustainable buildings and improve quality of life.. There is no available updates of the system since the first issuance in 2010. As such, the studied version is 2010.

Out of selected 12 WHO recommendations, 5 were found in line with PBRS documentation. No actions were taken as a response to COVID-19. The system is aligned with some local tools and programs (e.g. green key).

International adaptability is considered very low. (Rahim et al. 2015), The system combines performance and evidence-based methodologies. See Appendix 1 for detailed scores.

GREENGLOBES as per the official website that is is identifing opportunities and provides effective tools to achieve success. A nationally recognized green rating assessment tool, guide e and certification system, Green Globes® works with stakeholders to achieve the sustainability goals for newly constructed projects, existing buildings and interiors. The studied version is "New Construction 2019".

Out of selected 12 WHO recommendations, 4 were found in line with GREEN GLOBES documentation. No evidence of training or awareness campaigns promotion as a response to COVID-19 Crisis.

Actions were taken as part of COVID-19 response including webinars and courses but added or amended credits. The system is aligned with other systems (e.g. ANSI). The system is a prescriptive bases evaluation tool. See Appendix 1 for detailed scores.

GREEN STAR, Launched by Green Building Council of Australia (GBCA) in 2003, Green Star is Australia's only voluntary and truly holistic sustainability rating system for buildings, fit outs and communities. No evidence of versions updates on the website.

Out of selected 12 WHO recommendations, 5 were found in line with GREEN STAR documentation. The system responded to COVID-19 crisis and issued a report addressing the changes in certification system. However, no added or amended criteria were introduced.

The system is aligned with other tools (like NABERS, BASIX, GEMS, ECS certification).The system is prescriptive and evidence-based tool. See Appendix 1 for detailed scores.

6 Results and Findings

The different rating systems were evaluated and compared based on the former discussions. More details are presented in Appendix 1.

The results are illustrated in Fig. 1.

The scores were concluded to be as follows (Tables 10 and 11).

LEED (90.3/100), BREEAM (74.0/100), GSAS (72.8/100), ESTIDAMA (PBRS) (55.8/100), GREENGLOBES (77.7/100) and GREEN STAR (73.5/100).

Fig. 1 Final scores/percentages

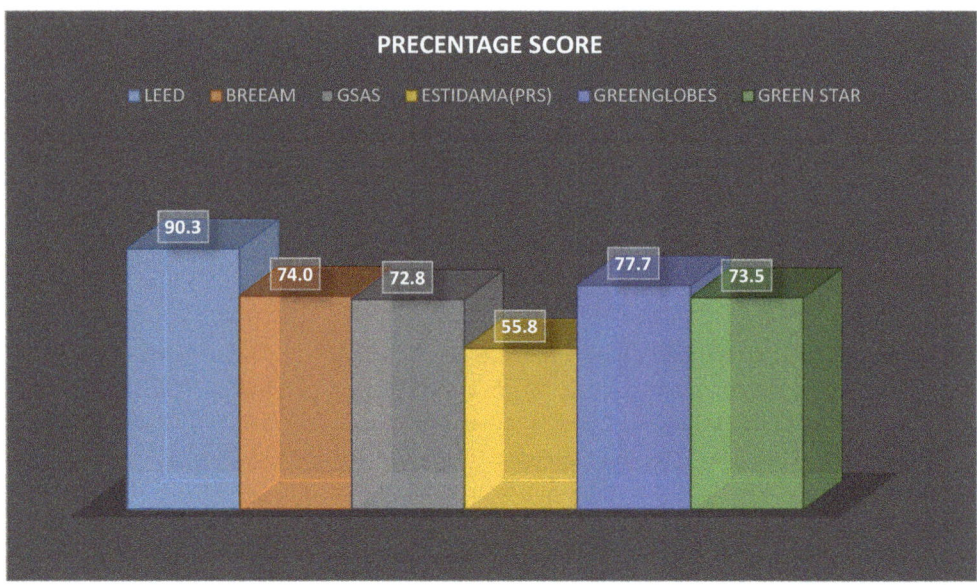

7 Conclusion

After the previous presentation and discussions, the following points could be concluded.

Firstly, all prominent sustainability rating systems responded to COVID-19 situation with different approaches and outputs. Some systems like LEED showed quicker response in changing some credits content while some other systems were stuck in providing guidance to maintain the certification process or promote the wellbeing of their employees.

Secondly, without exceptions, all systems showed deficiency in providing the required actions to protect the built environment against future pandemics. A further study is to be prepared on the feasibility of introducing a new tool dedicated for pandemics and biological threats.

As such, there are three recommended courses of actions to tackle the shortcomings of the current sustainability rating systems. These actions can be listed as follows:

1. To update the current versions of rating systems to include protective precautions of pandemics.
2. To update the policies to make health and safety tools (like WELL or Fitwell) mandatory as part of the assessment process.
3. To introduce a new tool or system which will be dedicated for pandemic situations, the tool shall combine sustainability, health and safety, resilience, wellbeing and environmental aspects in one comprehensive tool that addresses the current biological threats. The tool should be "solutions based".[1]

Acknowledgements Special thanks to Dr. Mohamed El Masry for his contribution in this paper.

---

[1] Further study to be made on the impact of solutions-based tools to improve the certification process.

Table 10 Summary of findings

Rating system	Country	Rsponsive	Adaptable	Additional measures taken	Additiona credits added	Methodology
Leed	US/International	Yes	Yes	Yes	Yes	Prescriptive Preformance based
Bream	UK/International	Yes	Yes	Yes	No	Prescriptive Preformance based
GSAS	Qatar	Yes	No	Yes	No	Prescriptive Preformance based
Estidama (PBRS)	UAE	No	No	No	No	Prescriptive Preformance based
Greenglobes	Canada/US	Yes	Yes	Yes	No	Prescriptive
Green star	Australia	Yes	Yes	Yes	No	Prescriptive

Table 11 Final scores

System	Version	Score (%)
Leed	Version 4.1	90.3
Breeam	2016 version	74.0
GSAS	Version 4	72.8
ESTIDAMA(PRS)	Version 1.0	55.8
Greenglobes	New construction 2019	77.7
Green star	Version 4.6	73.5

Appendix 1. Detailed Review Scoring

Number	Aspect			Score	LEED		BREEAM		GSAS		ESTIDAMA(PBRS)		GREENGLOBES		GREEN STAR	
					Key	Score	Key	Score	Key	Score	Key	Score	Key	Score	Key	Score
1	Science Base/Technical content (Pandemic Precautions inclusion)			20 pt												
	1.A	Alignment with WHO recommendations.		10												
		1.A.1	Number of Credits/Criteria aligned.		B	4.2	B	5.0	B	4.2	B	4.2	B	3.3	B	4.2
		1.A.2	Accuracy and effectiveness of application.		B	3.0	B	3.0	A	5.0	A	5.0	A	5.0	B	3.0
		1.A.3	Applicability to certfied or existing buildings.		A	2.5	A	5.0	A	5.0	C	0.0	A	5.0	A	5.0
	1.B	Credits/Criteria contradicting with WHO recommendations or		10												
		1.B.1	Number of Credits/Criteria not aligned.		B	1.5	A	2.5	A	2.5	A	2.5	A	2.5	A	2.5
		1.B.2	IMPORTANCE (Mandatory/optional)		A	2.5	A	2.5	A	2.5	A	2.5	A	2.5	A	2.5
		1.B.3	Number of Credits/Criteria not covered.		A	2.5	A	2.5	A	2.5	A	2.5	A	2.5	A	2.5
		1.B.4	IMPORTANCE (Mandatory/optional)		A	2.5	A	2.5	A	2.5	A	2.5	A	2.5	A	2.5
2	Transparency/Accessibility			20 pt												
		1.C.1	Easy to access/Material available		A	5.0	A	5.0	A	5.0	A	5.0	A	5.0	A	5.0
		1.C.2	Cost of system/Certification fee		A	5.0	A	5.0	B	3.0	A	5.0	A	5.0	A	5.0
		1.C.3	availability of user interaction/support		A	5.0	A	5.0	C	1.0	C	1.0	A	5.0	A	5.0
		1.C.4	Availability of case studies		A	5.0	A	5.0	B	5.0	C	1.0	A	5.0	A	5.0
	Total			40	38.7		43.0		38.2		31.2		43.3		42.2	
3	Objectivity/focus/Proactivity/Responsive actions.			20 pt												
	2.A	Action Taken		10	A	10.0	B	5.0	B	5.0	C	0.0	B	5.0	B	5.0
	2.B	Number of Credits/Criteria added.		5	A	5.0	C	0.0	C	0.0	C	0.0	C	0.0	C	0.0
	2.C	Addendums or chapters published.		5	A	5.0	C	0.0	C	0.0	A	5.0	C	0.0	A	5.0
	Total			20	20.0		5.0		10.0		0.0		10.0		5.0	
4	Progression/development/Adaptability/diversity			20 pt												
	3.A	Synchronisation with other tools		10	B	5.0	B	5.0	B	5.0	B	5.0	B	5.0	B	5.0
	3.B	Cultural & Geographical zone adaptation		5	A	5.0	B	3.0	C	0.0	C	0.0	B	3.0	B	3.0
	3.C	different building types coverage		5	A	5.0	B	3.0	A	5.0	A	5.0	A	5.0	A	5.0
	Total			20	15.0		11.0		10.0		10.0		13.0		13.0	
5	Miscellaneous aspects (number of projects, assessment method, weighting).			20 pt												
	4.A	Number of countries using the system.		5	A	5.0	A	5.0	B	3.0	B	3.0	B	3.0	A	5.0
	4.B	number of projects.		5	A	5.0	A	5.0	A	5.0	A	5.0	A	5.0	A	5.0
		4.B.1	Registered		A		A		A							
		4.B.2	Certified.		A		A		A							
	4C	different building types coverage		10		6.7		5.0		6.7		6.7		3.3		3.3
		4.C.1	Prescriptive based		A		A		A		A		A		A	
		4.C.2	Performance based		A		B		A		A		C		C	
		4.C.3	solutions based.		C		C		C		C		C		C	
	Total			20	16.7		15.0		14.7		14.7		11.3		13.3	
	Grand total			100	90.3		74.0		72.8		55.8		77.7		73.5	

Key	Explanation
A	Fully Meets Criteria
B	Partially Meets Criteria
C	Does not meet Criteria
(blank)	Information not availble
n/a	Not applicable

Appendix 2. Different Approaches of Evaluation with Colour Codes of Similar or Related Criteria

RSMeans		K.Nguyen/Hasim Atlan		E.Bernardi/S.Carlucci/C.Cornaro/R.Bohne		H.M.Karmany	
Science-based	Results and decisions must be reproducible by others using the same standard	Popularity & Influence	Well Known Importance Numer of Countries Involved Number of projects Involved Versatility	Focus	an inclusive focus on buildings	Technical Content & Sustainability aspect metrics	
Transparent	Standards and process for awarding the certification should be transparent and open for examination	Availability	1- Availability to the system itself: Easy to access System's format Information available for public Cost of system Certfication fees 2- vailability of References: Vailability of on-line information vailability of non on-line information availability of case studies avilability of user's interaction	Scientific interest	Cited in at least 20 papers reflected in the Elsevier Scopus database; the search was excuted on article titles,abstracts and keywords	Access to Rating system	
Objective	Certification body should be free of conflict	Methodology	Methodology Summary Weighting Rating Levels Standardization Quantitative Criteria Qualitative Criteria Whole Life cycle Assessment Complexity Efficiency of assessment method	Widespread Adoption	More than 500 Certfied projects	Cost of assessment	
Progressive	Standards should advance industry practices, not simply reward business as usual	Applicability	1-Stages of Building lifecycle incfluenced: Predesign/Planning/Site selection Design/Procurement Construction/Post construction review Existing building management/operations/Maintenance tenant fit-out/refurbishment Demolition. 2- Technical Contents:	A Consolidated Development State	More than 5 years of service	Local Context (regional Priority)	
		Data Collection Process	Data Gatherer Data collection method Documentation measurability Convenience			Weighting method	
		Accuracy & Verification	1-accuracy of data processing stage 2-Accuracy of data outputting stage 3-Verification: Assessor qualaifications Level of details to check thrird party results aknowledgement			Registration & Assessment Process	
		User-Friendliness	1-Ease of use: 2- Product support availability of responsive assisstance FAQ & Record inquiries Training & courses available instructions or helps			Maturity	
		Development	Systems Maturity Systems stability Update Development approach Future development			Life cycle stage coverage	
		Results Presentation	Presentation Method Clarity comparability Result usability			Validity	
						Adaptability	

References

Rahim, A., Mohamed, A., & Baqutayan, S. (2015). Sustainability achievement and estidama green building regulations in Abu Dhabi Vision 2030. *Mediterranean Journal of Social Sciences* July 2015. https://doi.org/10.5901/mjss.2015.v6n4s2p509. W.-K. Chen, (1993) *Linear Networks and Systems (Book style)*. Belmont, CA: Wadsworth, pp. 123–135.

Nguyen, B. K., & Altan, H. (2011a). Comparative review of five sustainable rating systems. In *2011a International Conference on Green Buildings and Sustainable Cities, Energy Procedia*, Vol. 111 (2017), pp. 41–50.

Nguyen, B. K., & Altan, H. (2011b). TPSI—tall-building projects sustainability indicator, international conference on green buildings and sustainable cities. *Engineering, 21*(2011), 387–394.

Bernardi, E., Carlucci, S., Cornaro, C., & André Bohne, R. (2017). An analysis of the most adopted rating systems for assessing the environmental impact of buildings. MDPI.

Karmany, H. M. (2016). Evaluation of green building rating systems for Egypt. Department of Architecture, American University in Cairo, Cairo, Egypt.

Green Buidling Council of Australia. (2020) Managing the impacts of COVID-19 on green star ratings, version 1. https://gbca-web.s3.amazonaws.com/media/documents/managing-the-impacts-of-covid-19-on-green-star-projects-200407.pdf.

Fowler, K. M., & Rauch, E. M. (2006). Sustainable building rating systems summary. Pacific Northwest National Laboratory.

Politia, S., & Antoninib, E. (2016). An expeditious method for comparing sustainable rating systems for residential buildings. In *8th International Conference on Sustainability in Energy and Buildings, SEB-16*, 11–13 September 2016, Turin, Italy.

Multiplying Effects of Urban Innovation Districts. Geospatial Analysis Framework for Evaluating Innovation Performance Within Urban Environments

Jeremy Burke, Ramon Gras Alomà, and Fernando Yu

Abstract

Urban innovation performance is the main driving force behind sustained economic growth. There have been many attempts to quantify innovation performance in order to guide decision-making, promoting investment and innovation projects. Most studies do not accurately describe innovation in terms of how knowledge-intensive activities develop over multiple phases, how they are distributed geographically, and the ways in which agglomerated urban innovation efforts can enhance economic growth. Primarily, they fail to effectively combine urban practice and economic findings through data science efforts to design quantitative innovation performance metrics. This paper presents a novel database and analytical methodology to measure and describe the nonlinear benefits of geographic aggregation of knowledge-intensive activities within urban environments. The results of this research empirically demonstrate that innovation districts, characterized by their geographic concentration of knowledge-intensive activities, benefit from the superlinear growth of innovation, both in terms of innovation output per employee (new patents, new products, new services, R&D, scientific papers) and in terms of innovation-related employment creation per resident. The geospatial, analytical framework has been applied to the study of 50 notable Innovation Districts to benchmark them against a baseline of all districts in the United States. We have extracted the most salient features of these districts that illustrate the value of investing in the geographic concentration of innovation activities. The analytical framework can then be applied to any geographical area to evaluate the economic performance of knowledge-intensive activities within urban environments. The work expands on general knowledge of how cities operate as complex systems and how they shape the collective knowhow of urban communities. Further research may identify the key factors, features, and dynamics underlying the success of innovation districts, such as urban design criteria and smart specialization strategies, and apply them to specific communities to support the economic growth of urban environments.

Keywords

City science • Geospatial analysis • Urban innovation • Complexity economics • Database creation • Innovation districts • Superlinear growth • Urban development • Territorial analysis • Network science • Innovation performance • Complex systems analysis

The original version of this chapter was revised, the author's name "Ramon Gras (Given name) Alomà (Family name)" has been changed to "Ramon (Given name) Gras Alomà (Family name)". The correction to this chapter can be found at https://doi.org/10.1007/978-3-030-98187-7_24

J. Burke · R. Gras Alomà (✉) · F. Yu
School of Engineering and Applied Sciences, Harvard University, Cambridge, USA
e-mail: rgras@mde.harvard.edu

J. Burke
e-mail: jburke@mde.harvard.edu

F. Yu
e-mail: fernandoyu@hks.harvard.edu

1 Introduction

Recent research in the study of economic development and urban phenomena has applied insights of economic complexity to provide a reliable methodology to describe industry comparative advantage at the national level. This strand of research applies network theory to study the linkages between knowledge-producing agents in an economy (Hidalgo & Hausmann, 2009). Such methodology enables a systematic understanding of collective knowhow advancement and knowledge diffusion at the national scale, as well as the identification of smart specialization and diversification strategies. However, the current literature presents two major limitations: on the one hand, it primarily focuses on international trade for physical goods, thus lacking the analysis of high value-added intangible services; on the other hand, the national level of aggregation does not permit a deeper understanding of geospatial dynamics, hence precluding a detailed understanding of territorial dynamics and the nonlinear benefits of geographic aggregation of knowledge-intensive activities. As a result, the economic complexity models tend to come short in terms of identifying sub-national and city-level *collective knowhow* dynamics, and illustrating urban development recommendations at the regional, urban, and local levels (Hausmann et al., 2014).

There is a research gap in the urban innovation literature to understand the links between good practices in urban design and the main factors, features, and underlying dynamics of successful cities. By combining city science techniques with insights from economic complexity, we can provide a higher resolution methodology to measure, evaluate, and better understand innovation systems at the urban scale. The geospatial analysis of innovation and knowledge-intensive activities within urban environments will enable a deeper understanding of (1) the dynamics of the nonlinear benefits of strategic geographic aggregation of knowledge-intensive activities; to (2) identify the key ingredients and dynamics facilitating economic growth and stable employment creation; thus (3) propelling regional growth, and distributed prosperity, by means of providing urban development decision-making recommendations to increase urban economic performance.

Previous research by Bettencourt et al. (2010) as well as Barabási (2017) shows that the scale of cities has a superlinear or sublinear impact on social measures such as patents, crime, and sustainability. The purpose of the present paper is to quantify the superlinear effects of the geographic aggregation of knowledge-intensive activities within innovation districts and provide quantitative measures of the multiplying effects and the derived economic surplus in terms of knowledge advancement, wealth, and employment creation for the surrounding communities. The novel framework used in this paper will allow readers to understand these relationships by modeling knowledge-intensive activities from a network theory perspective.

Further research will subsequently evaluate best practices in urban design (topology, morphology, entropy, and scale) enhancing fruitful human interaction to propel the knowledge economy, as well as smart specialization and diversification strategies for sustainable economic growth within successful innovation districts.

2 Literature Review

Our work contributes to four main strands of the literature linking economic development and innovation. First, the literature of economic complexity and collective knowhow tries to model the relation between the stock of knowledge in a region and economic outcomes (Hidalgo & Hausmann, 2009). Our work incorporates these intuitions but places the scope of analysis at the city and district levels and incorporates higher quality data such as firm-level data. Second, the study of power laws in urban scenarios, as proposed by Bettencourt et al. (2010), which study the universal relation between scale and urban phenomena by using graphical descriptions of sublinear or superlinear effects with larger and larger scales of aggregation. We contribute to this literature by focusing the analysis on innovation districts in urban settings, third, the study of agglomeration economies by Ellison and Glaeser (2009). This strand of the literature tries to disentangle the effect of local comparative advantage and endogenous spillovers to explain the geographic distribution of economic activity. The literature finds that intra-industry externalities play an important role, and local natural advantages have a limited explanatory power. We build upon these findings to concentrate on the powerful externalities driven by innovative activities in urban settings. Finally, the city-level evidence-based approach to increase urban performance, developed by Kent Larson and Andres Sevtsuk, serves as a source of inspiration to bridge the gap between economic geography, economic complexity studies, and urban design (Ekmekci et al., 2016).

Economic complexity as a measure of collective knowhow

The study of how people collaborate to add value to the economy dates back to the writings of Adam Smith (1776), who studied the division of labor. People and firms specialize in different activities, increasing economic efficiency and the impact of the interactions between them. Hidalgo and Hausmann (2009) applied this insight at a national scale to study the relation between the human and physical capital resources in a given country and the type of goods that they export. Basing their work on the study of scale-free networks by Barabási (2016), they modeled the structure of an economy as a bipartite network in which countries are connected to the products they export and showed that it is possible to quantify the complexity of a country's economy by characterizing the structure of this network. Furthermore, this measure of complexity is correlated with a country's level of income, and deviations from this relationship are predictive of true growth (Barabási, 2017). This suggests that countries tend to converge to the level of income dictated by the complexity of their productive structures. The level of complexity is modeled as the combination of capabilities available in a given country or more broadly as a measure of collective knowhow.

This body of work spurred further research including Hartmannet et al. (2017), which expands the scope of analysis of economic complexity to study implications on institutional design and income distribution. Youn et al. (2016) use similar motivation to investigate how the diversity of economic activities depends on city size.

The limitation of this approach as usually applied is that it lacks the level of detail to be implemented at an urban scale. In addition, it is mainly tailored for the analysis of export data, which tend to lack measures of service industries. Our goal is to contribute in these two dimensions, by focusing the analysis of collective knowhow at the urban scale and in addition making use of firm-level data to also incorporate the production of services. In addition, we believe that a less indirect and more precise approach to measuring collective knowhow is through the output of the innovation process, as measured by patents and innovation-related metrics.

Superlinear effects of urban scaling in economic development

In an influential body of research, Geoffrey West and collaborators have studied the remarkable similarity between cities and living creatures based on the scale of their body and total residents (Bettencourt et al., 2006). There are simple and universally applicable laws that link the size of mammals to their fundamental biological functions such as metabolism and energy consumption. This implies that regardless of size, all mammals are a scaled manifestation of a single, idealized mammal. Could this be true as well for human cities and agglomerations?

West and his collaborators showed that this is in fact the case, by analyzing a myriad of urban phenomena such as infrastructure, crime, pollution, wealth creation, and innovation. The research finds that there are universally applicable power laws across different geographies and economic dimensions that translate the size of a city as measured by population and other measures to the urban phenomena listed above. These laws can be classified as linear, superlinear or sublinear and the exponent of the power law and have interesting implications (Bettencourt et al., 2006).

Other papers by the same group of researchers focus on related topics involving the network structure of urban infrastructure. In particular, Kühnert et al. (2006) study the supply patterns for energy, fuel, medical, and food supply. Bettencourt et al. (2010) study the superlinearity of urban growth, and Bettencourt et al. (2007) study the increasing

returns to patenting as a scaling function of metropolitan size. In follow-up work, Bettencourt et al., (2008, 2009) focus on the self-similarity implications of these networks, and in 2013, they propose a production function for cities that illustrate these superlinear patterns. Finally, Bettencourt et al. (2013) study the implications for innovation and crime patterns within these networks. Lane et al. (2009) study the patterns of self-organization of urban agglomerations that motivate the modern analysis of urban networks.

Innovation Districts: superlinear effects of geographic concentration in urban innovation

In recent years, a rising number of innovative firms and talented workers are choosing to congregate and co-locate in compact, amenity-rich urban settlements in the cores of central cities. These patterns are described in Katz and Wagner (2014). These districts are geographic areas where leading-edge anchor institutions and companies cluster and connect with start-ups, business incubators, and accelerators. They are also physically compact, transit-accessible, and technically wired and offer mixed-use housing, office, and retail.

The intensity of social interactions is a key metric in the evolution of innovation districts. The scaling exponent of urban infrastructure networks with population is usually estimated below one, therefore suggesting economies of scale in the use of buildings, transportation, and communication networks. On the other hand, social interactions, which culminate in innovation and wealth creation, present a scaling exponent above one, therefore implying increasing returns to aggregation in human interaction. The latter effect can be so powerful that it creates jumps in innovation with cycles shorter than a typical human lifespan, in contrast with biological phenomena where large innovation jumps occur sporadically and in a timeline longer than many lifespans of a living creature.

Previous research has expanded on this topic. One critical reason to understand the social dynamics that align with the scale of a city includes research from Moretti (2012), who states that for every innovation job there are five service jobs being created. Arbesman et al. (2009) provide a theoretical model to understand the superlinear effects of urban innovation. The network model proposed is not tested on empirical data but could serve as a solid ground for future tests with urban innovation empirical patterns. Schläpfer et al. (2014) study superlinearity of communication networks in urban environments by analyzing mobile-phone interactions within European cities.

Our contribution to this body of literature is to narrow the focus of the study of innovation from the city to the district level and to study in particular the superlinear effects of agglomerations of knowledge-intensive activities around innovation districts. This is an area of analysis that deserves a set of data sources and methodological approach of its own.

Innovation and Economic Geography

Within the economic geography literature, there has been a large body of work focusing on agglomeration economics. Why do industries agglomerate? How much of this agglomeration is explained by local advantages and how much is a result of endogenous intra-industry spillovers? Ellison and Glaeser (2009) try to tackle this fundamental question by disaggregating the effect of economic agglomeration between natural advantages and intra-industry spillovers in a sample of four-digit manufacturing industries in the United States. They study the determinants of agglomeration based on cost of inputs (electricity, gas, coal, agricultural products), cost of labor inputs (relative wage differences), relative price of skill, transportation costs, and unobserved spillovers. They find that natural advantages have a limited explanatory power, being able to explain only 20% of observed agglomeration. The implication is that agglomeration effects are an important force driving the geographic distribution of economic activity. As noted by the authors, this effect is particularly extreme in manufacturing industries, in particular, the automobile manufacturing sector.

We build upon these insights to carry the question of agglomeration of manufacturing industries to the more general understanding of innovation activities. How much of the agglomeration in innovation activities can be explained by local comparative advantage and how much is an endogenous spillover effect that can be replicated in different regions? This is a fundamental question since as we will show below the spillover effects of innovative activities appear to be large.

Based on the body of work described above, in the present paper, we wish to expand the analysis to narrow the scope from a regional level to a city and district level, incorporate new granular data on innovation activities, and provide theoretically based measures to analyze the agglomeration of innovation efforts. In the following sections, we aim to explain how and to which extent innovation districts outperform other urban areas in knowledge and wealth creation and how to measure the impact.

3 Research Goals

The purpose of this paper is to define a conceptual and methodological framework to perform geospatial analysis of innovation performance metrics within urban environments. By building an urban economic performance database describing geographically distributed innovation variables, we can design and compute geospatially detailed urban economic variables across the United States, as well as tracking their evolution over time. Building the database requires knowledge and testing of different methods for combining data using specific joining keys, tagging methods, and geospatial join and merging commands to create the specific Key Performance Indicators (KPIs). The brand new KPIs will enable the measurement of economic performance metrics describing the nonlinear benefits of strategic aggregation of knowledge-intensive activities.

The main hypothesis is that certain types of urban development and geographic clustering of innovation-related activities, usually observed in Innovation Districts, tend to generate sustainable prosperity, and superlinear economic growth patterns. The methodology will enable the identification of strengths and weaknesses in the geospatial network of urban infrastructure and collective knowhow, hence illustrating both the local comparative advantage of specific industries and knowledge areas, and the economic growth prospects. The empirical research will describe the analysis of the entire US territory, as well as highlight the economic performance metrics of 50 notable innovation districts with respect to the rest of the national territory.

The expected contribution of the novel theoretical and methodological framework is to create a geospatial model illuminating best practices in economic development to evaluate the impact of urban design, and knowledge-intensive economic activity concentration. Applications of the model may include the evaluation of the adequateness of different types of urban topology, morphology, and entropy for a given city and region.

4 Methodology

The current stream of work integrates city science and network theory and applies this framework to urban design and the knowledge economy. On the empirical side, there is heavy emphasis on the measurement of innovation activity in knowledge-intensive urban ecosystems. These data points are combined to identify the main trends and relevant patterns. The theoretical framework and empirical analysis are then combined to evaluate the performance of alternative urban design and economic activation strategies.

The analytical rationale is based upon the assumption that cities operate as dynamic complex systems, susceptible to be analyzed by means of network science. Thus, by analyzing the nature of the underlying networks supporting knowledge creation, we can compare, contrast, and extract best practices in terms of economic and innovation agglomeration dynamics, and city design (urban topology, morphology entropy). Hence, by analytically identifying the nature and potential of a specific city or region, we can extract insights and guidelines to increase urban performance by determining the key ingredients and dynamics to propel economic growth and prosperity.

4.1 Scope of Analysis

The territorial scope of the analysis covers the entire contiguous United States and consists of the 48 adjoining US states (plus the District of Columbia) for the year 2017. Within this geographic area, 50 of the most well-known innovation districts were identified and tagged within our database, as depicted in Fig. 1. The tagging process identified census block groups, in which businesses, known to be significant contributions to innovation within the neighborhood, were located. Due to their irregular shapes, the block groups were chosen as close to the center of research or business activity of the neighborhood as possible. It was important to choose innovation districts based on the geographic area of the block group so that the data could be combined with social demographic information from the US Census. It was also a relevant criterion that block groups could be compared to one another in an apples-to-apples comparison and describes how businesses and industries cluster together in high-performing areas.

4.2 Definition of Key Performance Metrics for Measuring Urban Innovation

The Key Performance Metrics for measuring Urban Innovation enable quantitatively describing key factors concerning knowledge-intensive activities across the US territory, at the smallest census unit; the census building group, as well as illuminating the correlations and potential causation between them. Innovation Intensity describes the societal effort in supporting innovation, in terms of the percentage of employees per geographic unit working on knowledge-intensive activities; Innovation Performance describes the business revenue generated as a result of knowledge-intensive activities (patents, new products, new services, scientific articles, R&D); and Innovation Impact serves as a proxy for the societal benefits of the multiplying effect of knowledge-intensive activities, in terms of accessibility to innovation-related employment per unit of resident, as a proxy for attractive, well paid, stable employment opportunities and the indirect benefits derived therefrom.

Notation:

$i \in [1, n]$; domain of values describing each geo-located business in the United States.

$\beta_i = BusinessEmployment(i)$; Number of employees working in the business location (i).

$\rho_i = BusinessRevenue(i)$; Annual revenue for each business location (i).

$\kappa_i = InnovationClassification$; Boolean value describing whether the business is knowledge-intensive.

$g \in [1, m]$; domain of values describing each Building Group in the United States.

$\delta_g = BuildingGroupResidents(g)$; Number of residents within a Building Group (g).

Innovation Intensity: describes the percentage of employees working on knowledge-intensive activities, broken down by each of the three innovation phases: research, technology transfer, and production. It is computed as the sum of all the employees working in geo-located businesses belonging to one of three innovation categories (research, technology transfer, and advanced production), divided by the total number of employees per building group.

$\alpha_{int}(g) = $ InnovationIntensity.ForeachBuildingGroupg :

$$\alpha_{int}(g) = \frac{\sum_{i=1}^{i=n}(\beta_i \cdot \kappa_i) \cdot 100}{\sum_{i=1}^{i=n}\beta_i} \tag{1}$$

Innovation Performance: describes the business revenue per building group, generated by measurable innovations: new patents, new products, new services, new processes, R&D, scientific papers.

$\alpha_{per}(g) = $ InnovationPerformance
$= $ TotalInnovationRevenueperEmployeeperBG(g).
ForeachBuildingGroupg :

$$\alpha_{per}(g) = \frac{\sum_{i=1}^{i=n}(\rho_i \cdot \kappa_i) \cdot 100}{\sum_{i=1}^{i=n}\beta_i} \tag{2}$$

Innovation Impact: describes innovation employment per resident for each BG, as a proxy for availability of knowledge-intensive employment opportunities within the community.

$\alpha_{imp}(g) = $ InnovationImpact.ForeachBuildingGroupg :

$$\alpha_{imp}(g) = \frac{\sum_{i=1}^{i=n}(\beta_i \cdot \kappa_i)}{\sum_{g=1}^{g=m}\delta_g} \tag{3}$$

Innovation Districts are defined as urban neighborhoods or districts purposely designed, nurtured, and supported, to concentrate and support innovation-related activities in dense urban areas, to leverage the nonlinear effects of geographic concentration of knowledge-intensive activities. By comparing representative statistics of the Urban Innovation Performance KPIs from those 50 notable Innovation Districts with those of the average block groups, we can provide a tangible understanding of the positive, superlinear effects

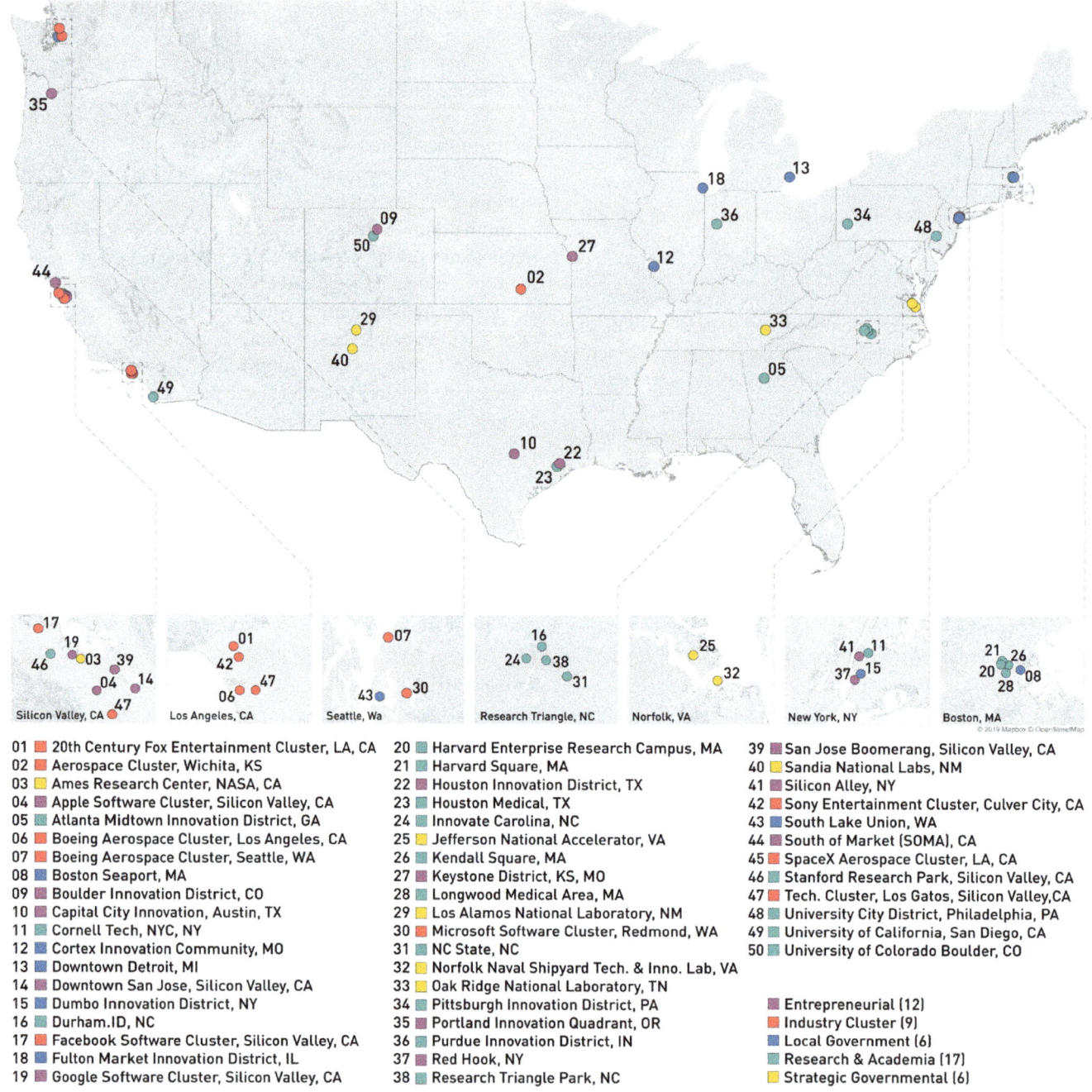

Fig. 1 Geographic distribution of 50 innovation districts

01 ▪ 20th Century Fox Entertainment Cluster, LA, CA	20 ▪ Harvard Enterprise Research Campus, MA	39 ▪ San Jose Boomerang, Silicon Valley, CA
02 ▪ Aerospace Cluster, Wichita, KS	21 ▪ Harvard Square, MA	40 ▪ Sandia National Labs, NM
03 ▪ Ames Research Center, NASA, CA	22 ▪ Houston Innovation District, TX	41 ▪ Silicon Alley, NY
04 ▪ Apple Software Cluster, Silicon Valley, CA	23 ▪ Houston Medical, TX	42 ▪ Sony Entertainment Cluster, Culver City, CA
05 ▪ Atlanta Midtown Innovation District, GA	24 ▪ Innovate Carolina, NC	43 ▪ South Lake Union, WA
06 ▪ Boeing Aerospace Cluster, Los Angeles, CA	25 ▪ Jefferson National Accelerator, VA	44 ▪ South of Market (SOMA), CA
07 ▪ Boeing Aerospace Cluster, Seattle, WA	26 ▪ Kendall Square, MA	45 ▪ SpaceX Aerospace Cluster, LA, CA
08 ▪ Boston Seaport, MA	27 ▪ Keystone District, KS, MO	46 ▪ Stanford Research Park, Silicon Valley, CA
09 ▪ Boulder Innovation District, CO	28 ▪ Longwood Medical Area, MA	47 ▪ Tech. Cluster, Los Gatos, Silicon Valley,CA
10 ▪ Capital City Innovation, Austin, TX	29 ▪ Los Alamos National Laboratory, NM	48 ▪ University City District, Philadelphia, PA
11 ▪ Cornell Tech, NYC, NY	30 ▪ Microsoft Software Cluster, Redmond, WA	49 ▪ University of California, San Diego, CA
12 ▪ Cortex Innovation Community, MO	31 ▪ NC State, NC	50 ▪ University of Colorado Boulder, CO
13 ▪ Downtown Detroit, MI	32 ▪ Norfolk Naval Shipyard Tech. & Inno. Lab, VA	
14 ▪ Downtown San Jose, Silicon Valley, CA	33 ▪ Oak Ridge National Laboratory, TN	▪ Entrepreneurial (12)
15 ▪ Dumbo Innovation District, NY	34 ▪ Pittsburgh Innovation District, PA	▪ Industry Cluster (9)
16 ▪ Durham.ID, NC	35 ▪ Portland Innovation Quadrant, OR	▪ Local Government (6)
17 ▪ Facebook Software Cluster, Silicon Valley, CA	36 ▪ Purdue Innovation District, IN	▪ Research & Academia (17)
18 ▪ Fulton Market Innovation District, IL	37 ▪ Red Hook, NY	▪ Strategic Governmental (6)
19 ▪ Google Software Cluster, Silicon Valley, CA	38 ▪ Research Triangle Park, NC	

of such strategies in terms of job creation, wealth creation, knowledge advancement and standards of living.

The three main innovation KPIs described in Fig. 2 are the building blocks of the analysis. They allow us to quantitatively measure the differential impact of Innovation Districts and benchmark different typologies of Innovation Districts against each other and other non-innovation-intensive districts. Finally, we analyzed the superlinearity of innovation-related activities in a regression framework against population, city size, and other variables to benchmark our results against previous studies.

5 Urban Innovation Performance: Database Building

The database is built by combining multiple different data sources, as described in Table 1.

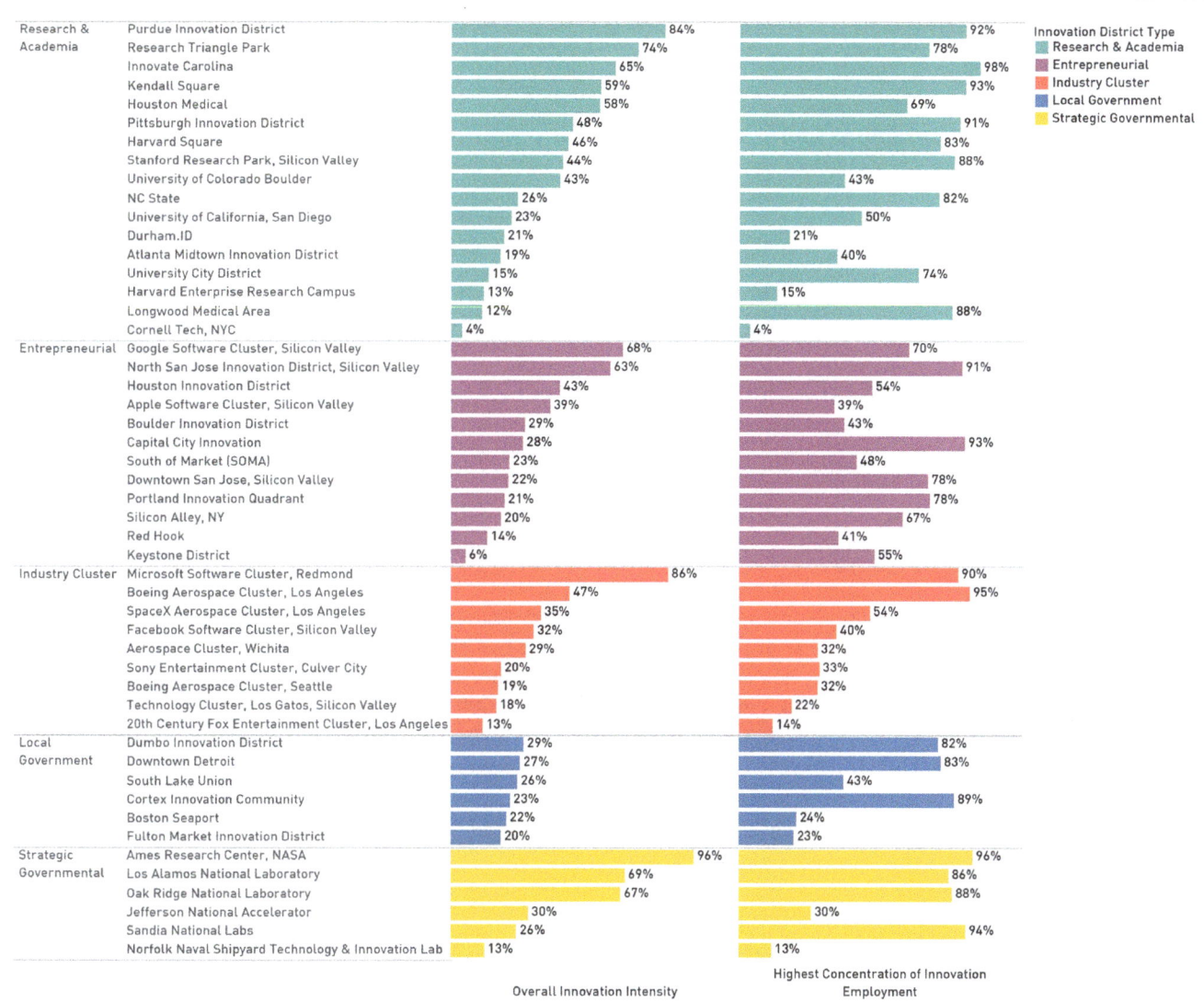

Innovation Intensity. The overall Innovation Intensity KPI describes the compounded percentage of knowledge-intensive jobs versus total employment in an Innovation District. The highest concentration Innovation Intensity is a complementary KPI that describes the peak percentage of knowledge-intensive jobs concentrated at the core US Census block group of each Innovation District.

Fig. 2 Nonlinear benefits of strategic aggregation: Innovation intensity, performance, and impact for the top 50 innovation districts within the United States

The business database covered the entire United States for the year 2017 with approximately 12.5 million businesses. It is understood that the largest, most well-reported firms were included in the dataset. The NAICS codes were six digits long and used for the industry/sub-industry and company classification. The dataset was cleaned with imputed where there were trivially incorrect or missing values. Addresses were already geolocated so there was no need to compute the latitude and longitude of each business. Key inputs include: employment, revenue, and industry code.

The US Demographic data came from ESRI's imputed database of the US Census for the year 2017. Data were used from the smallest geographic area of roughly 2000 people per building block group to provide the most granular level of analysis for the study. Key inputs for the model included variables that described employment, income, wealth, education, and employment density.

Industry innovation metrics were used from data gathered by a survey conducted by the National Science Foundation in 2014. The data were gathered through interviews and surveys, which covered topics such as the total number of

Table 1 Data sources to support the geospatial analytical framework for urban innovation performance

Data sources		
Title	Source	Type
Business data Summary and locations	Esri Business Analyst 2017 Sourced from Infogroup	Points
Population demographics	Methodology Statement: 2017/2022 Esri US Demographic Updates	ESRI shape files
Business enterprise research and development survey (BERD)	The National Science Foundation https://ncses.nsf.gov	PDF and Excel
America's advanced industries: new trends	The Brookings Institute	Tables
North American industry classification system (NAICS)	United States Census Bureau	Tables
Standard industrial classification (SIC)	United States Department of Labor	Tables
Harmonized system (HS)	International Trade Administration	Tables

patents per company and revenue generated from the patents; and total spent on R&D, new products, new services, and new processes. The data were then aggregated at various levels of industry codes from 2 to 4 digit NAICS codes, to provide an understanding of innovation activity per industry.

Classification methods of analysis were conducted by tagging specific industries as "innovative", "innovation related", and "other" based on the NAICS code per industry. The tagging originated from previous classification methods was determined by the Brookings Institute's Advanced Manufacturing Industry List and the BERD survey from the National Science Foundation. The final set of industry codes then became drivers of the innovation performance metrics used for further analysis, and the tagging process is continually updated and refined as new data and technology continuously change industry operations throughout the United States.

The final list of tagged NAICS Codes was translated to HS Codes through a crosswalk table. Through this industry code system, information about innovation performance was combined with each business entry from the INFO Group Database, thereby creating a proxy for the level of innovation for that specific business.

Urban Innovation Performance Dataset Creation

Each business entry was then tagged with the specific block group in which it was located, through a spatial merge command. With every business associated with its own block group, it was then possible to aggregate all businesses to provide highly accurate totals for business and innovation metrics at the block group level and associate the data with the US Census. Therefore, the US Census block group

dataset was expanded with private business and innovation metrics, which allowed each block group to be compared to one another because they all have the common feature of roughly 2000 residents. For example, if a block group has close to no businesses, it was clear that innovation activity is low, whereas block groups that contained high numbers of businesses, with strong innovation performance metrics from the National Science Foundation, were ranked highly in our model.

The block group level of detail was chosen given that it is the smallest administrative geographic area used by the US Census, therefore, allowing for the most granular analysis of high-density urban environments. In addition, it is possible to use the same attribution method and apply metrics at larger census areas such as census tracts, metropolitan statistical areas, counties, states, regions, and divisions of the United States. However, there is less and less actionable information at higher levels, due to the high amounts of aggregating business activity. With the standardization of variables across the United States, it is now possible to create mathematical distributions describing the ranges of possible values for urban performance.

New Boolean variables were also introduced that describe generally how new knowledge, which is then designed and engineered, is finally mass produced to the general public. We denote this process with the three phases of innovation, where businesses are tagged based on their NAICS code, according to their category: Research and Academia, Technology Transfer, and Production, based on previous research from research institutions, National Science Foundation surveys and analysis, as well as the advanced manufacturing industries listed by the Brookings Institution. In addition, another new variable allowed for the assessment of how central Innovation was for each industry. For each phase, a new variable column was created, where Boolean values were added, first if the industry code was innovative, innovation-related, or other. This tagging system allowed for specific innovation metrics to be appropriated to each business within the database.

Below are summary Tables 2 and 3, which provide a list of descriptive statistics for the new database.

6 Results

The first innovation KPI described in the Methodology section corresponds to Innovation Intensity and measures the share of people working in knowledge-intensive activities in a given geography, broken down by innovation phase. The second innovation KPI, Innovation Performance, tracks the revenue generated by measurable innovations in the form of new patents, products, services, processes R&D, scientific papers, and startup companies. Finally, the last set of indicators corresponds to the Innovation Impact, which encompasses the societal benefits, knowledge-intensive employment, wages, unemployment metrics, and housing affordability. Additional performance measures include quantitative surplus of good practices, urban design characteristics, urban typologies, and their associated positive and negative externalities. Figure 3 shows the distribution of the three innovation KPIs.

We observe that the Innovation Intensity follows a Lognormal distribution, Innovation Performance follows a Pareto/Power Law distribution, and the Innovation Impact follows a Gamma Distribution. The three distributions are characterized by a right skew and a long tail, which correlate with the geographical agglomeration patterns that we observe in Innovation Districts. These results are summarized in Table 4 and expanded in more detail in the section below.

The comparison between Innovation Intensity, Innovation Performance, and Innovation Impact depicts the amplifying effects of clustering knowledge-intensive activities: on average, innovation districts present 2.8 times higher concentration of knowledge-intensive activities per employee, 4 times higher innovation output per employee in terms of patents, new products, new services, new processes, and R&D, and 16 times higher creation and availability of knowledge-intensive employment opportunities per resident. These results reveal that Innovation Districts systematically benefit from structural nonlinear innovation patterns as a result of geographic aggregation of knowledge-intensive activities within urban environments. These results apply as well for medium-sized Innovation Districts such as Boston Seaport, Silicon Alley, and the Pittsburgh Innovation District illustrated in Fig. 4.

Municipality scale

We have determined above the importance of innovative employment and its effects on the broader economy through knowledge externalities. We now want to understand the determinants of this special type of employment. To this end, we analyzed the distribution of employment by type in all municipalities in the United States. After data cleaning, we obtained 14,342 municipalities. The descriptive statistics for all municipalities are presented in Table 5.

Average daytime residents and average total employment are very similar, and in addition, they both have a very large standard deviation. This large dispersion in both daytime residents and employment levels is depicted in Fig. 5. Both histograms, in logs, show a right skew of the distribution, which implies that daytime residents and employment opportunities are concentrated in a few very large municipalities.

As we might expect, this extreme concentration of daytime residents and employment tends to co-occur in the same

Table 2 Number of business classification categories within each innovation phase—United States

Businesses—innovation phases	NAICS	SIC
Research	9 (0.72%)	121 (1.12%)
Technology transfer	40 (3.19%)	692 (6.39%)
Advanced manufacturing	209 (16.65%)	1,747 (16.13%)
Innovation total	258 (20.56%)	2,560 (23.63%)
All businesses—total	1,255 (100%)	10,833 (100%)

Table 3 Number of businesses within each innovation category —United States

Businesses—innovation phases	NAICS and SIC businesses (%)
Research	2.08
Technology transfer	5.45
Advanced manufacturing	2.42
Innovation total	9.95

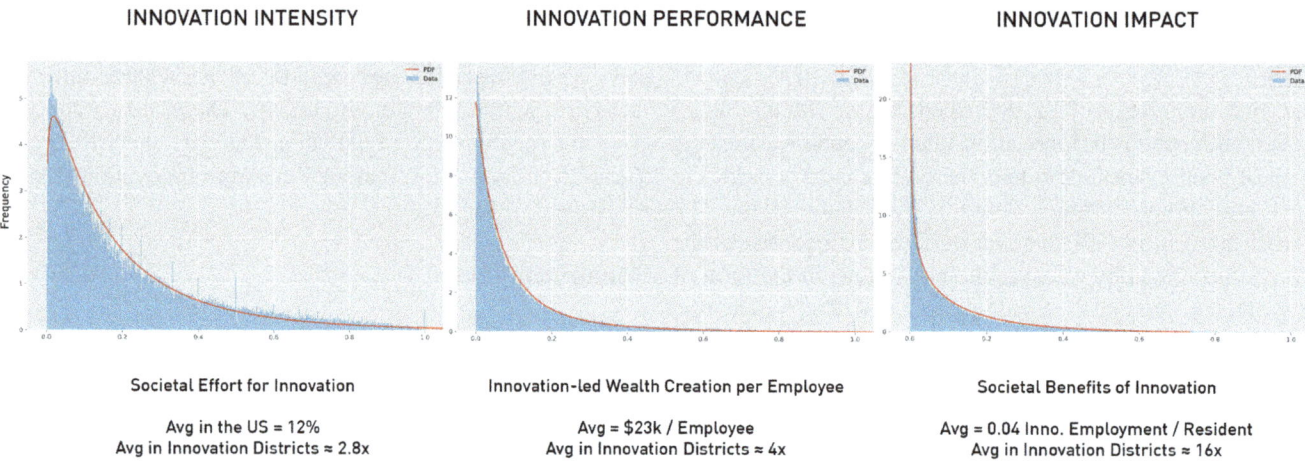

Fig. 3 Amplifying effects of concentration of knowledge-intensive activities

Table 4 Multiplying effects of knowledge-intensive activity concentration in innovation districts

Innovation KPIs	Multiplying factor
Innovation intensity	2.8
Innovation performance	4
Innovation impact	15

municipalities. Figure 5 confirms this intuition by showing the almost perfect correlation between our measure of day-time residents and employment.

More central to our analysis is to understand what fraction of these employment opportunities happen in innovative sectors. Our construction of innovation metrics described in the previous section allows us to study this relation. Figure 6 shows the distribution of innovative employment across all municipalities. We observe that the distribution is lognormal. What is the relation between the lognormal distribution of innovative employment and municipality size? To this end, we run a preliminary regression:

$$\log(innovative\ employment_{ct}) = \alpha + \beta \log(daytime\ residents_{ct}) + \varepsilon_{it}$$

where c denotes municipality or city and t represents the year. We begin by analyzing the year 2017. The parameter β is the Power Law coefficient. A value below 1 implies sublinear effects of scale on innovative employment.

A parameter above 1 implies superlinear effects. We find a parameter value of 1.17, which is a remarkable value for superlinear effects. Figure 7 describes the Lognormal distribution of innovative employment across building groups.

Finally, we want to decompose this finding between a scale component and an innovation component. As we observed above, municipality scale as measured by daytime residents also correlated strongly with total employment. We run the above regression again, this time taking into account total employment in addition to innovative employment. We find that the coefficient on total employment is 1.108, which explains 59% of the total effect found for innovation. This implies that there are two contributing factors to the superlinear result. The first factor is purely due to agglomeration economics of total employment: larger municipalities concentrate more employment opportunities. The second and most important factor is innovation externalities: even when we take into account agglomeration of total employment,

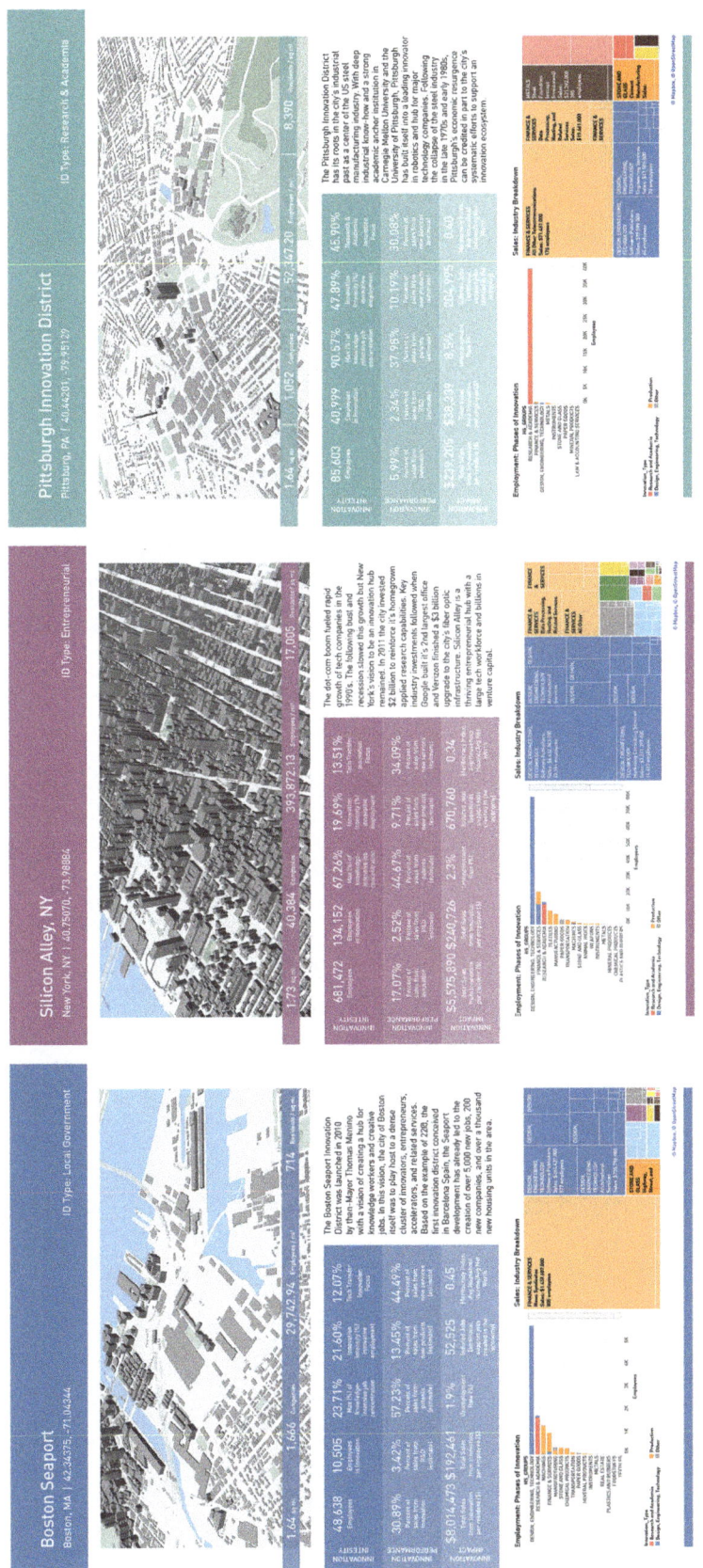

Fig. 4 Summary profiles for three innovation districts, succinctly describing the networks of talent, industries, and urban design

Table 5 Descriptive statistics of population, employment, and innovative employment in the United States

Variable	Mean	Standard deviation
Daytime residents	11,510.13	45,063.10
Total employment	10,391.14	49,615.91
Innovative employment	1,989.92	10,039.86

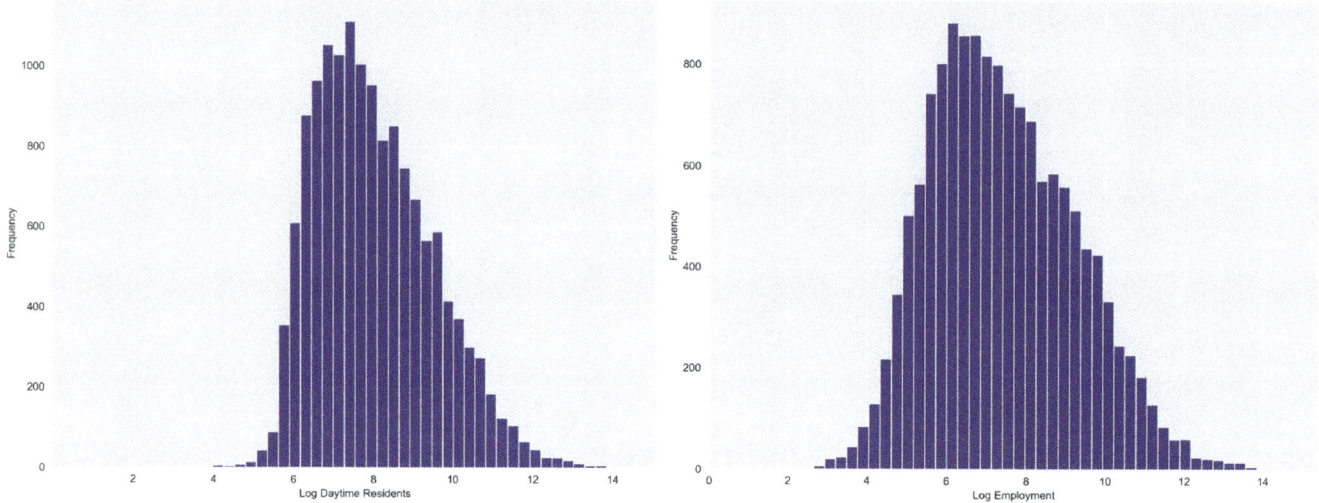

Fig. 5 Distribution of daytime residents and employment (in logs)

Fig. 6 The strong correlation between daytime residents and employment

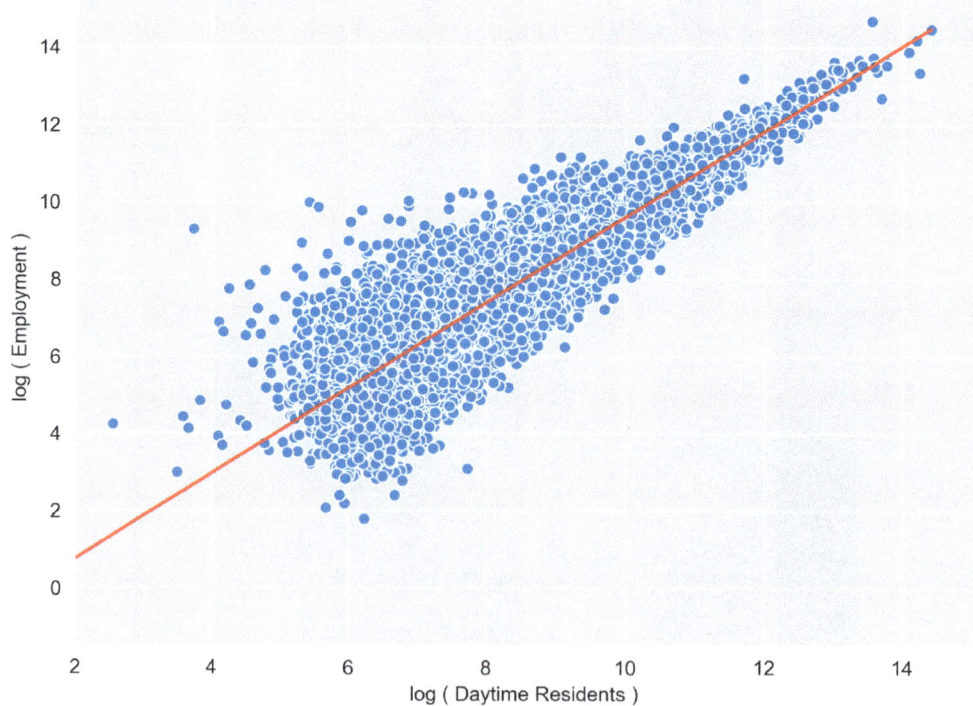

innovative employment tends to concentrate even more. We illustrate this decomposition in Fig. 8.

Measuring superlinear innovation in Innovation Districts

The analysis of the over 267 k building groups in the continental US reveals the amplifying effects of investing on knowledge-intensive activities. The goodness of fit analysis of the distributions describing innovation effort (intensity), tangible knowledge advancement (performance), and societal benefits (impact) present increasingly nonlinear mathematical distributions. By comparing the representative descriptive statistics of the Urban Innovation Performance KPIs of 50 notable Innovation Districts with those of the rest of the US territory, we observe that systematic amplifying effects of investment in knowledge-intensive activities take place.

• Innovation Intensity

The statistical distribution describing the best goodness of fit for the Innovation Intensity (%) PDF follows a Lognormal (Power Lognormal) distribution, where c = 2.14 (scaled), and s = 0.446 (scaled):

$$f(x, c, s) = \frac{c}{xs}\Phi(\frac{log(x)}{s})(\Phi(\frac{-log(x)}{s}))^{c-1}$$
$$= \frac{2.14}{x \cdot 0.446}\Phi(\frac{log(x)}{0.446})(\Phi(\frac{-log(x)}{0.446}))^{2.14-1} \quad (4)$$

The Goodness of Fit to best describe the Probability Density Distribution for the Innovation Intensity metric

follows a Lognormal shape, as shown in Fig. 9. On average, the top 50 Innovation Districts in the United States concentrate 2.8 times more innovation intensity (34%) than the national average (12%), as described in Table 6.

• Innovation Performance

The statistical distribution describing the best goodness of fit for Innovation Performance PDF follows a Pareto/Power Law distribution, where b = 2.62, loc = 0.21, scale = 0.21:

$$f(x) = \frac{b}{x^{b+1}} = \frac{2.62}{x^{2.62+1}} \quad (5)$$

The function describing the best Goodness of Fit for the Probability Density Distribution regarding the Innovation Performance metric follows a Pareto-shaped Power Law function, as described in Fig. 10 and Table 7. On average, the 50 notable Innovation Districts produce up to four times more Business Revenue per Employee than the national average, as a result of inventions and knowledge advancement in the form of patents, new products, new processes, scientific articles, and R&D projects.

• Innovation Impact

The statistical distribution describing the best goodness of fit for Innovation Impact PDF follows a Gamma Distribution, where: a = 1.99:

Fig. 7 The lognormal distribution of innovative employment (abs)

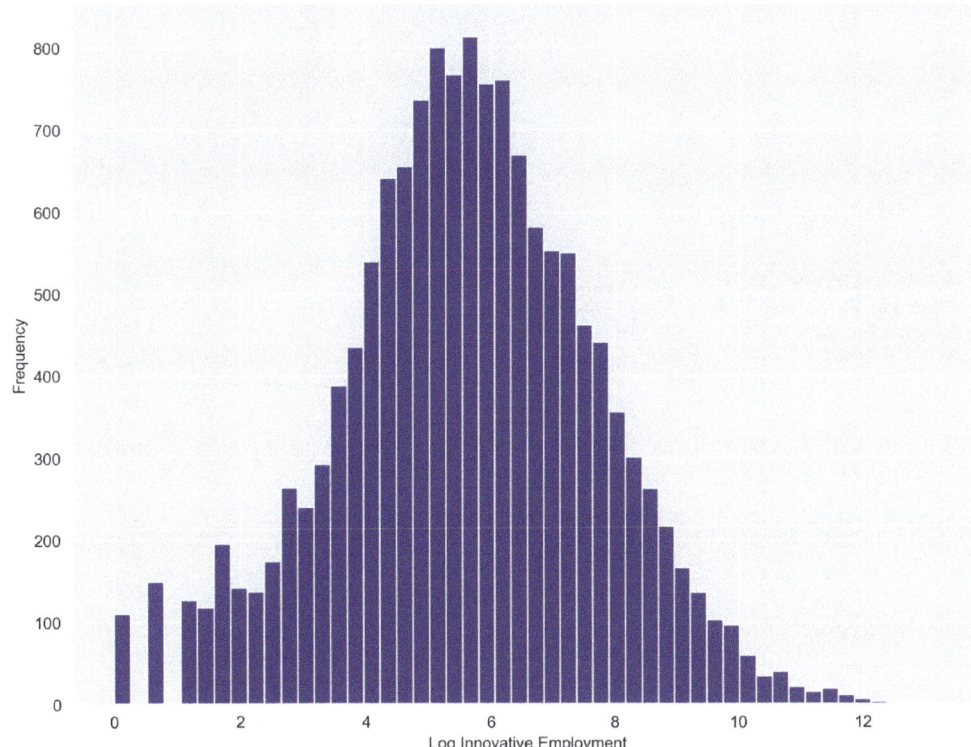

Fig. 8 Superlinear relationship between size of municipality and innovation-intensive employment

Fig. 9 Goodness of fit analysis of the innovation intensity PDF: lognormal distribution

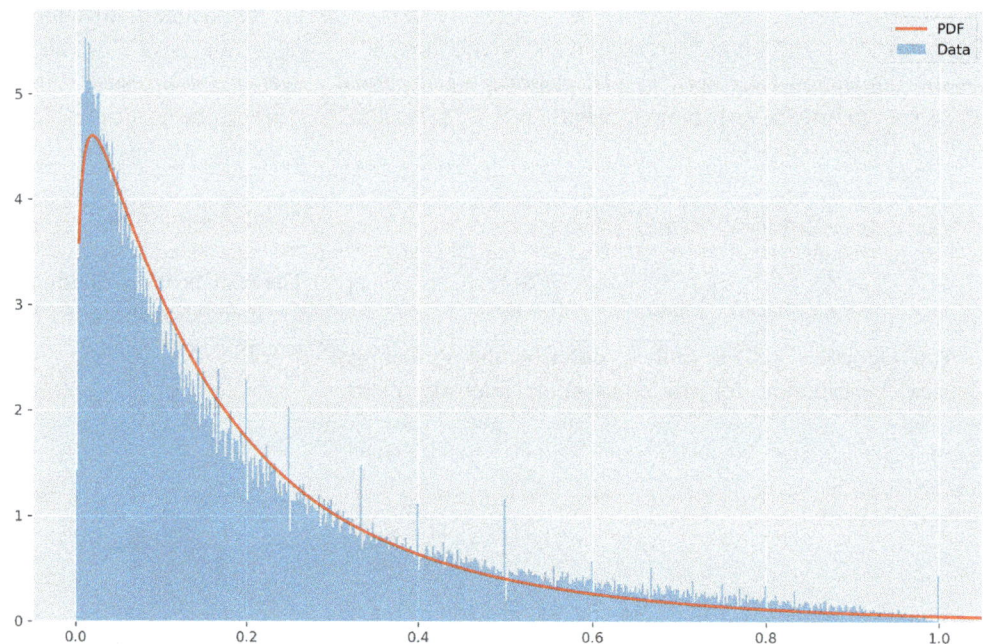

Table 6 Descriptive statistics goodness of fit analysis of the innovation intensity PDF

Concept	Values
KPI	Innovation intensity
Definition	Innovation employees/total employees
PDF distribution	Lognormal (power lognormal)
Parameters (scaled)	c = 2.14 and s = 0.446
Value range	0–100 (%) | 0–100,000 innovation employees (abs)
Average	12%
Average in IDs	34% (82th percentile)
Core building group in IDs	69% (95th percentile)

Fig. 10 Goodness of fit analysis of the innovation performance PDF: power law distribution

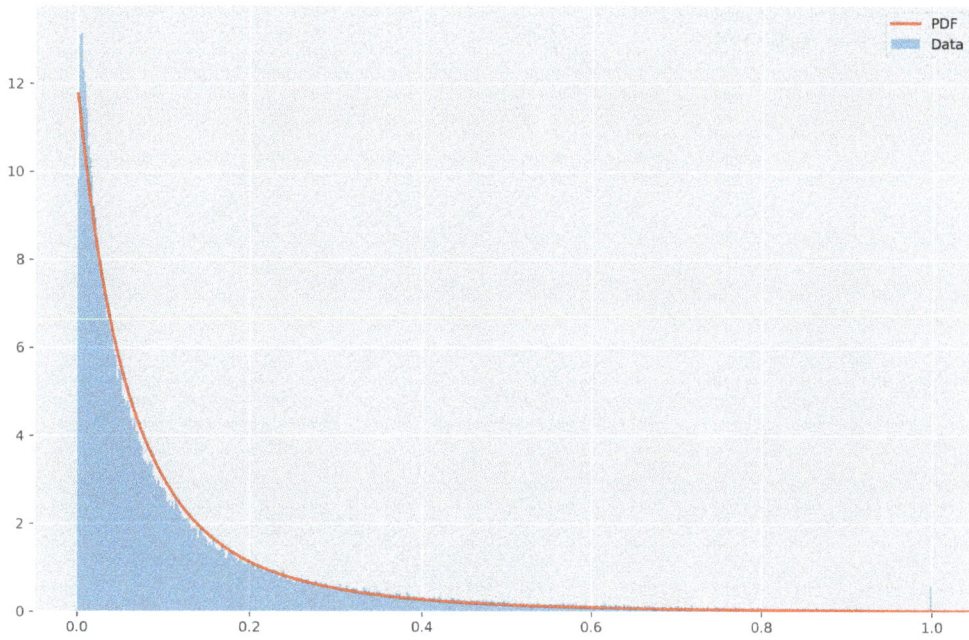

Table 7 Descriptive statistics goodness of fit analysis of the innovation performance PDF

Concept	Values
KPI	Innovation performance
Definition	Innovation sales/employee
PDF distribution	Pareto/Power law
Parameters	b = 2.62
Value range	0–100 (%): 0–$40 M (abs, per BG)
Average	$21 k Innovation revenue/employee
Average in IDs	$81 k Innovation revenue/employee

$$f(x) = \frac{x^{a-1}e^{-x}}{\Gamma(a)} = \frac{x^{1.99-1}e^{-x}}{\Gamma(1.99)} \quad (6)$$

The function describing the best Goodness of Fit for the Probability Density Distribution regarding the Innovation Impact metric follows a Gamma-shaped distribution, as depicted in Fig. 11 and Table 8. The 50 notable Innovation Districts outperform by a factor of 16 in terms of availability of knowledge-intensive job opportunities per resident.

7 Conclusions and Further Research

This paper presents a geospatial analysis framework to model and measure urban innovation performance, enabling the evaluation at the smallest census unit scale: the building group. The novel methodology allows for measuring the nonlinear benefits of geographic aggregation of knowledge-intensive activities. Three new Key Performance Metrics have been designed to

measure the societal effort or investment supporting knowledge-intensive activities (Innovation Intensity), the tangible outcomes of such innovations (Innovation Performance), as well as the societal benefits derived from knowledge advancement activities (Innovation Impact).

The goodness of fit adjustment for the Probability Density Function of the three KPIs reveals the increasingly nonlinear nature of the amplifying effects of investing in knowledge-intensive activities. The Innovation Intensity (societal effort) metric follows a Lognormal Distribution, the Innovation Performance Metric (knowledge advancement and wealth creation) follows a Scale Free/Power Law/Pareto distribution, and the Innovation Impact Metric (societal benefits) follows an extremely skewed Gamma distribution.

The comparisons between 50 notable innovation districts and the average descriptive statistics for building groups across the United States enable the extraction of revealing insights regarding the amplifying effects derived therefrom. The superlinear effects of geographic concentration of

Fig. 11 Goodness of fit analysis of the innovation impact PDF: Gamma distribution

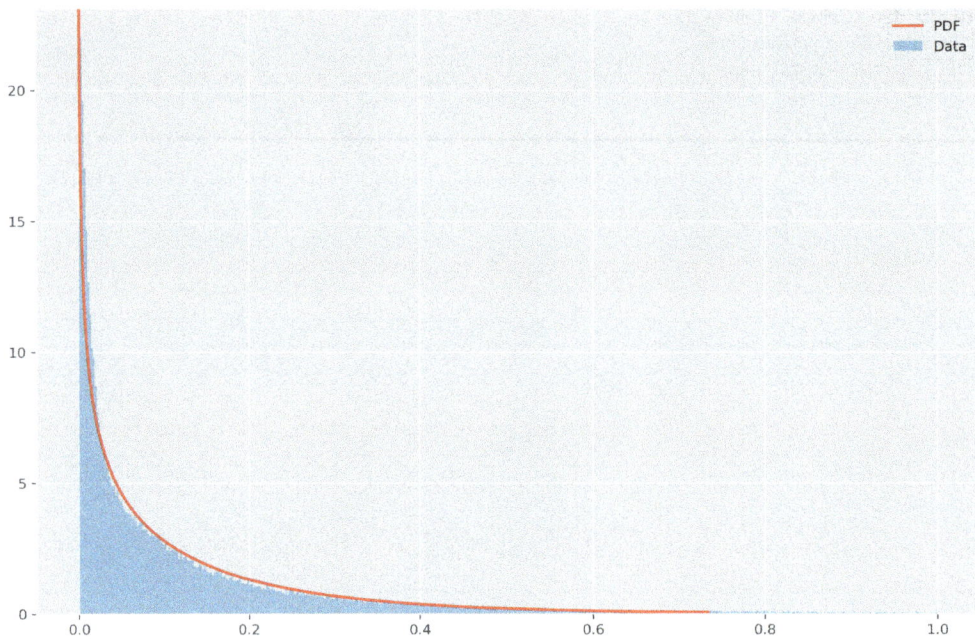

Table 8 Descriptive statistics goodness of fit analysis of the innovation impact PDF

Concept	Values
KPI	Innovation impact
Definition	Innovation employment/residents
PDF distribution	Gamma
Parameters	a = 1.99
Value range	0–250 innovation employee/residents | 0.04–19 in IDs (abs)
Average in the United States	0.04 innovation jobs/resident
Average in IDs	0.68 innovation jobs/resident

knowledge-intensive activities can be described when observing that Innovation Districts systematically outperform regular districts. The multiplying factors of 2.8 (Innovation Intensity), 4 (Innovation Performance), and 16 (Innovation Impact) on average. This illustrates the strong agglomeration economies present in urban innovation and reinforce the current efforts in understanding innovation patterns and their implication for urban performance. Finally, we observe that agglomeration in innovation activity is due to two factors. The growth of total employment is linear but the growth of innovative employment is superlinear, which is amplified in Innovation Districts.

Further Research

A further application of the geospatial analysis methodology is a building group-level clustering analysis to identify the five types of anchor institutions behind the success of

innovation districts, and how they shape the networks of talent, industries, and urban infrastructure: city government led, research led, industrial cluster, entrepreneurial and strategic governmental agencies. The current methodology can be applied to different geographies but is currently restricted to the United States due to data limitations. Future work will expand its applications to other geographies and make use of new sources of information about innovation intensity, economic activity, and urban design.

The ability to measure, compare, and contrast the superlinear growth effects of investing on geographically clustering knowledge-intensive activities will enable to study the urban design typologies and economic activation dynamics fostering fruitful human interaction behind the success of urban innovation performance communities. Future applications of the model may include a network theory-driven evaluation of the adequacy of different types of city design (urban topology, morphology, and entropy) for a given city and region. One model for such an application is the CityScope described in Alonso et al. (2018) data-driven platform.

Opportunities for further learning will enable raising the level of understanding regarding discerning the difference between urban development malpractice and best practices, location intelligence of innovation centers, urban infrastructure, and real estate investments, and the causal mechanisms between exogenous shocks in the urban system, economic performance levels, and economic growth prospects. The sensitivity analysis may help identify best practices in terms of urban design criteria, smart specialization, and innovation pipeline shaping.

Further research may investigate what conditions (urban design, organizational structures, incentive schemes, smart specialization strategies, entrepreneurship support resources, academic and professional training, among others) facilitate and foster the successful development of knowledge creation, transfer, and advanced production, as well as the associated benefits for the different urban communities.

References

Alonso, L., Doorley, R., Elkhatsa, M., Grignard, A., Noyman, A., Larson, K., Sakai, Y., & Zhang, Y. (2018): CityScope: A data-driven interactive simulation tool for urban design. Use case: Volpe Center.

Arbesman, S., Kleinberg, J. M., & Strogatz, S. H. (2009). Superlinear scaling for innovation in cities. *Physical Review E, 79,* 016115.

Barabási, A. L. (2016). Network science: The scale free property. Retrieved from: https://barabasi.com/f/623.pdf

Barabási, A. L. (2017). The elegant law that governs us all. *Science, 357*(6347), 138.

Bertaud, A., & Malpezzi, S. (2003): The spatial distribution of population in 48 world cities: Implications for economies in transition.

Bettencourt, L. M. A., Lobo, Helbing, D., Kühnert, C., & West, G. B. (2006). Growth, innovation, scaling and the pace of life in cities. *PNAS, 104*(17), 7301–7306. https://doi.org/10.1073/pnas.0610172104

Bettencourt, L. M. A., Lobo, J., & Strumsky, D. (2007). Invention in the city: Increasing returns to scale in metropolitan patenting. *Research Policy, 36*(1), 107–120. ISSN 0048-7333. https://doi.org/10.1016/j.respol.2006.09.026

Bettencourt, L. M. A., Lobo, J., & West, G. B. (2008). Why are large cities faster? Universal scaling and self-similarity in urban organization and dynamics. *The European Physical Journal B., 63,* 285–293.

Bettencourt, L. M. A., Lobo, J., & West, G. B. (2009). The self similarity of human social organization and dynamics in cities. In D. Lane, D. Pumain, S. E. van der Leeuw, & G. West (Eds.), *Complexity perspectives in innovation and social change.* Springer.

Bettencourt, L. M. A., Lobo, J., Strumsky, D., & West, G. B. (2010). Urban scaling and its deviations: Revealing the structure of wealth, innovation and crime across cities. *PLoS ONE, 5*(11), e13541.

Bettencourt, L. M. A., Lobo, J., Strumsky, D., & West G. B. (2013). Urban scaling and the production function for cities. *PLoS ONE, 8* (3), e58407.

Ellison, G., & Glaeser, E. (2009). The geographic concentration of industry: Does natural advantage explain agglomeration? *American Economic Review.*

Ekmekci, O., Kalvo, R., & Sevtsuk, A. (2016). Pedestrian accessibility in grid layouts: The role of block, plot and street dimensions. *Urban Morphology, 20.*

Hausmann, R., Hidalgo, C. A., Bustos, S., Coscia, M., Simoes, A., & Yildirim, M. A. (2014). *The Atlas of economic complexity. Mapping paths to prosperity.* MIT Press.

Hartmann, D., Guevara, M. R., Jara Figueroa, C., Aristarán, M., & Hidalgo, C. A. (2017). Linking economic complexity, institutions, and income inequality. *World Development, 93,* 75–93.

Hidalgo, C. A., & Hausmann, R. (2009). The building blocks of economic complexity. *PNAS. Proceedings of the National Academy of Sciences of the United States of America, 106*(26), 10570–10575.

Katz, B., & Wagner, J. (2014). The rise of urban innovation districts. *Harvard Business Review.*

Kühnert, C., Helbing, D., & West, G. B. (2006). Scaling laws in urban supply networks. *Physica a: Statistical Mechanics and Its Applications, 363*(1), 96–103.

Lane, D., Pumain, S., van der Leeuw, S. E., & West, G. B. (2009). Power laws in urban supply networks, social systems, and dense pedestrian crowds. *Complexity perspectives in innovation and social change.* Springer.

Moretti, E. (2012). *The new geography of jobs.* Houghton Mifflin Harcourt.

Schläpfer, M., Bettencourt, L. M., Grauwin, S., Raschke, M., Claxton, R., Smoreda, Z., West, J., & Ratti, C. (2014). The scaling of human interactions with city size. *Journal of the Royal Society Interface, 11* (98), 20130789.

Smith, A. (1776). *The wealth of nations.*

Youn, H., Bettencourt, L. M. A., Lobo, J., Strumsky, D., Samaniego, H., & West, G. B. (2016). The systematic structure and predictability of urban business diversity. *Journal of the Royal Society Interface, 13.* arXiv:1405.3202v1

Understanding Challenges of Urban Regeneration

Examination of the Population Density Impact on Major Air Pollutants: A Study in the Case of Germany

Kamyar Fuladlu and Haşim Altan

Abstract

Background: Today, there is a profound shift in the study of atmospheric pollutants at an urban scale especially due to their negative effects on humans and the environment. Atmospheric pollutants are emitted as a result of rapid urbanization, burning fossil fuels, deforestation, and agricultural intensification. The majority of studies believe cities have both the highest pollution rates and the highest impact targets. Unfortunately, the alteration of atmospheric pollutant patterns was faster than planners were able to come up with solutions. Major Air Pollutants (MAPs) are considered one of the important groups of atmospheric pollutants. Nitrogen Dioxide (NO_2) and Sulfur Dioxide (SO_2) are the main MAPs, which are released from road transport and are mostly exhaust emissions arising from fuel combustion. **Aim and Method**: Until a few years ago, monitoring MAPs, including NO_2 and SO_2 pollutants, was not as easy as today. Recently, the Sentinel-5 Precursor (S5P) satellite was successfully launched and scenes were released. Therefore, monitoring MAPs with the use of Remote Sensing (RS) and Geographical Information System (GIS) became possible. Accordingly, the current study aims to monitor NO_2 and SO_2 pollutants in the case of Germany as a country scale. Besides, this study has examined the relationship between the population density; NO_2 and SO_2 pollutants with the use of RS and GIS on the same scale as the use of a regression model. **Result**: Obviously, the concentration of the NO_2 and SO_2 pollutants over the cities is a result of mutual complex factors like meteorological situation, geographical location, population density, and land-use land-cover. Therefore, it cannot be stated that the population density contributes to the increase of pollutants.

Keywords

Germany • Major air pollutants • Nitrogen dioxide • Remote sensing • Sentinel-5 precursor satellite • Sulfur dioxide

1 Introduction

The potential of the urban area over the past centuries made it a living destination for many people. The United Nations estimated two-thirds of the world population by 2050, which will be living in urban areas (United Nations, 2019). Rapid urban development threatens sustainability, and to obtain environmental sustainability further challenges will be required (United Nations, 2015). At the beginning of the twenty-first century, rapid urban growth resulted in the transformation of the natural environment to the built-up area. An increase in population and demand for housing puts pressure on land cover. Hence today, the urban areas occupied 2% of the world's land, and those amounts of urban lands are responsible for consuming 75% of the world's energies and resources (Pacione, 2009). The main reasons could be addressed to unplanned, single-use, low-density, less accessible, and rapid urban development (Fuladlu, 2019, 2020; Fuladlu et al., 2018b, 2021). Cities become complicated as there are a variety of problems, including traffic congestion, urban pollution, waste and water management, resource management, social inequality, human health, and public security (Chourabi et al., 2012; Khan et al., 2013; Marceau, 2008; Neirotti et al., 2014). In microscale analysis, there are seven important environmental parameters that are effective; namely, "geographical location, meteorological situation, urban form, surface materials, amount of

K. Fuladlu (✉)
Department of Architecture, Faculty of Architecture, Eastern Mediterranean University, 99628 Famagusta, Cyprus
e-mail: kamyar_fuladlu@yahoo.com

H. Altan
Department of Architecture, Faculty of Design, Arkin University of Creative Arts and Design, 99300 Kyrenia, Cyprus

vegetation and watershed, and anthropogenic pollution" (Fuladlu, 2021).

Besides, burning fossil fuels, deforestation, and agricultural intensification are threatening the environment, and are known as the source of anthropogenic pollutants. Unfortunately, the alteration of pollutant patterns was faster than planners were able to come up with a solution (Marsh and Smith 1996). Fundamentally, the nature of cities is shaped by concentrations of humans, materials, and activities. Therefore, they have both the highest pollution rates and the highest impact targets (Fenger, 1999). Additionally, the gradient concentration of the particulate increases according to the land-use density from the countryside toward the core of the urban area. In this sense, the particulate levels in the urban areas are usually two to three times greater than that of the suburban areas, and five times greater than that of the countryside (Marsh & Grossa, 1996). There are several reasons for the accumulation of pollutants in cities. According to Zabalza et al. (2007) at the local level atmospheric pollutants are a major urban problem due to the following factors; (a) majority of the anthropogenic sources are within the urban area, (b) the effect of Urban Heat Island (UHI) to cities (Fig. 1), and (c) concentration of the human in the urban area. Therefore, there is a profound shift in the study of atmospheric pollutants at an urban scale especially due to their negative effects on humans and the environment.

For instance, Weng and Yang (2006) "in the case of Guangzhou city in southern China, by using the Geographic Information System (GIS) approach, examined the spatial pattern of air pollution and land use, and the results showed that the spatial patterns of air pollutants probed were positively correlated with urban built-up density". Feizizadeh and Blaschke (2013) in the case of Tabriz city in Iran found a positive correlation among land-cover, land surface temperature, and air pollutants. Kaplan et al. (2019) in the case of Turkey found a significant correlation between population density and maximum tropospheric Nitrogen dioxide (NO_2). Fuladlu and Altan (2021) in the case of Tehran province with the use of Remote Sensing (RS), GIS, and Moving Average (MA) showed that Sulfur dioxide (SO_2) and NO_2 are mainly concentrated in the Tehran metropolis and the core urban area. The air pollutants are categorized into three groups; (a) Major Air Pollutants (MAPs) include "SO_2, NO_2, Carbon monoxide (CO), Particular matter (PM) lead (Pb), and Ozon (O_2), which are frequently used as indicators of air pollutants in ambient air. MAPs are target species due to their negative impact on human health and vegetation" (European Union Council Directive 96/62/EC). (b) Hazardous Air Pollutants (HAPs) include physical, chemical, and biological agents. Generally, HAPs in comparison to MAPs are in smaller concentrations and seem regularly more localized (Fenger, 1999; Flemming et al., 2005; Wiederkehr & Yoon, 1998; Zabalza et al., 2007). (c) Heat could be considered as a pollutant since the built-up area tends to be warmer than the adjacent area. The generation of heat in an urban area is the climatic reaction to disruptions caused by urban development known as UHI (Fuladlu et al., 2018a; Marsh & Grossa, 1996). The urbanization process transforms the natural material into artificial with the minimum ability of evapotranspiration. This results in a partitioning of energy into sensible fluxes rather than latent heat fluxes. The UHI effect was studied at two distinct layers, Urban Canopy Layer (UCL) and Urban Boundary Layer (UBL). The UCL extends vertically up to the roof level, where the UBL is situated directly above the UCL (Fig. 1) (Oke, 1973, 1976, 1982, 2011).

Among the air pollutants studied, majority have focused on SO_2 and NO_2 due to the environmental consequences, for instance, SO_2 is a major acid rain precursor, which contributes to acidic deposition in terrestrial ecosystems as dry-deposited gas (Cox, 2003; Meng et al., 2008). SO_2 oxidation product, Sulfate, plays a significant role in the radiative forcing of the climate. Similarly, NO_2 is of particular concern, as it plays a crucial role in atmospheric chemistry by participating in the formation of tropospheric O_2, peroxyacyl nitrate, and nitrate aerosols (Meng et al., 2008). SO_2 and NO_2 are short-lived trace gases that have key

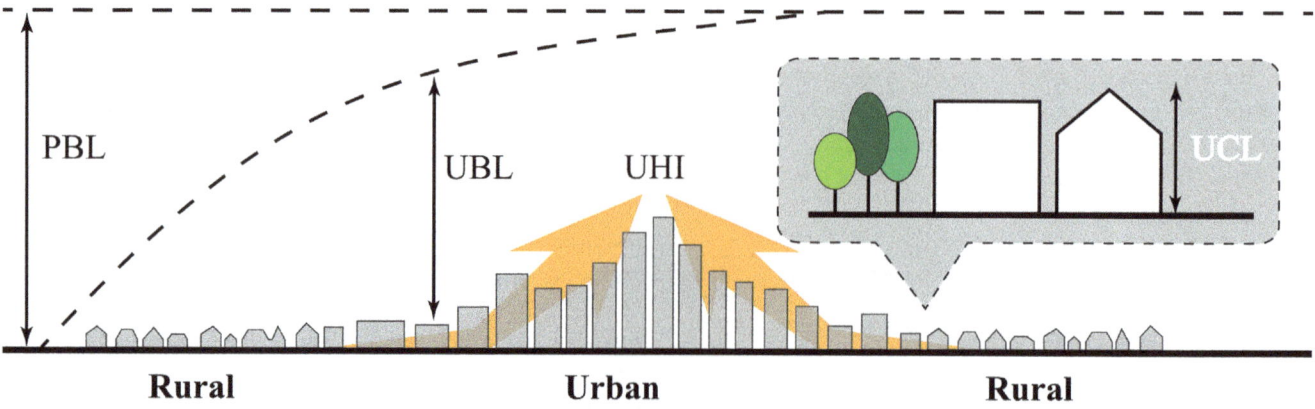

Fig. 1 Planetary Boundary Layer (PBL), Urban Boundary Layer (UBL), Urban Canopy Layer (UCL), and Urban Heat Island (UHI)

roles in tropospheric chemistry. Besides, SO_2 and NO_2 are involved in the formation of secondary aerosol particles (Krotkov et al., 2016; Xue et al., 2020). "Major sources of NO_x ($NO_x = NO + NO_2$) include fossil fuel combustion, mediated biomass burning, industrial and transportation emissions, and lightning" (Krotkov et al., 2016; Xue et al., 2020). "NO_2 affects atmospheric oxidation rates participating in the formation of surface ozone" (Krotkov et al., 2016; Meng et al., 2008). "The natural source of SO_2 is volcanic eruptions, which can release a large amount of SO_2 above the Planetary Boundary Layer (PBL). The anthropogenic source of SO_2 was burning sulfur-contaminated fossil fuels and the refinement of sulfide ores, which mainly concentrated in the PBL" (Fig. 1) (Brunelli et al., 2007; Krotkov et al., 2016; Xue et al., 2020).

For decades, new dimensions of RS and its ability to integrate with GIS have created new opportunities for monitoring and interpretation of the spatial dataset. Besides, RS and GIS, when used together, are recognized as a strong and outstanding tool for monitoring alterations in the spatial dataset. The advantage is the cost efficiency of RS and GIS as a method of monitoring. Moreover, RS and GIS are derived from the accessibility of data collected by thousands of satellites orbiting the planet. However, until a few years ago monitoring MAPs from a satellite was not as easy as today. Recently the Sentinel-5 Precursor (S5P) satellite was successfully launched by scene release. Accordingly, the S5P satellite aims to fill in the data gap and to provide data continuity between the retirement of the Envisat satellite and NASA's Aura mission and the launch of S5P satellite. The S5P satellite objectives are to provide operational space-borne observations in support of the operational monitoring Air Quality, O_2, Surface Ultraviolet, and Climate. Besides, it will provide measurements of O_2, NO_2, SO_2, CO, Formaldehyde, Aerosol, Methane, and Clouds.

Until now, a systematic review of the literature provides a focus on the atmospheric pollutants in the cities. Literature includes the main sources of the atmospheric pollutants, classification, and the S5P satellite as a tool for monitoring the pollutants, especially MAPs. The current study firstly aims to monitor NO_2 and SO_2 pollutants in the case of Germany as a country scale. Secondly, this study examines the relationship between the population density, NO_2 and SO_2 pollutants with the use of RS and GIS on the same scale with the use of a regression model. The reasons to choose Germany for a case in this study were:

A. Germany has the largest economy in Europe, on a global scale, it has leading several industrial and technological sectors.
B. Germany as a highly developed country provides census, statistical dataset, and temporal map on different levels, which make the base for any study.
C. Germany with an area of 357,022 km^2 is suitable for study with the S5P satellite resolution.

2 Materials and Methods

2.1 Case Study

"Germany, officially the Federal Republic of Germany, is a country in Central and Western Europe. It is located between the Baltic and North seas to the north, and the Alps to the south. It borders Denmark to the north, Poland and the Czech Republic to the east, Austria and Switzerland to the south, and France, Luxembourg, Belgium, and the Netherlands to the west". Germany comprises sixteen federal states, each state has its state constitution and is largely autonomous regarding its internal organization. By 2017 Germany is divided into 402 districts at municipal levels; these consist of 295 rural districts and 107 urban districts. According to Germany government estimates on 31st December 2018, the German population was about 83,040,944 (Wikipedia, 2020).

The satellite is a material of the study chosen according to the published population estimation by the German government on 31st December 2018. Therefore, the population date is considered as a groundwork for the current study. The S5P satellite RS was utilized to acquire data for NO_2 and SO_2 pollutants using the following means:

(a) The twelve Level 2 (level-two) senses for each pollutant were acquired from the S5P satellite; the twelve senses are good enough to present the overall condition of the year and to increase the study accuracy.
(b) The last day of each month was chosen for the period of 31st October 2018 to 30th September 2019.

The detailed steps are described under the next sections and the overall study methodology diagram is summarized in Fig. 2.

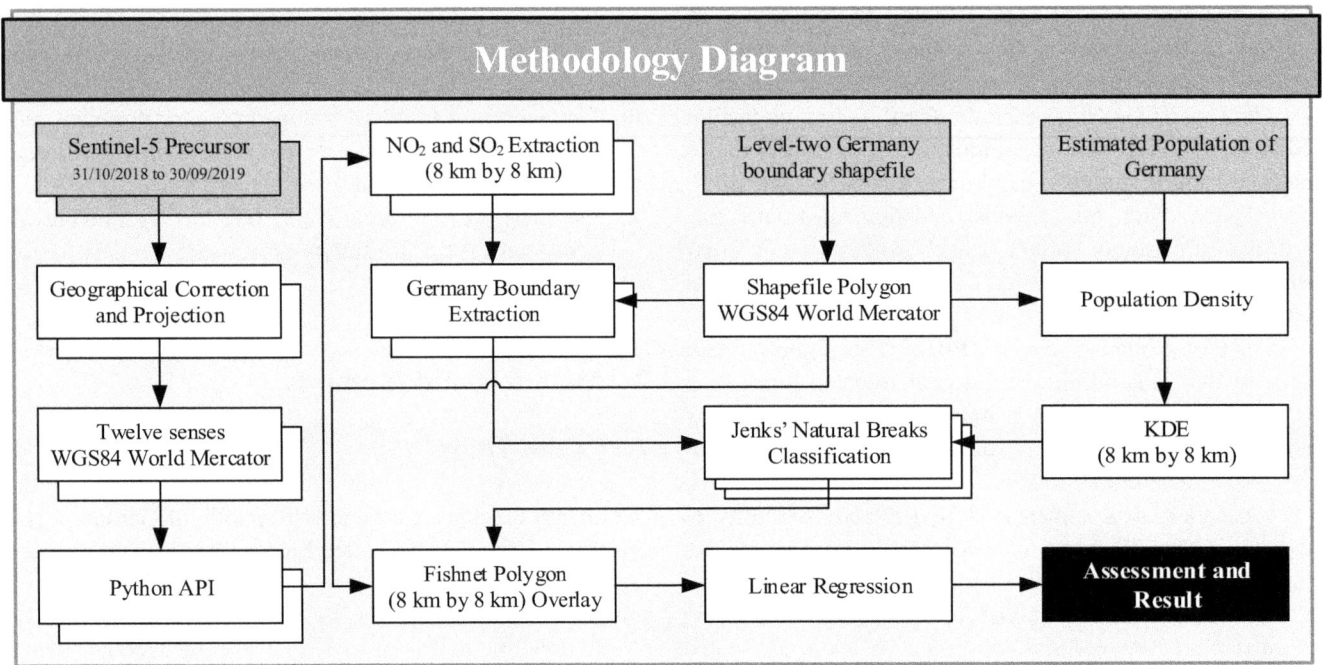

Fig. 2 Methodology diagram

2.2 Pre-Processing

The population density raster developed with the use of the Level 2 Germany boundary shapefile. The shapefile projection was WGS84 World Mercator and had 403 polygons including rural, urban districts, and water bodies. The estimated population dataset was integrated into the shapefile attribute based on the records. The population density was calculated in GIS for each district and rasterized according to the S5P satellite resolution (8 km by 8 km). The process was performed with the use of Kernel Density Estimation (KDE). The KDE is known as an outstanding method which facilitates a cell-by-cell means for the future correlation between rasters (Fotheringham et al., 2000; O'Sullivan & Unwin, 2010; Silverman, 1998). The KDE outcome is classified with the use of Jenks' natural breaks method as shown in Fig. 3a.

"Time-series take place in a wide range of phenomena" (Salcedo et al., 1999). "Air quality monitoring is an example of an environmental time-series study. The method commonly is used to estimate environmental parameters and based on classical descriptive statistics, but due to the high variability associated with air quality data and the low signal-to-tonnage ratio of existing measurements, this value is relatively limited. Time-series studies may be a good tactic to avoid these troubles by agreeing to the identification of hidden deterministic behavior and thus contributing to an understanding of origin and consequence associations in

environmental and urban problems" (Schwartz & Marcus, 1990). "Time-series still form the basis of decomposition methods; a smoothing technique is required to apply on time-series to remove the fine-grained variation between time steps. MA is a simple and common type of smoothing used in time-series analysis and is a calculation for analyzing data points by creating an average series of different subsets from a big dataset" (Fuladlu & Altan, 2021).

The NO_2 and SO_2 rasters developed with the use of the twelve senses to represent points of time for each of the pollutants. As stated earlier, all the required senses are acquired from the S5P satellite. The python API was integrated into the GIS to minimize human errors and save time during the big data interpretation. Accordingly, the geographical correction was applied to all senses to minimize the distortion, and later all the senses were projected to the WGS84 World Mercator. The NO_2 tropospheric (mol/m^2) and SO_2 total (mol/m^2) bands were used for calculating the MA of each pollutant. The MA of each pollutant is accurately calculated from twelve senses with the use of the spatial analyst function of GIS. As a result, two rasters were created for NO_2 and SO_2 pollutants, all the pixels beyond the Germany boundary shapefile were eliminated from the rasters and classified with the use of Jenks' natural breaks method as shown in Fig. 3b and c. About 18% of Fig. 3b is unusable (blank area) due to the no-data pixel of the sources, which is considered as a study bias.

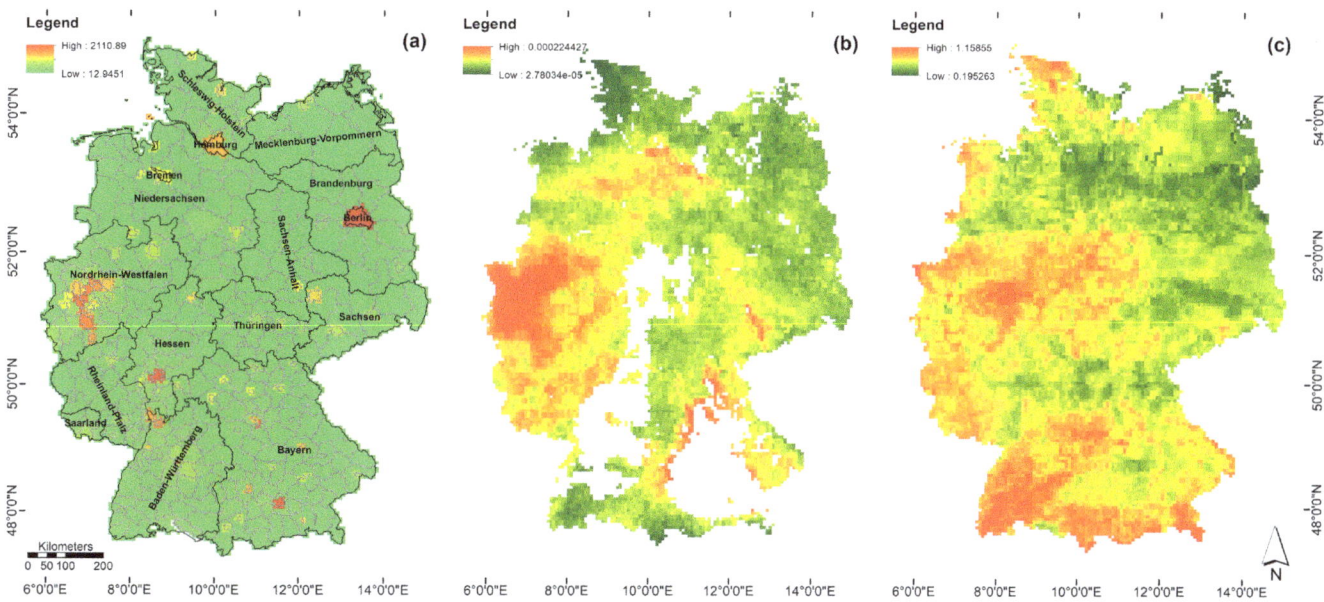

Fig. 3 Germany **a** population density ($\#/km^2$), **b** NO_2 (mol/m^2), and **c** SO_2 (mol/m^2)

3 Results and Analysis

Different approaches exist to quantify the raster's data. However, the current study experimentally puts its approach into practice. According to the methodology diagram (Fig. 2), all rasters are made according to the resolution of the S5P satellite and there is no need to resample it. Therefore, a fishnet is created according to the German boundary with the same resolution and projection in GIS. The required fields added to the fishnet attribute overlay the rasters of population density, NO_2, and SO_2 pollutant values. The records of those fields were used to develop the future correlation matrix between the population density, NO_2, and SO_2 pollutants. Accordingly, the cell-by-cell examination is performed with the use of the Pearson two-tailed correlation matrix for 22,750 cells and the results are presented in Table 1.

The correlation between population density, NO_2 and SO_2 pollutants visualized with the use of the scatter plot and the trend of each plot is shown with linear regression (Fig. 4).

The relationship between population density and pollutants is quantified with the use of linear regression and equation, which are written for each pollutant (Table 2).

Based on the correlation matrix (Table 1; Fig. 4), there is a correlation at the 0.01 level between the variables. Accordingly, a positive correlation was noted between the population density and NO_2 pollution ($r = 0.314$, $p < 0.01$). Likewise, the population density and SO_2 pollution had a positive correlation too ($r = 0.276$, $p < 0.01$). Besides, a positive correlation between both NO_2 and SO_2 pollutants is outstanding ($r = 0.663$, $p < 0.01$). A one-way ANOVA (Table 2) was conducted to compare the effect of the pollutants on population density. The variance analysis showed that the effect of NO_2 on population density existed, and the effect of SO_2 on population density existed. In summary, there is a positive correlation between population density, NO_2 and SO_2 pollutants. However, the mean of pollutants is relatively minimized in districts with a high population density (Fig. 3a, b, and c).

Table 1 Correlation matrix of the variables

Variables		Population density	Nitrogen dioxide	Sulfur dioxide
Population density	Pearson correlation	1		
	N	22,750		
Nitrogen dioxide	Pearson correlation	0.314**	1	
	N	22,750	22,750	
Sulfur dioxide	Pearson correlation	0.276**	0.663**	1
	N	22,750	22,750	22,750

**Correlation is significant at the 0.01 level (two-tailed)

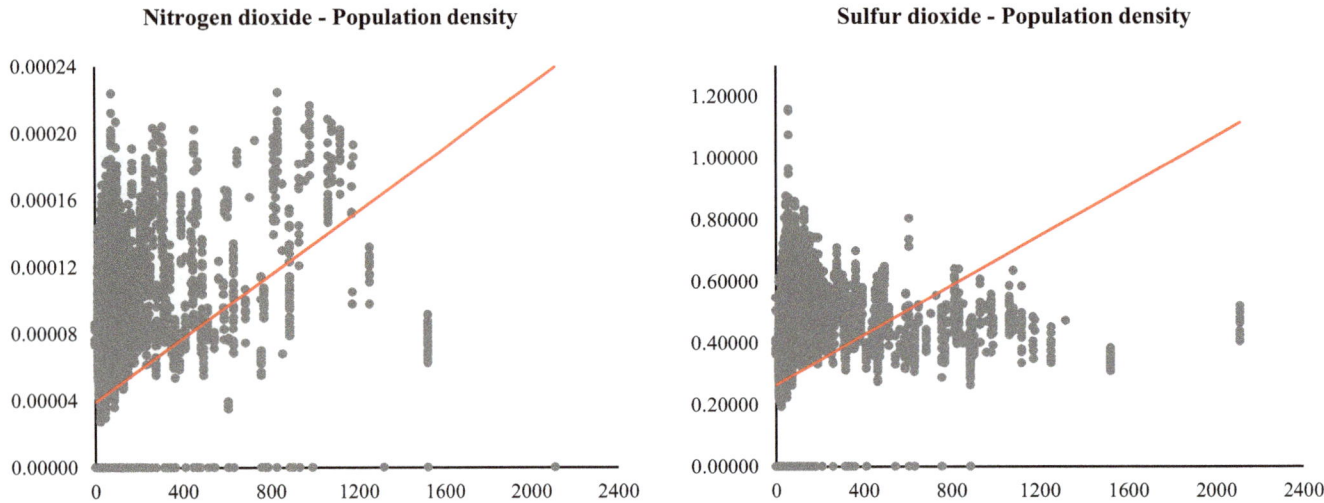

Fig. 4 Scatter graph for population density ($\#/Km^2$), NO_2 (mol/m^2), and SO_2 (mol/m^2)

Table 2 Linear regression equations for population density and each pollutant

Pollutants	Equation	R	R^2	df1	df2	F. change	Sig. F. change
NO_2	Population density = 20.495 × NO_2 1,038,659.509	0.314	0.099	1	22,748	2486.557	0.000
SO_2	Population density = 12.703 × SO_2 + 189.425	0.276	0.076	1	22,748	1878.813	0.000
Both	Population density = (772,155.898 × NO_2) (83.301 × SO_2) × + 8.355	0.327	0.107	2	22,747	1359.989	0.000

4   Discussion

According to the outcome (Fig. 3) München 2110.89 and after Berlin 1520.81 had maximum population density among the German districts. However, the maximum concentration of the pollutants was mainly located in the west and south-west of Germany (Fig. 3b and c). Likewise, NO_2 is seen at Nordrhein-Westfalen state while SO_2 is seen at Baden-Württemberg, Bayern, and Schleswig–Holstein state. Although Tables 1 and 2; Fig. 4 depict an overall positive correlation between the population density, NO_2 and SO_2 pollutants, this correlation was not significant as expected. The concentration of NO_2 and SO_2 pollutants over the cities as a result of mutual complex factors like meteorological situation, geographical location, population density, and land-use land-cover (Weng & Yang, 2006). Therefore, it cannot be stated that the population density contributes to the increase of pollutants.

As stated previously, the main source of SO_2 and NO_2 pollutants is fossil fuel combustion, which are released by the manufacturing industries. A quick look over Germany depicts that, majority of the manufacturing and industries such as Rolls-Royce, MTU Aero Engines, Cassidian, Astrium, Airbus, BMW, Daimler, ThyssenKrupp, Otto

Fuchs, OHB, and etcetera mainly located in Hamburg, Niedersachsen, Hessen, Baden-Württemberg, and Bayern states; whereas the pollutants are mainly concentrated (Fig. 3b and c). Therefore, proximity to manufacturing and industries more than population density can be attributed to the concentration of NO_2 and SO_2 pollutants. Similarly, NASA (2020) and European space agency pollution monitoring satellites by the end of February 2020 had detected significant decreases in NO_2 pollution over Wuhan, China. NASA scientists confirm the significant decrease of NO_2 pollution in eastern and central China in comparison to the previous years. Decrease of NO_2 pollution over Wuhan attributes to the Chinese quarantines by 23rd January 2020. Accordingly, Chinese authorities had shut-down transportation and local businesses of Wuhan to reduce the spread of the COVID-19 pandemic.

5   Conclusions

The study of atmospheric pollutants due to their negative impacts on humans and the environment becomes a topic for discussion. Until now, different studies confirm the role of population concentration on MAP emissions. Likewise, the current study was put forward to provide comprehensive reviews of atmospheric pollutants. Besides this study aims to monitor the MAPs including NO_2 and SO_2 pollutants in the case of Germany as a country scale. Moreover, the study tries to quantify the relationship between the population density, NO_2 and SO_2 pollutants with the use of RS and GIS on the same scale with the use of a regression model. The S5P satellite RS was utilized to retrieve the necessary data for each NO_2 tropospheric (mol/m^2), and SO_2 total (mol/m^2) over Germany.

The $\bar{x}$ of each pollutant was accurately calculated from twelve senses for the period of 31st October 2018 to 30th September 2019. The value is compared with the Level 2 population density of Germany. Thus, the results depict the positive correlations between the population density, NO_2 and SO_2 pollutants. However, this relationship is not as expected as the concentration of pollutants that was found mainly in the states where the manufacturing industries are located. Further studies are required to investigate the correlation of MAPs, and land surface temperature with population density and other factors, which are contributing to environmental pollution.

Acknowledgements The authors wish to acknowledge Canay Ataöz, Head of Technical Services of the Library at Eastern Mediterranean University, Arosha Architecture and Urban Research Development and Technology for providing the resources and materials that supported the author's execution of this study.

Funding
The authors received no financial support for the research, authorship, and publication of this manuscript from agencies in the public, commercial, or not-for-profit sectors.

Availability of Data and Materials
All data generated or analyzed during this research are included in this manuscript.

Competing Interests
The authors declare that they have no competing interests.

References

Brunelli, U., Piazza, V., Pignato, L., Sorbello, F., & Vitabile, S. (2007). Two-days ahead prediction of daily maximum concentrations of SO2, O3, PM10, NO2, CO in the urban area of Palermo Italy. *Atmospheric Environment, 41*(14), 2967–2995. https://doi.org/10.1016/j.atmosenv.2006.12.013

Chourabi, H., Nam, T., Walker, S., Gil-Garcia, J. R., Mellouli, S., Nahon, K., Pardo, T. A., & Scholl, H. J. (2012). *Understanding Smart Cities: An Integrative Framework*. Paper presented at the 2012 45th Hawaii International Conference on System Sciences, Maui, HI, USA. https://doi.org/10.1109/HICSS.2012.615.

Cox, R. M. (2003). The use of passive sampling to monitor forest exposure to O3, NO2 and SO2: a review and some case studies. *Environmental Pollution, 126*(3), 301–311. https://doi.org/10.1016/S0269-7491(03)00243-4

Feizizadeh, B., & Blaschke, T. (2013). Examining urban heat Island relations to land use and air pollution: multiple endmember spectral mixture analysis for thermal remote sensing. *IEEE Journal of Selected Topics in Applied Earth Observations and Remote Sensing, 6*(3), 1749–1756. https://doi.org/10.1109/jstars.2013.2263425

Fenger, J. (1999). Urban air quality. *Atmospheric Environment, 33*(29), 4877–4900. https://doi.org/10.1016/S1352-2310(99)00290-3

Flemming, J., Stern, R., & Yamartino, R. J. (2005). A new air quality regime classification scheme for O3, NO2, SO2 and PM10 observations sites. *Atmospheric Environment, 39*(33), 6121–6129. https://doi.org/10.1016/j.atmosenv.2005.06.039

Fotheringham, A. S., Brunsdon, C. F., & Charlton, M. (2000). *Quantitative Geography: Perspectives on Spatial Data Analysis* (1st ed.): SAGE Publications Ltd.

Fuladlu, K. (2019). Urban Sprawl Negative Impact: Enkomi Return Phase. *Journal of Contemporary Urban Affairs, 3*(1), 44–51. https://doi.org/10.25034/ijcua.2018.4709.

Fuladlu, K. (2020). Urban sprawl measurement with use of VMT Pattern: A longitudinal method in case of Famagusta. *International Journal of Advanced and Applied Sciences, 7*(5), 12–19. https://doi.org/10.21833/ijaas.2020.05.003.

Fuladlu, K. (2021). Environmental parameters for campus outdoor space: A microclimate analysis of the eastern mediterranean university (EMU) campus. *Journal of Green Building, 16*(3), 217–236. https://doi.org/10.3992/jgb.16.3.217

Fuladlu, K., & Altan, H. (2021). Examining land surface temperature and relations with the major air pollutants: A remote sensing research in case of Tehran. *Urban Climate, 39*, 100958. https://doi.org/10.1016/j.uclim.2021.100958

Fuladlu, K., Riza, M., & İlkan, M. (2018a). *The Effect of Rapid Urbanization On the Physical Modification of Urban Area*. Paper presented at the The 5th International Conference on Architecture and Built Environment with AWARDs S.ARCH 22–24 May 2018a, Venice, Italy.

Fuladlu, K., Riza, M., & İlkan, M. (2018b). *Impact of Urban Sprawl: The Case of the Famagusta, Cyprus*. Paper presented at the 1st Regional Conference: Cyprus Network of Urban Morphology CyNUM 16–18 May 2018b, Nicosia, Cyprus.

Fuladlu, K., Riza, M., & İlkan, M. (2021). Monitoring urban sprawl using time-series data: Famagusta region of Northern Cyprus. *SAGE Open, 11*(2), 21582440211007464. https://doi.org/10.1177/21582440211007465

Kaplan, G., Avdan, Z. Y., & Avdan, U. (2019). Spaceborne nitrogen dioxide observations from the sentinel-5P TROPOMI over Turkey. *Proceedings, 18*(1). https://doi.org/10.3390/ECRS-3-06181.

Khan, Z., Anjum, A., & Kiani, S. L. (2013). *Cloud Based Big Data Analytics for Smart Future Cities*. Paper presented at the 2013 IEEE/ACM 6th International Conference on Utility and Cloud Computing, Dresden, Germany. https://doi.org/10.1109/UCC.2013.77.

Krotkov, N. A., McLinden, C. A., Li, C., Lamsal, L. N., Celarier, E. A., Marchenko, S. V., Swartz, W. H., Bucsela, E. J., Joiner, J., Duncan, B. N., Boersma, K. F., Veefkind, J. P., Levelt, P. F., Fioletov, V. E., Dickerson, R. R., He, H., Lu, Z., & Streets, D. G. (2016). Aura OMI observations of regional SO2 and NO2 pollution changes from 2005 to 2015. *Atmospheric Chemistry and Physics, 16*(7), 4605–4629. https://doi.org/10.5194/acp-16-4605-2016

Marceau, J. (2008). Introduction: Innovation in the city and innovative cities. *Innovation: Management, Policy & Practice, 10*(2–3), 136–145. https://doi.org/10.5172/impp.453.10.2-3.136.

Marsh, W. M., & Grossa, J. M. (1996). *Environmental Geography: Science, Land Use, and Earth Systems* (1st ed.). John Wiley & Sons.

Meng, Z. Y., Ding, G. A., Xu, X. B., Xu, X. D., Yu, H. Q., & Wang, S. F. (2008). Vertical distributions of SO2 and NO2 in the lower atmosphere in Beijing urban areas China. *Science of the Total*

Environment, 390(2), 456–465. https://doi.org/10.1016/j.scitotenv. 2007.10.012

NASA. (2020). Airborne nitrogen dioxide plummets over China. https://earthobservatory.nasa.gov/images/146362/airborne-nitrogen-dioxide-plummets-over-china.

Neirotti, P., Marco, A. D., Cagliano, A. C., Mangano, G., & Scorrano, F. (2014). Current trends in smart city initiatives: Some stylised facts. *Cities, 38*, 25–36. https://doi.org/10.1016/j.cities.2013.12.010

O'Sullivan, D., & Unwin, D. (2010). *Geographic Information Analysis* (2nd ed.). New Jersey: John Wiley & Sons. https://doi.org/10.1002/9780470549094.

Oke, T. R. (1973). City size and the urban heat island. *Atmospheric Environment (1967), 7*(8), 769–779. https://doi.org/10.1016/0004-6981(73)90140-6.

Oke, T. R. (1976). The distinction between canopy and boundary-layer urban heat islands. *Atmosphere, 14*(4), 268–277. https://doi.org/10.1080/00046973.1976.9648422

Oke, T. R. (1982). The energetic basis of the urban heat island. *Quarterly Journal of the Royal Meteorological Society, 108*(455), 1–24. https://doi.org/10.1002/qj.49710845502

Oke, T. R. (2011). Urban heat islands. In I. Douglas, D. Goode, M. Houck, & R. Wang (Eds.), *The Routledge Handbook of Urban Ecology* (1st ed., pp. 120–131). Oxon: Abingdon: Routledge. https://doi.org/10.4324/9780203839263.ch11.

Pacione, M. (2009). *Urban Geography: A Global Perspective* (3rd ed.). London: Routledge. https://doi.org/10.4324/9780203881927.

Salcedo, R. L. R., Alvim-Ferraz, M. D. C., Alves, C. A., & Martins, F. G. (1999). Time-series analysis of air pollution data. *Atmospheric Environment, 33*(15), 2361–2372. https://doi.org/10.1016/S1352-2310(99)80001-6

Schwartz, J., & Marcus, A. (1990). Mortality and air pollution j London: A time series analysis. *American Journal of Epidemiology, 131*(1), 185–194. https://doi.org/10.1093/oxfordjournals.aje.a115473

Silverman, B. W. (1998). *Density Estimation for Statistics and Data Analysis*. Routledge. https://doi.org/10.1201/9781315140919

United Nations. (2015). *World Urbanization Prospects: The 2014 Revision*. (ST/ESA/SER.A/366). P. D. Department of Economic and Social Affairs. New York. https://population.un.org/wup/Publications/Files/WUP2014-Report.pdf.

United Nations. (2019). *World Urbanization Prospects: The 2018 Revision*. (ST/ESA/SER.A/420). P. D. Department of Economic and Social Affairs. New York. https://population.un.org/wup/Publications/Files/WUP2018-Report.pdf.

Weng, Q., & Yang, S. (2006). Urban air pollution patterns, land use, and thermal landscape: An examination of the linkage using GIS. *Environmental Monitoring and Assessment, 117*(1–3), 463–489. https://doi.org/10.1007/s10661-006-0888-9

Wiederkehr, P., & Yoon, S.-J. (1998). Air quality indicators. In J. Fenger, O. Hertel, & F. Palmgren (Eds.), *Urban Air Pollution - European Aspects* (pp. 403–418). Dordrecht: Springer Netherlands. https://doi.org/10.1007/978-94-015-9080-8

Wikipedia. (2020). Germany, https://en.wikipedia.org/wiki/Germany, Accessed 6 Mar 2020.

Xue, R., Wang, S., Li, D., Zou, Z., Chan, K. L., Valks, P., Saiz-Lopez, A., & Zhou, B. (2020). Spatio-temporal variations in NO2 and SO2 over Shanghai and Chongming Eco-Island measured by Ozone Monitoring Instrument (OMI) during 2008–2017. *Journal of Cleaner Production, 258*, 120563. https://doi.org/10.1016/j.jclepro.2020.120563

Zabalza, J., Ogulei, D., Elustondo, D., Santamaría, J. M., Alastuey, A., Querol, X., & Hopke, P. K. (2007). Study of urban atmospheric pollution in Navarre (Northern Spain). *Environmental Monitoring and Assessment, 134*(1–3), 137–151. https://doi.org/10.1007/s10661-007-9605-6

A Comparison Between Italian and French Case Studies on Urban Regeneration

Elena Gualandi, Tiziano Innocenzi, Chiara Pompei, and Alessandro Stracqualursi

Abstract

Urban regeneration is a process that aims to transform the existing city while improving its performance, adopting strategies for the recovery and enhancement of the building heritage. In the contemporary age, we are experiencing a phase of industrial crisis that is reflected in the consolidated European city themes of reactivation and rehabilitation, which are assuming a central role for both local and metropolitan communities. The aim of this paper is to highlight issues and strategies for sustainable urban regeneration from a socio-economic point of view. To achieve this, we will compare different case studies from Italy and France, all located in dynamic contexts, whose projects have been completed in the last ten years. The cases are gathered in two categories of brownfield land, each addressing a different perspective: ex-industrial areas and docks. Urban regeneration is a key challenge for the contemporary city. Restoring brownfield sites means putting the theme of sustainability, declined in all its forms, at the centre of the future city development process.

Keywords

Functional recovery • Brownfield land • Industrial area • Docks • Sustainable development

E. Gualandi (✉) · T. Innocenzi · C. Pompei · A. Stracqualursi
Planning, Design, and Technology of Architecture, Sapienza University, Rome, Italy
e-mail: elena.gualandi@uniroma1.it

T. Innocenzi
e-mail: tiziano.innocenzi@uniroma1.it

C. Pompei
e-mail: chiara.pompei@uniroma1.it

A. Stracqualursi
e-mail: alessandro.stracqualursi@uniroma1.it

1 Introduction

The disposal of industrial sites is a complex, large-scale process that has occurred since the second half of the twentieth century, with the decline of some traditional production sectors and the consequent progressive transition from an industrial society linked to the Fordist model to a post-industrial society characterized by a marked outsourcing. The resulting areas have become the ideal field of intervention for setting in motion and realizing the development of dynamics of contemporary cities and metropolises, with the aim of redefining their image and functioning in a sustainable key (Donnarumma, 2013; Sennett, 2018). This process contributes to the sustainable development of cities through the reuse of soil and buildings, saving on demolition waste and new construction materials as well as reducing peripheral urban growth. Furthermore, it facilitates the densification of existing urban areas. Urban regeneration strategies involve economic, social and environmental improvement measures for the dismissed areas. These areas are facing periods of decline due to: compounding pressures from major economic problems, demographic changes, underinvestment, infrastructural obsolescence, structural issues, political disenfranchisement, social tensions and physical deterioration in urban areas (Czischke et al., 2015). Brownfield regeneration, together with the reuse of dismissed industrial areas, is a common action included in local urban plans to reduce land consumption as recommended by European Environment Agency (EEA, 2019). It is important to realize public space for citizens, especially in suburban areas where the industrial districts were built. Regeneration can be expensive or complicated for several reasons cause its economic return, its management by municipalities, and its social fruition derive from how the regenerated space is accepted by local communities. Besides, considering the actual situation of Covid-19 health emergency, the current use of the public space is significantly reduced (Gehl, 2020) and its future will strictly depend on measures that will be capable of re-thinking it in a pliable and adaptive way (Honey-Rosés et al., 2020). At a time when our built environment has been altered by a pandemic, constantly changing day by day, speaking about major investment might seem incongruous. Otherwise, brownfield reuse, resilience and adaptability are undeferrable necessities, as assumed by the 2030 Agenda for Sustainable Development by United Nations. In particular, the 11th Sustainable Development Goal (SDG 11) "*Sustainable cities and communities*" underlines the importance of "*make cities and human settlements inclusive, safe, resilient and sustainable*" (UN Habitat, 2020). All this considering, the purpose of the paper is to investigate the effects of a brownfield regeneration process on the public spaces' fruition and their integrability within the city, meeting the targets of inclusiveness, accessibility and sustainability.

2 Methodology

Case studies are used to investigate the effects of brownfield regeneration on inclusiveness, accessibility and sustainability of the city where it takes place, showing modes and types of intervention. The selection of the cities was made among European Metropolitan and Large-Metropolitan Areas (OECD, 2012, 2013), following a two stages process based on familiarity (in terms of industrial heritage) and variation (in terms of geographical and structural context). Representative cases were collected for Most Different and Most Similar Systems, covering a broad investigation framework and highlighting extremes and points of contact between the issues addressed (see Sect. 2.1). Moreover, the authors have direct knowledge of the cases through visits and study trips.

To meet the research objectives, a semi-qualitative analysis was carried out comparing the case studies with each other and with the goals. The evaluation was supported by the use of SDG 11 Indicators (UN, 2020), which provide a

common and internationally relevant standard of comparison. Indicators have been revised on the basis of the available data and shaped to merge with macro-categories of analysis (Table 1).

2.1 Selection of the Case Studies

Case studies selection is based on three different criteria: characteristics of the place, regeneration process and type of intervention. Following the first criterion, we chose cities with an industrial tradition, which suffered a halt in their development during the immediate post-war period. We highlighted different industrial areas typologies from each city. The second criterion considers the time of the interventions, as we decided to consider projects realized in the past twenty years, notwithstanding the fact that they have older roots. Finally, the third criterion relates to the type of intervention, as we preferred large-scale urban transformations, consisting of several different design areas that respond to a unitary vision.

Combining the previous criteria, we frequently encountered two types of regeneration projects: ex-industrial areas and docks. In both places, the industrial heritage and infrastructures—terrestrial or maritime—became a starting point for a renewal that was strongly linked with the local environment. According to the EU "*Nomenclature des unités territoriales statistiques*" (*NUTS*), we chose to compare Italian and French cities for their resemblance in the process methodology and in their urban-planning regulation. In Italy and France, the urban regeneration process responds to national economic policies and municipal strategies at different scales. The processes are managed by planning tools, implemented through strategic guidelines and municipal master plans (PRG for Italy and PLU for France). At a local scale, regeneration interventions are individuated as strategic areas that can be transformed according to a public–private partnership programme.

Table 1 Reference indicators for the evaluation of the case studies

Macro-category	SDG target	SDG indicator	Proposed indicator
Inclusiveness	11.3	11.3.2	1. Percentage impact of direct participation of civil society in urban planning and management
	11.7	11.7.1	2. Surface extension of open space for public use for all
Accessibility	11.2	11.2.1	3. Proportion of population with convenient access (15 min' walk) to public transport
	11.a	11.a.1	4. Implementation of urban and regional development plans based on population projections and resource needs
Sustainability	11.3	11.3.1	5. Land consumption rate
	11.6	11.6.2	6. Adoption of mitigation and adaptation measures to reduce the environmental impact

Each indicator is associated with a SDG 11 Indicator and target, and with a macro-category of analysis

3 Case Studies on Urban Regeneration

Among the most representative examples, we propose an analysis of ex-industrial areas as Italian case studies, and of docks in the French context: *FICO Eataly World* in Bologna, *Porta Nuova* district in Milan, *Euromèditerranèe* in Marseille and *Les Docks* in Saint-Ouen.

3.1 Ex-industrial Area: Bologna, FICO Eataly World

Bologna is a city in central Italy, the first city of the Emilia-Romagna region. Bologna experienced a period of economic and cultural splendour in the late Middle Ages (Dondarini & Borghi, 2014), growing with textile and food goods sector trade (Sicari, 2004). The commercial spirit of the city, combined with a passion for cookery, makes the traditional Bolognese cuisine one of the cornerstones of Italian gastronomy, emphasizing the common perception of San Bologna as the 'city of Italian food' (Crinò, 2020). This perception inspires many projects where the city proposes itself as a new tourist attraction in central Italy, getting over the traditional role as a university city, political and cultural centre. FICO Eataly World, an agri-food park, is an urban project made in 2017 in the outskirts of Bologna that should strengthen the touristic dimension of Bologna, and create new jobs and activities in the metropolitan area.

FICO was built on a public area of about 100,000 m^2 in *San Donato Nuovo,* in the city outskirts, among *Terza Bologna* areas. Bologna is a historically left-wing city: since 1960, the Municipality has accomplished several urban and social policies focused on housing rights, giving rise to the cultural season of reformist town planning in Italy. The reformist model was based on a gradual reform of urban planning, providing a social dimension to the urban policies, as the priority role accorded to social housing, the protection of green areas, the reuse and renovation of the existing building stock (Oliva, 2010). The reform of urban planning was mainly carried out in the outskirts, towards planning tools such as PEEP (Plan for economic and social housing) that offered a significant number of green areas and public services (at least 30 m^2 per inhabitant) and public transport connections with the historic centre. *Terza Bologna* was built between 1964 and 1990, and today holds a good standard of quality of life, commonly recognized by the same inhabitants (Municipality of Bologna, 2020). However, in *San Donato Nuovo* and *Pilastro* neighbourhoods—historic parts of *Terza Bologna*—in recent years, it has been reported a growing demand for new jobs and activities, particularly in the *Pilastro* district. *Pilastro,* which was built in 1966, today is one of the greener suburbs of the city

(*Arboreto del Pilastro* park covers 10 hectares). The district suffers the distance from Bologna's centre, where most of the jobs and cultural activities are located. Besides, social problems and weaknesses in public transport generated critical issues and requests for a greater connection with the city and more employment opportunities. The demand for new jobs, joint with other objectives, such as zero soil consumption and building stock reuse, constitute a priority strategy of the PUG (General urban development plan). Within the PUG, FICO World Eataly represents a touristic pole for the city and a new centrality for *San Donato Nuovo* and *Pilastro* neighbourhoods, through the creation of new public areas and new jobs in the touristic, gastronomic and agricultural sector.

The FICO (Fig. 1) project started in 2013 as an agreement for the redevelopment of a large, dismissed part of the CAAB (Agri-Food Centre of Bologna), a public logistic and trade centre for advanced services dedicated to the agri-food sector. FICO has been designed to respond, with a form of public–private partnership, to the regeneration needs of *San Donato* and *Pilastro* area, through the creation of new public spaces and new jobs in the touristic and agricultural sector. The project occupies one-fifth of the CAAB, born in 1991 as a consortium of cooperatives with participatory processes, typical of the farming sector in the Emilia-Romagna region (Menzani, 2007). The CAAB is a mainly public company, owned by the Municipality and the Chamber of Commerce of Bologna. It covers an area of 500,000 m^2 including structures such as the Vegetables Market, platforms and frozen logistics, offices, retail outlets and related services (CAAB, 2020). In 2013, the Municipality decided to allocate a dismissed part of the Market of 70,000 m^2, and a management centre of 40,000 m^2, to a shared project with *Eataly,* the Italian food distribution company. The project concept was based on the idea to reuse the spaces of Vegetables Market to host a "Citadel of food and sustainability", reconstructing the route of food from "field to table", in the words of Andrea Segrè, the President of FICO Foundation (Garuti, 2017). The concept was shared with the Municipality, the majority shareholder of the CAAB, who approved the project on 1st July 2013 (Metropolitan City of Bologna, 2017). The actual project is quite similar to an ordinary Eataly point, as a commercial centre of typical and luxurious Italian food. Nevertheless, FICO joined this character with an education will. The FICO Foundation, which manages the project with FICO World Eataly, declares that it will provide customers a shared knowledge on the agri-food sector and Italian gastronomy. FICO, beyond the part of the retail trade, offers the discovery of food production chains, from farming to final processing of the food products. The vision which aims the project is based on teaching nutrition issues, using an educational farm

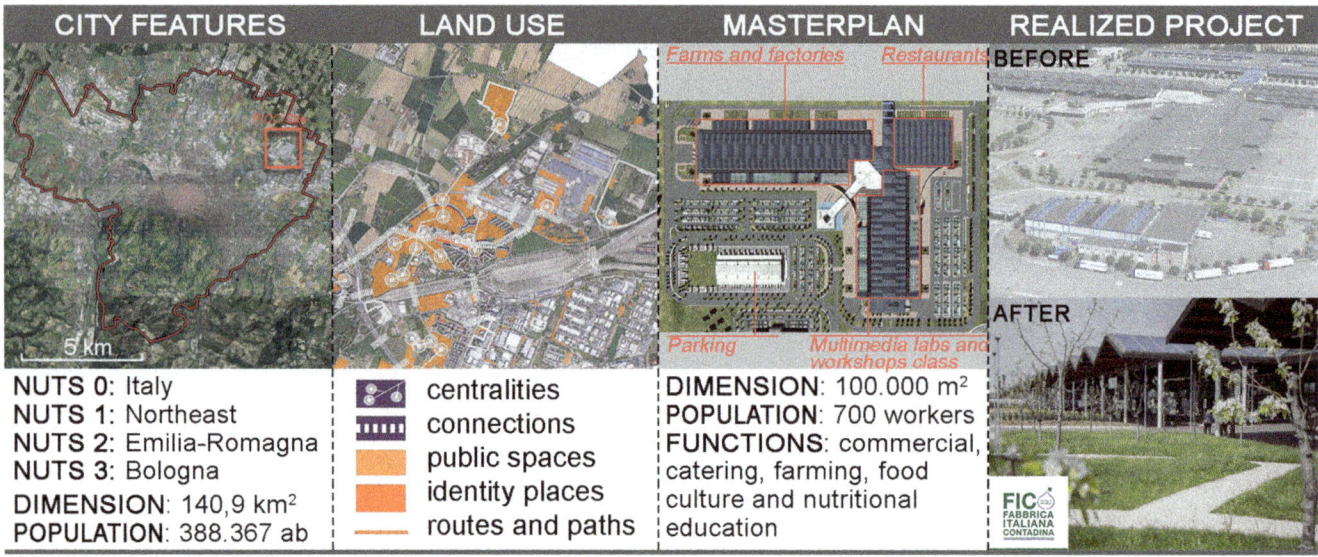

Fig. 1 FICO Project. The sheet shows project's location and land use in the urban context of Bologna with a focus on the Masterplan area (Authors own elaboration, 2020)

and multimedia paths and combining in a single circuit the food cultivation, breeding, processing and consumption. For the production part, the project includes farms, orchards and vegetable gardens, with an enhancement of Italian biodiversity. Regarding the transformation and goods distribution, the structure includes ovens, an oil mill, a brewery and a fishmonger. The educational section consists of practical cooking classes, gastronomic sciences and nutrition science. The aim is to inform visitors both about Italian agricultural offers and on the importance of the Mediterranean diet, endorsing healthy food and a correct lifestyle. A series of dining options allow the consumption of locally produced goods, as well as retail or wholesale. FICO collaborates with the CAAB companies, which supply products and materials for the sale, the educational farm and the food-experience courses; at the same time, FICO advertises companies in the tourism and catering sector, thanks to the international relevance of the project. Professor Segrè said that FICO will prove to be "a formidable promotional tool for Italian food, and the country will have to learn to take advantage even in the export" (Garuti, 2017). FICO constitutes an attraction for the city of Bologna, assuming culinary tourism as a strategic line of city development. To reach this, the project also includes a dedicated bus line, an ecological shuttle service that connects the city centre to the suburbs, as well as a special infrastructure—the 'People Mover'—that links the Bologna Airport to FICO. The complex was also designed respecting environmental sustainability issues, such as zero soil consumption, due to the redevelopment of dismissed areas and energy self-sufficiency, thanks to the cover structure made entirely with solar panels.

3.2 Ex-industrial Area: Milan, Porta Nuova

Milan can be defined as a "sponge city that absorbs and releases, it is the engine of the infinite city", a set of smart city, social city and smart land. It is structured as a city that enhances its features and resources in a continuous process of technological innovation, and social and economic development (Bonomi, 2012; Bonomi & Masiero, 2014). Developed as an industrial hub of Italy, between the XIX and XX centuries, the city grows on the model of two Town Plans. The 1889 Beruto Plan defined an infrastructure system to connect the inner and the external city, guarantee the city's industrial development and improve the city's living conditions. The Pavia-Masera Plan of 1912 instead planned to expand the city prescribing the road network and the buildings construction rules, thus giving the current structure of the city. In both plans, industrial areas were underlined as the key assets for the city's development. However, during the war and in the post-war period, the city underwent a halt in its development like many other European cities (Marseille, Hamburg, Manchester, Liverpool) until it tries to revive its fortunes with the 1952 Town Plan, then with the 1980 Variant to the Town Plan and with the 1984's Loop Municipal project. The urban regeneration begins in this period, with the 1988 "*Director Document for Decommissioned Industrial Areas*", which classified industrial areas, including Pirelli alla Bicocca and the Porta Nuova area as "*Urban industrial sites of predictable disposal and reuse*" (Denti & Mauri, 2000; Oliva, 2002).

In these last twenty years, the city has turned into a metropolis, divided between a historic industrial city and its widespread metropolitan territory. Thereby the municipality redirected the industrial area's regeneration on different guidelines focusing on innovation and creativity, poverty phenomena and hinterland integration. The city, therefore, changed, despite the crisis, during the crisis. The regeneration processes have been promoted by private stakeholders who transformed the city with their managerial skills through public policies, such as the Bicocca, Expo and Porta Nuova (Andreotti, 2019; D'Ovidio, 2009; Gibelli, 2016; Goldestein & Bonfantini, 2007). In particular, *Porta Nuova* is a large-scale urban and architectural regeneration project within the Milan Business Center, the tertiary district that extends from *Porta Garibaldi* train station to *Piazza della Repubblica*, from *Porta Nuova* to *Palazzo Lombardia*. After the disposal of the *Varesine* railway yard, in the 1950s, the area has been highly regarded by all public and private stakeholders of the economic, financial, social and cultural sectors. Thus, they began to develop planning hypotheses for an urban transformation, which were not implemented for legal disputes. Since 1999, Milan Municipality has developed a unitary urban project extended to a territorial area of over 26 hectares, in collaboration with private stakeholders. The operation aims to give vitality to the site, which is divided into three autonomous areas—*Porta Nuova Garibaldi, Porta Nuova Varesine* and *Porta Nuova Isola*—to create a place of great prestige for the entire Lombard community. Its main objective is to mend the three different districts through the business centre and the new public park *Biblioteca degli Alberi*.

The operation starts between 2003 and 2006, with the approval of the Integrated Intervention Plans (P.I.I.) as planning tools: the P.I.I. Varesine by Kohn Pedersen Fox Architects, the P.I.I. Garibaldi-Repubblica, by Pelli Clarke Pelli Associates, and the P.I.I. Isola-Lunetta, by Boeri Studio. All PIIs were presented by Hines Italia, a global real estate investment, development and management firm, founded in 1957, which makes these interventions very similar to the major regeneration interventions in Hamburg, Liverpool and Manchester, (Anselmi & Vicari, 2020; Comune di Milano, 2000, 2004; Dragone, 2007; Molinari & Catella, 2015).

Porta Nuova Isola neighbourhood represents a part of the intervention that seeks to synergistically reconcile the historical characteristics of the area with the most advanced ideas of the green city model, representing an abandoned industrial areas regeneration model, negative effects included. The area is located in the northern part of the Porta Nuova Project and it is a 31,500 m^2 area with residential (22,000 m^2), commercial (850 m^2), offices (6,300 m^2) and leisure (2,360 m^2) land uses. It has been realized by Boeri Studio, between 2006 and 2014. Its historical features are closely connected to the name of the neighbourhood. The name *Isola* (island) derives from the construction of the railway in 1865, which limited connections, dividing and isolating the neighbourhood from the city. At the beginning of the 1900s, it was a working-class neighbourhood, due to its proximity to the factories of Tecnomasio-Brown-Boveri, Pirelli and Helvetica. According to the first Town Plan, the blocks were regular of about 120 × 100 m and the houses included shops and craft shops with warehouses on the ground floor and residences with a balcony or landing on the upper floors, with projects by rationalist architects such as Terragni, Lingeri and Giò Ponti. The innovative feature instead is represented by the urban renewal strategy. It is based on the architectural re-evaluation to develop a neighbourhood that maintains its origins but evolves into a futuristic district, joining and raising many questions on financial, social and environmental sustainability. Office and residential buildings are reducing the solar heat with energy control and production systems. The most representative residences building is the *Bosco Verticale* by Boeri Studio, which is a new green architecture prototype building. The "vertical forest" is the main character, with condominium maintenance independent from the tenants. This is a metropolitan forestation project that allows to regenerate the environment and urban biodiversity without implying an expansion of the city in the territory (Distretto Isola, 2020; Ordine e Fondazione degli architetti, pianificatori, paesaggisti e conservatori della provincia di Milano, 2015; Porta Nuova, 2013; Stefano Boeri Architetti, 2014). Nowadays, the Isola district area is served by several bus and tram lines, underground lines and stations, suburban railway lines and regional, national, international lines and Malpensa Express in the Porta Garibaldi area. The new connections aim to retrain the negative aspect that gave the neighbourhood its name into a positive one, making it one of the most connected neighbourhoods of the city. Therefore, all the area of Porta Nuova Project has been individuated as one of the fourteen "Unitary districts for commerce", due to its financial, commercial and real estate influence and attractiveness (Bianchini, 2019; Unione Confcommercio imprese per l'Italia, 2020). As result, the project defines a regeneration process to achieve great urban transformations by supporting and promoting private stakeholders' initiatives. This aspect brings negative effects too since the price to pay for the economic attractiveness is the increase in housing rental costs and all the district activities. Thirdly, the results draw attention to the issue of environmental sustainability, realized through prototype buildings (Bosco Verticale) which, despite countless criticisms of its feasibility in the context, today represent the landmark of the entire urban planning and real estate operation (Fig. 2).

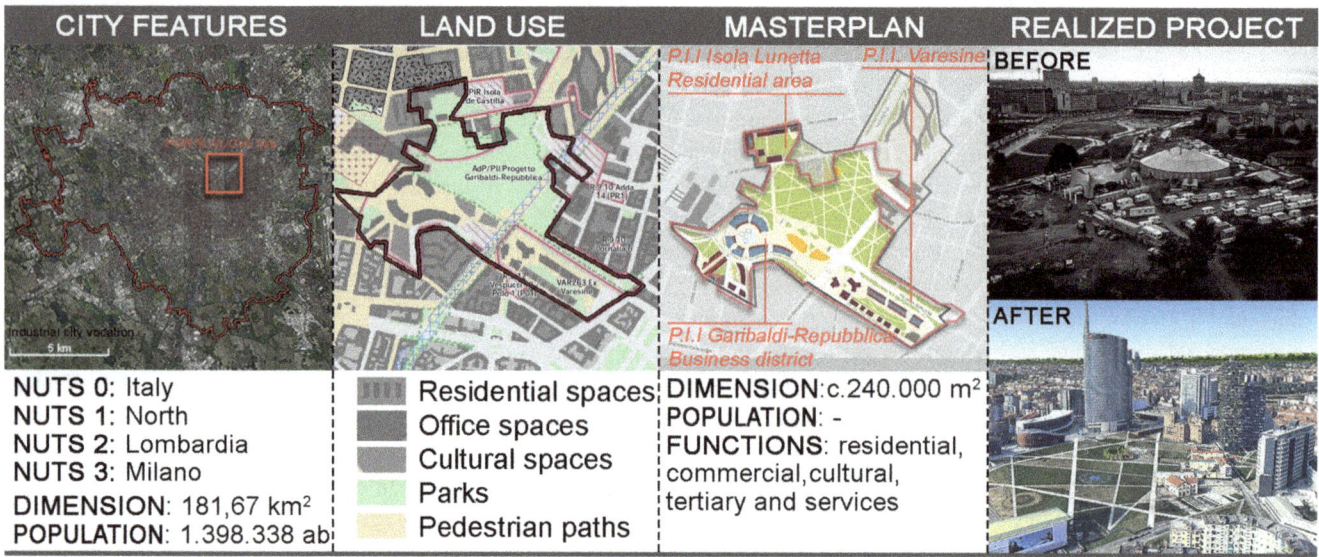

Fig. 2 Porta Nuova. The sheet shows project's location and land use in the urban context of Milan, with a focus on the Masterplan area (Authors own elaboration, 2020)

3.3 Docks: Marseille, Euroméditerranée

Marseille, a strategic historical pole of the Mediterranean with a deeply rooted maritime tradition, is the first French port for size and flows and the second city for inhabitants (826,700) (INSEE, 2013) of southern France. We can determine that four main areas give identity to the city: the old town with its Vieux-Port, the popular disadvantaged districts to the north, the south with its high-class residential neighbourhoods and its eastern part, where the interventions of industrial rehabilitation concentrate. From the administrative point of view, the city is divided into 16 arrondissements that group 111 neighbourhoods (AGAM, 2020). The city model of Marseille is a virtuous example of how to invest in the culture to give a new and attractive image of a city and how creativity can be a leading force in the sustainable regeneration of waterfronts and port areas.

The transformation process of Marseille began at the end of 1999 with the design competitions announced by the Ministry of Culture, under the name of "Operation of National Relevance". The administration has chosen to use contemporary architecture as a vehicle to promote the new image of the city, by choosing some of the most prominent studios on the international scene as the ones of Zaha Hadid or Kengo Kuma. Since 2000, Marseilles has been associated with the 18 neighbouring municipalities to form the Communauté urbaine Marseille Provence Métropole, recognizing the Grande Marseille as an urban community (Esposito, 2011). The human settlement in this coastal area can boast a history of several thousand years and today extends over fifty-seven kilometres. In 2013, Marseille Provence became the capital of culture receiving funds to rehabilitate the city with the project Marseille Euroméditerranée (Euroméditerranée, 2020). The project has been linked to the improvement of cultural infrastructure throughout the territory including new locations, restoration projects in pre-existing industrial and commercial areas, as well as numerous renovations and extensions of existing facilities. The chosen strategy rejects the need for punctual interventions, focusing on a system of widespread quality, more suited to the specific reality of the city. Many architectural interventions were realized with different languages and functions, allowing better continuity to the urban system.

The area of the project covers 480 hectares with an increase in housing of 18,000 units, a million square metres dedicated to offices, 200,000 square metres for trade, as well as for public equipment. The green space covers 60 hectares and the entire project involves an investment of seven billion euros with a return of 35,000 new employees and 38,000 inhabitants (McAteer et al., 2014). Mobility remains closely linked to its maritime tradition as the development of the urban route stretches towards the sea, with its modern buildings serving the Port and showing off different historical phases, distinguishing the value of an improved port–city interface. The project "Marseille-Euroméditerranée" had a clear European dimension through the theme Partage des Midis (sharing the south), which crossed the entire cultural programme, highlighting the historical role of the city as a meeting point between European and Mediterranean cultures. Thanks to this high-profile international dimension of the project, public funds were supplemented by large amounts of private sector sponsorship, which sought to build on the urban renewal in the heart of Marseille, while promoting greater integration in the surrounding area. Marseille

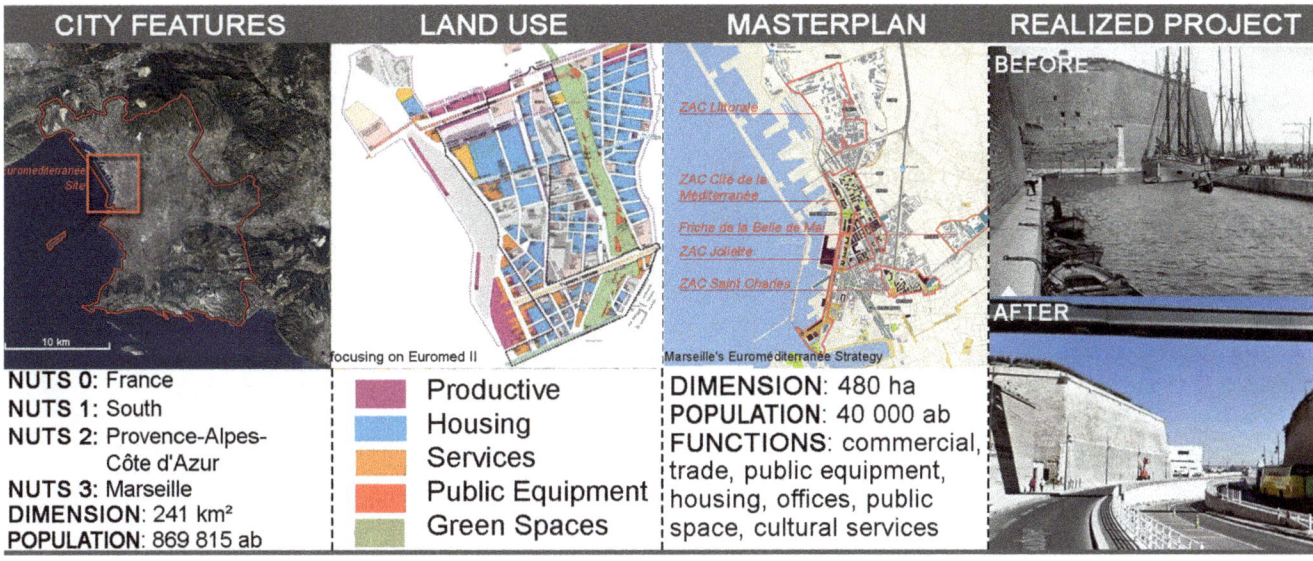

Fig. 3 Euroméditerranée. The sheet shows project's location and land use in Marseille, with a focus on the Masterplan area and a focus on the land use of Euroméditerranée II (Authors own elaboration, 2020)

Provence has surpassed its private sponsorship target, with €16.5 million raised by 207 companies (McAteer et al., 2014). This success reflects a consistent strategy to create partnerships with corporate sponsors and an understanding of the need to generate mutual benefits. One of the main goals of the project was to cancel the great social and geographical contrast between the North and the South of the city, (Buslacchi, 2013) an attempt already put in place at different times in Marseille (Anselme & Peraldi, 1990; Tarrius, 1987; Roncayolo, 1993; Donzel, 1998; Ascaride, 2001), actively involving locals and artists, called to live and reinvent the degraded areas to make it the aggregating heart of the city. From the economic point of view, the project has created several impacts, especially in terms of total cultural audiences and the increase in the number of tourist visits.

Among the different operations, the "Friche de la Belle de Mai" in the east area of the city is interesting to mention. This area is comparable in dimension to a real neighbourhood. In 1860, the factory, then located in Rue Sainte, near the Vieux-Port, was the city's first employer and the second manufactory of France after Paris. The factory has undergone various transformations and extensions linked to cigarette consumption and the evolution of production modes. The transformation process begins in 1992 when the twelve hectares inside the district are acquired by the Société Française d'Assainissement (French Sanitation Company) from the private company SEITA (La Friche, 2020). From 1992 to 1997, thanks to the inclusion of the area in the scope of the Euroméditerranée project, a commission was appointed (to which architect Jean Nouvel was added) which has the task of developing an overall plan of rehabilitation.

The rehabilitation aimed to create a local and international centre for artists and the locals, which will improve the quality of the city's social spaces. From a conceptual point of view, the operation proposal is based on education and awareness of cultural activities as an opportunity for the local territory. The former factory covers an area of 12 hectares with a centre of 45,000 square metres area. One of the most catching elements of the project is the relationship between the activities of the association Friche and the work carried out by the Commission chaired by Jean Nouvel. The Commission has focused on the artistic presence as a driving agent for the urban development of the district. Accessibility of the public space was not only a key point, but also the value of the *mixité* of the area, the relationship between public and private and the definition of a programme for sports and cultural activities to be carried out over time. In this specific case, the Commission played the role of facilitator and intermediary between the Friche project and the Public Administration. The majority of the estate belongs to the City of Marseille which has agreed with the association a free lease of the spaces, on a condition to carry out activities for the neighbourhood. (Fig. 3).

3.4 Docks: Saint-Ouen (Paris), Les Docks de Saint-Ouen

The *Les Docks de Saint-Ouen* (Fig. 4) is an *EcoQuartier* under construction in Saint-Ouen-sur-Sein, a municipality north of Paris of 51,108 inhabitants. The plan is located on the old logistic site of the fluvial docks on the Seine, a brownfield previously occupied by industrial buildings. As

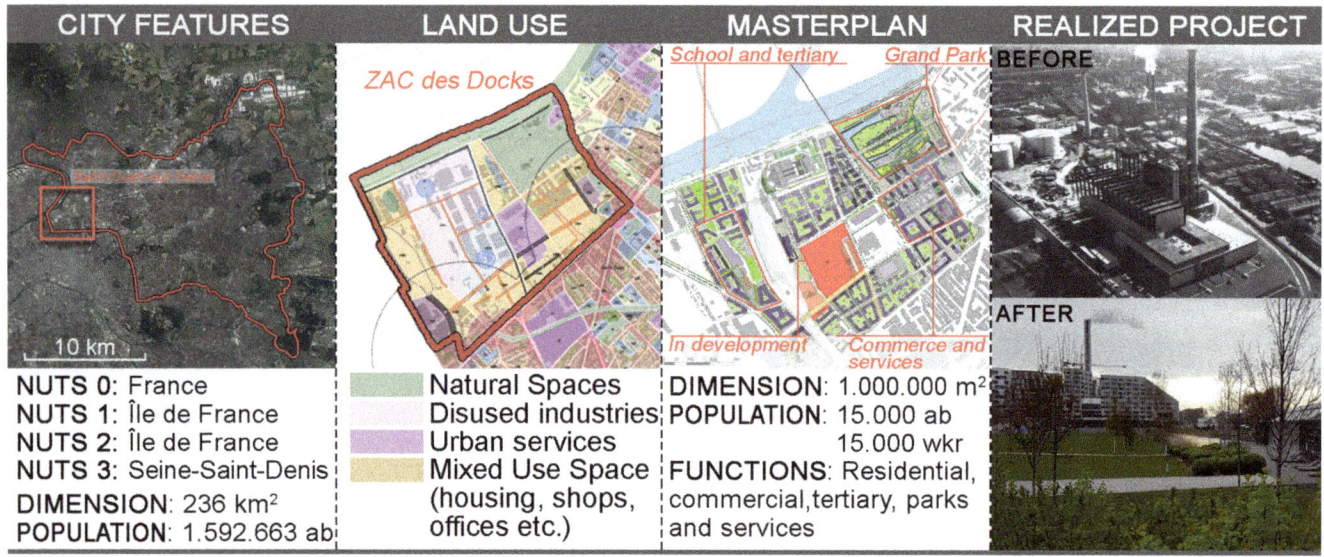

Fig. 4 Les Docks de Saint-Ouen. The sheet shows project's location and land use in the urban context of Saint-Ouen with a focus on the Masterplan area (Authors own elaboration, 2020)

in many places in the metropolitan area of the *Grand Paris*, Saint-Ouen is a place of profound transformation conceived and implemented under the supervision of the Ministry of Ecological and Solidarity Transition. Public domain agencies and private agencies participated in the project.

The programme started in 2004 to regenerate the areas of *Total* and *Alstom* factories and transform them into an eco-neighbourhood. Saint-Ouen has a long industrial tradition, but it clashed with the deep conversion of the labour market, which brought visible effects on the economic and demographic decrease (Commune de Saint-Ouen, 2017; Guironnet, 2017). The culture of the *banlieue rouge* (red suburb), strictly connected with the principles of communism, lapsed when national spatial planning policies encouraged deindustrialization, until the 1970s, with the emersion of new liberal logic of land profitability (Albecker, 2015; Séquano Aménagement, 2020). This transition culminated at the beginning of the twentieth century when the last remaining industries left the site, *Total* in 2003 and *Alstom* in 2004, leading to the definitive abandonment of the area and the decision to create an urban regeneration plan. In 2007, this programme took shape through the development of a *Zone d'Aménagement Concerté* (ZAC), born through the first feasibility studies and the involvement of the stakeholders in a partnership between Paris and Saint-Ouen municipalities. In 2010, the *Plan local d'urbanisme* (PLU), or local urban plan, was revised and updated several times until 2017, while the *Plaine Commune* was deposited in 2011, starting the first excavations. The first phase of construction, which included 2,000 houses, the *Grand Parc* and school facilities, took place between 2012 and 2016. The second phase started in 2015, including housing, offices,

shops and more, and ended in 2019. In the same year, the third and final construction phase, for the completion of the infrastructures by 2025, began.

The district develops on a quadrangular plan of about 100 ha, of which 12 ha are destined for the local park "*Le Grand Parc de Saint-Ouen*". Within the eight project sectors, it is planned to establish 15,000 new inhabitants and create 15,000 new jobs through the construction of 878,600 m^2 of built-up areas destined for various activities, with an intervention budget of €666 million. The project includes 443,000 m^2 of residences (about 50% of the total and comprehensive of 40% of social housing), 300,000 m^2 of offices and economic activities (34%), 68,000 m^2 of commercial structures (8%), 52,000 m^2 of collective structures (6%), and 15,600 m^2 of public structures (2%) (Ekopolis, 2020).

The active participation of the stakeholders (local elected representatives, developers, inhabitants, associations) allowed the structuring of an advanced plan able to satisfy requests from different directions. The city of Saint-Ouen and project management teams used two tools to engage citizens in the local planning process: public meetings and the Docks website. Citizens presented themselves in public meetings through the *Mon voisin des Docks* association, created in March 2014, fuelling the debate on structures and services, which continued from 2017 through telematic meetings on the Docks website. In addition to housing, central was the idea of creating new jobs for residents, preventing them from living in the neighbourhood only when they came home or went out to work in central Paris or in other peripheral areas of the capital, and configuring the intervention as a fully pedestrian-friendly neighbourhood.

The short distance (<1 km) from the two main existing public transport hubs of RER C and line 13, with the new addition of line 14, ten bus lines and modal shift incentives are aimed at developing a sustainable mobility model that makes the neighbourhood accessible and directly connected to a large part of the city without using the car. Water is a constant feature in outdoor public spaces. The high concentration of infrastructures close to the Seine has prevented urban development towards the river over the years, not only marking a great production opportunity, but also a strong break with the territory. The neighbourhood is planned as a natural extension of the river network, including a water canalization system for phytoremediation. Stormwater management promotes drainage and treatment, according to the Integrated Urban Water Management approach, with bioswales along the roads and green roofs on buildings, moving the runoff to permeable surfaces or the Seine, and reducing discharge into underground municipal sewage conduits. Part of the rainwater is collected in two storage basins, respectively of 3,000 and 10,000 m^3, treated and reused for irrigation. Furthermore, riparian wetlands serve as a transit corridor for the fauna along the Seine, protecting biodiversity (also insects and mosquitoes findable in these places). Inside the park, the greenhouse and large urban gardens for the cultivation of vegetables (about 5,000 m^2) provide food for the inhabitants and become a place of multiform activities. The *Île des partages* (Island of sharing), is a protected area for pesticide-free cultivation that offers space for family gardens (former Alstom workers' gardens created in 1928), shared gardens (for local citizens), educational gardens (with pedagogical activities for the children of Le Petit Prince school), therapeutic gardens (for people with a handicap or illness) and integration gardens (for the reintegration of people in social or occupational difficulties) (Toura, 2019). In Saint-Ouen, physical and mental wellbeing takes on the same relevance: an inter-partnership cooperation platform has been created with the *Centre Local de Santé Mentale* to identify and help young people, the elderly over 65, young mothers, parents at home, single-parent families, unemployed, to offer them the opportunity to confront other people and talk about their daily lives (Brisse, 2017). Finally, energy resources management takes place through production from renewable sources. A district heating network provides heat and domestic hot water generated through heat recovery from Seine's waters, steam from local production processes and waste incineration. Refurbishment of former industrial complex is realized with the project of the *Etoile verte*, an old disused chimney which will be an integral part of a future double filtration system for energy production from biomasses in 2021, flanked by 5,800 m^2 of vegetated land, 7,700 m^2 of green roofs and 600 new trees planted. The Docks operation was designated "New Urban District" by the Region in 2009, obtained ISO 14001 certification (environmental management) in 2012 and won the Ile-de-France programme "100 innovative and ecological neighbourhoods" in 2016. The second phase of development, which involves the completion of line 14, the development of a cultural and commercial centre that extends to the rehabilitated Alstom room, is underway. This programme is supported by the installation of the hotel in the Ile-de-France region and by the development of an important pole, mainly tertiary, west of the ZAC. Actually, 2,920 houses, the entire Grand Parc and various public structures such as a school complex, a gym, a nursery, 74,000 m^2 of offices and 2,300 parking places are completed (Séquano Aménagement, 2020).

4 Discussion

Each city, in its own way, has benefited from brownfield regeneration:

- Bologna tries to become a new touristic pole in central Italy, choosing to reuse the spaces of a logistic and agri-food centre to create the first theme park on healthy food and nutrition education.
- Milan, as a financial district, focuses heavily on the technological-environmental sustainability of buildings, aiming to reduce the heat island and the energy needs in the city and to realize a prime neighbourhood.
- Marseille, as a creative city, encourages wider participation in culture through free and public events, specific events for young people and activities taking place in disadvantaged neighbourhoods or showing the diversity of cultures in the territory.
- Saint-Ouen is implementing a path of economic development and repopulation of the city through an urban intervention characterized by inclusiveness and a variety of public spaces, ecological management of natural sources, wastes, mobility and renewable energy.
  Positive aspects have been recognized, as the attention to the climate change issues, the zero-consumption soil goal and the wealth produced by tourism, according to the Agenda 2030:
- Innovative design approach. The futuristic approach to the city realized some international landmarks. In Milan, the *vertical urban forest* is the prototype to bring all over the world to promote urban forestation. In Marseille, *les docks* represent a modern approach to historical buildings, transforming a huge industrial area into a multicultural dynamic living neighbourhood.
- Social gathering. In all the case studies, due to the economic and cultural activities, the opportunity for the meeting of the social classes is enhanced through the public spaces.

- Economic attractiveness. The private companies' investments in the regeneration process have promoted the development of an autonomous economic system related to different sectors—from food tourism to tertiary activities, as in Milan and Marseille. Therefore, the private funds allowed to achieve faster and better interventions providing 700 new different job positions in the outskirts, as in Bologna.
- Land reuse and reclamation. The attention to environmental sustainability is the core of all the regeneration processes analyzed. In particular, the predominant aspect is the attention to minimize the soil consumption, against the current high trend in Milan, Bologna and Marseille. Negative aspects can be traced along with all these processes, divided into the morphological, social, economic and environmental dimensions. They share problems related to the transformation of a circumscribed area, such as brownfields, influencing the entire urban context:
- Outsized-out of scale- self-referred morphology. The oversized scale of intervention, as in Milan, Bologna and Saint-Ouen, clashes over the surrounding area and sometimes appear in a self-referential stylistic form. In Saint-Ouen, the urban plot dispersity has led to petty crime spread. Also, in absence of a stratification process over time, situations of rejection by citizens often occur.
- Social inequalities. The marked distinction between city districts, as in Marseille and Milan, has been increased with the transformation. Access disparities in housing policies, health, education and transport, are not cancelled but strengthened despite the regeneration. The plan benefits only a narrow segment of citizens and homelessness is constantly increasing, despite the high number of social housing.
- Gentrification. The incomplete partnership between public and private stakeholders, as in Bologna, Milan and Marseille, has generated projects that appear unbalanced in favour of an entrepreneurial vision. In both cases, the aspects of community development seem to be faded in the background. This led to the gentrification issues due to the change of the surrounding living costs, caused by the realization of prime housing, slowly moving away from the regeneration process target, that is the middle-working class.
- Environmental ordinary standards. Despite the environmental aspects are the core of the projects, not all the cases have promoted significant environmental standards.

The strengths and weaknesses of the case studies, declined in the social, economic and environmental dimensions, highlight the points of contact that can be collected in three common issues, such as inclusiveness, accessibility and sustainability, as proposed by SDG 11 (Table 1).

As regards inclusivity, it is understood as the active involvement of all citizens, in all its forms, at different levels in the regeneration processes, and it is directly connected with the presence of public spaces.

While in the French cases, active participation is a central theme of the intervention, in the Italian cases, the aspect is less relevant. Despite FICO was realized from a collaboration between public and private stakeholders, the project has a marked entrepreneurial vision. The project is mainly targeted at a paying clientele, and this prevents spontaneous participation. In Milan, the lack of commitment between experts and citizens has strengthened the financialization of the project's results excluding the citizens' role. In Saint-Ouen the constant commitment between experts and citizens has encouraged social resilience. The urban renewal process development allows redefining roles, establishing the citizens as participants of equal dignity. In the case of Marseille, the constant communication between citizens and administrations in a participatory development model has allowed the balancing of private interests with individual needs. Also, the local cultural planning strategy offered the opportunity to transform local structures and social ties into "third places", offering new services and uses to the population.

The involvement of citizens is also related to the use of public spaces, a very important element in most of the cases studied. In Fico, 0.5 ha is dedicated to green areas, but mainly intended for private use or linked to the needs of the agri-food park (fruit and vegetable cultivation, educational gardens). In Milan, 10 ha are dedicated to public spaces and pedestrian paths. In particular, the new urban park *Biblioteca degli alberi* and *Gae Aulenti* square became new public centralities for the entire neighbourhood. In Saint-Ouen, 12 ha are occupied by *Le Grand Parc de Saint-Ouen*, which is a reference point for the entire neighbourhood as green infrastructure and which also offers various activities related to integration (see *Île des partages*). In Marseille, one of the main aspects of the Euromediterraneé II plan, in extension to the recovery of the port of the first phase, is that of providing equipped spaces and public green areas. The equipped spaces area is 60 ha on 480 ha, in total. Located in the north of the city, the Aygalades envisages a total covered area of approximately 14 ha by 2025 as an extension of the Parc François Billoux.

As far as accessibility is concerned, it is to be understood as the project connected the whole city. In all the projects, public transport is highly valued, especially as a link to the city.

There is an eco-bus service which connects the city centre and the railway station of Bologna with FICO. Furthermore, is shortly expected the realization of a new public mobility service, the People Mover, which should directly connect the airport with FICO, like a monorail service. In the other cases, there was an approach to the "15 min city". In Milan, the district is accessible to everyone from everywhere. In less than 15 min by foot, citizens can reach the main public transport hubs: five metropolitan line stops (on three lines), three train stations and the tramlines. The same approach is valued in Saint-Ouen, there are not any kind of barriers and child safety is guaranteed by avoiding the cars' circulation or limiting their speed (30 km/h). In less than 15 min by foot and 5 min by bike, citizens can reach the main public transport hubs: three metropolitan line stops (on two lines) and ten bus stops. Marseille changed from the infrastructural nets and mobility (TGV, new subway lines and tram lines, waterways, tunnels, parking areas under all the squares of the city centre, underground rerouting of motorways) up to the delocalization of the harbour equipment (merchant and tourist), to the arrangement of the Old Port and the entire urban waterfront. The main objective was to abolish the physical and psychological barrier between the city and its seaport.

In general, all the case studies were developed according to the urban and regional development plans guidelines. The interventions join a network of interventions, promoting general accessibility to the rest of the city. FICO enhanced the third urban strategy, *attractiveness & work*, representing a strategic hub, and creating 700 new jobs and an economic induced for local production. Porta Nuova Project accords to the city's industrial and infrastructure development. The transformation guaranteed a high-density settlement, with respect to soil consumption and it increased the public spaces areas, lining up to the new Milan Vision 2030. *Les Docks des Saint-Ouen* realize the PLU inclusion of social and cultural *mixité*, respecting ecological principles. In Marseille, the *Euromediterraneè* has been able to track the changes in the "urban thinking" of recent decades. The project creates a city, certainly tertiary, but first marked by the diversity of its functions.

As far as sustainability is concerned, it is primarily to be understood as environmental sustainability and how projects affect the urban ecosystem with deep attention to climate mitigation and adaptations solutions.

The *project FICO* envisages a total of 10 ha area. Among these, 7 ha are covered by building, 0.5 ha with green areas for gardens and farming and 2.5 ha are covered by parking lots, with around 85% of land rate consumption. It includes advanced technology to reduce energy consumption: magnetic levitation refrigeration units have been used which are characterized by high yields at partial loads, and a large photovoltaic system is installed on the roof, covering more than 100,000 sqm, producing 15 million kWh. The provided energy is used for an efficient LED lighting system, with an advanced regulation system, which optimizes energy consumption based on real requests. The project—*Porta Nuova*

garibaldi Repubblica, Isola and Varesine—envisages a total 24 ha area. Among these, 10 are covered by building, with around 40% of land rate consumption. The project guarantees sustainable energy production, mitigating greenhouse gas (GHG) emissions and solar heat. Especially the *Bosco Verticale* has started a transition to reduce the CO_2 emission and fine dust by planting around 800 trees of different sizes—equivalent to that of 30,000 square meters of forest and undergrowth, concentrated on 3,000 square meters of urban surface. Irrigation is also centralized: the needs of the plants are monitored by a remote digitally controlled probe system and the water is drawn from the filtering of the grey drains of the towers. In *Saint-Ouen*, the project envisages a total covered area of approximately 18 ha by 2025. In parallel, 8 ha of land occupied by *Alstom,* with asbestos in soil, were decontaminated along with the demolition of 50,000 m^2 of workshops. Approximately, 60–75% of energy district demand is covered by renewable sources, moving to more sustainable production and mitigating GHG emissions. The *Syctom* waste treatment plant has started a transition to reduce the CO_2 from the incineration system and to produce biomaterials (bioplastic, biofuel), together with the planting of 600 trees. The adaptive research for eco-urbanism is also reflected in the integration of the urban water cycle in open spaces and in buildings, with particular attention to wastewater management. In Marseille, the project is made up of a perimeter of 310 ha. The city recovers places of memory, integrating social, economic and cultural objectives, as well as environmental rehabilitation, pursued through the transformation of hectares of industrial areas and volumes. The Euro-Mediterranean eco-city focuses on bioclimatic design to exploit natural resources, in order to produce houses that are energy efficient. In 2016, the first geothermal power plant was inaugurated to feed the Docks area.

To summarize the various aspects of the different case studies, it is proposed a table of synthesis and analysis (Table 2).

5 Conclusions

The paper's primary proposition is to analyze some relevant case studies of urban regeneration reusing brownfield sites, in order to understand how urban processes of functional recovery worked in major European metropolitan cities. The paper seeks to individuate common and innovative analysis issues, referring and extending the study both to the project's authors and to all stakeholders and shareholders directly or indirectly involved. From the selection of the case studies, it has been possible to define recurring types of dismissed industrial areas and different regeneration models emerged with the common goal of redefining the image of the city in a

Table 2 Comparison between the case studies in relation to the Issues and the indicators described in paragraph 2. Methodology (Authors own elaboration)

Issues	Indicator	FICO Eataly World 10 ha	Porta Nuova 24 ha	Euroméditerranée I e II 480 ha	Les Docks de Saint-Ouen 100 ha
Inclusiveness	1	General lack of participation: the projects have been designed without considering or taking little account of the needs of Pilastro and San Donato Nuovo population	Lack of participation of citizens, because of the private property of land and the real estate company management	Active involvement of locals and artists called to live and reinvent the degraded areas of the Vieux-Port. Process open to all local residents, political representatives and civic associations	Direct participation of citizens through meetings, also online, open to all local residents, political representatives and civic associations
	2	0.5 ha	10 ha of public green spaces	60 ha	12 ha of public green spaces
Accessibility	3[*]	90%	95–100%	95–100%	95–100%
	4	The project respects the need for regional plans and laws (Regional Law 24/2017)	New urban functions and sustainability issues according to the PGT (town plan) vision of Milano 2030	The main objective was to abolish the physical barrier between the city and its seaport, through its infrastructural net and mobility	Only strategic goals are included in the 2015 revision of the PLU
Sustainability	5[**]	85%	40%	35%[*]	15–20%
	6	Solar energy Water reuse system	Renewable energy Vertical urban forestation Mitigation of the solar heat with energy control and production systems	Renewable energy Reduction in GHG emissions and 65% reduction in water using thermo refrigerating pumps and high energy-efficient refrigerated units	Mitigation of GHG emissions from industrial processes and waste management Adaptation in rainwater and wastewater management, and promotion of renewable energy

[*] The percentage derives from the *15 min city approach*
[**] The percentage of land use is calculated by estimating the soil coverage of the new building projects.

sustainable key, with different economic development guidelines. The four case studies show how industrial areas are fragile territories that frequently clash with a necessity of change, imposed by productive and economic trends but hampered by the citizen's identity will and needs. In fact, both physically and symbolically, there is still a gap between top-down and bottom-up regeneration initiatives, which have almost never successfully linked the needs and the wills of the stakeholders and shareholders involved. In conclusion, these projects are innovative and sustainable as advertised by their authors, but nevertheless through their weaknesses. It is to us to underline the necessity to re-think the brownfield regeneration process in Italy and France, like in similar European cities, embracing aspects of social inclusion, local communities' involvement and gentrification restraint. The new city model should be oriented on Inclusiveness, Accessibility and Sustainability, looking at the opportunities available in its lands and its heritage.

References

AGAM (2020). *Agence d'Urbanism de l'Agglomeration Marsellaise.* https://www.agam.org/data/

Albecker, M. F. (2015). *Banlieues françaises/la banlieue parisienne, périphérie réinvestie?* http://www.revue-urbanites.fr/la-banlieue-parisienne-peripherie-reinvestie

Andreotti, A. (2019). *Governare Milano nel nuovo millennio.* Il Mulino.

Anselmi, G., & Vicari, S. (2020). Milan makes it to the big leagues: A financialized growth machine at work. *European Urban and Regional Studies, 27*(2), 106–124. https://doi.org/10.1177/0969776419860871

Anselme, M., & Peraldi, M. (1990). Marseille et ses soeurs: notes sur ladynamique urbaine de quelques métropoles méditerranéennes, Aix-en-Provence: Cerfise.

Ascarides, G., Condro, S. (2001). La ville précaire: the isolés in the center-ville de Marseille. LA VILLE PRÉCAIRE, 1-288. ISBN: 9782296269446, https://www.digital.casalini.it/9782296269446

Bianchini, R. (2019). *Il Bosco Verticale di Boeri: da fenomeno ad archetipo?* Inexhibit. https://www.inexhibit.com

Bonomi, A. (2012). *Milano. Le tre città che stanno in una.* Bruno Mondadori.

Bonomi, A. & Masiero, R. (2014). *Dalla smart city alla smart land.* Marsilio.

Brisse, M. (2017). La prise en compte de la santé mentale dans la dynamique des projets de renouvellement urbain. Étude de cas sur Saint-Ouen et L'Ile-Saint-Denis. *Mémoire de Master 2 APTER* (pp. 79–107).

Buslacchi, M. E. (2013). *Cambio. Anno III*, Numero 6.

CAAB. (2020). *What is CAAB.* https://www.caab.it/en/what-is-caab-3/

Commune de Saint-Ouen. (2017). *Plan local d'urbanisme—Saint Ouen/1. Rapport de présentation.* Saint-Ouen. https://www.saint-ouen.fr/services-infos-pratiques/urbanisme/145-plan-local-d-urbanisme-plu-plui.html

Comune di Milano. (2000). *Ricostruire la grande Milano.* https://www.comune.milano.it/

Comune di Milano. (2004). *Accordo di Programma Progetto Garibaldi Repubblica.* https://www.comune.milano.it/

Crinò, L. (2020). *Bologna la Saporita: viaggio nella città del cibo.* La Repubblica. https://www.repubblica.it/sapori/2020/01/30/news/itinerario_bologna_food_town_2020-246500856/

Czischke, D., Moloney, C. & Turcu, C. (2015). Setting the scene: raising the game in environmentally sustainable urban regeneration. In URBACTII, *Sustainable Regeneration in urban areas, Urbact II capitalisation April 2015*, 6–15. https://urbact.eu/sites/default/files/04_sustreg-web.pdf

D'Ovidio, M. (2009). Milano, città duale. In C. Ranci (Ed.), *I Limiti Sociali della Crescita: Milano e le Città d'Europa, tra Competitività e Disuguaglianze* (pp. 9–72). Maggioli.

Denti, G. & Mauri, A. (2000). *Milano, l'ambiente, il territorio, la città.* Alinea.

Distretto Isola. (2020). *L'Isola.* http://www.distrettoisola.it/lisola/

Dondarini, R. & Borghi, B. (2014). *Bologna. Storia, volti e patrimoni di una comunità millenaria.* Minerva Edizioni.

Donnarumma, G. (2013). Il fenomeno della dismissione dell'edilizia industriale e le potenzialità di recupero e riconversione funzionale. In *Proceedings of the V International Conference on History of Engineering* (pp. 1345–1361). Naples. ISBN: 9788887479805.

Donzel, A. (1998). *Marseille, l'expérience de la cité.* Anthropos.

Dragone, R. (2007). *Milano: nasce l'area Porta Nuova. Il maxi intervento che riunifica le aree Garibaldi, Varesine e Isola.* Archiportale. https://www.archiportale.com/news/2007/06/architettura/milano-nasce-l-area-porta-nuova_10034_3.html

EEA. (2019). *Land and soil in Europe. Why we need to use these vital and finite resources sustainably.* Publications Office of the European Union.

Ekopolis. (2020). https://www.ekopolis.fr/operation-amenagement/ecoquartier-des-docks#target-plus-loin

Esposito, G. (2011). Marsiglia: Città Euromediterranea tra storia, multiculturalismo e innovazione. In M. Clemente (Ed.), *Città dal Mare, l'arte di navigare e l'arte di costruire la città* (pp. 169–176). Editoriale Scientifica s.r.l.

Euromediterranée. (2020). *Tous les projets.* https://www.euromediterranee.fr/projis

Garuti, M. (2017). *Fico a Bologna: l'intervista ad Andrea Segrè, ideatore del progetto.* Il Giornale del Cibo. https://www.ilgiornaledelcibo.it/fico-bologna-apertura/

Gehl, J. (2020). *Public space & public life during COVID-19.* https://gehlpeople.com/announcement/public-space-public-life-during-covid-19/

Gibelli, M. C. (2016). Milano: da metropoli fordista a mecca del "real estate". *Meridiana*, *85*, 61–80. https://www.jstor.org/stable/43840171?seq=1

Goldstein, M. B., & Bonfantini, B. (2007). *Milano incompiuta: interpretazioni urbanistiche del mutamento.* Franco Angeli.

Guironnet, A. (2017). *La financiarisation du capitalisme urbain: Marchés immobiliers tertiaires et politiques de développement urbain dans le Grand Paris et le Grand Lyon, les projets des Docks de Saint-Ouen et du Carré de Soie.* Doctoral dissertation, Université Paris-Est.

Honey-Rosés, J., Anguelovski, I., Chireh, V., Daher, C., Konijnendijk van den Bosch, C., Litt, J., Mawani, V., McCall, M., Orellana, A., Oscilowicz, E., Sánchez-Sepúlveda, H., Senbel, M., Tan, X., Villagomez, E., Zapata, O., & Nieuwenhuijsen, M. (2020) 31 July 2020. The impact of COVID-19 on public space: An early review of the emerging questions—design, perceptions and inequities. *Cities & Health.* https://doi.org/10.1080/23748834.2020.1780074

INSEE (Institut national de la statistique et dec études économiques). (2013). http://www.insee.fr/en/default.asp

La Friche. (2020). *La Friche La Belle de mai.* http://www.lafriche.org/fr/agenda

Iachello, E., & Roncayolo, M. (1993). The image of Marseille. Port, ville, pôle, Marseille, Chambre de Commerce et d'Industrie de Marseille. Tome V: History of commerce and industry of Marseille, XIXe-XXe siècles, 1990, III 368 p. Annales. *Histoire, Sciences Sociales, 48*(4), 974–976. https://doi.org/10.1017/S0395264900060182

McAteer, N., Rampton, J., France, J., Tajtáková, M., & Lehouelleur, S. (2014). *Ex-post evaluation of the 2013 European capitals of culture.* Publications Office of the European Union.

Menzani, T. (2007). *La cooperazione in Emilia-Romagna. Dalla Resistenza alla svolta degli anni Settanta.* Il Mulino.

Molinari, L. & Catella, K. R. (2015). *Milano Porta Nuova: l'Italia si alza.* Skira.

Metropolitan City of Bologna. (2017). *Progetto FICo—Accordo di Programma per la realizzazione del progetto.* https://www.cittametropolitana.bo.it/pianificazione/Pianificazione_del_territorio/Accordi_di_Programma/

Municipality of Bologna. (2020). *PUG—Piano Urbanistico Generale: Proposta di Piano febbraio 2020—Leggere il Piano.* Comune di Bologna.

OECD. (2012). *Redefining "Urban": A new way to measure metropolitan areas.* OECD Publishing.

OECD. (2013). *Third ministerial meeting of the territorial Development policy committee, Summary Report.* https://www.oecd.org/cfe/regionaldevelopment/Summary-Aix-Marseille.pdf

Oliva, F. (2002). *L'urbanistica di Milano. Quel che resta dei piani urbanistici nella crescita e nella trasformazione della città.* Hoepli.

Oliva, F. (2010). *Giuseppe Campos Venuti. CITTÀ SENZA CULTURA: Intervista sull'urbanistica.* Laterza.

Ordine e Fondazione degli architetti, pianificatori, paesaggisti e conservatori della provincia di Milano. (2015). *Milano che cambia.* https://www.ordinearchitetti.mi.it/it/mappe/milanochecambia/aree

Porta Nuova. (2013). *Porta Nuova Isola.* http://www.porta-nuova.com/

Sennett, R. (2018). *Building and dwelling: Ethics for the city.* Farrar Straus & Giroux.

Séquano Aménagement. (2020). *Rapport d'activité 2019.* Seine-Saint-Denis:Abaca Créa et Com.

Sicari, D. (2004). *Il mercato più antico d'Italia. Architetture e commercio a Bologna.* Compositori.

Stefano Boeri Architetti. (2014). *Bosco Verticale.* https://www.stefanoboeriarchitetti.net/project/bosco-verticale/

Toura, V. (2019). *Désindustrialisation et décroissance urbaine. L'interdépendance de politiques publiques intercommunales et l'enjeu local dans deux projets de renouvellement urbain en France: l'Ile-de-Nantes et les Docks-de-Seine à Saint-Ouen.* Paper presented at the 55ème colloque de l'Association de Science Régionale de Langue Française (ASRDLF), Caen, France. https://halshs.archives-ouvertes.fr/halshs-02861234

Tarrius, A. (1987). L'entrée dans la ville: migrations maghrébines et recompositions des tissus urbains à Tunis et à Marseille. *Revue européenne des migrations internationales, 3*(1), 13–148. https://doi.org/10.3406/remi.1987.1131

UN. (2020). *Global indicator framework for the sustainable development goals and targets of the 2030 agenda for sustainable development.* https://unstats.un.org/sdgs/indicators/Global%20Indicator%20Framework%20after%202020%20review_Eng.pdf

UN Habitat. (2020). *The 2030 agenda for sustainable development.* https://sdgs.un.org/

Unione Confcommercio Imprese per l'Italia. (2020). *DUC Distretti Urbani del Commercio.* https://www.confcommerciomilano.it/

Relevance of Urban Ecosystem Services for Sustaining Urban Ecology in Cities-A Case Study of Ahmedabad City

Vibha Gajjar and Utpal Sharma

Abstract

Rapid urbanization and changing dynamics of cities in a developing nation like India are experiencing its impact in the form of degraded environmental conditions. Apart from various other factors, the spatial growth of cities is also associated with the process of land management techniques deployed for its development. An alternative approach to holistic development suggests integrating ecological perspective in the decision-making process of spatial growth and pattern of cities. A global perspective on such a topic tries to formulate agendas to ensure the environmental health of the planet. The field of planning is associated with spatial development considering the physical, social, economic, and environmental concerns. The interrelation of the above aspects and their adopted weightage results in the spatial pattern generated in the cities. The field of urban ecology makes a notable attempt to bridge the gap between spatial planning and environmental concern, simultaneously fostering the resilience of cities. The urban ecology field knowledge having integrated consideration of ecosystem service could be promoted as a tool for ensuring future development planning process in cities. This research investigates the scenario and status of ecological concern in the case of the rapidly growing metropolitan city of Ahmedabad using a response survey and LULC change over a temporal scale of 25 years. The secondary data from the USGS site has been used to conduct spatial analysis of remote sensing data. The decadal change in land cover and its effects on ecosystem services have a direct link with the environmental degradation in the study area. This decadal change in LULC is used to analyze the overall environmental quality of urbanizing areas. The inferences from the case study are used to formulate the suggestive future policy framework for a city development plan.

Keywords

Urbanization • Urban ecology • Urban ecosystem services • LULC • Ahmedabad

1 Introduction-Urbanization Trend

The growth of the human population is a phenomenon that contributes to the rise in the concentration or density of population in urban areas. Keeping in mind the total population rise at the global level, the main concentration continues to be in urban areas or cities. As per the UN report, the global population rise of more than 25 million for the period of coming 35 years from 2014 to 2050 (United Nations, 2020) will bring more challenges related to urbanization.

In the upcoming decades, most of the future population growth is anticipated to occur in the developing world; as shown in Fig. 1, which indicates the concentration of population in India. Also, as per the census of India, the urban population in India was 11% in 1991, which increased to 27.81% in 2001 and 31% in 2011. As per the estimation given by the UN report, 50% of the population will be urbanized by 2046 in India. Along with population rise, the concentration of population in urban areas is bringing major challenges to the development processes. The graphical representation of the map shown in the report—'Atlas of Human Planet 2016' (Pesaresi Martino, 2016), indicates the concentration of population in developing nations like India. The population growth is not a challenge, but the density of population concentration in urban areas is putting pressure on land conversion from agricultural to urban use. The resultant degradation in environmental and climatic modification along with pollution of land, water, and air has become a major challenge. The spatial growth of cities is associated with the process of development plan preparation, along with land management techniques deployed for its

V. Gajjar (✉) · U. Sharma
Institute of Architecture and Planning, Nirma University, Ahmedabad, India
e-mail: vibha.gajjar@nirmauni.ac.in; vibha.gajjar.29@gmail.com

© The Author(s), under exclusive license to Springer Nature Switzerland AG 2022
C. Piselli et al. (eds.), *Innovating Strategies and Solutions for Urban Performance and Regeneration*, Advances in Science, Technology & Innovation, https://doi.org/10.1007/978-3-030-98187-7_18

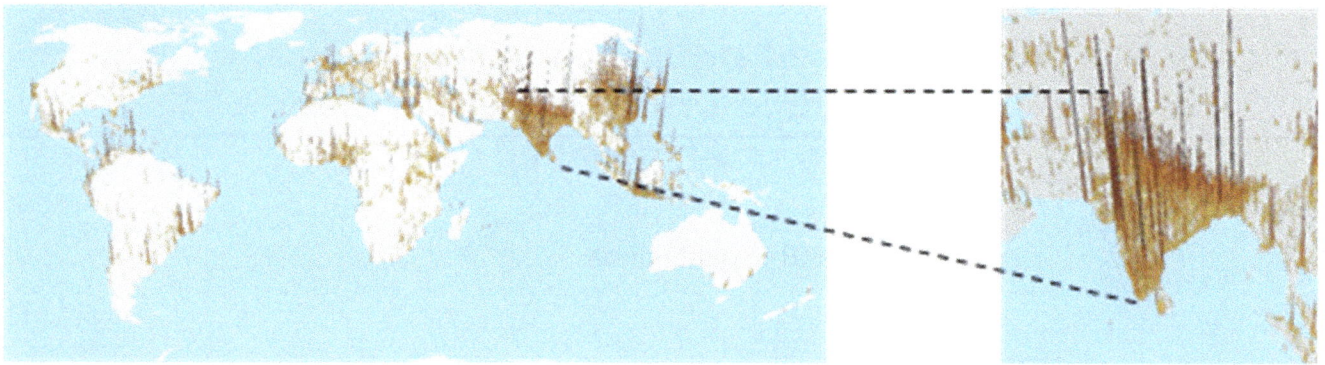

Fig. 1 Concentrations of population distribution across the world and in India. *Source* Atlas of the Human Planet 2016 (Pesaresi Martino, 2016)

development, which are critical factors in deciding the quality of the living environment for the people.

The influx of population in large cities of a developing country, like India, has a direct relation to the availability of employment opportunities. The aspirations, job opportunities, and lifestyles also draw most of the rural population toward urban areas. Besides natural growth, a high level of rural immigration has put pressure on the already rising demand for serviced land in urban areas. The land management technique is practiced in the Indian context as a part of the regular development plan-making process that involves the development plan for providing appropriate land use by various spatial planning standards as also by governing authorities like the municipal corporations of the area/urban development authority. The primary objective is to fulfill the basic need of serviced land along with physical and social infrastructure. The gradual shift of focus in making the cities sustainable has developed the need in the concerned authorities to develop environmentally friendly cities. The alternative method of incorporating environmental concerns in the planning exercise is through ecological approaches. The ecological aspects consider the interrelationship with the surrounding environment, and thus can be considered in the planning exercise, to ensure the sustainability aspect of city planning. One of the alternative approaches to ensure sustainability aspect in the urbanizing area is to promote ecosystem service accountability in land-use planning. The ecosystem functions are altered by land-use planning as it changes the ecology of the place. Thus, making the data of available stock for ecosystem service can help in making an informed decision on land-use planning.

2 Urban Ecology and Cities: Ecological Models and Their Relevance in Land-Use Planning

The definition of urban ecology can be explained by separating the two words—'urban' and 'ecology'. The meaning of urban takes the accountability of the degree of change in land use from natural to built-up environment and ecology means the study of the surrounding natural environment. The generic definition could be said as the study of manifested environment existing within the complexity of human settlement in the name of city structure. Many variants of definitions have justified the meaning of urban ecology from various lenses. Therefore, it could also be defined as —'Urban Ecology is the study of ecosystems that include humans living in cities and urbanizing landscapes' (Cary institute of ecosystem studies, 2019). In the context of an urban environment, the scientific study conducted for studying the relationship of a living organism with its surroundings constitutes the major part of urban ecology. Also, the human and ecological processes coexist in cities under settlements in the form of societies having built as well as unbuilt or natural settings. Thus, the major aim in the field of urban ecology is to understand the process occurring in such interactions and their resultant effects (Grove, 2015). 'It has deep roots in many disciplines including sociology, economics, geography, urban planning, landscape architecture, infrastructure planning, anthropology, climatology, public health, and environmental planning. Because of its interdisciplinary nature and unique focus on humans and natural systems within urbanized areas, 'urban ecology' has been

Fig. 2 A view of Urban Ecology that emphasizes the coupled relationships between humans and nature proposed by Maria Alberti in "Integrating Humans into Ecology: Opportunities and Challenges for Studying Urban Ecosystems". *Source* figure from Book Advances in Urban Ecology (Alberti, 2008)

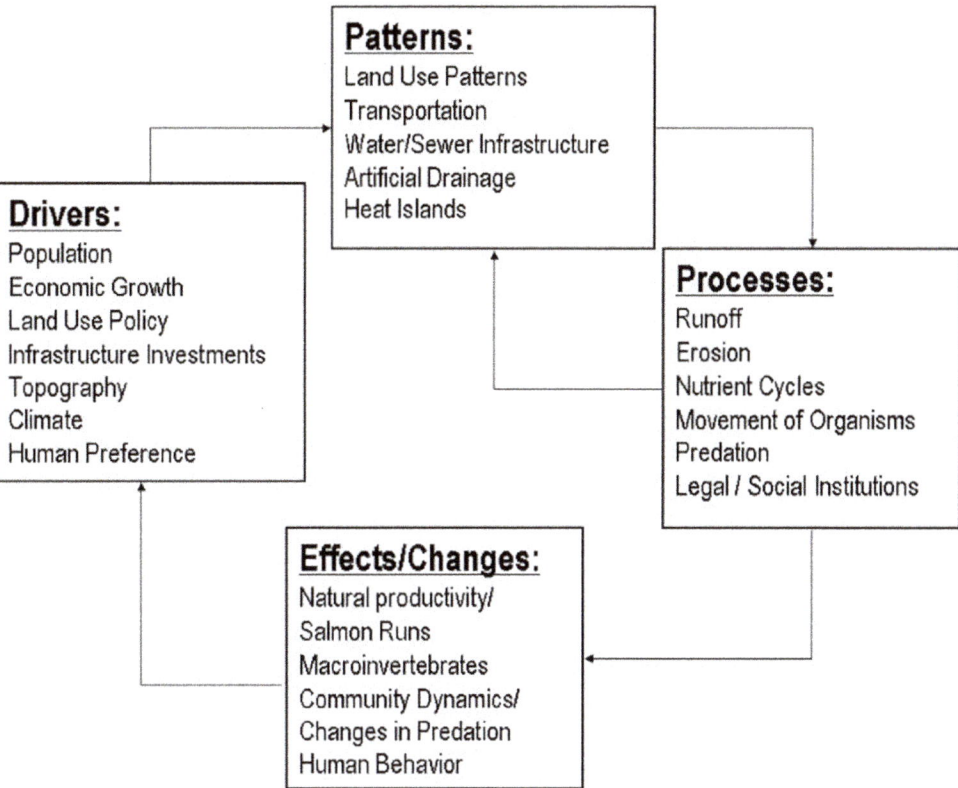

used variously to describe the study of humans in cities, nature in cities, and the coupled relationships of humans and nature' (Alberti et al., 2003).

This concept can be better explained through the pattern, process, effects, and driver as shown in Fig. 2 developed by Marina Alberti (2008). Based on the evolution of urban ecology as a branch of knowledge gathered and collated over 70 years, it has been considered an important aspect for study through various parameters. As urban ecology incorporates the larger concern of survival of all biotic and abiotic components present on earth, the subject can be studied from macro to micro level. Subsequently, urban ecology can be defined as 'the study of spatial and temporal patterns, environmental impacts, and sustainability of urbanization with emphasis on biodiversity, ecosystem processes, and ecosystem services' (Zhi, 2016). The trend of evolution in the urban ecological perspective can be understood through different paradigms. One such aspect is a detailed inquiry for ecosystem service which, when considered for land-use planning for urbanizing areas, can show better urban ecological concern for rapidly developing cities.

The change in viewing ecological science from a purely theoretical to a highly scientific-based knowledge is a significant shift in understanding urban ecology perspectives from many angles. The precise ecological information catering to individual benefit to mass benefit is a commendable shift and extension in paradigms for urban

ecology. The most popular paradigms for any urbanized area can be studied through three main anecdotes, namely 'ecology in, ecology of and ecology for city' (Steward and Pickett, 2016), wherein the city is a referential term for any urban dominant settlement. The three popular paradigms are shown in Fig. 3.

All three paradigms represent a distinctive area of interest and approach and can be represented as the evolution of three possible fields of detailed study. The 'ecology in city' approach is a bio-ecological approach and the 'ecology of

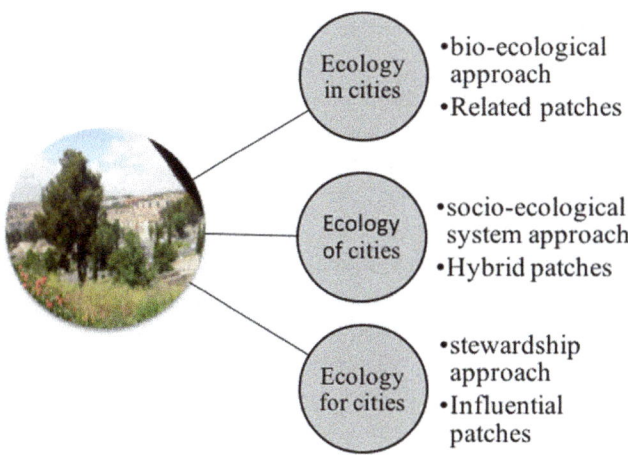

Fig. 3 Paradigms of urban ecology. *Source* Author

the city' is a socio-ecological system approach, whereas 'ecology for the city' is a combination of both the perspectives and more of a stewardship approach.

With the development of cities and accelerating population growth in urban areas, the natural ecosystem is steadily being replaced by an urban ecosystem. The urban ecosystem can also be called a hybrid ecosystem as it consists of a network of many systems. It consists of the natural environment as well as layers of economic, social, cultural, political, and many more aspects. Ecosystem health and diversity are essential to maintain flows of beneficial services and the survival of cities. So developing a better understanding of ecosystem service plays a pivotal role in assessing the real asset and decision-making process of land-use planning. The recognition and relevance of the term ecosystem services came in with the publication of Millennium Ecosystem Assessment (MA), which involved more than 1300 scientists worldwide. 'Ecosystem services' (ES) are the ecological characteristics, functions, or processes that directly or indirectly contribute to human wellbeing: that is, the benefits that people derive from functioning ecosystems (Robert Costanza, 2017).

All-natural ecosystems are capable of delivering free and unconditional valuable services. The basic needs of all organisms, including humans, are satisfied by nature and its services, such as the provision of productive soil and clean water, cultivation of crops for food, maintaining climate regulation, etc. So the importance of ecosystem service is core to the survival of humans including social and economic development. Thus, it is important to acknowledge the ecosystem services and know their relevance in detail. A healthy environment and the natural capital benefit received by people is referred to as ecosystem services (ES) (Haines-Young, 2016). ES support life (e.g., by providing air, water, food, raw materials, medicines), security (e.g., by mitigating extreme weather events, the spread of vector-borne diseases), and quality of life (e.g., by supporting mental and physical health, cultural identity, recreation), among many other things (Preston, 2017). The emerging discipline of ecology examines the interactions between organisms and the human-dominated ecosystems in which they reside, and can provide additional solutions to urban environmental problems. This takes into account incorporating ecosystem services into spatial planning exercises, promoting an ecological way of thinking in the formulation of policies, introducing ecological functions in modifying and operating land-use planning, and fostering scientific basis and consequences of ecological thinking in land management. Strategies aimed at restoring and enhancing urban ecosystem services should play a significant role in the transition toward healthier, resilient, and

sustainable cities (Gómez-Baggethun, 2013). The study of LULC changes can offer a better insight into the transformation process that happened over a temporal scale on any landscape patch. Thus, it can become a useful tool in investigating the ecosystem service at selected patches of landscape.

Land conversion is a resultant effect of the rapid urbanization process. The transformation of the natural landscape and modification in the biogeochemical cycle to cater to the growing need for urbanization affects the structure and function of the natural ecosystem of cities (Niemela, 1999). 'Land development and human activities in urbanizing regions alter land cover and the availability of nutrients and water, affecting population, community, and ecosystem dynamics' (Alberti, 2007). These altered ecosystems of cities, in turn, affect the ecosystem's capacity to deliver important services to inhabitants of the city and human wellbeing. The moment a city extends its limit beyond the capacity of regeneration, the multiplication effects on the surrounding environment need urgent attention. It is important to safeguard the ecosystem services by managing the land management tool, which is used for the urban development process efficiently. The only way to ensure a sustainable approach to city planning and land management tool is by incorporating the relevance of ecosystem services at various scales of planning from regional to city and local area planning. Urban ecosystem sustainability assessment serves as a tool that helps policy and decision makers in improving their actions toward sustainable urban development (Yigitcanlar & Dizdaroglu, 2015). The need for generating database of existing ecosystem conditions must be prioritized for decision- making of the land-use planning process. One of the alternative means to safeguard the ecology of urbanizing cities is to analyze the LULC change and its effect on the ecosystem service. Direct links can be established between the ecosystem service and a healthy environment by checking the effects on existing environmental conditions. The normal functioning of ecosystem service will result in a healthy living environment.

One of the major contributions of maintaining the ecosystem service is to warrant a healthy environment in cities. Out of four ecosystem services as per the Millennium Ecosystem Assessment (MA), namely, Provisioning service, regulating service, Cultural Service, and Supporting services, the development plan processes must investigate them in detail before taking the land-use planning decision. Different land parcels located in the core and peri-urban areas of cities are contributing to various ecosystem services like agricultural activities for the provision of food, parks and garden for recreational activities, trees and dense vegetation cover for temperature regulation. The available stock of different services needs to be mapped in order to identify the areas for potential retention. It will help in taking decisions on land management. Also, the data collection of the available stock of ecosystem should be used for decision-making of land management. The incorporation of such a concept will not just improve available ecosystem service but contributes toward the better environmental condition. Further, Green Infrastructure (GI) concept is one of the better approaches for ensuring a healthy environment in an urban context. GI has emerged as a concept that may be employed to operationalize an ecosystem services-based approach within spatial planning policies and practices (Lennon & Scott, 2014). The GI approach moves beyond traditional site-based approaches of 'protect and preserve' toward a more holistic ecosystems approach, which includes not only protection but also enhancing, restoring, creating, and designing new ecological networks characterized by mufti-functionality and connectivity (Lennon & Scott, 2014). Thus, it becomes necessary to take the stock record of available vegetation cover along with the urbanization process of built forms for analyzing the sustainability aspect. The availability of ecosystem services in urbanizing areas has a direct relationship with well-being and better environmental quality of cities. Thus, the inclusion of ecosystem service assessment by the local authority before processing any development process and land-use plan becomes necessary. Taking the accountability of Land Use Land Cover (LULC) change over the temporal scale can give insights for decision-making regarding the present situation. The land cover change is studied using remote sensing data for a selected case study area. The change in land cover has a direct impact on immediate surroundings and its detailed investigation can be used for assessing the status of the available ecosystem. In particular, mapping and assessing ecosystem conditions can help prioritize where green infrastructure could be best deployed and degraded ecosystems need to be restored (Maes, 2018). Various conversions happening in peri-urban areas are gradually changing the existing land parcel. This phenomenon is responsible for altering the ecosystem function. Depending upon the land-use change, it may result in positive as well as negative impacts on ecosystem performance and ecosystem services. Thus, it is a greater concern to study and take accountability for the land-use change effect over a temporal scale in rapidly urbanizing areas.

In the Indian context, the Urban and Regional Development Plan Formulation and Implementation guidelines (URDPFI) (Ministry of Urban Development, 2014) published in 2014 are available to ensure balanced development activities at the urban level. It mentions the need for ecological aspects by conserving the Eco-sensitive zones identified during the development plan process to safeguard overall natural resources and environmental conditions. These guidelines also mention the carrying capacity assessment for potential urban development pockets, as shown in Table 1, but do not provide further details for investigating the mechanism.

The local authority responsible for land management at the city level thus has to choose between the interpretation of carrying capacity as per the expertise available. In the case of Indian cities, the local development authority does address the land suitability analysis, depending upon the scale of urbanization. But, it fails to address the concept of ecosystem services as a major attribute in the land-use decision-making process. Also, the impact created by the depletion of environmental conditions in highly urbanized areas needs urgent attention. The need for ensuring environmental sustainability is also being addressed through Environmental Impact Assessment process. But the comprehensive data mapping, evaluation, and analysis regarding the Ecosystem service is not a part of the land-use decision-making process. Thus, through this paper, the change in LULC is studied at a temporal scale to understand the land cover change over 25 years (1991–2016) along with a change in 2019. The case study is located in a highly urbanizing setting in the state of Gujarat, India, known as Ahmedabad.

Table 1 Level of evaluating capacity for urban areas based on URDPFI Guidelines

Level of evaluation	Infrastructure capacity level	Environmental capacity level	Sustainable capacity level
Focus	Main purpose is infrastructure development or intensity of usage based on the demand of an area	Main purpose is accountability toward the environment and ecosystems of an area	Main purpose is safeguarding resources for long-term sustenance of an area
Indicator	Development of water supply, sewerage, transportation, etc	Status of environmental condition and pollution level like air, water, soil, etc	Available stock of resources and its status of exhaustion like resource flow process

Source Author

3 Introduction to Case Study: Overview and Location of Ahmedabad City

Ahmedabad is one of the rapidly growing cities located in the western part of the country (India) and the central part of the state of Gujarat as shown in Fig. 4. The Sabarmati river divides the city into the eastern and western parts, showing a lot of disparity in terms of development right since the year of its establishment in 1411 AD. The old city of Ahmedabad is situated on the eastern bank of the river Sabarmati. The core city of Ahmedabad has the distinct characteristic of the built form and dense fabric having mixed-use development. The city has grown manifold over 600 years of establishment, possessing a rich confluence of cultural, social, and economic dynamics. It was known as Manchester of East with the textile industry as its economic backbone during the British era. It was the state capital till the establishment of new a capital city in 1976 at Gandhinagar. Since the establishment of Ahmedabad Municipal corporation in 1950, the city has seen progressions and amendments in land management techniques to safeguard the growing needs of the city.

The Ahmedabad municipal corporation has a 466 sq km area having a historic core and urban development authority beyond its limit. The city has outgrown manifold, and the new limit proposed by the Ahmedabad urban development authority for 2021 is 1851 km^2. For the purpose of this study, the Ahmedabad Urban Development Authority's new DP proposal is taken into consideration having an area of 1851 km^2, which also includes the area under the jurisdiction of Ahmedabad Municipal Corporation. Ahmedabad could be judged from its population growth as it crossed the one million population in 1961 as per the census of India. Over another 50 years, and as per the census record of 2011, the city of Ahmedabad had more than five million population.

The city allowed various segments of society to fulfill the aspiration of city life and keep pace with the growing urbanization trend. Ahmedabad city has witnessed many phases of development in each era and they have left traces of existence in the overall fabric of this urban landscape. The city has transformed culturally, socially, economically, and environmentally over the years.

3.1 Land Management Technique

In the Indian context, right since the Indus valley civilization, the development process exercising land management techniques has a long history of evolution. With the advent of urbanization, there was an increase in population growth as well as in the opportunities available after the post-industrialization phase in cities. Due to this, a drastic change was seen in the conversion of land from rural to urban. It also resulted in the necessity for the formation of a city custodian to ensure the development planning process in urbanized areas. The administrative setup in cities in the form of municipal corporations and urban development authorities was established to ensure proper land management of cities through a mechanism of Town Planning Schemes (TPS). It is a land pooling and readjustment process to ensure adequate space for roads, open spaces, and physical and social infrastructure. A town planning scheme is a hybrid form of land adjustment in which agricultural landowners in the urban fringe are required to give away 40% of the land parcel in exchange for compensation to the government (Sanyal, 2012). After the readjustment of plots and deduction from the original plot area, the original owner gets the serviced land parcel. The deducted portion of the land parcel is used for the development of roads, social infrastructure, and reservation as a land bank for the

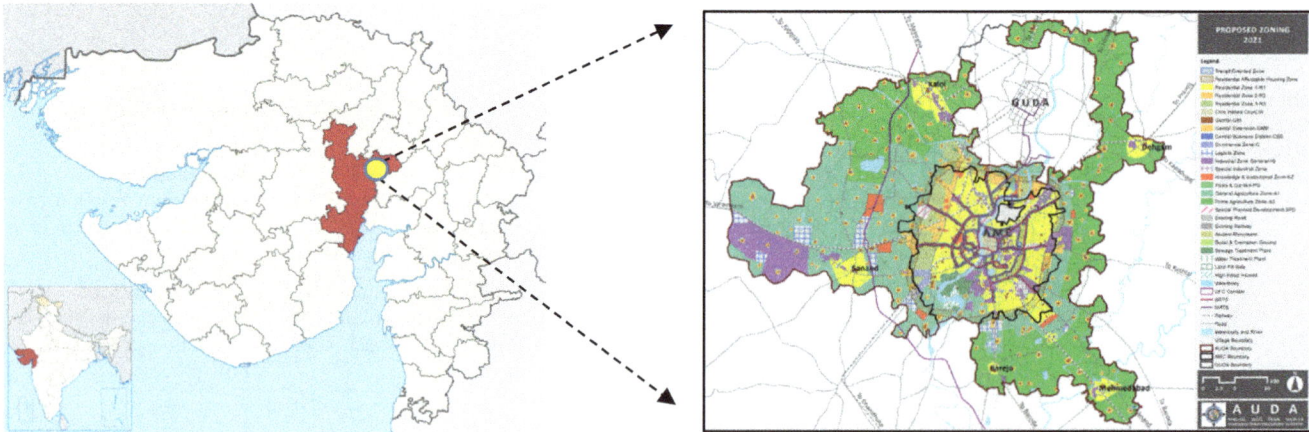

Fig. 4 Map of Gujarat showing location of Ahmedabad District and further City of Ahmedabad proposed development plan for 2021 by AUDA. *Source* Developed by Author from different map sources (https://en.wikipedia.org/wiki/Ahmedabad_district and AUDA)

Fig. 5 Land management
process in the state of Gujarat.
Source Developed by Author
based on AUDA website

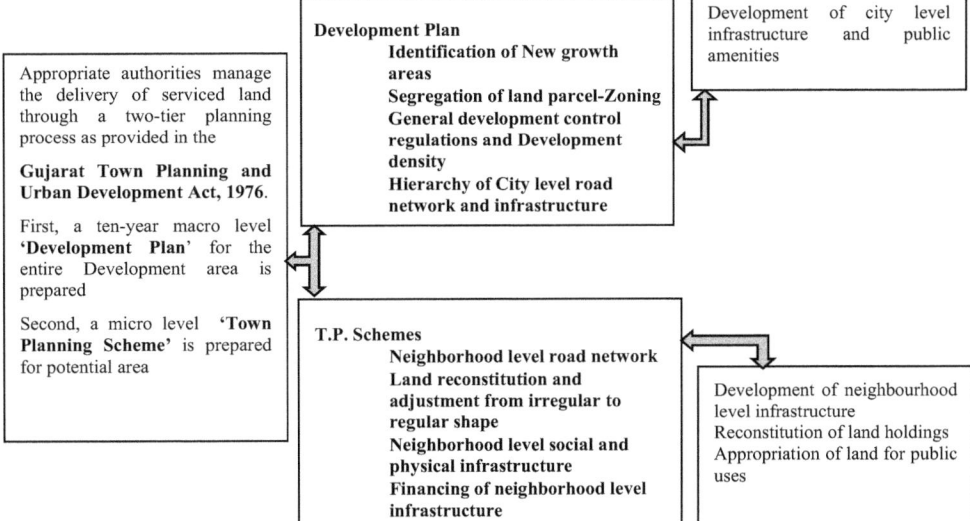

respective sale for generating revenue. In the case of Gujarat state, where the case study is located, the practice of land management is exercised under the provision of the GTPUD Act, enabling the making of macro-level development plan (DP) process and micro-level Town Planning schemes by Ahmedabad Municipal corporation (AMC) and Ahmedabad Urban Development Authority (AUDA) as shown in Fig. 5.

AUDA is responsible for preparing the statutory development plan for AMC and peri-urban areas that are likely to be urbanized in the future (Ballaney, 2012) The 1981 draft development plan was the first master plan prepared under the Gujarat Town Planning and Urban Development Act, 1976 and sanctioned in 1987. Doubts were expressed about the validity and authenticity of the information based on which proposals were formulated. The mismatch related to ground reality and plan was reported in newspapers. The city also experienced leapfrog development and haphazard growth with a rising population. But with the new development plan of 2002, AUDA showed the use of technological advancement and the use of remote sensing data for the first time. A land suitability analysis using remote sensing data and GIS was jointly carried out by the Space Application Centre (ISRO) and AUDA in order to assess the urban sprawl since 1972 and land suitability for development. It shows the use of technology but could not address the issue related to environmental degradation. Rather, land-use decisions should have been based on environmental land suitability study, to prevent damage to the environment and protect the environmentally sensitive zones (e.g., agricultural land, forests, water bodies, and rivers). The new upcoming draft development plan of 2021, apart from land suitability analysis, considers the preservation and conservation of natural resources like water bodies and the

promotion of vegetation cover. But the missing component is the ecosystem service assessment approach, fostering an ecological concern for making resilience and sustainability aspects. The vision of this DP is 'To make Ahmedabad a liveable, environmentally sustainable and efficient city for all citizens; a city with robust social and physical infrastructure, vibrant economy and a distinct identity; a globally preferred investment destination' (AUDA, 2014). This draft report has identified the inadequacy of green spaces in the city; lack of city-level public open spaces and uneven distribution of community-level parks. It has also identified the inadequate green cover, and hence the need for green streets in the city. Therefore, taking up the accountability of change in vegetation cover will give better insight into the course of development happening in the case study area.

3.2 Method of Data Collection

The overall framework follows an inductive research method to understand the development process in urbanized areas. To understand the ecosystem service functioning and its impact on the environmental condition, LULC change is studied on a temporal scale and in spatial paradigms. LULC change is a common phenomenon observed across the globe to satisfy the growing need for urbanization and also to study the resultant effect of environmental degradation. On the other hand, survey data is collected through a perception-based study of industry experts and Government officials who are working in the field of planning. Information is collected through interviews, questionnaire surveys followed by their analysis. The remote sensing data is acquired from the USGS site having Landsat images of years selected for the case study purpose (1991–2019). The Tiff

Table 2 Data used for LULC (Tiff file downloaded from USGS site)

Year	Landsat data	Date of path 148	Date of path 149
1991	Landsat -5 (TM)	23-Jan-91	30-Jan-91
1996	Landsat -5 (TM)	21-Jan-96	28-Jan-96
2001	Landsat 7 (ETM+)	10-Jan-01	01-Jan-01
2006	Landsat 7 (ETM+)	Landsat 7 Data for 2006 have stripping error and there is no data available of Landsat 4–5 TM in required time period, therefore data of three dates from Jan. to Feb. are used each for both the Path, so total 6 tiles are used to create data for 2006 by mosaicking all database (15-Jan, 24-Jan, 02-Feb-2006)	
2011	Landsat -5 (TM)	30-Jan-11	21-Jan-11
2016	Landsat 8 OLI and TIRS	28-Jan-16	19-Jan-16
2019	Landsat 8 OLI and TIRS	20-Jan-19	27-Jan-19

Source Author

files for path 148 and path 149 were collected for processing. Mosaicking helped in combining and merging the two files into one single file. The shape file for the latest administrative limit in the case of Ahmedabad was obtained from the AUDA office and clipped for extracting the required images. Table 2 summarizes the data used for the analysis of LULC in the case study area. These remote sensing data acquired for the selected geographical area for temporal analysis of Land Use Land cover (LULC) was analyzed systematically using GIS software. One of the advantages of using GIS in site suitability analysis is the capability of GIS in the development of alternative scenarios for urban development (Parry, Ganaie, & Sultan Bhat, 2018).

3.3 Data Analysis Using Response Survey

Data was also collected using a response questionnaire survey method with industry experts, academicians, and city custodians, who work in the field of planning. Out of 70 respondents, 15 were officials from urban local bodies working in the department who are directly linked with the decision-making process of urban development activities. Some of the questions asked in the survey were related to environmental degradation over a temporal scale in the case study area. The intention of asking such a direct question was to get clear feedback on the performance of ecosystem service that is directly linked with available vegetation cover and degraded environment. The performance of ecosystem services like regulation and cultural services is linked with the available land parcel in the city fabric having green vegetation cover and open spaces. Based on previous experience, the most direct question for pollution level came out to be Air pollution as shown in Fig. 6a. Also, 96% of the response was in favor of enrichment of green cover in

improving the environmental quality and overall condition as shown in Fig. 6b. Further, according to the data from the website of air quality index and the Central Pollution Control Board (CPCB) Report, the air quality of Ahmedabad city usually ranges from moderate (101–200) to very poor (301–400) (Central Pollution Control Board, 2019). The graphical presentation published in the new bulletin of CPCB in January 2019 is shown in Fig. 7. The first column represents the case study area.

The city has been experiencing lockdown due to the pandemic situation of COVID-19 since March 23, 2020. This means that no pollution has been created by vehicular traffic and industries. Both of which are considered to be the major contributors to air pollution in the city for almost one month. Still, the air quality index of the city as recorded on April 24, 2020, was moderate (103 US AQI) which is considered to be unhealthy for breathing (IQAIR, 2020). Hence, it supports the reason for degradation in the air quality of Ahmedabad due to the lack of green cover and inadequate provision of open spaces in DP and TPS. It is much less than URDPFI guidelines of 9 mt per sq. capita. Even open spaces and roads do not have adequate tree plantation. The Development plan of AMC (1965) had the provision of the green belt but while preparing the first DP of AUDA 1987, it was changed for another purpose. The response survey also validated the importance of green cover for addressing air pollution and got 96% of weightage as shown in Fig. 6b. From the views of the respondents, if the green belt would have been protected, it would have resulted in better air quality in the city. A similar response can be looked upon in the response survey as shown in Fig. 8. The buffer zone would have created a better environmental condition although the width, size, and other qualitative initiatives on the green belt zone would have made a difference.

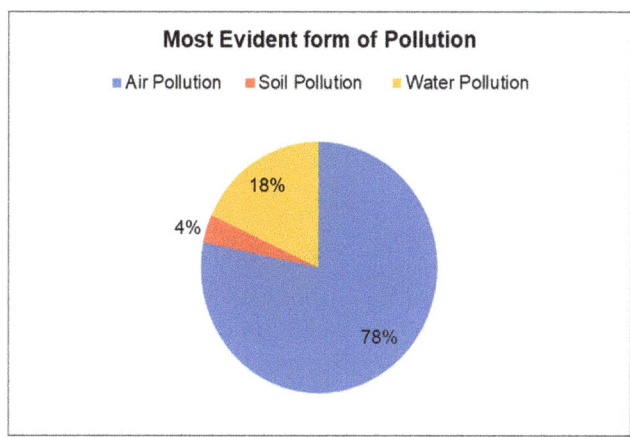
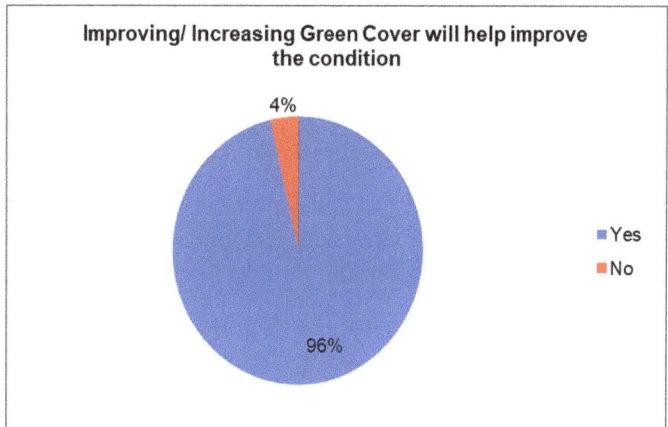

Fig. 6 **a** and **b** Response survey of questionnaire. *Source* Author

Fig. 7 AQI of Ahmedabad by CPCB. *Source* GPCB Report (Central Pollution Control Board, 2019)

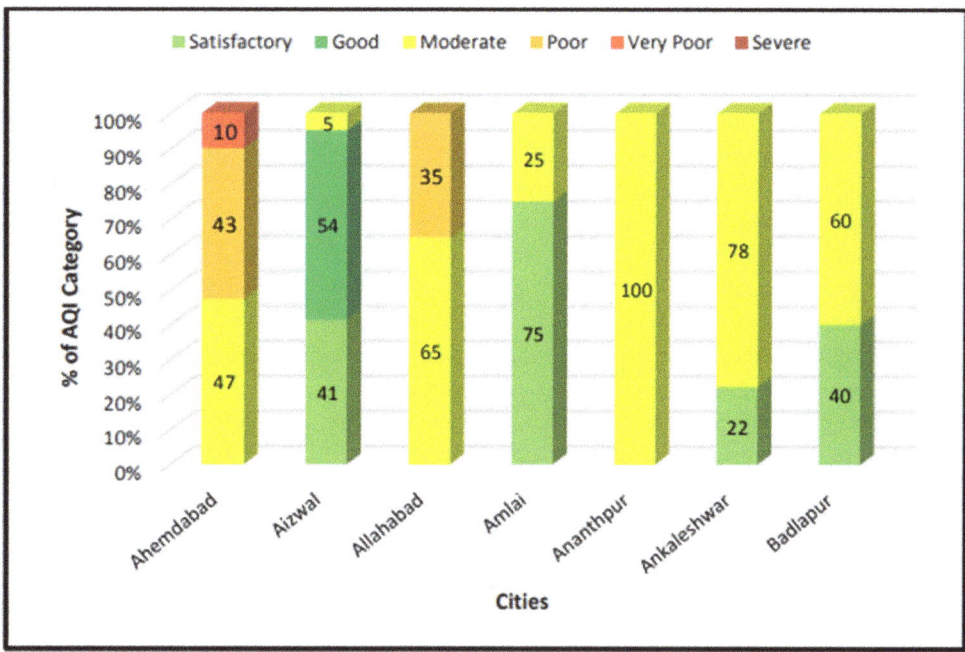

3.4 Data Analysis of Decadal Change in LULC

Figure 9 displays the LULC of Ahmedabad city from 1991 to 2016 and for 2019. The Land cover change from 1991 to 2019 is also depicted in Table 3. Looking at the LULC classification for the city of Ahmedabad (AMC + AUDA) over the years from 1991 to 2019, there has been a significant increase in the built-up area. It shows a visible growth of the city in an outward direction and is also evident in Fig. 9. The maximum growth in terms of built up is observed from the year 2001 to 2006. The reason for this is that the government had started laying down physical infrastructure in the outer parts of the city post the 2001 earthquake to reduce the pressure in the core city. Due to this, the population of the city started to move toward the outer areas of the city. Along with the increase in the built-up area, there has also been an increase in the population over the years. The population of the Ahmedabad Urban Development Authority area has increased from 3.31 million in 1991 to 4.5 million in 2001 to 5.8 million in the year 2011 (Census of India, 2011).

The vegetation in the area has been decreasing over the years except for two scenarios wherein there is a visible increase in vegetation. There has been an increase in the vegetation from the year 2006 to 2011, the reason for which can be the first objective of the 2011 Ahmedabad Development Plan that was 'To create a good environment and to minimize the environmental pollution with green spaces, open spaces and places of public activities with recreational areas' (Adhvaryu, 2015). To meet the objective of the

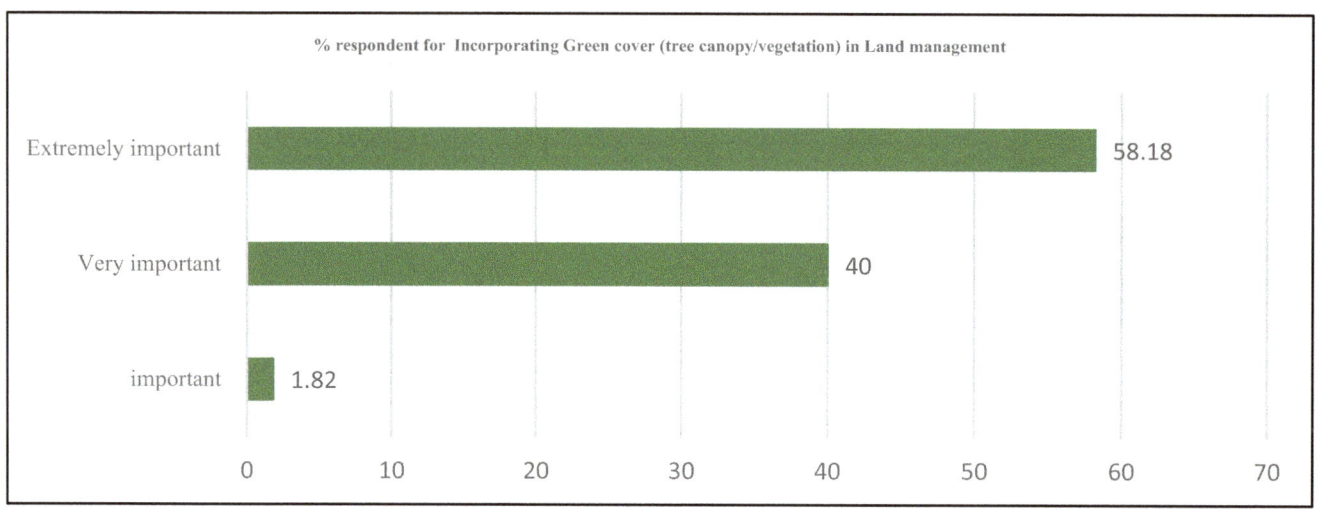

Fig. 8 Response survey of questionnaire. *Source* Author

Development Plan for the year 2011, the government may have taken necessary steps to increase the vegetation of the site area. But it was a temporary measure as the farmland converted to urban use and the green cover was gone. The vision documents prepared by the authority during the different phases of development plans preparation have never mentioned the importance of green cover and its link to the performance of ecosystem services. The resultant effect, thus, is being observed in the form of varieties of pollution and degraded environmental condition in the study area.

In the case of vacant land, there is an increase in the share of the area from 1991 to 2006. This is because the area under vegetation has been converted into vacant land in the fringe areas, which later converts into built-up areas. Here, it is observed that there is a reduction in the vegetation to cater to the increasing population due to which there is an increase in the city limits. Post-2006, the share of vacant land decreased for a decade. From the decade 2006 to 2016, it is noticed that a major share of the vacant land has been converted into vegetation cover. The rise in the built area against available green cover has now become a reason for in-depth inquiry. In 2019, the overall proportion of vacant land to vegetation cover is almost equal and the built up has also increased simultaneously. Thus, creating an imbalance between available vegetation cover for ecosystem service performance in relation to increasing built up. The detailed inquiry regarding the quantity and quality of total green cover for sustaining the ecosystem service is a greater concern now in the case study area.

There has also been a decrease in the number of water bodies over the years. The share of the water body increased through the years 2006–2011 due to the construction of a riverfront along the Sabarmati river in the city of Ahmedabad. This change is of ecological concern as the decision makers have altered the natural river edge into the riverfront

development project. The notion of green infrastructure is not considered by the municipal authorities who do not acknowledge natural systems as infrastructure (Nicola Dempsey, 2017). The riparian ecological concern as described by the researcher is a missed opportunity in the case of Sabarmati riverfront development.

The area distribution of change of various land cover like built up, vegetation cover, vacant land, and water are as shown in Table 4 based on data extracted for LULC as shown in Table 2. It is clearly indicating the increasing trend of built-up areas but variation in other components like vegetation, vacant land, and water bodies.

4   Inferences and Discussion

As seen from the analysis, there has been a variation in the share of vegetation cover in the study area as shown in Fig. 9 and Table 4. Apart from other factors, the decrease in vegetation cover leads to concern-worthy air quality in the study area. The poor air quality for the city of Ahmedabad has been depleting steadily, which directly affects the health and quality of life of its citizens. Also, the continuous changes within the various land covers lead to a depletion in the quality of the soil. It is evident from the data analysis that there has been a continuous conversion of vegetation to vacant land and eventually built-up area. The growth of the city limits to cater to the increasing population, there has also been a change of land cover under vegetation and vacant land to the urban built up. Due to these continuous changes, the soil loses its ability to renew itself. Due to this, there is a decrease in the soil quality of the site over the years. It has also become the reason for various levels of pollution. The natural cycles are pressurized in an urbanizing

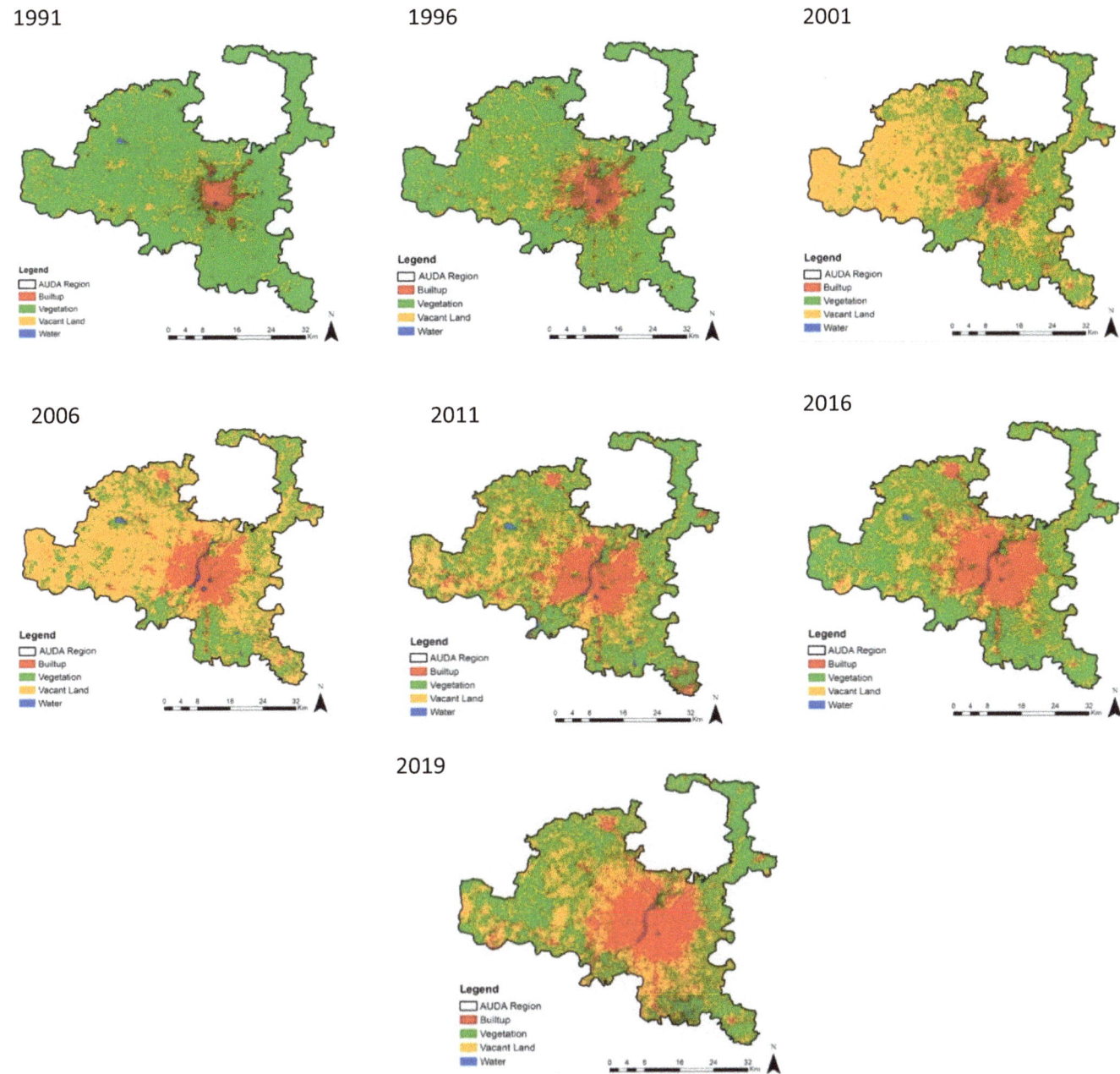

Fig. 9 LULC change (1991–2019). *Source* Author

context, where the focus is on physical land-use planning and not the ecological planning approach. There has been no provision in the development plan to allocate separate land parcels to ecosystem services. Land-use allocation is a priority for roads, open spaces, and amenities as the standard format. Except where there are natural features like rivers, lakes, there is no provision of open space, plantation area or vegetation cover. Peripheral agricultural land is something that provides the space for green cover. But as the city grows, its agricultural land gets converted to urban land- use.

In the case of water, the city is dependent upon the Narmada River for catering to the daily requirements of its population. The city is dependent upon the Narmada River for its water needs as it does not have any source of water of its own. Also, the government has invested a huge amount in the beautification of the city in the construction of the Sabarmati Riverfront which is also a city-level recreational space. But, the water in this Sabarmati river is stagnant and has been manually filled up for the creation of this city-level recreational space.

Table 3 LULC change % distribution in Ahmedabad (1991–2019)

Land cover	% distribution of Land cover change in Ahmedabad (AMC + AUDA)						
Year	1991	1996	2001	2006	2011	2016	2019
Built-up	3.68	5.95	7.49	11.78	14.89	16.16	19.18
Vegetation	82.03	68.72	32.41	21.68	39.00	47.26	37.19
Vacant Land	13.72	24.67	59.80	65.64	44.98	35.98	43.13
Water	0.58	0.66	0.30	0.90	1.14	0.59	0.49

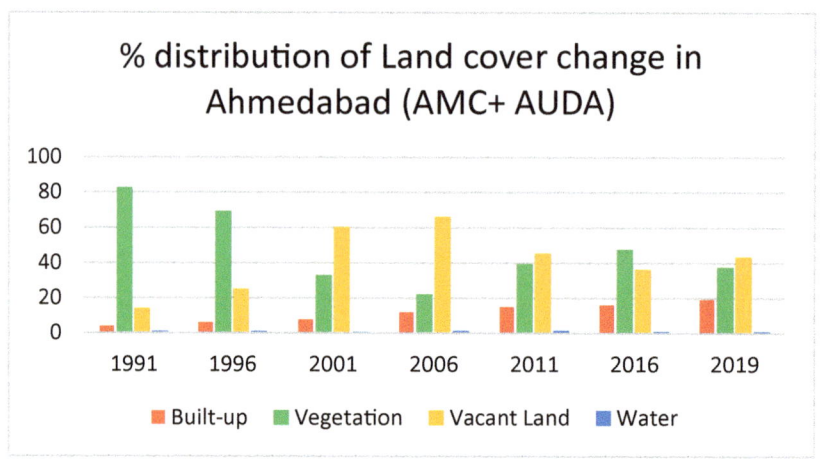

Source Author (based on USGS data extracted for analysis)

Table 4 LULC area distribution (Area in km^2) for AUDA area from 1991 to 2019

Land cover	1991	1996	2001	2006	2011	2016	2019
Built-up	68.15	110.14	138.67	218.12	275.71	299.31	355.13
Vegetation	1518.84	1272.50	600.09	401.42	722.06	875.14	688.66
Vacant Land	254.00	456.76	1107.33	1215.47	832.84	666.24	798.71
Water	10.67	12.26	5.57	16.66	21.05	10.96	9.16
Total Area	1851.66	1851.66	1851.66	1851.66	1851.66	1851.66	1851.66

Source Author (based on USGS data extracted for analysis)

Air, water, and soil pollution have been focused upon as they are the key aspects that contribute to the better functioning of the ecosystem services if looked upon at the macro level. Also, these three characteristics directly affect the quality of life of the people living in the city of Ahmedabad. The change in the LULC in case of study area shows the increasing rate of built up and falling rate of vegetation cover. The altered land use can negatively affect the overall performance of ecosystem service in urban areas. Also, one notices similar results regarding the ecosystem services in the study area resulting from altered LULC. In order to assess environmental performance, examine ecological limits as well as provide the long-term protection of environmental quality, urban ecosystem sustainability assessment is a potential planning tool for policy and decision-making (Yigitcanlar & Dizdaroglu, 2015). The LULC conversions should be monitored as they can be used for evaluating the overall environmental quality of

urbanizing areas. The effects created on the environmental condition have a direct link to the performance of urban ecosystem which is simply an outcome of LULC change over a period of time. Thus, a detailed investigation of LULC can become the alternative means for assessing urban sustainability.

Also, the decadal variation of LULC shows the patch formations in the land parcels. Such patches should be analyzed on a detailed scale, as they can become an answer for sustaining the ecosystem at the micro level. The fragmented patches and uneven distribution of built, vacant, and vegetation cover also negatively impact ecosystem services. The land management techniques adopted to make a development plan using the Town Planning Scheme mechanism can offer clues from a detailed study of LULC. Again, in the TP Scheme mechanism, the minimum amount of area allocated for roads, open spaces, and amenities is much lower than the URDPFI guidelines. In addition, additional space in the form of ecosystem services, open space by natural drainage channels, enhancement of water bodies, and the surrounding ecologically sensitive areas are not considered. Therefore, at the micro-scale, the ecosystem service approach has not been included for a public purpose. Even the provision of open spaces is negligible.

The fragmented land parcels could be studied by monitoring and analyzing the patch using decadal change in LULC. Thus, a detailed investigation of potential patches of development in rapidly urbanizing cities becomes necessary. It is, therefore, recommended to go for a detailed vegetation cover analysis from the quantitative and qualitative aspects of the case study area. The vegetation species with quantitative and qualitative presence shows a better understanding of available ecosystem service for a given population density at a particular location. Thus, analysis of vegetation density on a patch undergoing a development process is recommended to policymakers for considering its relevance.

Based on the analysis and findings, we can conclude that ecosystem services are a vital part of the better functioning of cities. Hence, it is recommended that the ecosystem services be considered during the preparation of the development plan of the cities so that the cities move toward more sustainable and liveable cities with better quality of life for their citizens. The current scenario of the planning approach is focused on spatial planning according to the perception response surveys conducted. Survey analysis also brings a higher rating for environmental planning which directly suggest the assessment of ecosystem service as a major concern. There is a need for a paradigm shift from the existing planning process toward an integrated planning approach which is a combination of physical, social, economic, ecological, and environmental planning. In this manner, we could move toward making our cities resilient as well as self-sufficient. Furthermore, if this approach is taken up and followed there will also be an enhancement in the air, water, and soil quality of the future cities.

Similarly, the urban ecology field knowledge could be promoted as a tool for ensuring future development planning process in cities. This will increase awareness regarding the importance of ecological services among the citizens and will help the government in moving toward a combined planning approach focusing on the environment and ecology. The inclusion of urban ecosystem sustainability is the only alternative for policy and decision makers in rapidly urbanizing cities. Thus, the relevance of urban ecosystem service can become a primary investigation tool for sustaining urban ecology in case of rapidly urbanizing Ahmedabad city. It is strongly recommended that the decision makers of urban planning in case of Ahmedabad takes the cognizance of ecosystem service for achieving sustainable development goal.

References

Adhvaryu, B. (2015). The Ahmedabad urban development plan-making process: A critical review. *Planning Practice & Research, 26*(2), 229–250. https://doi.org/10.1080/02697459.2011.560463

Alberti, M. (2008). Advances in urban ecology. *Springer, US.* https://doi.org/10.1007/978-0-387-75510-6

Alberti, M., Marzluff, J., Shulenberger, E., Bradley, G., Ryan, C., & Zumbrunnen, C. (2003). Integrating humans into ecology: Opportunities and challenges for studying urban ecosystems. *BioScience, 53,* 1169–1179

Alberti, M. (2007). Advances in urban ecology: Integrating Humans and ecological processes in development (1st ed.). In K. C. Seto, & D. Satterthwaite (Eds.). Springer US. https://doi.org/10.1007/978-0-387-75510-6

Auda. (2014). *Ahmedabad urban development authority.* Retrieved 20 May 2020, from http://www.auda.org.in/RDP/

Ballaney, S. B.-A. (2012). *Inventory of public land in Ahmedabad, Gujarat, India.* The World Bank. Retrieved 05 27, 2021, from SSRN: https://ssrn.com/abstract=2342426

Cary institute of ecosystem studies. (2019). https://www.caryinstitute.org/news-insights/definition-ecology. Retrieved from cary institute of ecosystem studies: https://www.caryinstitute.org/

Census of India. (2011). *Census of India: District census handbook, Ahmedabad* (2011th ed.). Government of India.

Central Pollution Control Board. (2019). *National air quality index.* Ministry of Environment & Forest.

Erik Gómez-Baggethun, Å. G. (2013). *Urban ecosystem services.* https://doi.org/10.1007/978-94-007-7088-1_11

Grove, J. M. (2015). *The Baltimore school of urban ecology* (1st edn). Yale University Press.

Haines-Young, M. P. (2016). Defining and measuring ecosystem services. In M. H.-Y. Potschin (Eds.), *Routledge handbook of ecosystem services* (pp. 25–44). Routledge.

IQAIR. (2020). *IQAIR: Ahmedabad City.* Retrieved April 13, 2020, from IQAIR Website: http://www.iqair.com/india/gujarat/ahmedbad

Lennon, M., & Scott, M. J. (2014). Delivering ecosystems services via spatial planning: reviewing the possibilities and implications of a green infrastructure approach. *Town Planning Review, 85*(5), 1–16. Retrieved from https://researchrepository.ucd.ie/handle/10197/7845 : https://www.researchgate.net/publication/283079402_Delivering_ecosystems_services_via_spatial_planning_reviewing_the_possibilities_and_implications_of_a_green_infrastructure_approach

Maes, J. (2018). *Mapping and assessment of ecosystems and their services: An analytical framework for ecosystem condition.* European Union. https://doi.org/10.2779/055584

Marina Alberti, J. M. (2008). Integrating humans into ecology: Opportunities and challenges for studying urban ecosystems. In E. S. John & M. Marzluff (Eds.), *Urban ecology—An international perspective on the interaction between humans and nature.* Springer.

Ministry of Urban Development. (2014). *URDPFI guidelines.* Retrieved 05 27, 2021, from MInistry of Housing and Urban Affairs: http://mohua.gov.in/link/urdpfi-guidelines.php

Nicola Dempsey, S. R. (2017). *Landscape research.* Retrieved 27 May 2021, from There's always the river: social and environmental equity in rapidly urbanising landscapes in India: https://www.tandfonline.com/action/showCitFormats?doi=10.1080%2F01426397.2017.1315389

Niemela, J. (1999). Is there a need for a theory of urban ecology? *Urban Ecosystems, 3,* 57–65. https://doi.org/10.1023/A:1009595932440

Parry, J. A., Ganaie, S. A., & Sultan Bhat, M. (2018). GIS based land suitability analysis using AHP model for urban services planning in

Srinagar and Jammu urban centers of J&K, India (A. Elsevier, Ed.). *Journal of Urban Management, 7*(2), 46–56.

Pesaresi Martino, M. M. (2016). *Atlas of the human planet 2016. Mapping human presence on earth with the global human settlement layer.* Joint Research Centre (JRC), the European Commission's. https://doi.org/10.2788/889483

Preston, S. M. (2017). *Ecosystem services toolkit.* Her Majesty the Queen in Right of Canada,Aussi disponible en français. Retrieved from www.biodivcanada.chm-cbd.net/documents/ecosystem-services-toolkit

Robert Costanza, R. D. (2017). Twenty years of ecosystem services: How far have we come and how far do we still need to go? *Ecosystem Services,* 1–16. https://doi.org/10.1016/j.ecoser.2017.09.008

Sanyal, B. (2012). Town planning schemes as a hybrid land readjustment process in Ahmedabad, India: A better way to grow? In G.K.-H. Hong (Ed.), *Value capture and land policies* (pp. 149–183). Lincoln Institute of Land Policy.

Steward T. A., & Pickett, M. L. (2016). Evolution and future of urban ecological science: Ecology in, of, and for the city. *Ecosystem Health and Sustainability, 2*(7), 16. https://doi.org/10.1002/ehs2.1229

United Nations. (2020). Retrieved from Department of Economic and Social Affairs: https://www.un.org/en/development/desa/population/publications/trends/concise-report2014.asp#:~:text=The%20report%20indicates%20that%20the,in%20the%20less%20developed%20regions

Yigitcanlar, T., & Dizdaroglu, D. (2015). Ecological approaches in planning for sustainable cities-A review of the literature. *Global Journal of Environmental Science and Management,* 159–188. https://doi.org/10.7508/gjesm.2015.02.008

Zhi, X. W. (2016). Impact of shared economy on urban sustainability: From the perspective of social, economic, and environmental. *Energy Procedia, 104,* 191–196.

Increase in Neighborhood Development and Urban Water Erosion in Tropical Cities from Central Brazil: The Case of Goiânia, Goiás

Lizandra Ribeiro Cavalcante and Selma Simões de Castro

Abstract

The literature on urban water erosion in Brazil emphasizes its installation of gullies mainly in the 1980s, in cities that received large migratory contingents, but without timely human settlement plans and without an urban drainage network. Goiânia, the capital of the state of Goiás, located in the Midwest region of the country was created in 1933 to receive 50,000 inhabitants. It has undergone a great transformation associated with the expansion of the agricultural frontier since the 1960s, by attracting countless immigrants seeking better living conditions, but without timely urban planning. The erosive process in Goiânia shows that the city had not yet been evaluated in space–time using detailed/ultra-detailed scales to identify critical areas and their causes. Also, it has been pointed out that the urban occupation created numerous spontaneous neighborhoods, initiated by intensive deforestation and implemented without technical assistance, in areas of springs and valley bottoms. It Follows in these areas are greater erosive water potential, due to shallow soils and greater declivity, under a tropical climate with intensive rainy season during the summer. These studies have demonstrated that the city of Goiânia is a recent example in Brazil of the urban water erosion process, due to the lack of adequate urban planning within tropical regions. This lack of planning leads to an over occupation of protected areas. The results emphasized that spontaneous and accelerated urbanization from 1982 to 2016, which led to a notable increase in the population of more than 1 million people, also promoted the intensive process of urban erosion, accompanying the direction of implantation of spontaneous neighborhoods, in coverage areas degraded natural resources such as springs and valley bottoms. Hypsometry, slope, and landforms showed that the preferred direction of the concentrated/subsurface runoff converges toward the springs and valley bottoms, often disagreeing with the design of streets and the layout of lots, but agreeing with the occurrence of erosive gullies. The analysis of the space–time quantitative evolution and distribution of erosive gullies has shown that the process has drastically reduced since 2007. This may have resulted from public control policies such as urban land use, housing deficit, drainage systems, and many others were implemented. In conclusion, Goiânia is a recent Brazilian example of an urban erosion process related to the lack of urban planning in tropical cities with a high number of migratory movements, which are subsequently controlled through policies to regularize spontaneous neighborhoods, but with high socio-environmental cost.

Keywords

Accelerated urbanization neighborhoods • Urban water erosion process • Intensive occupation urban planning

1 Introduction

Brazil's urban population increased from 31.2% in 1940 to 81.2% in 2000 (Goiânia, 1944; IBGE, 1952; Goiânia, 1969; Santos, 2009; Silva & Travassos, 2008), this growth is mainly associated with the industrialization process that intensified in the last century. From the 1970s, the medium and large cities, especially in the south and southeast regions of Brazil, were subject to high migratory contingents from intra-state (countryside-city) and inter-state. This was essential to promoting rapid demographic growth and to increasing the housing deficit, especially in the areas that were being occupied in peripheries and areas located nearly of the urban border expansion. These areas were the preferred target for spontaneous occupation, proliferated irregular subdivisions, creating new neighborhoods without territorial planning, and poor infrastructure.

L. R. Cavalcante (✉) · S. S. de Castro
Institute of Socioenvironmental Studies, Federal University of Goiás, Goiânia, Brazil
e-mail: lizandra.geo93@gmail.com

© The Author(s), under exclusive license to Springer Nature Switzerland AG 2022
C. Piselli et al. (eds.), *Innovating Strategies and Solutions for Urban Performance and Regeneration*,
Advances in Science, Technology & Innovation, https://doi.org/10.1007/978-3-030-98187-7_19

In this sense, there was an absence or insufficiency of regulatory instruments for urban land use, such as the *"Plano Diretor Urbano"* (In Brazil, the abbreviation PD). Although when these instruments existed, they prioritized only the architectural and landscape aspects of cities and not compensatory measures for socio-environmental problems, nor installed the urban infrastructure. Regarding the year 2000, for example, Cruz and Tucci (2008) state that out of the 5,507 Brazilian municipalities, only 841 (15.3%) had *Plano Diretor Urbano* (PD), while only 489 (8.9%) had the last version after the 1990s. It was prior to the implementation of the *"Estatuto da Cidade"* (in English, Statute of the City) by Law n°. 10.257 on 10 July 2001 (Brasil, 2001). Among several environmental impacts, many of these municipalities have developed linear hydraulic erosive processes.

Selby et al. (1993), Guerra and Jorge (2013), and Bertoni and Lombardi Neto (2017) described the concept of water erosion as a process of interaction between sediment load, disintegration, detachment, transport capacity, and deposition of soil particles caused by the impact of raindrops on the soil. The most important variable of water erosion is characterized by the runoff of rainfall not infiltrated in the soil of areas between the channel's drainage and the watershed divides. These areas are the overland-flow portions of the landscapes (Morgan, 2005; Selby et al., 1993; Young, 1972). It means that water flows can only be from upstream to the valley bottoms (downstream) of a watershed (drainage basin, catchment).

The water erosion analyzed in the spatial context of a watershed consists of two types of conduits by water flow, open channels and pipes (Toy et al., 2001). The cross-sectional shape of open channel flow is linked only by a concentrated flow, in which the conduits under the surface (called a subsurface or interflow process) are characterized by the development of pipe and interception of the water table (Morgan, 2005; Toy et al., 2001). Salomão (2015), based on a study of the erosion control by *"Instituto de Pesquisas Tecnológicas & Departamento de Água e Energia"* (In Brazil, the abbreviation IPT/DAEE) in the State of São Paulo (1989), differentiates rill from gullies. Salomão (2015) states that rills result only from the runoff being descended on the slope, while gullies result from the subsurface which develops piping and interception of the water table or suspended.

The main natural conditions for water erosion are: erosivity which measures the impacts caused by the energy, mass, and velocity of the raindrops on a surface (Bogaart & Troch, 2006; Carvalho Junior et al., 2010); erodibility based on variables of texture, permeability and density of the soil (Salomão, 2015); slope steepness that represents the effect of flow accumulation on transport capacity, while that slope length is a function of the product of distance along the slope (Lal, 2014); Landforms commonly used to classify the shape of slope as convex, concave or flat (Valeriano, 2003); and

fragmented or absent vegetation cover. Otherwise, anthropic influences such as urban areas are characterized by the addition of new conditions, in which the linear erosion (that is known as a rill flow) quickly erodes to the deeply ephemeral gullies. This is common in tropical countries due to high rainfall levels and the frequent lack or insufficiency of territorial planning.

As for Brazilian urban erosion, IPT/DAEE (1989) Almeida Filho and Ridente Júnior (2001) and Faria (2007) emphasize that: (a) flow velocity and volume of runoff in the impermeable surface of urban soil are much higher, and thus the transport capacity of sediment is induced; (b) the implementation of housing developments for low-income population in susceptible areas of geological risk, added to the streets perpendicular to the contour lines, without paving and the absence or inadequacy of the micro drainage system; and (c) the frequent occurrence of these subdivisions in areas with concave curvatures, with a high probability of events such as erosions, landslides, and floods. In summary, these factors clearly reveal the lack of a housing policy that contemplates urban territorial planning, especially in areas of accelerated urban expansion.

The urban gully erosion of Goiânia began to be studied in more detail in the 1980s by *"Instituto de Planejamento e Habitação de Goiânia - IPLAN"* or "Institute of Planning and Housing of Goiânia" (IPLAN, 1983, 1987) and later by Nascimento in the 1990s (1993, 1994, 2002). Cavalcante (2019) and Cavalcante and Castro (2021) updated the mapping of erosive gullies based on aero photography and high-resolution satellite images which registered 106 erosive gullies in 1982, 114 gullies in 1983, 112 in 1992, 84 in 2002, 75 in 2006, 30 gullies in 2011, and only 29 gullies in 2016. According to Cavalcante and Castro (2021), this spatial variability allowed the identification of three phases: at first, between 1983 and 1992, the number of erosive gullies oscillated much higher than 100; secondly, they are reduced to 84 and 75 gullies in 2002 and 2006; and in the third phase, the gullies reduce much more to 30 and 29 in the years 2011 and 2016.

The neighborhoods most affected by the water erosion were those from irregular and clandestine subdivisions without the infrastructure that are regularized years later by the local government (Cavalcante, 2019). After 2006, the municipal government of Goiânia started to promote better control of urban land use by interventions applied to all those impacts caused in the past. It explains the fact that the number of erosive gullies has reduced from 114 to 29 in the last thirty years. This is a common situation in Goiânia, where the areas of springs and valley bottoms on convexo-concave and convex-rectilinear-concave slopes are occupied by the low-income population due to the high housing deficit and not being covered by public policies for urban settlements (Soares Neto & Faria, 2014; Cavalcante, 2019).

The Brazilian pattern of urban occupation by low-income populations in susceptible areas of geological risk is constantly disordered, whose mean character is intensive deforestation followed by land movement to delimitate the lots (IPT/DAEE, 1989). As identified by Cavalcante (2019) and Cavalcante and Castro (2021), these areas are often exposed to the impact of soil erosion until the implantation of infrastructure, which in some cases can be more than 10 years later. It is common to observe intensive and indiscriminate post-deforestation, a close spatial–temporal correlation between spontaneous, dense, and disordered occupation to reduce the housing deficit by low-income populations, in susceptibility lands that concentrate flow and subsequently process water erosions.

The objective of this work is to demonstrate that the urban gully erosion process in Goiânia is a recent example of a well-known urban expansion dynamic. This is where there is no planning for human settlements in susceptible areas of geological risk in tropical regions. These urban expansion areas are recognized as unplanned and the result of the intense migratory flow. This is known to be spontaneous, disordered and rapid, and without adequate infrastructural

planning. In these areas, the erosive gullies are linked to rapid occupancy due to the lack of application of technical standards. These gullies are only controlled after the (consolidated) urban expansion, and with that come high socio-environmental costs.

Study Area

The municipality of Goiânia, capital of the State of Goiás, is located in the geoeconomic region of Central-West Brazil. This area was previously dominated by the phytophysiognomies ecosystem of the "*Bioma Cerrado*" (Brazilian Neotropical Savannah). With the "*Lei Complementar do Município de Goiânia*" (Municipality law of Goiânia) n°. 171, from 29 May 2007, of the last PD, the municipality of Goiânia was divided into two *Macrozones*: The "*Macrozona Rural*" (Rural area) that corresponds to 284.9 km² (38.89%) of the total area which is subdivided into seven units corresponding to drainage basins; and the "*Macrozona Construída*" (Urban area) that includes the constructed and urban expansion area covering 441.6 km² (61.10%) of the municipality area (Fig. 1). The specific area of study corresponds to

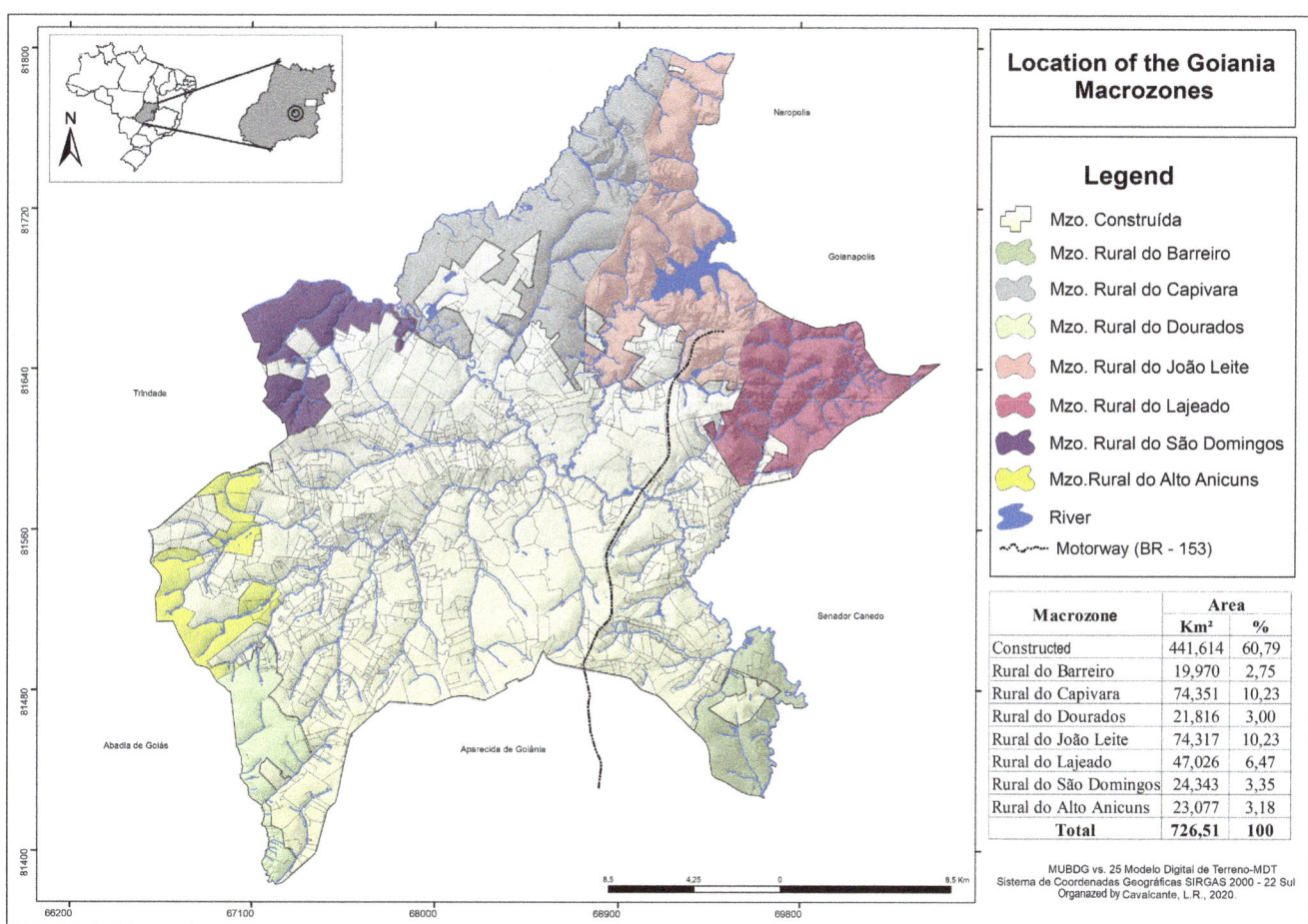

Macrozone	Area	
	Km²	%
Constructed	441,614	60,79
Rural do Barreiro	19,970	2,75
Rural do Capivara	74,351	10,23
Rural do Dourados	21,816	3,00
Rural do João Leite	74,317	10,23
Rural do Lajeado	47,026	6,47
Rural do São Domingos	24,343	3,35
Rural do Alto Anicuns	23,077	3,18
Total	**726,51**	**100**

MUBDG vs. 25 Modelo Digital de Terreno-MDT
Sistema de Coordenadas Geográficas SIRGAS 2000 - 22 Sul
Organazed by Cavalcante, L.R., 2020.

Fig. 1 Territorial division in Macrozonas of the municipality of Goiânia

the "*Macrozona Construída de Goiânia*" (Abbr. MZCG) and is the actual urban area (Goiânia, 2007).

Overview of Physical Characteristics

Goiânia's climate is tropical, therefore, hot and humid with two well-marked seasons, a dry season in autumn and winter from April to September, and another humid season in spring and summer from October to March which concentrates about 80% of the annual precipitation (Nimer, 1979). Convective rains are frequent and concentrate 94% of the annual erosivity of the State of Goiás, with average pluviometry indexes of 247.67 mm and 4.312 MJmm ha-1 h-, respectively, these are erosive rains typical of tropical regions (Galdino, 2015). Average monthly temperatures vary between 24 and 28 °C, but in more urbanized and vertical areas, the maximum temperatures can reach 32 °C constituting islands of heat (Nascimento & Barros, 2009).

The geomorphology of the MZCG is considered 80% flat with slope height above 720 m and gradient between 0 and 3% in surfaces formed by the "*Planaltos Dissecados, Chapadas e Planaltos Embutidos*" of Goiânia (In English, Dissected Valleys, Mountains and Plateau Edges) (Casseti, 1992). These formations are classified as stable slopes with very low susceptibility to gully erosion. The exceptions applied to the "*Planaltos Dissecados e Embutidos*" over those convexo-concave and convexo-rectilinear-concave slopes appear to be greater than 8%. Although, these developed slopes seem most likely to be in areas of high to moderate concentration potential of surface flows, inducing runoff and often water erosion in the flow lines (Lopes & Romão, 2006).

The soil, classified as "*Latossolos Vermelhos Distróficos*" (Red Latosols), abbreviated as LV in Portuguese, is usually found in the higher part of Goiânia. This means that 85% of the surface is characterized by low erodibility in a natural situation as described by Salomão (2015). Otherwise, soils classified to be moderate to high erodibility is only found in 9% of MZCG on surfaces of "*Cambissolos Vermelhos (CX), Neossolo Litolico (RL),* and *Argissolo Vermelho (PV)*" identified as Cambisols, Leptsols, and Acrisols by FOOD AND AGRICULTURE ORGANIZATION OF THE UNITED NATIONS—FAO and ITPS (2015).

Soils of low erodibility "*Gleissolos Haplico (GX) and Neossolo Flúvico (RY)*" known as Gleysols and Fluvisols (FAO and ITPS, 2015) are located in slope heights between 700 and 720 m and equivalent slope lengths lower than 10% on flat surfaces of terraces and floodplains of the Meia Ponte River. *Gleissolos* (Gleysols) is found in slope forms characterized by convexo-concave and convexo-rectilinear-concave at a height lower than 700 m and slope gradient greater than 10% (Casseti, 1992). Despite this, studies on urban gully erosion in Goiânia showed these areas as having a concentration of erosive gullies, even though they are located in soils of very low erodibility. This is corroborated by the literature on urban erosion by Nascimento and Podestá Filho (1993), Nascimento (1994), and Nascimento and Sales (2002).

Urban expansion

Goiânia is a planned city that was founded in 1933 to house 50 thousand inhabitants. But by the year 1960, it already had 130 thousand inhabitants, by 1970, it had 363 thousand, and 703 thousand in 1980 (Goiânia, 1944; 1969; 2007). This increased to 913 thousand in 1991 and then to 1,085,806 in 2000 (IBGE, 2010; SEGPLAM, 2018; SEPLAM, 2018). In 2020, it became the 10th most populous city in Brazil with an estimated 1,516,113 inhabitants (IBGE, 2019). Such demographic growth and regional role in little more than 80 years is the result of policies for the interiorization of the country implemented since 1950, the expansion of the agricultural frontier and the switching of the federal capital from Rio de Janeiro to Brasilia in the 1960s (Nascimento & Oliveira, 2015). In 1999, Goiânia was named as the "*Região Metropolitana de Goiânia*" TRANS "Metropolitan Region of Goiânia" (in Brazil, the abbreviation RMG) (Goiânia, Lei Complementar nº 27/1999).

Since Goiana's foundation, it has had several *Plano Diretor Urbano (PD)*. These were in the years 1938, 1962, 1971, 1992, 2000, and 2007, although those PD of 1962 and 2000 were not implemented due to the national political situation during those periods. Nascimento and Oliveira (2015), when carrying out a synthesis on the urban expansion of Goiânia between 1930 and 1980, based on documentary research and mapping on digital orthophoto and Landsat ETM satellite images, emphasized that PDs had a significant role in defining the vectors of urban expansion, but were neglected in several periods. This favored disorderly growth, evidenced by the presence of urban voids and disjointed vertical buildings, did not adequately consider territorial planning (Fig. 3).

The same authors rescued the work of Moraes (1991) (Fig. 2) and his proposal to periodize the expansion, which was also adopted in the *Plano Diretor* of 1992, namely: (1) 1933 to 1950—the creation of the place; (2) 1950 to 1964—the expansion of the place; (3) 1964 to 1975—the concentration of new places in space; (4) 1975 to 1992—the intensification of urban expansion; (5) 1992 to the present day—the segregation of urban spaces.

Nascimento and Oliveira (2015) also pointed out that the first PD of Goiânia (1933) was proposed according to the concentric radial urban model, which has been redesigned by the PD of 1938. For Oliveira (2005) and Nascimento and Oliveira (2015), the first PD in fact, indicated the preferred axes for urban expansion—initially the central and northern sectors, after six years the southern sector, and finally the

Fig. 2 Urban expansion of Goiânia from 1933 to 1983. *Source* Nascimento and Oliveira (2015) apud Moraes (1991)

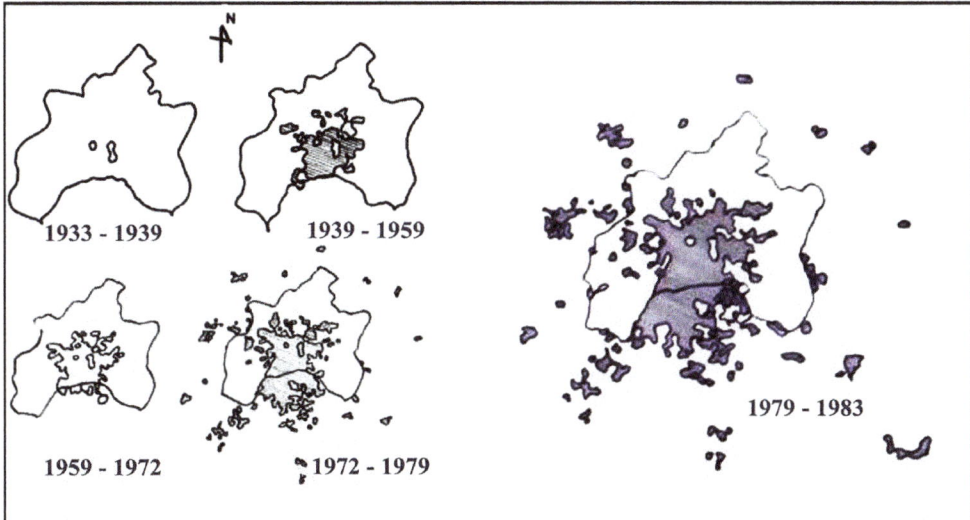

west. However, the northern region was not included due to physical barriers (railroad, sharp relief, and supply sources) which drove the expansion to the south. Figure 3 shows the accelerated growth between 1972 and 1983 and the preferential axis to the west and northwest.

The authors also emphasized the fact that the *"Código de Edificações de Goiânia"* of 1947 (its translation means Goiânia Building Code) allows the implementation of new subdivisions, as long as the entrepreneurs implanted the infrastructure. But later the *"Decreto nº. 16 de 1950"* (In Brazil, local law) started to require only the lease and opening of roads, which led the municipal government to lose control over the urban land use. This favored an increase in the number of irregular subdivisions in this decade, although mainly in the east and west regions that were destined for the low-income immigrant population (Cavalcante, 2019). Thus, numerous irregular subdivisions were created between the 1960s and the 1970s in the center and south sectors, in addition to the periphery of the north and west regions of the city.

Nascimento and Oliveira (2015), claim that the PDs (1947, 1971) were not respected and allowed the implementation of subdivisions without the obligation to implement the infrastructure and only the unpaved road system. Not even the axes of urban expansion were followed and urbanization continued spontaneously and disorderly to the south. Much later, in 1992, the axis of urban expansion took place to the southwest accompanied by various control measures. Those control measures, in the 1992 PD, considered the impact caused by the previous expansions, including irregular subdivisions in geological areas of risk of occupation such as springs, valley bottoms, and floodplains.

Only in 1991, the *"Carta de Risco de Goiânia"* (IPLAN, 1991) was drawn up, which determined the plains, terraces,

valley bottoms, and the entire north and northeast portion of the municipality to be susceptible areas of geological risk. Also, this document suggests areas that are designated for environmental preservation, to recharge water sources and it is also accompanied by the implementation of urban drainage works and measures to contain erosive areas. After the *"Carta de Risco de Goiânia", the* number of gullies reduced from 112 to 75 erosive gullies from 1992 to 2006, and then decreased further to 30 and 29 gullies in 2011 and 2016, when the *Plano Diretor* was updated in 2007 (GOIÂNIA, 2007) which determined priority actions for erosion control.

2 Material and Methods

The purpose of this study was to examine the relationship between the axes of urban expansion and the distribution of water erosion processes in the urban expansion area of Goiânia, specifically the macrozone *"Macrozona Construída de Goiânia—MZCG"*. To achieve this, the search was based on procedures and techniques applied to spatially distributed models of events structured for geographical information systems (GIS). These are described in six main steps below in this section, under the headings preliminary procedures, the identification of urban linear gullies, the delimitation of critical areas, the urban land use, the flow direction, and the data analysis and interpretation.

Preliminary procedures

At the beginning, the preliminary procedures involved documentary research and data information. Three data sources: (i) Aerial photographs of high-quality resolution (0.6, 0.3, and 0.1 m) for the years 1992, 2002, 2006, 2011,

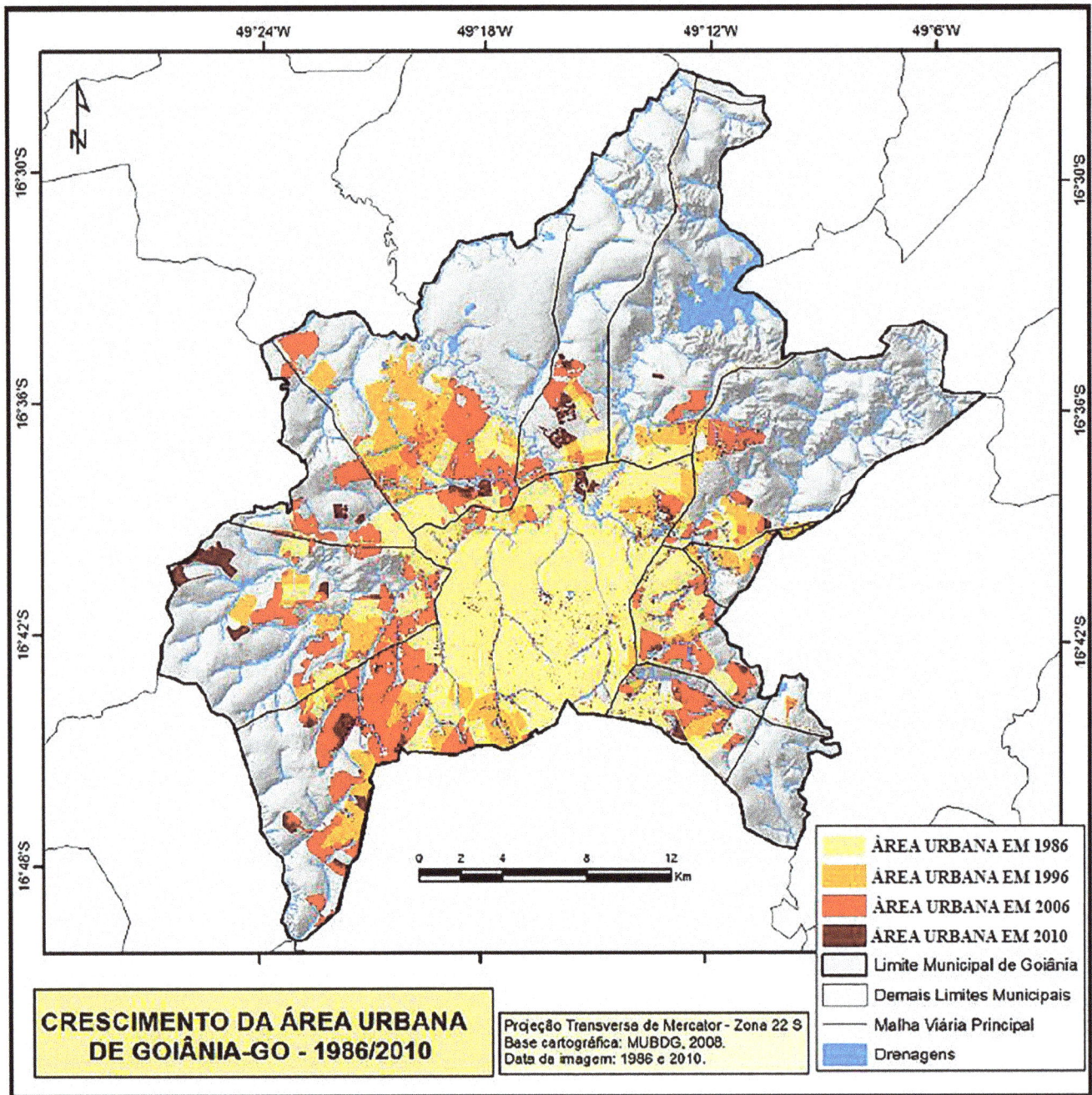

Fig. 3 Growth of the urban area of Goiânia from 1986 to 2010. *Source* Nascimento and Oliveira (2015)

2016, (ii) the Digital Terrain Model (DTM) with 4.79 m of resolution, and (iii) a geodatabase file, were officially acquired from the Department of Geographic Information Territorial from Goiânia City Government (MUBDG, 2017). The geodatabase included shapefiles of the municipality border, neighborhood, blocks, lots, drainage network, and contour lines with intervals of 5 m. With the use of ArcMap software, the spatial reference was defined for every dataset in the geodatabase.

Identification of urban linear erosive gullies

In this phase, the goal was to identify the number of water erosion processes over space using techniques of temporal and spatial variability distribution of gullies for each year analyzed, i.e., 1992, 2002, 2006, 2011, 2016, and 2019 in the urban expansion area of Goiânia. For this analysis, the ArcGis 10.4 software was used to analyze, interpret, and create geodatabases from those images of high resolution visible on the 1:2.000 scale, resulting in six specific point vector layers that were based on the manual vectorization method with a single point per gully positioned up the head. These input values were based on the same definitions and measurement techniques as those used to develop the model of spatial interpolation by estimating the value of a field anywhere from a limited number of random sample points.

Delimitation of critical areas and density of erosive gullies

This index is important to describe the density of erosive gullies and delimitation of critical areas, corresponding to the concentration of the water erosion process in the study area. Following the method used by Ferreira (2014), apud Willians (1972), Taylor (1977) and Maarseeven et al. (2019) the statistical index Rn is defined by the ratio between the average distance calculated from each point to the nearest neighbor (Ln) and the expected average distance between all points on the layout (Lc). For it, the point vector layer for each year was used to input data to the Kernel Density tool of the software ArcGis 10.4, which allowed the construction of the density modeling of gullies in the positioning of the 5 intervals for each year.

These intervals were based on the relationship between erosion rates and the nearest neighbors (Rn) as a function of density and distance within the entire urban area "*Macrozona Construída de Goiânia— MZCG*", which showed a representative spatially variable distribution of gullies for this area and in each year. This allowed the interpretation of the obtained intervals values by Rn, that varied from 2.15 for the maximum aggregation of the gullies and from 0.0 for the maximum dispersion of the points. Next, the five intervals classified in the same values were converted from raster to vector following the aggregate features within specified attributes in a particular area by the aggregation/concentration of erosive gullies/area identified. These were, namely: High criticality for areas with a higher concentration of gullies; High to Average for aggregates but less than the previous one; Average for dispersed erosive gullies; Medium to Low are attributed to random patterns and Low criticality, the occurrence of erosive gullies are isolated or null.

Use and occupation of urban land

In this phase, one of the most important objectives was to produce data of sufficient quality and quantity to permit confident analyses and interpretations in addressing the cause-and-effect of gullies over the urban area of Goiânia. For it, an exploratory analysis was made from databases described in the preliminary stage such as images of high resolution for each year and the official delimitation of neighborhoods, which allowed a better understanding of the area object and the classification that should be adopted for those critical areas.

After making observations from the previous exploratory analysis, a new classification of urban land use for Goiânia was created using variables from the methodologies of Canil (2001), Nascimento and Oliveira (2015), and Cavalcante (2019). The geodatabase created on a scale of 1:1.000 resulted from manual vectorization of the classes from aerial photography for each year. This classification identified the different areas within the space and allocated those fragmented areas as vegetal cover, neighborhoods, and blocks by number of lots with construction and occupancy of buildings.

The urban land use is classified under the following conditions: (1) the vegetation areas of the native forests with a density higher than 80%; (2) the vegetation areas with high level degradation with shade of tree with density between 10 and 80%; (3) low or absent vegetation in areas of exposed soil; (4) the consolidated urban use defined by neighborhood with high rates of constructed building; (5) unconsolidated use that is characterized by high numbers of buildings constructed per block with no more than 20% of lots available for a new building; (6) unconsolidated use that is linked by a new neighborhood with rates between 40 and 80% of constructed buildings; (7) unconsolidated areas in a new neighborhood with rates of construction lower than 40% per block; and (8) local traffic route that can be (a) paved and (b) unpaved.

The road network was manually vectored for single and double lines specifically between the limits for each block, with a segment classified as paved and unpaved. The following step was applied to the Buffer tool in the ArcGIS 10.4 software using the minimum dimensions of public roads such as the arterial roads (30 m) and local roads (13 m) given by the parameters described in annexes I, II, and V (Goiânia, 2007). The Erase tool was used at this stage to subtract from the urban land use layer and other features such as the road and the drainage network vector.

Flow Direction

In this phase, the methodology used in studies by Rieke-Zapp and Nearing (2005), Bogart and Troch (2006) Oliveira et al. (2012), and Jardim (2017) was strongly followed for each step of the hydrology spatial analysis. The main product of that analysis was flow accumulation and flow direction, which was extracted by the Digital Terrain Model (DTM) which has 4.79 m resolution. The flow direction was extracted using the Flow Direction tool in the ArcGIS 10.4 software, which is the input data of the DTM. This method is known as the eight directions, in which a single cell of the grid is linked by the eight adjacent cells.

While working with the same software, was extracted the flow accumulation using the Flow Length tool. This processing set as the flow direction raster and direction of measurement from Upstream. This setting was to calculate the longest upslope distance along the flow path, from each cell to the top of the drainage divide to obtain the flow concentration (O'Callaghan & Mark, 1984) and Maarseeven et al. (2019). It is important to note that the D8 method has been used frequently in studies of water erosion, especially to estimate concentrated flows by a particular contribution area from upstream of the terrain (Valeriano, 2003).

Data Analysis and Interpretation

The erosion was estimated by calculating simple indices of gully density, using aero photography of different dates to spatial variability distribution. For this, the density modeling of gullies classified in five intervals was an essential survey to identify the highly critical areas for each year. In this way, changes to the density of gullies could be examined in relation to the water erosion process over changing urban land use and increasing population pressures on a particular area.

The density modeling of gullies from 1992 with two highly critical areas was chosen as a procedure for data analysis and interpretation by the cause-and-effect relationship between erosion processes and existing conditions of topography and urban land use. In this analysis, the spatial correlation was linked by variables that influence these processes, such as: existing problems in different phases of urbanization, that was divided before and after the basic document of environmental, urban, and territorial planning policies; the reactivation of old erosive gullies; layout of the road system, topography and concentrated flows that were based on parameters of the Cartesian plane in the transversal, parallel, and perpendicular space in Euclid's foundations in geometry.

3   Results and Discussion

The dynamics of gullies, from year to year (Table 1), were based on simple analysis and interpretation made by the frequency of an event in the past and future. Also, the same data information shows that the number of erosive gullies

decreased remarkably from 112 in 1992 to 29 in 2016, as already explained. Although many gullies have been controlled, it is observed that new gullies have emerged and others were reactivated from one year to another. Thus, in 1992, 112 gullies were registered, of which only 9 (8%) were identified in previous registrations by IPLAN (1983, 1987), with 103 (92%) being new gullies. In 2002, 84 gullies were identified, of which 2 (2.4%) already existed in 1982 and 17 (20.2%) in 1992, and thus 65 were new gullies. Of the 75 gullies identified in 2006, 65 (86.7%) were new and only 9 gullies (13.3%) registered since 1982 were repeated. In 2011, of the 30 identified gullies, 26 (86.7%) were new and only 4 (13.3%) repeated since 1992. In 2016, only 29 gullies were registered, of which 19 (65.5%) were new and 10 (34.4%) had existed since 1982.

The methodology applied by density estimation of gullies identified five intervals of critical areas, with spatial variability for each year given by the highest and lowest number of gullies aggregated in the *MZCG*. Figure 4 allowed us to see that the high criticality identified by red areas showed spatial changing from the eastern region in 1992 to the southwest in 2016, passing through the northwest and west, developing a script in a counterclockwise direction. In these areas of high criticality, the density decreased from 2.3 gullies/km^2 in 1992 to 1.06 gullies/km^2 in 2002, followed by 0.28 gullies/km^2 in 2011. Exceptions to this were 2006 in which there was an increase to 1.44 gullies/km^2, and in 2016 an increased to 1.43 gullies/km^2.

The areas of medium to high criticality decreased year after year, starting with 1.14 gullies/km^2 in 1992, moving to 0.57 gullies/km^2 in 2002, 0.46/km^2 in 2006, 0.17/km^2 in 2011, and finally to 0.16/km^2 in 2016. Average densities decreased from 0.67/km^2 in 1992 to 0.07/km^2 in 2011, although in 2016 they increased to 0.11/km^2. Figure 4 also shows that the medium to low and low densities decreased significantly from year to year, where the average to low density, which in 1992 was 0.30/km^2, in 2016 reduced to 0.03/km^2; and the low density, which in 1992 was 0.01/km^2, decreased to 0 gullies/km^2 in 2016.

Despite the reduction in the number of erosive gullies per critical area between 1992 and 2016, it is important to understand the factors that conditioned the high concentration of erosions over specific areas, as well as why and how these areas changed from high to low criticality over the years. As an illustration, only the results of the high critical

Table 1 Spatio-temporal dynamics of erosive gullies of MZCG

Total	Erosive dynamic/Year				
	1992	2002	2006	2011	2016
Erosion/Year	112	84	75	30	29
New erosions	103[a]	65	65	26	19

[a] Extracted from previous registration (IPLAN, 1983, 1987)

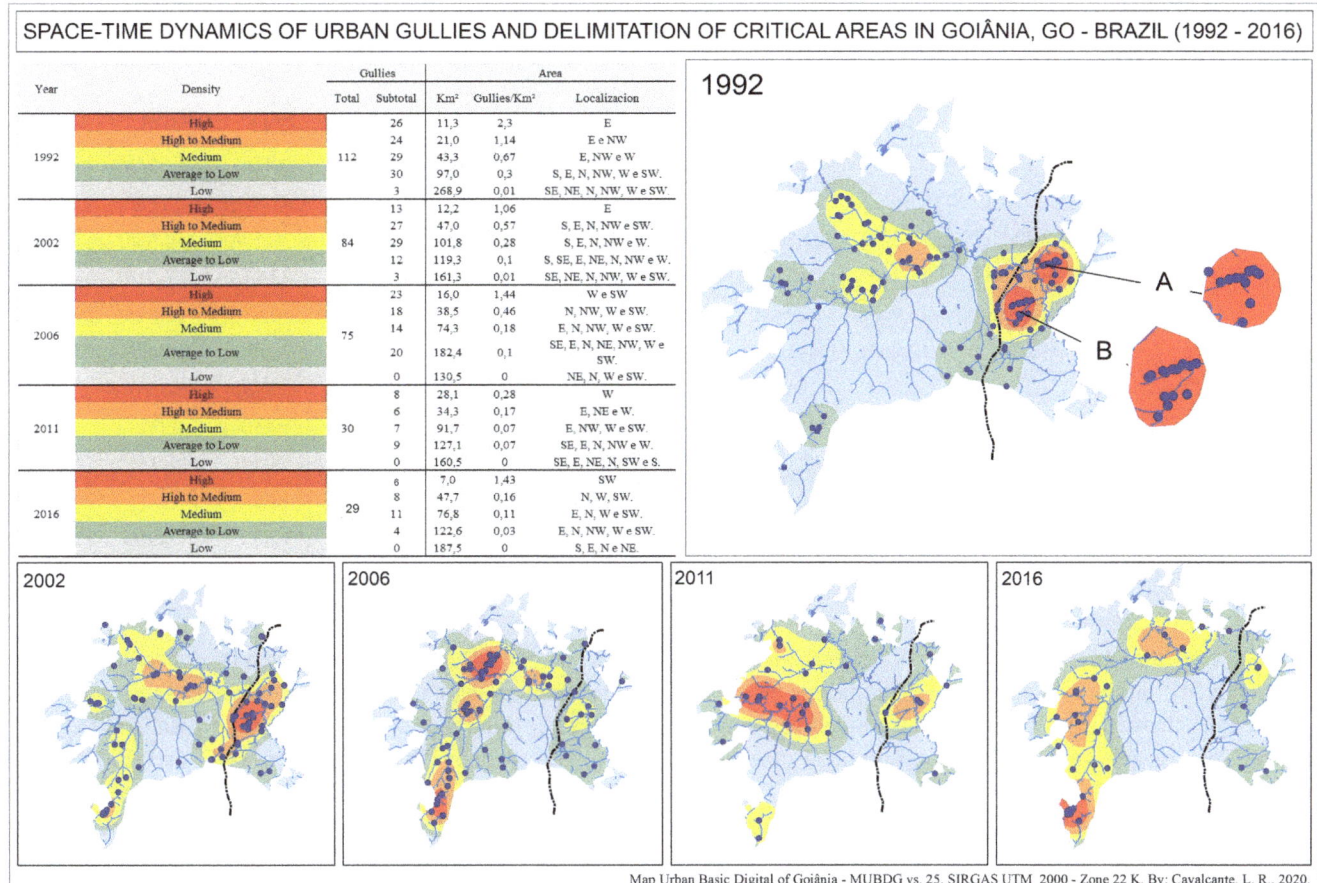

Fig. 4 Density of linear erosive gullies in the "*Macrozona Construída de Goiânia*"

areas of 1992 are presented below in the following results. These are in order to understand the relationship between concentrated areas of gullies and urban land use, physical environment, and axes of concentration of the flow.

3.1 Temporal and Spatial Variability Urban Land Use

The great changes that occurred in the period from 1992 to 2019 were especially in the beginning, due to the intense suppression of the vegetal cover for accelerated implantation of housing developments. These are characterized by irregular subdivisions that were resulting in a discontinuous urban network with poor infrastructure, where the occupation process is still recent and appears to be unconsolidated, as shown in Table 2. Based on the limits of the high critical areas *A* and *B* of 1992 delimited in Fig. 4, similar indices were found regarding the predominance of erosive gullies in certain classes of anthropic land use interventions, but different phases of the urbanization process. The results of interpretive areas *A* and *B* were divided into three phases,

before and after the basic document of environmental, urban, and territorial planning policy such as the "*Plano Diretor*" and the "*Carta de Risco de Goiânia*" since 1991.

1st Phase

The use of land and the distribution of erosive gullies in 1992 from areas *A* and *B* show two different models of the urbanization process of *MZCG*. The first (area *A*) is characterized by the absence of infrastructure with 82.3% of open roads with exposed soil and the absence of a micro drainage system. In these, most areas are recently occupied subdivisions with low occupancy of lots and irregular neighborhood close to the springs and valley bottoms started in the 1980s.

While the second area shows the urban process is more accelerated, with numerous irregular subdivisions created between 1950 and 1960 to serve the low-income immigrant population. It often involves infrastructure installation and regularization after occupation. Clearly, the first phase corresponds to the old urban policy implemented since the "*Decreto* nº16 de 1950" in 1992. It is recognized by the

Table 2 Use and coverage of urban land in an area of high criticality of erosive areas in Goiânia

Use	A						B					
	Area (%)/Erosion						Area (%)/Erosion					
	1992	2002	2006	2011	2016	2019	1992	2002	2006	2011	2016	2019
1	1.5/0	0.98/0	1.11/0	2.29/0	4.21/0	8.9/0	0.41/0	0.96/0	1.41/0	3.41/0	5.85/0	4.33/0
2	**8.8/1**	7.64/0	9.11/0	9.46/0	10.19/0	4.7/0	**8.72/2**	**11.19/3**	13.01/0	11.94/0	7.12/0	6.01/0
3	**47.3/5**	**34.7/1**	29.26/0	26.52/0	21.8/0	**23.8/2**	**24.27/12**	**17.44/3**	**16.34/2**	10.59/0	10.48/0	**13.09/6**
4	2.4/0	7.37/0	12.96/0	17.23/0	20.5/0	22.6/0	7.96/0	11.19/0	13.95/0	22.11/0	27.11/0	44.32/0
5	12.2/0	**16.97/1**	20.08/0	25.96/0	26.99/0	29.6/0	34.91/0	33.49/0	38.95/0	40.67/0	36.87/0	**25.47/1**
6	**15.5/2**	21.56/0	20.37/0	12.26/0	9.41/0	5.4/0	18.14/0	19.85/0	11.75/0	7.3/0	7.06/0	**5.68/2**
7	**12.4/2**	10.76/0	7.1/0	6.28/0	6.9/0	5.0/0	5.59/0	5.87/0	4.6/0	3.97/0	5.51/0	1.1/0
Total	100/10	100/2	100/0	100/0	100/0	100/2	100/14	100/6	100/2	100/0	100/0	100/9
8[a]	17.7/0	72.9/0	92.5/0	96.2/0	98.1/0	98.5/0	87.1/0	91.4/0	96.8/0	98.4/0	98.1/0	99/0
8[b]	**82.3/2**	27.1/0	7.5/0	3.8/0	1.9/0	1.5/0	12.9/0	8.6/0	3.2/0	1.6/0	1.9/0	1/0
Total	100/2	100/0	100/0	100/0	100/0	100/0	100/0	100/0	100/0	100/0	100/0	100/0

*Paved (**a**) and unpaved (**b**) local roads

continuous process of occupation of peripheral areas and respectively the phase with the highest number of erosions. In those areas there is exposed soil with degraded vegetation, followed by neighborhoods of unconsolidated use with unpaved roads.

The values presented in Table 2 allowed us to see that the area A is represented by 47.3% of exposed soil and vegetal cover of less than 10%. It also shows 27.9% of unconsolidated use with occupation/building less than 80%. 82.3% of the roads are unpaved which is linked to two erosion processes. While area B is represented by 34.9% of unconsolidated use with occupancy building higher than 80 and 24.2% exposed soil with vegetation cover degraded. This also shows area B having 87.1% of local paved streets and a concentration of erosive gullies in areas of low vegetation close to the drainage channel. Also, the consolidated urban use has the lowest percentages, with only 2.4% for area A and 7.9% for area B. Areas A and B has no erosion processes in 1992.

2nd Phase

The areas described in 1992 as having vegetation covered lower than 10% and the highest rate of erosion (Table 2), were found to have reduced in 2002 and 2006 to 34.7% and 29.2%, respectively, for area A. Area B showed reductions up to 17.4 and 16.3% for those same years. These areas are characterized by a concentration of gullies in peripheral neighborhoods, valley bottoms, and springs. In 2002, there was only one erosive occurrence in areas of unconsolidated use with high occupancy within area A. This area was previously defined by the "*Carta de Risco de Goiânia*" (Fig. 5) in 1991 as at risk of erosion by regressive evolution of the

headwaters of watercourses, promoted by concentrated flows downstream toward the drainage channel. Area A showed the greatest difference in terms of *paved* local traffic roads. These increased by 55.2% between 1992 and 2002, and a further 19.6% up to 2006 following the implementation of infrastructural developments. Overall local *paved* roads totalled 92.5% by 2006.

3rd Phase

The third phase corresponds to the period following the update of the PD (Goiânia, 2007) and the "*Carta de Risco de Goiânia*" (ITCO, 2008). This period is characterized by the inclusion of the criterion of continuity to another neighborhood that had already been implemented and also the requirement of having at least 30% occupancy, built and occupied. This inclusion contributed to the continuity of public policies already implemented since the 1990s, which regarded the continuous process of filling urban voids and consolidation of existing subdivisions. That allowed better control and ordering of land use in terms of filling the urban grid by 62.5% for Area A and 76.6% for Area B, especially with infrastructure installed on average 98% in 2019.

The results also point out that area A has a higher rate of degraded vegetation with tree coverage below 10%. Also, its unconsolidated use with average occupation/building, showed it was an area supposedly less urbanized when compared to area B. This would be expected since both areas had seen occupation in different periods. However, areas A and B are similar in terms of the evolutionary dynamics of land parceling, whose occupation/construction is less than 40% and located mostly in risk areas such as springs and valley bottoms (Fig. 6). The areas are characterized by

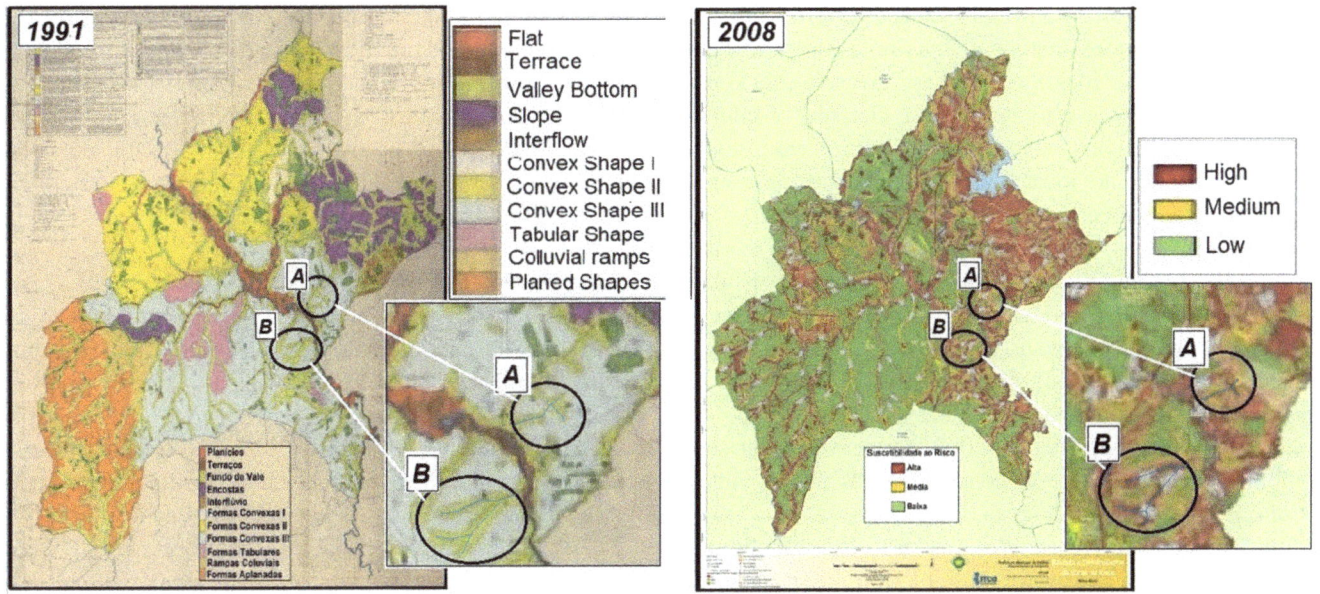

Fig. 5 "*Carta de Risco de Goiânia*" in 1991 and updated in 2008. *Source* "*Carta de Risco de Goiânia*" (1991, 2008)

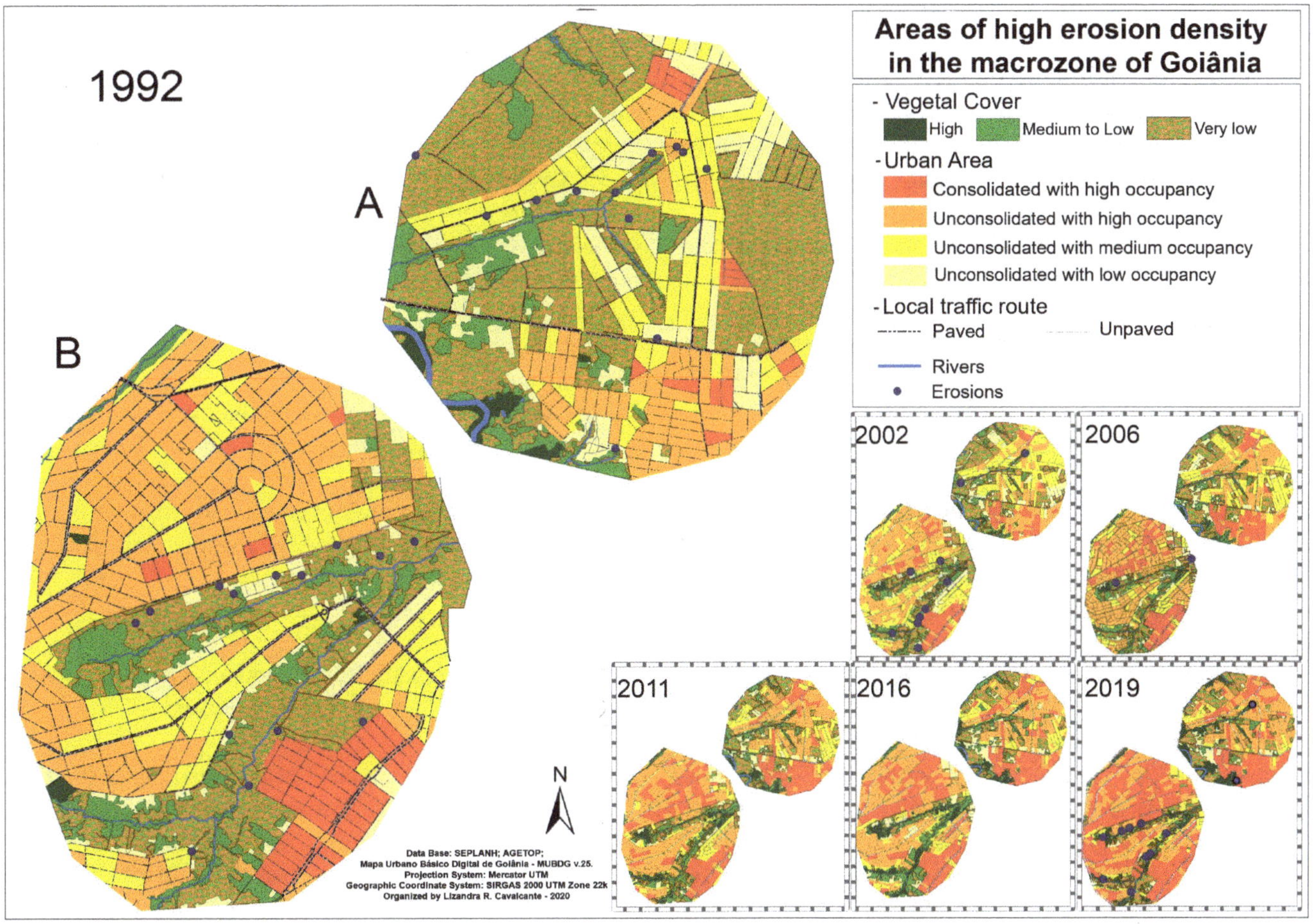

Fig. 6 Urban land use of areas *A* and *B* between 2002 and 2019

decreasing rates from 1992 to 2019 except in 2016 when the number of occupations increased. It is noteworthy that no erosive gullies were identified in these areas in 2011 and 2016, based on aerial photography of Goiânia. While in 2019, in the image of GoogleEarth, 11 gullies were registered, of which two are located in area A drainage headland, which had already been identified in 1992 and 2002.

Therefore, it was found that the highest rate of erosion occurred in the 1st phase when the urban area is still unconsolidated and a large amount of exposed soil susceptible to rain erosivity prevails (Tucci, 1995; Cruz & Tucci, 2008; Goudie, 2006). This was especially in the areas of occupation restriction defined by the "Carta de Risco de Goiânia" in 1991—updated in 2008, as areas of high susceptibility of erosion due to concave slopes in the valley bottom, as expected (see Fig. 5).

Still as predicted, the categories of land use that did not show water erosion rates were those of high-density vegetal cover, such as forests, parks, APPs, areas of consolidated use, and also unconsolidated use with occupation/building greater than 80%. This is probably due to the role of intercepting raindrops by vegetation in 40% of total rainfall for tropical climates (Prandini et al., 1976; Morgan, 2005). This is in addition to the low exposed soil index, the high topography, but with a straight and convex surface, as well as the presence of a road system connected to the micro drain network (Nascimento & Faria, 2007; Nascimento, 1994; Podestá Filho, 1993).

It is also worth noting the fact that urban voids and unconsolidated areas decrease as they are consolidated, as well as the data indicating that the number of erosive gullies reduced, between the first and second phases, from 26 in 1992 to 8 in 2002, followed by only 2 gullies in 2006 for area A and B. This suggests at the outset that the problems pointed out in the diagnoses of IPLAN (1983, 1987), of the Plano Diretor (1992) and of the "Carta de Risco de Goiânia" (1991), and later by Nascimento and Sales (2002) helped as a strong instrument in the control of erosion processes as well as for territorial reordering. Other public policies were developed and implemented since the 90s, after the creation of the Municipal Council for Urban Policy (COMPUR in 1991), the Municipal Environment Agency (AMMA in 1992), the Construction and Housing Company of the Municipality of Goiânia (COMOB in 1995), of Basic Sanitation (ARG—Regulation, Control and Inspection Agency for Public Services in Goiânia, in 2016), served as a contribution to the priority occupation of peripheral areas, the filling of intra-urban voids, the structuring of the basic road network and the urban rainwater drainage system (Goiânia, 2007).

However, at the same time, the additional costs generated by the government, due to the delay in services, regarding the provision of water, energy, paving services and, above all, the micro drainage system in remote subdivisions, resulted in speculative profits for land owners, the suppression of vegetation cover for the opening of new subdivisions, and irregular occupations in valley funds, as shown in Fig. 7.

Thus, as already pointed out by Nascimento and Podestá Filho (1993), Nascimento (1994), Nascimento and Sales (2002), and Nascimento and Oliveira (2015), the problems existing in different phases of urbanization are the result of old public administrations that did not control the use of urban land. Although control measures were subsequently employed, through the expropriation of families in risk areas and the recovery of vegetation for the years after 2006 (Fig. 7), the reactivation of the erosive process in 2019 has not yet been verified. This reinforces the idea of associating environmental degradation with the spontaneous and disorderly occupation process.

It draws attention to the failure to identify erosive gullies in the aerial photographs of 2011 and 2016—shown in Table 2 and Fig. 6, which leads to the supposition that the policies applied to combat erosion were effective. However, when observed in a second temporal analysis based on Google Earth images, this interpretation was discarded when the reactivation of old erosive gullies in the different months over the years was verified. This was probably due to the use of ineffective and temporary corrective measures, e.g., as shown by the erosion reactivated in area B (Fig. 8).

Figure 8 allows us to correlate the evolution of the erosive process with the pluviometric indexes, whose representative months from October to April correspond to the rainy season. As a consequence, a greater concentration of surface flows, due to the absence of the micro drainage system and the direct release of rainwater into the drainage channel, intensified after the opening of roads and irregular occupations in a short period of time. A fact clearly noticeable after 2014.

3.2 Contour, Urban Road System, and Routes Concentration of Surface Flows

In Table 3, the values obtained at the crossing of the level curves, the directions of surface runoff and the accumulation of rainwater flows with the occurrence of erosive processes are presented in general for the years 1992, 2002, 2006, and 2019. It should be noted that the results were based on the Digital Terrain Model for the hydrological analysis of

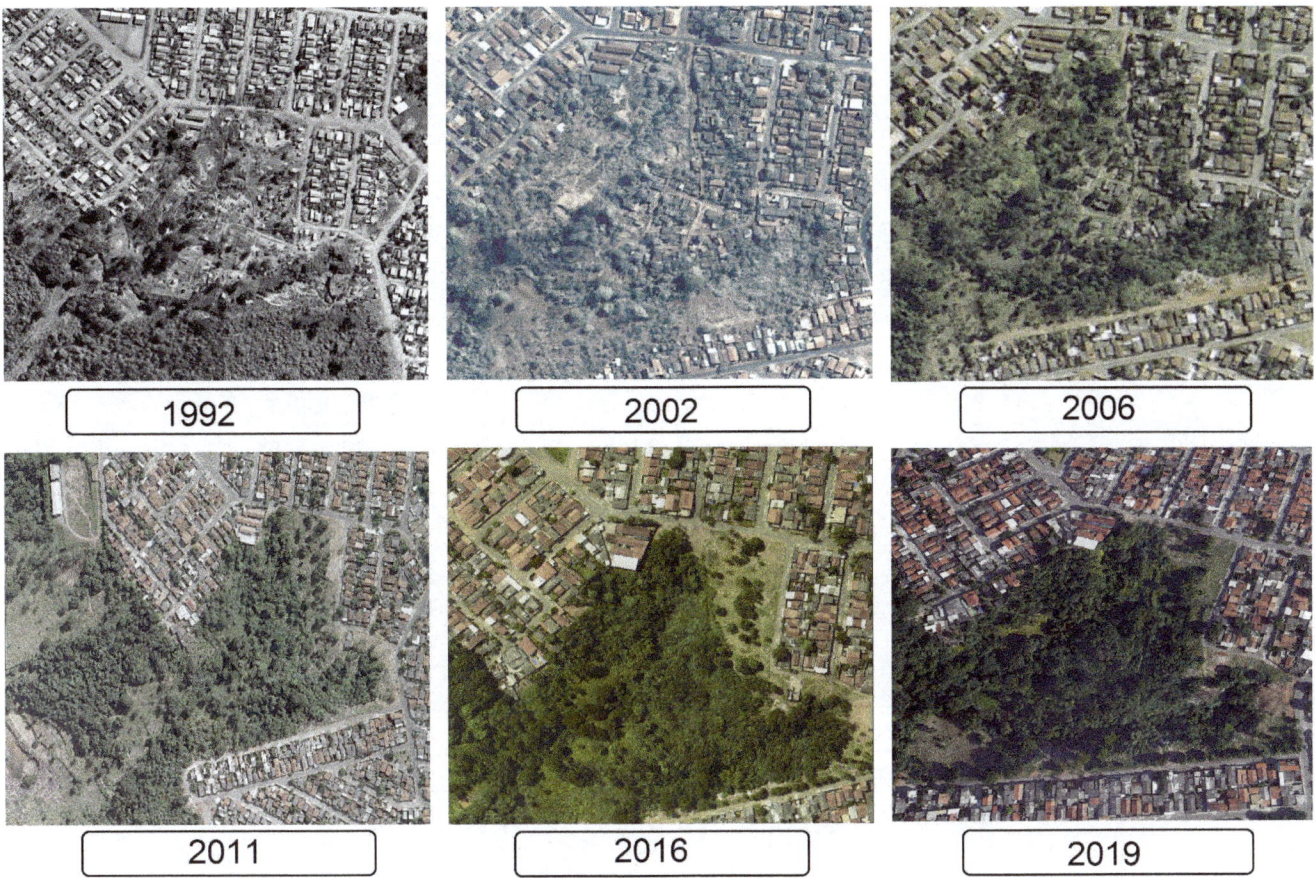

Fig. 7 Example of an area degraded by erosion due to the accelerated process of clandestine occupation in area *A*. *Source* Aerial Photographies from 1992 to 2016 (MUBDG, 2017) and GoogleEarth (2019). Geographic coordinates 16°39′48″S and 49°11′38″W

interpretive area *A* and *B*, as already mentioned in the methodology (item 2).

The interpretation of urban morphology enabled us to relate the layout of the road system to the problems of water erosion, especially in irregular installments or with the project not adjusted to the physical (geotechnical) characteristics of the terrain. Examples of this are poor infrastructure, wide local streets, very straight streets, and areas of great slope. Still, the results obtained revealed a different behavior regarding the occurrence of erosive processes in areas of low to high concentration of surface flows in interpretive areas *A* and *B*. Though they reflect similarities in terms of the physical characteristics of the terrain.

Figure 9 shows that the majority of the area *A* and *B* road system has a total width of 13 m for local roads, corresponding to the streets created for internal displacement. These may or may not have a direct connection with wide arterial roads of the 30 m, corresponding to long and perpendicular to contour lines, whose implementation project took into account the viability and mobilization of local traffic (Goiânia, 2007). The analysis of the indices presented

in Table 3 for the values obtained from contour lines and erosive gullies with the morphology of the urban road system, as well as the adoption of parameters based on the Cartesian plane in the transversal, parallel, and perpendicular space in Euclid's foundations in geometry, allowed the following observations:

1. The arterial road is perpendicular to the contour lines, where the points of intersection occur preferably on a concave surface and connected to the drainage channel. These arterial roads are located in a segment from 810 to 720 m altitude. This fact may condition the increase in the magnitude of the volume and speed of the concentrated flows due to the impermeabilization of the urban soil;

2. Arterial road perpendicular to mid-slope level curves with intersection in local roads within 90° angle. This particularly occurs in areas having a concave surface form and a high susceptibility to erosion due to high sediment transport by concentrated flows. This is especially when the system of galleries intended to receive

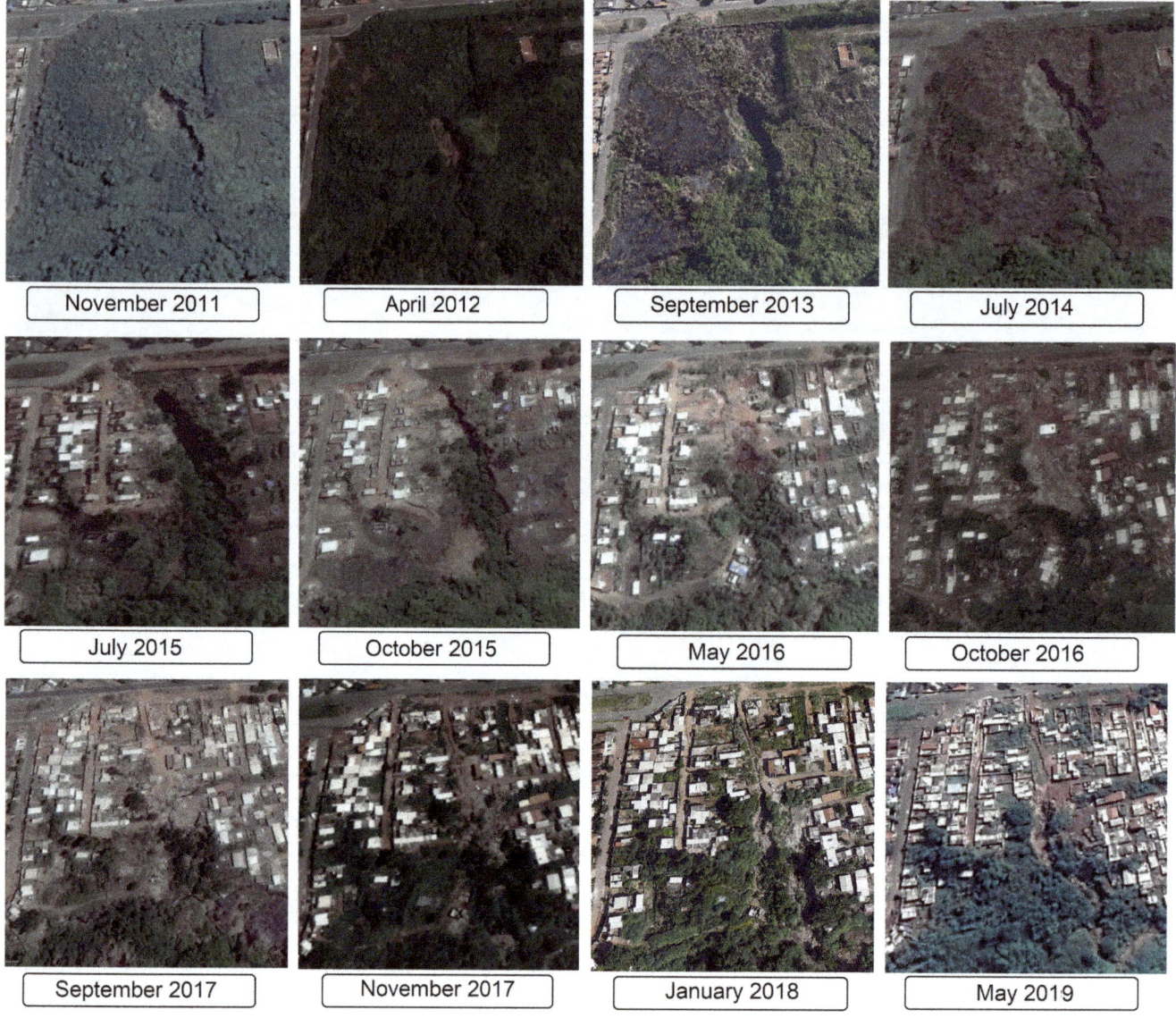

Fig. 8 Example of erosion process in area *B* reactivated in representative months between 2011 and 2019. *Source* GoogleEarth (Geographic coordinates 16°40′49″S and 49°13′13″W)

rainwater captured on the surface is ineffective or non-existent. The case of erosions located at 715–725 m altitude on unpaved roads as shown in Tables 2 and 3.

3. Asymmetrical linear road system in areas located between the foot slope and the half slope. This is represented by the open roads in irregular subdivisions that were later regularized by the local government and were paved perpendicularly to the contour lines and to the drainage system. This was at 90° angle with the intersection of roads in the shape of '*T*' and '*H*', without meeting the geotechnical standards of the terrain.

Although the expropriation of areas at risk of erosion had already been foreseen in the 1992 PD.

Still, the values presented in Table 3 indicate that the highest erosion rates occur in areas with low accumulation of superficial flow for *A*, while in area *B,* erosive gullies predominate in low and medium axes and moderate flow accumulation for all years analyzed, as shown in Fig. 10.

In addition, the results indicate that area *A* has a greater number of routes with a low accumulation of surface flows, corresponding to 87.36% of the total area, being an area

Table 3 Contour lines, flow accumulation, flow direction, and erosion in the interpretive area *A* and *B*

Synthesis *A* and *B*					
Contour lines (m)		1992	2002	2006	2019
695–700		0/0	**1/0**	0/0	0/0
700–705		**1/0**	0/0	0/0	0/0
705–710		**0/1**	**0/1**	**0/1**	0/0
710–715		**0/1**	0/0	0/0	**0/1**
715–720		**1/0**	0/0	0/0	**0/1**
720–725		**2/3**	0/0	0/0	0/0
725–730		**2/2**	**2/4**	0/0	**0/1**
730–735		0/0	**0/1**	0/0	**0/2**
735–740		**0/1**	0/0	0/0	**0/1**
740–745		**1/1**	**0/1**	0/0	0/0
745–750		**0/1**	0/0	0/0	**2/2**
750–755		**3/2**	**1/0**	**0/1**	0/0
755–760		**1/2**	0/0	0/0	0/0
760–765		**1/0**	**0/1**	0/0	**0/1**
Total		12/14	2/8	0/2	2/9
Flow accumulation (m)		1992	2002	2006	2019
Low	0–266,604	**12/6**	**2/0**	0/0	**2/3**
Reasonable	266,6041–283,7618	0/0	0/0	0/0	0/0
Average	283,7619–550,3658	**0/7**	**0/5**	**0/2**	**0/4**
Moderate	550,3659–4.692,965	**0/1**	**0/3**	0/0	**0/2**
High	4.692,966–69.062,32	0/0	0/0	0/0	0/0
Total		12/14	2/8	0/2	2/9
Flow direction		1992	2002	2006	2019
East	67.5–112.5	0/0	0/0	0/0	**0/1**
Southeast	112.5–157.5	**3/8**	**0/2**	**0/2**	**0/7**
South	157.5–202.5	**3/2**	0/0	0/0	**1/0**
Southwest	202.5–247.5	**1/0**	**1/0**	0/0	**1/0**
West	247.4–292.5	**3/0**	**1/0**	0/0	0/0
Northwest	292.5–337.5	**1/3**	**0/5**	0/0	0/0
North	337.5–360 and 0–22.5	**1/0**	**0/1**	0/0	0/0
Northeast	22.5–67.5	**0/1**	0/0	0/0	**0/1**
Total		12/14	2/8	0/2	2/9

Values separated by (/) are equivalent to interpretive areas (A/B)

supposed to have a flatter topography when compared to *B* (80.69%), slightly lower. Still, it was found that 8.82% of the total area *A* and 13.33% of area *B* are located on routes of medium concentration of flows and moderate flow with 2.77% of area *A* and 13.33% for *B* between upstream and downstream of the slope. There was also the identification of high values of flow accumulation only in the areas, corresponding to the total of 0.11% of area *A* and 0.04% for *B*,

downstream between the confluence of the tributaries with the main river Meia Ponte—illustrated in Fig. 10.

Attention is drawn to the hydraulic characteristics of water flow initiated at a critical distance from the top of the slope, which reveals that the routes of concentration of flows indicate the existence of four stages in cases where there is an erosion process, namely: 1. Diffuse runoff; 2. Runoff with some concentration in preferred points; 3. Flow concentrated

Fig. 9 Urban road system and the distribution of erosive gullies in areas *A* and *B*

in microchannels, without defined headlands; and 4. Concentrated flow in microchannels, with defined headlands. As expected, the areas of contribution of the gullies in the slope, for the most part, do not present unique patterns as to the flow directions at points of occurrence of erosive processes. Although they are still in agreement with the areas with the highest declivity and the concentration of surface flows, as shown in Fig. 10.

4   Conclusions

In summary, techniques of temporal and spatial variability distribution of gullies showed that qualitatively address issues described and alleged planning errors in the past due to the rapid growth of the Goiânia city. Those issues are associated with the reduction of the density vegetal cover, the increase of the unconsolidated neighborhoods with low occupancy, built and occupied lots, the occupation of the spring and valley bottoms, the poor infrastructure with the absence of the micro drainage system and unpaved streets and the street project not adjusted to the physical characteristics of the terrain. These issues resulted from three

phases of the urbanization process that was started in the 1950s, and also correspond to the highest number of erosion.

In the first phase, the municipal government lost control over the urban land use that favored the high number of erosion more than 100 and the increasing of irregular subdivisions without infrastructure. The second phase, after 1992, with the implementation of the public policies, decreased the rates of erosion to 84 and 75 and the absence of infrastructure in the following years. The third phase, after updating the PD in 2007, contributed to the continuity of public policies already implemented since the 1990s, which favored reducing the number of erosion and increase of consolidated neighborhoods and unconsolidated use with high occupancy.

The morphology of the urban road system showed a straight relationship between gullies and routes the concentrated flows. These routes of water flow are initiated at a critical distance from the top of the slope, increasing the volume and speed of the concentrated flows as low and moderate flows accumulate from the variable direction. These concentrated flows are associated specifically with the arterial roads perpendicular to the contour line, the streets with intersections crossed in the mid slope, and the

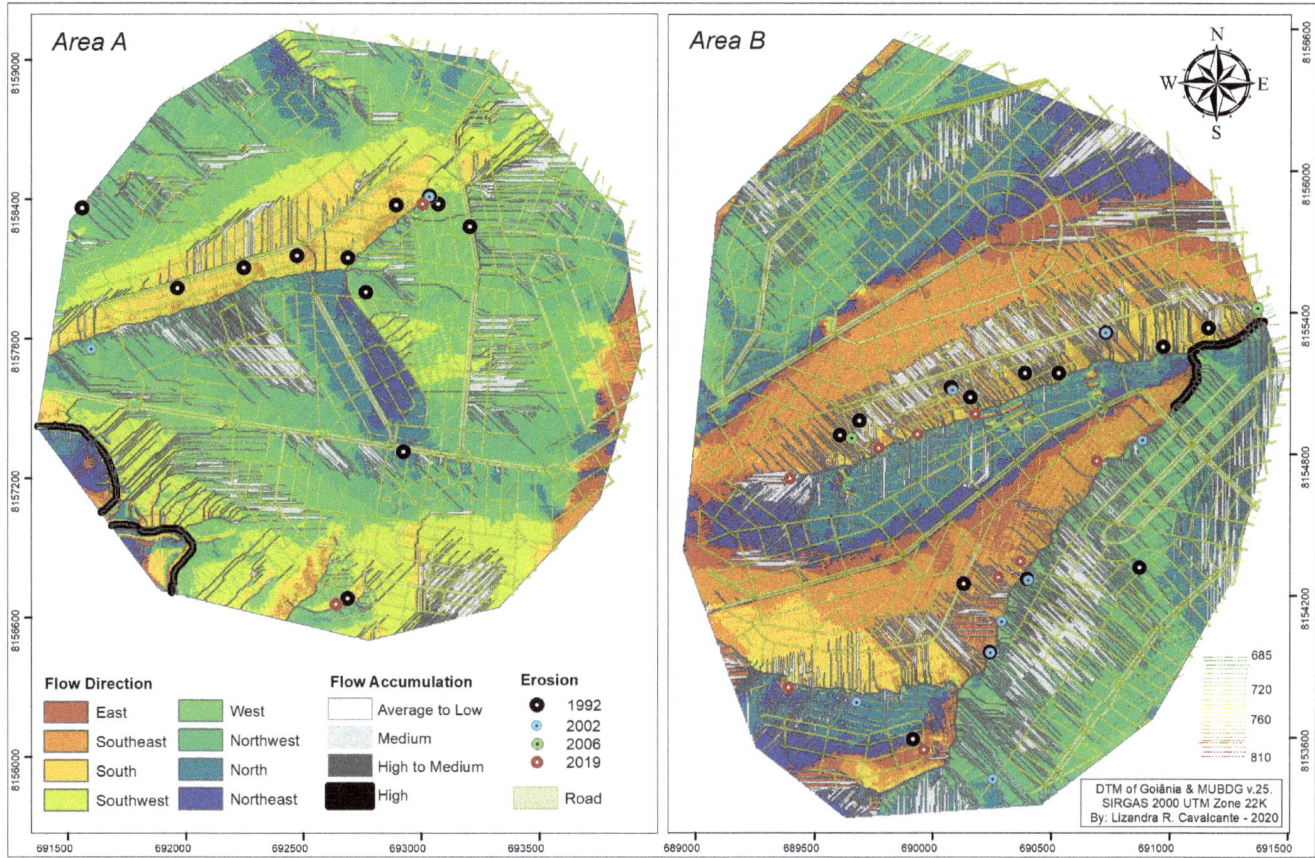

Fig. 10 Contour lines, flow accumulation, flow direction, and erosion in areas *A* and *B*

asymmetrical linear road system, which were implanted with wide and straight streets in areas of the great slope with projects that were not adjusted to the physical characteristics of the terrain.

The spatial data information with high quality resolution from the different months and the years before 2016, was a limitation found during the analysis of the water erosion process. This restriction resulted in no erosive gullies being identified for those representative areas in 2011 and 2016, based on aerial photography of Goiânia. While working with images of Google Earth in 2019, these results showed to be not representative. This limitation of the spatial data information also showed that it is necessary to use more than one image to analyze the water erosion process in urban areas. Otherwise, it can be covered by ineffective and temporary measures to control the erosion.

References

Almeida Filho, G. S., & Ridente Júnior, J. L. (2001). *Diagnóstico, Prognóstico e Controle de Erosão: Noções básicas para controle e prevenção de erosão em área urbana e rural*. VII Simpósio Nacional de Controle de Erosão.

Bertoni, J., & Lombardi Neto, F. (2017). *Conservação do Solo* (10th ed.). Ícone.

Bogaart, P. W., & Troch, P. A. (2006). Curvature distribution within hillslopes and catchments and its effect on the hydrological response. *Hydrology and Earth System Sciences, 10*(6), 925–936. https://doi.org/10.5194/hess-10-925-2006

Brasil. (2001). *Estatuto da Cidade Lei n. 10.257*. REGULAMENTA OS ARTS. 182 & 183 DA CONSTITUIÇÃO FEDERAL, ESTABELECE DIRETRIZES GERAIS DA POLÍTICA URBANA E DÁ OUTRAS PROVIDÊNCIAS. https://www.planalto.gov.br/ccivil_03/leis/leis_2001/l10257.htm

Casseti, V. (1992). Geomorfologia do Município de Goiânia-GO. *Boletim Goiano De Geografia, 12*(1), 65–85. https://doi.org/10.5216/bgg.v12i1.4377

Canil, K. (2001). *Indicadores para monitoramento de processos morfodinâmicos: aplicação na Bacia do Ribeirão Pirajuçara (SP)*. Tese de Doutorado – Programa de Pós-Graduação em Geografia Física. Universidade de São Paulo. https://doi.org/10.11606/t.8.2007.tde-04062007-141138

Carvalho Junior, O., et al. (2010). Urbanization impacts upon catchment hydrology and gully development using multi-temporal digital elevation data analysis. *Earth Surface Processes and Landforms, 35*(5), 611–617. https://doi.org/10.1002/esp.1917

Cavalcante, L. R. (2019). *Análise da evolução da paisagem urbana de Goiânia (GO) e a distribuição de focos erosivos hídricos de 1992 a 2016*. Dissertação de Mestrado. Universidade Federal de Goiás, Instituto de Estudos Socioambientais (IESA). Programa de Pós-Graduação em Geografia, Goiânia. http://repositorio.bc.ufg.br/tede/handle/tede/9631

Cavalcante, L. R., & Castro, S. S. (2021). Space-time dynamics of urban water erosive features in Brazil: The case of Goiânia, Capital of Goiás State. In *Soil conservation: strategies, management and challenges*. Nova Science Publishers.

Cruz, M. A. S., & Tucci, C. E. M. (2008). Avaliação dos Cenários de Planejamento na Drenagem Urbana. *Revista Brasileira de Recursos Hídricos RBRH, 13*(3), 59–71. https://doi.org/10.21168/rbrh.v13n3.p59-71

Faria, K. M. S. (2007). *Processos Erosivos Lineares no Município de Goiânia – Goiás*. Agência Municipal de Meio Ambiente de Goiânia. Gerência de Contenção e Recuperação de Erosões (GRECRE). http://www.geomorfologia.ufv.br/simposio/simposio/trabalhos/trabalhos_completos/eixo11/054.pdf

FAO & ITPS. (2015). *Status of the World's Soil Resources (SWSR)— Technical Summary*. Food and Agriculture Organization of the United Nations and Intergovernmental Technical Panel on Soils, Rome, Italy. http://www.fao.org/3/i5126e/i5126e.pdf

Ferreira, M. C. (2014). *Iniciação á análise geoespacial: Teoria, Técnicas e Exemplos para Geoprocessamento* (1st ed.). Editora Unesp.

Galdino, S. (2015). Distribuição espacial da erosividade da chuva no Estado de Goiás e no Distrito Federal. *Boletin de Pesquisa e Desenvolvimento*. Embrapa Monitoramento por Satélite. Campinas (SP). https://ainfo.cnptia.embrapa.br/digital/bitstream/item/138337/1/4668.pdf

GOIÂNIA. (1944). *Decreto – Lei n°11 de 06 março de 1944*. Proíbe a criação de novos loteamentos por um período de cinco anos.

GOIÂNIA. (1947). *Código de Edificações de Goiânia Lei 574 de 1947*. Instituiu a Regulamentação de Parcelamento, Zoneamento e Uso do Solo.

GOIÂNIA. (1950). *Decreto Municipal n.16 de junho de 1950 –* Regulamenta a implantação de infraestrutura básica em loteamentos, que exigia apenas a abertura de vias.

GOIÂNIA. (1969). *Plano de Desenvolvimento Integrado de Goiânia (PDIG)*. SECRETE Eng. S/A.

GOIÂNIA. (1971). *Lei Municipal n° 4.526 de dezembro de 1971*. Dispõe sobre loteamentos urbanos e remanejamentos.

GOIÂNIA. (1999). *Lei Complementar n.27, de 30 de Dezembro de 1999*. Região Metropolitana de Goiânia. http://fnembrasil.org/regiao-metropolitana-de-goiania-go/

GOIÂNIA. (2007). *Lei Complementar n°. 171, de 29 de maio de 2007*. Dispõe sobre o Plano Diretor e o processo de planejamento urbano do Município de Goiânia e dá outras providências. Lex: Legislação Municipal, Goiânia.

Goudie, A. (2006). *The human impacto on the natural environment: Past, presente, and future* (6th edn.). Oxford.

Guerra, A. J. T., & Jorge, M. C. O. (2013). *Processos erosivos e recuperação de áreas degradadas* (1st ed.). Oficina de Textos.

IBGE. (1952). *Recenseamento Geral do Brasil, Série Regional parte XXI – Goiaz*. INSTITUTO BRASILEIRO DE GEOGRAFIA E ESTATÍSTICA. Censo Demográfico Estado de Goiás. Rio de Janeiro. https://biblioteca.ibge.gov.br/visualizacao/periodicos/65/cd_1940_p21_go.pdf

IBGE. (2010). *Censo Demográfico de Goiânia*. INSTITUTO BRASI-LEIRO DE GEOGRAFIA E ESTATÌSTICA. https://cidades.ibge.gov.br/brasil/go/goiania/panorama

IBGE. (2019). *População estimada*. INSTITUTO BRASILEIRO DE GEOGRAFIA E ESTATÌSTICA. http://www.agm-go.org.br/noticia/1359-ibge-divulga-as-estimativas-da-populao-dos-municpios-para-2019.

IPLAN. (1983). *As erosões de Goiânia: Diagnóstico e Indicação de Alternativas para Controle e Combate*. Instituto de Planejamento Municipal de Goiânia.

IPLAN. (1987). *Levantamento das Erosões existentes em Goiânia*. Instituto de Planejamento Municipal de Goiânia.

IPLAN. (1991). *Carta de Risco do Município de Goiânia*. Goiânia: 1 mapa, color., 107 cm x 140 cm. Escala 1: 40.000. Instituto de Planejamento Municipal de Goiânia.

IPLAN. (1992). *Plano de Desenvolvimento Integrado de Goiânia (PDIG)*. Instituto de Planejamento Municipal de Goiânia.

IPT/DAEE. (1989). *Controle de Erosão: bases conceituais e técnicas, diretrizes para o planejamento urbano e regional, orientações para o controle de boçorocas urbanas* (2nd edn.). INSTITUTO DE PESQUISAS TECNOLÓGICAS, DEPARTAMENTO DE ÁGUAS E ENERGIA ELÉTRICA. São Paulo.

ITCO. (2008). *Revisão e detalhamento da carta de risco e planejamento do meio físico do município de Goiânia*. Instituto de Desenvolvimento Tecnológico do Centro-Oeste.

Jardim, A. C. (2017). *Direções de fluxo em modelos digitais de elevação: Um método com foco na qualidade da estimativa e processamento de grande volume de dados*. Tese de Doutorado. Instituto Nacional de Pesquisa Espaciais - INPE. São José dos Campos.

Lal, R. (2014). Soil conservation and ecosystem services. *International soil and Water Conservation Reserch, 3*(2).

Lopes, L. M., & Romão, P. A. (2006). Geomorfologia urbana da região metropolitana de Goiânia. *Guia de Excursões Centro - Oeste, Simpósio Nacional de Geomorfologia*.

Maarseveen, M. V., et al. (2019). GIS in Sustainable Urban Planning and Management: *A global perspective*. Taylor & Francis Group. U. S. Government works. ISBN 9781138505551. 364p.

Moraes, S. (1991). *O empreendedor imobiliário e o Estado: O processo de expansão de Goiânia em direção sul (1975–1985)*. Universidade de Brasília.

Morgan, R. P. C. (2005). *Soil erosion and conservation*. Silsoe College, Cranfield University. Longman, 3°edn.

MUBDG. (2017). *Mapa Urbano Básico Digital de Goiânia*. Mapa Geral. Secretária de Planejamento de Habitação de Goiânia. Prefeitura de Goiânia

Nascimento, M. A. L. S., & Podestá Filho, A. (1993). *Carta de Risco de Goiânia* 1(13). *Boletim Goiano de Geografia*, 95–107.

Nascimento, M. A. L. S. (1994). Erosões Urbanas em Goiânia. *Boletim Goiano de Geografia*, 77–101.

Nascimento, M. A. L. S., & Sales, M. M. (2002). Diagnóstico do Processo Erosivo em Goiânia. *X Simpósio Brasileiro de Geografia Física Aplicada*.

Nascimento, D. T. F., & Barros, J. R. (2009). Identificação de Ilhas de Calor por Meio de Sensoriamento Remoto: Estudo de Caso no Município de Goiânia – GO/2001. *Boletim Goiano de Geografia*. Goiânia, Goiás. 29, 119–134. https://doi.org/10.5216/bgg.v29i1.7112

Nascimento, D. T. F., & Oliveira, I. J. (2015). Mapeamento do processo histórico de expansão urbana do município de Goiânia-GO. *Geographia., 17*(34), 141–167.

Nimer, E. (1979). *Climatologia do Brasil*. IBGE, Série Recursos Naturais e Meio Ambiente.

O'Callaghan, J. F., & Mark, D. M. (1984). The extraction of drainage networks from digital elevation data. *Computer vision, graphics, and image processing* (Vol. 28, pp. 323–344).

Oliveira, M. das M. B. (2005). O padrão territorial de Goiânia: um olhar sobre o processo de formação de sua estrutura urbana. *Arquitextos*, 065.07, ano 6. https://www.vitruvius.com.br/revistas/read/arquitextos/06.065/419

Oliveira, A. H., et al. (2012). Consitência Hidrológica de modelos digitais de elevação (MDE) para definição da rede de Drenagem na sub-bacia do Horto Floresta Terra Dura, Eldorado do Sul, RS. *Revista Brasileira De Ciência Do Solo., 36*, 1259–1267.

Prandini F. L., et al. (1976). *Atuação da cobertura vegetal na estabilidade de encostas: uma resenha critica*. Instituto de Pesquisas Tecnológicas.

Rieke-Zapp, D. H., & Nearing, M. A. (2005). Slope shape effects on erosion: A laboratory Study. *Soil Science Society of America Journal, 69,* 1463–1471.

Santos, M. (2009). *Metrópole Corporativa Fragmentada: O caso de São Paulo* (2nd ed.). Editora da Universidade de São Paulo.

Selby M. J., et al. (1993). *Hillslope materials and processes.*

SEGPLAN. (2018). *Caderno de Entrega e Resultados do Déficit Habitacional.* SECRETARIA DE ESTADO DE GESTÃO E PLANEJAMENTO DE GOIÁS. Aliança Municipal pela Competitividade.

SEPLAM. (2018). *Goiânia em Dados.* SECRETARIA DE PLANEJA-MENTO E DESENVOLVIMENTO URBANO SUSTENTÁVEL. https://www.goiania.go.gov.br/shtml/seplam/dados/biblioteca.shtml

Silva, L. S., & Travassos, L. (2008). *Problemas ambientais urbanos: desafios para a elaboração de políticas públicas integradas, vol 1.* In: Cadernos Metrópole. Editorial Pontifícia Universidade Católica de São Paulo. 27–47

Soarez Neto, G. B., & Faria, K. M. S. (2014). Geomorfometria e planejamento urbano: Padroões de vertentes para ocorrência de erosões urbanas no município de Goiânia/GO. *Revista Geonorte,* Edição Especial *4*(10). 458–462.

Toy, T. J., et al. (2001). *Soil erosion: Processes, prediction, measurement and control.* Wiley.

Tucci, C. E. M. (1995). *Inundações Urbanas.* In C. E. M. Tucci & R. L. Porto (Org.). Drenagem Urbana. ABRH/Editora da Universidade/UFRGS (pp. 15–36).

VALERIANO, M. M. (2003). Curvatura Vertical de Vertentes em microbaciais pela análise de modelos digitais de elevação. *Revista Brasileira de Engenharia Agrícola e Ambiental* (Vol. 7, pp. 539–549). Campina Grande. PB. DEAg/UFCG.

Young, A. (1972). *Slope. Geomorphology texts.* University of East Anglia. Longman.

The Dimension of Urban Morphology on Placemaking Theory: A Comparative Typomorphological Analysis of a Self-built Neighbourhood and Its Regeneration Project in Ankara

Merve Okkalı Alsavada

Abstract

The informal settlement as a self-urbanity is a widespread phenomenon that causes many political, social and environmental dilemmas in Turkey. These settlements are considered as problematic environments that need to be urgently transformed into more healthy spaces by the public authorities, and are not considered as an integral part of the city environment even by dwellers. However, the rapid urban regeneration significantly affects the physical and social quality of built form and urban life in Turkish cities. The main problem driving this research is the continuum of built environment identity and sustainability in Turkish cities due to the urban regeneration model. Therefore, this study attempts to draw attention to the potential of the concept of morphology and typology in evaluating spatial characteristics of urban life in architecture. The main purpose of the study is to perform a comparative analysis of self-built neighbourhoods and their regeneration projects from the perspective of placemaking theory to determine the spatial dynamics of self-built neighbourhoods. The study has a methodology that consists of a typomorphological analysis which reveals the impact of spatial changes on the neighbourhood environment. In accordance with this, Feridun Çelik neighbourhood in Altındağ, Ankara was selected as a case study to reveal both typological and morphological regenerations in terms of the continuity of the living patterns of its inhabitants. The analyses of "Urban Form", "Plan Type", "Housing Unit-Street Relationship" and "Climate-Responsive Design and Material" are documented via mapping, housing plan drawings and photographs. The main findings of the research are that the design quality of self-urbanity can be acknowledged as recognizable wholes and the destruction of existing spatial patterns in regeneration projects results in the loss of urban identity and environmental quality. Observing the built environment with regard to places means observing their richness and complexity as a complicated interplay between individuals and their environment.

Keywords

Urban morphology • Placemaking • Urban regeneration • Squatter neighbourhood • Self-urbanity • Sustainability

1 Introduction

Housing production through urban regeneration projects in Turkey is significantly affects the quality of built form and urban life in the country's cities. Urban regeneration projects in Turkey have destroyed the existing urban texture and leaving high density high-rise settlements in their aftermath. Hence, they lack in terms of the continuum of urban identity and their typological characteristics do not maintain on the advantages of the older squatter neighbourhoods' spatial properties as places. From this perspective, this study reflects on placemaking theory with regard to the concept of urban regeneration via a typomorphological approach in the case of Ankara. Therefore, this study aims to draw attention to the potential of the concept of morphology and typology, together with evaluating the quality of the urban form and characteristics of urban life in a sustainability context. An urban form and its produced types determine the role of urban identity in the city environment. Accordingly, the study attempts a comparative analysis of self-built environments and their regeneration projects to reveal the typomorphological change on city milieu. The main issue driving the research is the loss of urban identity in Turkish cities due to the current urban regeneration model. There is a considerable amount of research investigating urban regeneration projects and squatter settlements that consider both their

M. Okkalı Alsavada (✉)
University College London, London, UK
e-mail: merve.alsavada.20@ucl.ac.uk

deficiencies and advantages from various perspectives. However, the study presented herein deals with both self-built and planned urbanity, combining the typomorphological approach with placemaking theory to offer a methodological approach to architectural analysis techniques.

The informal settlement as a self-urbanity is a widespread phenomenon which includes many political, social and environmental dilemmas in Turkey. These settlements are not identified or addressed as an integral part of the city environment, even by their residents. Also, these areas are considered problematic environments that need to be urgently transformed to more healthy spaces by the public authorities. Hence, the characteristics of their form and architectural grammar are seen as irrelevant and context-based (Dovey & King, 2011). According to this perspective, the main motivation of this study is to conceptualize squatter settlements as everyday phenomena and make them visually distinguishable as formal settlements in their own right. However, we intend to avoid aestheticizing these settlements by just emphasizing their positive aspects comparing to their ultimate regeneration as high-density mass housing projects in Turkey. Squatter residents can be affected to their detriment, by the otherwise reasonable intentions behind glorifying such settlements. Therefore, the study presents both positive and negative aspects of squats as ordinary places and analyses their environmental qualities in terms of their social and design dimensions. Methods of analysing the morphological characteristics of self-urbanity include the considerable potential for site-specific design responses in the city transformation. Hence, the study proposes a conceptual framework with which to examine the urban form, and architecture of self-urbanity and their regenerations. The works of Kellet and Napier has been useful as a foundation to this study to elaborate on the above. This approach explores the design variety of self-urbanity and spontaneous settlement (Kellet & Napier, 1995) and their spatial continuum in the city context in terms of urban identity. Hence, the study aims to explore typomorphological patterns of neighbourhood spaces to reveal the spatial impact of social, economic and site factors on residents' live. The main purpose of the study is to conduct a comparative analysis between self-built neighbourhoods and their associated regeneration projects from the perspective of placemaking theory to determine the different morphological properties and spatial dynamics of self-built neighbourhoods as places. Accordingly, the main research question addressed by this study is: "What changes occur during the regeneration process in squatter neighbourhoods in Ankara?". The sub-questions are formulated as follow: "How do self-built settlement types produce place identities?" and "What are the socio-spatial characteristics of a squatter settlement and its regeneration as a process?".

The study has a methodology that constitutes a typomorphological analysis. The literature review develops a theoretical framework on explorations of "place" in terms of human geography. The concept of "place" is defined by Tim Cresswell and Rephl provides a useful starting point for this study, and that further indicates three main influential approaches to the consideration of "place" that provide the analytical prospect for the study. These are phenomenological and critical social geographic approach to place in the context of a self-built environment and place-making theory. The phenomenological and critical social geographic approaches to "place" emphasize the creative spatial production of society. The idea of placemaking provides an alternative lens through which the advantages of these two approaches can be considered. The typomorphological analysis reveals the impact of spatial changes to neighbourhood environments from the perspective of placemaking. In accordance with this, Feridun Çelik neighbourhood in Altındağ, Ankara has been selected for consideration as a case study area to reveal both typological and morphological regeneration in terms of the continuity of place identity and the living patterns of residents. The analysis is documented via mapping, housing plan drawings, and photography (Fig. 1).

Analysed plan types were produced by the author during site visits between 2017 and 2020 when all the squats had not yet been demolished, and it is assumed that these plan types show typical squatter plan schemes in the settlement. As a conclusion, approaching squatter areas as places and analysing their regeneration period from the perspective of the typomorphological process provide for the growing visibility of squatter areas and addresses not only their socio-political dimensions in the city environment (such as land tenure, policymaking, etc.) but also reveals the properties of living environments, patterns of everyday activities and lived experiences of residents over time.

2   Research Methodology and Case Study

2.1   *Typomorphology*

This study adopts the methodology of a typomorphological process analysis that includes a case study of such Turkish city. The methodology will not only determine the typomorphological changes with regard to the physical characteristics of squatter neighbourhoods but also analyse the socio-spatial process through the regeneration of these squats as places. An associated literature review will attempt to identify relevant theory on squatter settlements and their regeneration processes, focusing on the relationship between self-urbanity and placemaking. The review establishes a conceptual framework that guides the research with regard to

Fig. 1 Analytical framework of the study. Prepared by the author

	Theory	Context
Aim	To review the theory of place and a squatter neighbourhood as a self-urbanity	To analyse the transformation process of a squatter settlement in Ankara, Turkey

Methods & Instruments	Literature Review
	• Self-Urbanity
	• Placemaking
	• Socio-Spatial Characteristics
	  - Physical Setting
	  - Social Inquiry
	  - Spatial Appropriation
	  - Lived Experience
		Typomorphological Analysis
		• Site Observation
		  - Mapping
		  - Drawing
		  - Photography

Outputs	Conceptualization of self-urbanization as a morphological process to guide site analysis	Identification of the socio-spatial process of Feridun Çelik neighbourhood in different scales and Analytical framework to the interpretation of results according to the theoretical context

analysing the parameters of self-urban settlements and regeneration processes, analysis of the presence and changes within the case study area, and data collected about the socio-spatial characteristics of the case study area. In a site context, the study typomorphology evaluates the built environment in relation to location, time and scale to clarify the production and regeneration process of the urban form and its design quality (Chen & Thwaites, 2018). The typomorphological analysis is a useful method to map and document the existing living patterns of inhabitants, and to produce data available for academics and students. Also, the determination of changes in urban form by building types is important to explore different approaches and design solutions for the transformation of problematic areas. Location is an important aspect due to form being heavily influenced by external factors, and reflects socio-economic and the cultural values of the residents. Time is another essential factor with regard to typomorphological analysis because the durability of types and morphologies can be assessed over time, whilst the process of adaptation can be explored according to specific time periods within the regeneration process itself. Hence, the study analyses housing forms and types of different morphologies in the same location, that is a squatter settlement and its associated regeneration project. Typological changes can take place to different extends, which can be defined as continuity, partial continuity and mutation. Continuity indicates the continuous development of form with regard earlier types; partial continuity demonstrates partial changes to typologies compared to the previous

types; and mutation refers to complete typological change compared to any previous types (Gokce & Chen, 2018). In this case, the typomorphological changes resulting from the urban regeneration projects in question can be defined as the latter. Urban regeneration occur to the urban form, street patterns, land division, positioning of buildings and neighbourhood patterns. Hence, the research conducted during the case study considers the building, street and neighbourhood scales. Data collection was achieved via observation with regular site visits during the daytime over a period spanning 2017–2020. Photos, drawings and video records were used to collect information about the socio-spatial characteristics of the environment over this time.

2.2 Site Selection for Case Study

TOKI (the Housing Development Administration of the Republic of Turkey) has developed mass housing regeneration projects in order to solve the housing problem currently being experienced by the low- and middle-income groups who cannot afford housing under the current market conditions. Its general approach of urban regeneration projects and social housing production for low-income social groups has previously been to build affordable high-rise high-dense housing in a relatively short time (Fig. 2). However, the administration has more recently changed its approach to housing, and has started to develop housing in a way as to meet social needs, including the "Neighbourhood

Fig. 3 Mass housing regeneration project in North Ankara

Fig. 2 Squatter settlement in North Ankara (*Source* https://www.toki.gov.tr/sosyal-konutlar)

Concept" as one of its basic space production approaches to achieve social sustainability in the city environment. Hence, as stated by the administration website, one of its main objectives is not just to build qualified affordable housing units but also to create neighbourhood environments that have genuine identities in Turkish cities (TOKI, 2020). Feridun Çelik regeneration project was run with this very intention. Therefore, the Feridun Çelik area was selected as our case study area to allow for a typomorphological analysis of socio-spatial changes of self-urbanity from the perspective of urban identity and placemaking theory.

3 Literature Review on Self-built Urbanity, Placemaking Theory and Urban Regeneration in Turkey

The study of places is one of the main fields of human geography. The concept of place is central to people's everyday live. Cresswell points out the use of 'place' in everyday speech as that of a location. It is not only part of academic terminology, in other words, containing both simple and complicated meaning. In one sense, it is common and familiar term to a majority; in another sense, it goes beyond a common sense-level understanding and gains a socio-geographical basis. Observing the world in terms of places is indicative of observing its richness and complexity, as to consider a space is to consider the complicated interplay between individuals and their environment. The connection between a person and location, area or building is also indicative of the associated privacy, belonging and memory. "Your place" is not "my place"—everyone has their own definition or understanding of meaningful places (Cresswell, 2004). This understanding has developed over the last two decades in human geography. The initial

positivist understanding of place within the human geography discipline had a functional perspective which is descriptive in terms of explaining the incomparable characteristics of a given region. During the 1950s and 1960s, spatial science affected the geography, and behaviours of human beings, which became abstract symbols within mapping and modelling platforms. In response to this idea, different ways of understanding the interactions between people and places were developed, and human centred geographies were created (Holloway & Hubbard, 2001) (Fig. 4). However, environmental image research was prevalent as an initially positivist approach.

In the 1960s and 1970s, the interaction of people with their environment and their responses to such became important in the field. Their representations were produced at different scales. During 1970s in particular, the notion of place became defined as a container of culture. This

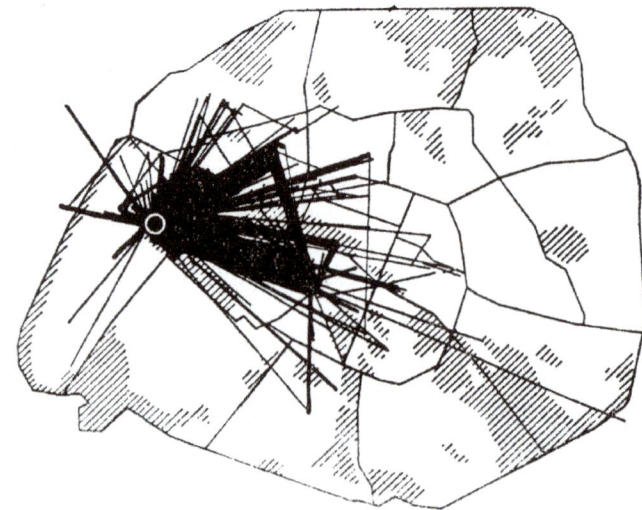

Fig. 4 Paul-Henri Chombart De Lauwe, map of a young woman's movements in Paris, 1957

increased the emphasis of social processes on human geography. However, this understanding was criticized due to the limitations about its definition of culture. Culture is defined as "a total way of life in common by a group of people" (Shurmer-Smith, 2002 in Lombard, 2010). Recently, place theory has begun to infer a reciprocity between people and their environment, according to which place has become a socio-spatial construct that means a location and social status within society (Cresswell, 1996 in Lombard, 2010). Hence, the notion of place does not only refer to a stationary meaning, and goes beyond a location on a map, but also means a material setting for social and economic relations. Agnew explains the three fundamental aspects of this notion, which are location, locale and sense of place. In addition to the meaning of a location, Agnew explains that the notion of "locale" can represent a place where people and the events associated with them take place. In this sense, place contains a material visual form and has a real connection with humanity. Place also contains a human capacity for production and consumption. With the term "sense of place" Agnew indicates the appropriation of and attachment of people to a place. Appropriation of a place is a way of seeing, knowing and understanding the world. It is also subjected to the critical reflection of individuals (Agnew, 2014). This means that places also construct people in the same manner that people construct places. Therefore, place is not only related to the quality of the things it contains but it is also an aspect of the approach we choose to consider, designate and emphasize it by. A fundamental dualism of place and space exists in geographic thinking which conceptualizes place as both part of and different to space; in this context, Tuan points out that the ideas of "space" and "place" need each other to allow for their complete clarification. Hence, they are interrelated concepts, both literally and experientially in terms of investing in a definition of either. He defines space in a more abstract concept which has volume, geometry and movement. On the other hand, place is defined with a pause in the movement. The perception of a location is a consequence of a pause in the movement (Tuan, 1977). Cresswell defines place as a narrative in this context. According to him, place often refers to a happening and a meaning, and a space becomes a place when individuals give meaning to it (Cresswell, 2004). In relation to the urban form, the structure of space affects the experience of place. At urban scale, the urban grid and its spatial properties affect how people move within it, which spaces interrelate to each other, how individual's travel within urban space is shaped, how much time they spend travelling, etc.

Urban morphology, as a study of the spatial structure, identifies and examines the patterns of urban component parts, the process of its development and its transformation. Whitehand defined urban morphology as a field of historical

and geographical knowledge towards the end of the 1970s (Whitehand, 1977, 1981). His studies became a root for the utilization of the theory of urban form in different geographical context, and offered a theoretical and methodological framework that combines the concepts of morphological periods, geography of cultural areas, typological processes and place characteristics. He also supervised a large number of Ph.D. studies which combined place characteristics and the study of urban form to understand the placemaking process inherent to different geographies (Oliveira, 2019). Accordingly, the urban space can be defined as an accumulation of actions of individuals and social groups. Hence, analysis of an area at different scales to identify place properties is not only undertaken with regard to the physical structures but also patterns of land use, movement, connectivity, space types, nature, etc. Despite the study of urban form having emphasized the many different fields and broader city, there are not enough studies of this nature to allow the characteristics of urban forms of poor settlements as a place to be determined (Duarte, 2009). The conceptual division of formality and informality in architecture and urban studies prevents the integration of various branches of similar research areas in the literature. This polarized understanding has resulted in the marginalization of squatters or informal settlements from morphological studies (McCartney & Krishnamurthy, 2018). Hence, there is a little documentation or morphological representations of informal settlements, especially with regard to the socio-spatial manifestations and transformation of urban poverty. In this context, this study investigates these settlements and their regeneration processes, not only to integrate planning strategies for the entire city but also to construct an urban narrative within the city environment. The determinants of urban identity can be social, geographical, environmental and architectural. These are defined in different ways in different studies, but the main elements of the city identity can be defined as follows: the physical structure of the settlement, the historical development of the city, socio-cultural accumulation of the urban environment, spatial characteristics of the milieu and urban typology. Hence, the huge change on urban form can result in the alteration of social structures and the collective identity that is formed by the values of cultural life (Yaldız et al., 2014).

There is a clear relationship between the morphological characteristics of the built environment and social practices. The urban form is shaped by the movement of people in organically grown settlements, and the configuration of the spatial structure consolidates this movement. Hence, the analysis of the urban patterns of cities, their historical and transformation processes provides a powerful tool to describe and interpret the social characteristics of communities (Kubat, 1999). The analysis of the urban structure and its socio-economic values with the historicity of cities could be a significant source for architects and urban planners while shaping the built environment. In that sense, Kubat made a quantitative analysis of Anatolian fortified towns and investigated Turkey's urban planning agenda through historical periods including the Romans, Byzantines, Turks and Ottomans. The main results show that Anatolian fortified towns have segregated urban layouts comparing to the other cities' analysis conducted in London, Athens and Gassin. In addition to that, the lack of openings on the street facades and the courtyard housing units with high garden walls increases the complexity of urban patterns in Anatolian cities shaped organically through the different morphological periods (Kubat, 1997). More recently, Unlu (2019) discusses the Turkish planning practice from the morphological perspective in relation to the urban change, the part-to-whole relationship and the historicity of urban form. The built environment functions the life of society, and the city acts as a dynamic unity regenerated through the production of ever-changing forms in the city. Therefore, he offers a framework to study the urban patterns and environmental characteristics of areas in their historicity and complex interactions between city parts and the whole. It is revealed that the successful management of changes to urban form needs to contain an analysis of the nature of urban growth and the roles of agents in the reproduction of urban forms. According to the consequences of his studies on different Turkish cities' context, the planners were concerned with the history and hierarchical nesting of urban form during the early Republican period while suggesting a new street system, block patterns and block types. However, this morphological perspective has changed after the second half of the nineteenth century. The identical freestanding apartment block became a dominant building type, and the designing the urban form was putten into the background as a second phase of shaping the built environment (Unlu, 2019). This understanding is still prevalent, especially in the regeneration process of poor areas in Turkey. Likewise, Yucel and Aksumer (2019) analyse the spatial changes to Selcuk province, Izmir through the five different morphological periods, which are until the end of the 1950s, between 1957 and 1980, between 1980 and 1994, between 1994 and 2008, and after 2008. Although the research area has unique spatial characteristics and natural resources, some of the results show the breakpoints of Turkey's urban planning understanding in terms of shaping the urban form. Researchers emphasize the radical changes in the third morphological period that is between 1980 and 1994. The squatter settlements were demolished, and the density of the built structure increased without any reference to the history of the area. This process affected the expansion of the city in this direction. Hence, this regeneration period is a significant point in shaping Selcuk's urban form by affecting physical characteristics, the social structure of the community and the collective memory of the society (Yucel & Aksumer, 2019).

Hence, the subject of urban regeneration in the context of urban form has an associated, and rather comprehensive literature on the theory of urbanization and its practical applications in Turkey. The rapid urbanization process and increased migration from rural to urban areas has been resulted in the explosion in number of squatter settlements in Turkish cities. These settlements have been important units in shaping the urban form of cities, especially at city peripheries. These self-solution emergent environments become problematic due to their poor living conditions and socio-spatial segregation. Their regeneration projects have been an essential constituent of Turkey's urbanization agenda since the 2000s. The evaluation of the Turkey's urban regeneration strategies and projects on the macro-, meso- and micro-scales has been researched from different perspectives by various scholars. While the scopes of the associated researches have diversified with regard the content of such studies, there are nevertheless a number of basic common principles underlying these critical analyses and consensus on their methods and aims of them. Korkmaz and Balaban (2020) discusses the sustainability of urban regeneration projects by the competition-based approach to describe what urban regeneration looks like in practice, and to conceptualize an "emergent epidemic community" in the world's regeneration agenda. After the analysis of five high-impact initiatives from different countries including Turkey, the improvement in access to economic and leisure activities for all inhabitants—the elderly, disabled people, children, etc., has increased the benefits to local businesses and shifted people's perceptions and lived experiences from that of decay to an inclusive urban milieu (Korkmaz & Balaban, 2020). On the other hand, Dündar (2001) discusses the regeneration projects outcomes and their impacts on the physical and social topography of Ankara. She emphasizes the rapid and extensive change of the urban form instead of applying a rehabilitation process to sustain flexible characteristics of and social relations with squatters in the city. The enlargement of block sizes and the unqualified dense construction inflict a resultant damage on the internal dynamics of these neighbourhoods. The rapid increase in population, and thus, the loss of neighbourhood relations, has damaged the fabric of society (Dündar, 2001). Eren (2014) conceptualizes the parameters of urban identity in order to follow its characteristics in regeneration projects. She discusses how the change in physical environment can have an impact on urban identity via two case study areas in the cities of Istanbul and Bursa. The main findings of her analysis show that the squatter settlements' urban forms were shaped over time by the impact of the social environment and history of the area as an example of self-urbanity. The inhabitants had a collective memory which provided a shared perception of the environment and place attachment. However, the huge change in the building types, the relationship between building units and green space and the block–parcel relationship leads to social breakage, and a loss of place attachment and socio-spatial identity within the urban milieu. Besides, the active street life of squatter settlements can be damaged due to the physical change of the urban layout (Eren, 2014). Uzun and Simsek (2015) examined the upgrading of squatter settlements through the case study of North Ankara. They revealed the principal strengths and weaknesses of the projects through a questionnaire study that included a sample of 115 people, analysis of ownership and building structures, development of parcel and receivable residence size and architectural projects of housing units. According to this study, the unhealthy infrastructures and service areas of squatter settlements were transformed into quality spaces, and large number of social, leisure and cultural facilities were provided for the residents' use. However, the main weak points of the regeneration project were the residential density and new high-rise building blocks. Hence, the project had weaknesses in terms of conserving the natural characteristics of the area and regional architectural identity of the old neighbourhood milieu. It also resulted in demographic changes and imposed an additional pressure on the original inhabitants of the area (Uzun & Simsek, 2015).

Unlu and Bas (2017) states that few studies investigates the morphological analysis of changes through generation, degeneration and regeneration process in Turkish cities. His study on the morphological process and the residential forms in the case of Çamlıbel, Mersin presents an analysis method consisting of three main principles: First, the formation and transformation process of the urban form results from the interaction of building blocks, plots and streets; second, the hierarchy of urban form can be studied in different scales starting with the city to an individual plot scale and ending with the buildings' material characteristics; third, the change of urban form is subject to the social and cultural contexts, and new emergent building types are part of the formation of urban form. After the analysis, Unlu concluded that the plot pattern metamorphosis is a result of division and amalgamation processes of planning practices. The single-family building blocks were converted to the apartment blocks in the area, and these changes continued with the addition of floors and changes to elevation details of buildings. In that sense, the planning decisions and the alliances between landowners and local government resulted in the rapid development and popularization of apartment blocks. Consequently, the urban fabric and the character of a neighbourhood environment has changed through this process (Unlu & Bas, 2017).

According to the above, defining, analysing and transforming a squatter neighbourhood as a place is not only conducted with regard to the quality of the physical properties of the environment but also pertains to the way of understanding the diverse creation of places through morphological processes in Turkey. Hence, the study refers to squatter settlements as self-urbanity instead of informality to emphasize the living environment of their residents rather than to refer to land ownership. In this sense, the relationship between self-urbanity and place theory is discussed in detail in three chapters which describe the phenomenological, critical social geographic and placemaking approaches in relation to the spatial structure properties.

3.1 The Lived Experience of Self-built Urbanity: Phenomenological Approach to Place

Yi-Fu Tuan and Edward Relph examined the lived-world experiences of place from the phenomenological perspective, as influenced by the thoughts of Heidegger. According to Cresswell, they brought the subject of place to the consideration of geographers in a sustainable way (Cresswell, 2004). Hence, their ideas of place are discussed extensively with regard to their potential application to squatter neighbourhoods. This chapter examines the place, lived experiences and dwelling issues and their application to self-urbanity.

Heidegger's inquiry about spaces is related to the associated locations, buildings and dwellings. According to him, the locations and the site expression of dwelling provide the spaces in which we live. Hence, the act of dwelling is defined as an expression of being in the world (Heidegger, 1971). Norberg-Schulz also mentions the relationship between dwelling and human existence itself. The idea of being and sense of belonging are linked to the act of dwelling (Norberg-Schulz, 1971). Therefore, "building" and "dwelling" are inseparable from the issue of identity. From phenomenological perspective, the notion of home refers to an intimate place and has an important role in the formulation of place. Relph's definition of home is a complete expression of place incorporating the associated aspects of location, people, time, space and place appropriation (Relph, 1976). Accordingly, home is a significant place of human existence and with regard to the identity of individuals and social groups. This phenomenological emphasis on place as the centre of belonging offers an influential focus for squatter settlements. From this perspective, self-urbanity can be defined as "a rich and complicated interplay of people and environment" (Cresswell, 2004). Also, the phenomenological approach provides a focus on everyday life, the lived experience of dwellers and "the other architectural history" of developing countries from the perspectives of inhabitants involvement with the place. In the context of regeneration projects of these settlements, despite the focus on the purely physical changes of settlements, historico-geographical theories proposes the combination of spatial analysis, and the typological and morphological approach with agents, i.e., the inhabitants, local authorities, architects and planners. The analysis of urban form is readdressed with the appearance of the built form, material use, land use, building density, activity patterns, architectural periods, the process of development, the characters of proposals of change and the decision-making processes followed by developers (Oliveira, 2019). In that combination of placemaking theory and urban morphology, this study redefines the inhabitants of squatter settlements as shapers of urban form and regeneration project developers as agents of change.

Hence, the conception of dwelling for squatter neighbourhoods can be evaluated as a self-help movement and inhabitants' highest expressions of being (Lombard, 2010). The socio-economic and socio-spatial properties of a settlement are associated with the belonging and place attachment of inhabitants. Hence, human interaction with the environment, site specifications and elements of human activities are important parameters through which to focus on the lived experiences of inhabitants. In the context of these parameters, finding formal solutions to unhealthy squatter areas can be evaluated on the basis of identity and defined as "placeless" (Varley, 2008).

3.2 Socio-spatiality of Self-built Environment: Critical Socio-spatial Approach to Place and Urban Form

While the phenomenological approach to the squatters seek the sense of place and lived experience of such, the critical social geographer considers the concept of place through the lens of social and cultural conflicts (Harvey, 1996). According to this, place is seen as a social construct and its fundamental role in social life is scrutinized from the perspective of a progressive political agenda (Cresswell, 2004). Looking at places as social hierarchies does not only include social processes but also the creation, continuance and regeneration of the relations of domination, oppression and exploitation (Cresswell, 2004). Stating a place is a social construct is also a statement of its materiality. Cresswell explains that the fabric of a space is a product of society. The elements of urban form, which are the building units, the public gardens, the streets, the trees, etc., are the production of society. In the context of squatter neighbourhoods as self-urbanity, the housing units and gardens are not just buildings and nature, but are rather the tireless efforts of the local residents in the city. In this manner, one of the key

concepts by which to explore squatter settlements is their relation to power. Hence, squatter neighbourhoods are evaluated as areas of negative relations, especially between the state and the community. However, social constructionist approaches offer a response to this with a real focus on the complexity of power in place (Lombard, 2010). This provides an improved consideration of the entangled processes that occur as a part of in self-urbanity. In a relational sense, place reflects the expectations about the behaviours associated with the social and spatial concerns. Place is not a simple geographical aspect, it also coincides with both social and cultural expectations reflected in the urban form. The existing patterns of street blocks, the ratio of empty to built-up areas and the utilization of building blocks are determined by inhabitants as place shapers in squatter settlements. Hence, the urban form of this self-organization is a process of identifying building the associated fabric. In contrast to the planned regeneration projects, the set of rules for space division and place construction do not allow for a clear delineation of boundaries. The hierarchy between streets, the degree of form continuity over time and the hierarchy of boundaries between housing units are based on the socio-historical development of the area. Correspondingly, Shields explains marginal spaces within the framework of social spatialization. These spaces are not only evaluated within geographical peripheries, but are also located at the periphery of cultural systems and can be ranked relative to each other (Shields, 2002). Hence, the squats, as organic urban forms, are the result of social spatialization of cultural systems. Inhabitants are able to resist the construction of their expectations by building, using and giving meaning to their places. Therefore, the morphology of squatter neighbourhoods can be stated as the power of the expectations of what the place is for. However, from the perspective of acquisition of land tenure, inhabitants gradually become involved in conferring their own attachment to what was formerly agricultural land. They appropriate vacant lots via illegal or semi-legal activities to create housing areas, streets, public spaces, meeting areas, playground spaces, etc. This self-appropriation simultaneously initiates the formal processes to gain urban services and supply. In a long-term situation, the state makes the attempt to exert its power to bring order to and transform these places. Whilst the construction of squat inhabitants' own spaces, appropriation of the areas, and initiation onto the formal processes are long term in nature, and indeed contain complexity in power and place relationship, the regeneration of the area through formalization occurs over a relatively short term period. In fact, their regenerations also comprises complex socio-spatial relations in relation to power, resistance and identity. Hence, from the point of critical social geographical thought on place, the regeneration of these settlements should be contextualized not only as a regularizing and formulization process from the perspective of political inquiry but also as an integral part of its neighbourhood within the city environment in connection with time. Hence, their typomorphological process of regeneration becomes important to the determination of place meaning and behavioural patterns in space.

3.3 Self-urbanity as Process: Conceptualization of Placemaking for Self-built Environment and Its Regeneration in Turkey

The other fundamental experiential property of the concept of place is movement. The phenomenological geographer Seamon, who discusses place as a central concept within his work, refers to bodily mobility rather than rootedness in order to explain the key component of place theory. He follows the French phenomenologist Merleau-Ponty and focuses on the "everyday movement in space". This is the production of particular patterns and the habitual nature of a settlement with regard to the experiential character of place (Seamon, 1980). In this context, Pred also explains the creation and utilization of physical setting according to structuration theory. Place is a never-ending convergence and is the result of processes and practices. In this context, the proposed theory regards the material continuity of people as a participant in that process and of objects as being employed in time–space practices (Pred, 1984). Hence, place as a process enhances the material continuity and reciprocal influence between a place and its inhabitants. The materiality of places influences people's living patterns and is influenced by people's actions. Within this framework, squatter settlements as representations of self-urbanity can be described as a confluence of flows and squatter inhabitants can be conceptualized as agents who embody a living environment and its possible material existence. Self-urbanity as process rests upon feelings of belonging within the rhythm of life in social production. Hence, the design properties of squatter settlements contribute to the history of social production and reproduction. A socio-spatial process is a communicative process that includes individual and collective dimensions. In this context, Amos Rapoport mentions the design qualities of self-urbanity and acknowledges them as being "recognizable wholes", and as worthy of being considered traditional vernacular settlements. Because of this, he uses the term 'spontaneous' rather than squat to refer to the nature of the built environment rather than land tenure. From this perspective, he identifies spontaneous settlements as "cultural landscapes" that are representations of individuals' decisions and their socio-spatial processes (Rapoport, 1988). He gives a list of numerous processes and product characteristics to describe all traditional settlements. Process characteristics contain "the identity, intentions and anonymity of the designer", "the reliance on a model with variations", "the extent of sharing of single models", "the congruence of the chosen model with ideals of the users", "degree of congruence between environment and culture-life style", "degree of self-consciousness of the design process", "form of temporal change" and "extend of sharing knowledge about design and construction". Product characteristics contain "degree of cultural and place specificity", "specific models, plan forms and morphologies", "nature of relation to landscape", "effectiveness of response to climate", "efficiency in use of resources", "open-endedness allowing changes and regarding activities", "degree of multisensory qualities of environment and differentiation of settings", "effectiveness of environment as a setting for life-style and activity system" and "ability to settings to communicate effectively to users" (Rapoport, 1988). This process is also defined as a placemaking process as an active verb. Placemaking explores the socio-spatial construction of place and highlights the potentiality of human activity in their environment. It also offers an analytical focus on the complex relationship between individual, community, physical setting and power.

Through the idea of placemaking, the process of appropriating a space is also a mirror of self (Friedmann, 2007). The appropriation of space is related to the activities such as going to the local butcher and baker, knowing the manes of the streets and being part of local meetings, and other forms of everyday life in Turkey. Therefore, the spatial characteristics of neighbourhood designate these activity patterns and the identity of place. In this sense, placemaking offers significant grounds on which to analyse the multiple and complex relationships in social and spatial construction in squatter settlements as self-urbanity and its regeneration in a sustainability context. As stated by Eren (2014), the squatter settlements contains sociospatial opportunities with regard to urban morphology and typology in Turkey. The most important place factor on the urban form of squatter settlements in Turkey is the impact of topography on shaping street layout. Most of such settlements were placed at the periphery of cities and on the foothills of mountains. While this provides a panoramic view of the city or landscape, the urban forms were shaped according to the places' geographies. Streets are formed at diagonals or parallels to the slopes. Hence, this organic urban layout formation differs from the planned linear urban configuration in Turkish cities in general. The size of housing units changes with regard to the gradient of slope. There is width hierarchy between streets, as well. Some become widened over time to allow access by larger vehicles. These organically formed self-built settlements creates unique community relations within which the inhabitants know each other and between them shape their environment over a long time. Hence, streets and shared green spaces characteristics provide common spaces for daily routines. While a regeneration project upgrades the poor environmental conditions in a squatter settlement, it should not be a purely physical renewal project, but it also needs to conserve strong socio-spatial relations in the neighbourhood community and hierarchy of boundaries in the urban form because it determines the social fabric of community.

4 Case Study in Feridun Çelik Neighbourhood, Altındağ, Ankara

Feridun Çelik neigbourhood is located on the north-east side of Ankara. The area is on the edge of the periphery highway of Ankara (Figs. 5 and 6).

According to TUIK's record from 2019, the population of the neighbourhood is 7,466. It has gradually decreased from 17,720 in 2007 to 6,394 in 2015 due to the regeneration process in the area (Fig. 7). The neighbourhood is still undergoing urban regeneration and the area of Cinderesi has been designed as the last step of the regeneration project in the neighbourhood.

Dovey and King explain the typology of informal settlements in terms of eight different types of urban forms and the visibility of squatter settlements. These are "districts", "waterfronts", "escarpments", "easements", "sidewalks", "adherences", "backstage" and "enclosures" (Dovey & King, 2011). According to this classification, Feridun Çelik neighbourhood can be defined as a "backstage" which is formed through attachment to the existing settlement of the city (Figs. 8 and 9). It is largely hidden from the public gaze of the formal city development.

According to Dovey and King's description, such informal developments are commonly seen in strong states or countries where the visibility of squatter neighbourhoods can be described as politically sensitive (Dovey & King, 2011). Therefore, the squatter settlement's insertion and excrescences are placed at the interstices of an existing formal sector housing area (Fig. 10). Even though squatter neighbourhoods can be visible from a distance, they are often impenetrable and enclaves for different social classes in Turkey's context. Hence, they may be defined as informally gated, and a key transformation has also occurred in the production of new sites within this environment for middle

Fig. 5 Feridun Çelik neighbourhood area in Ankara. Prepared by the author

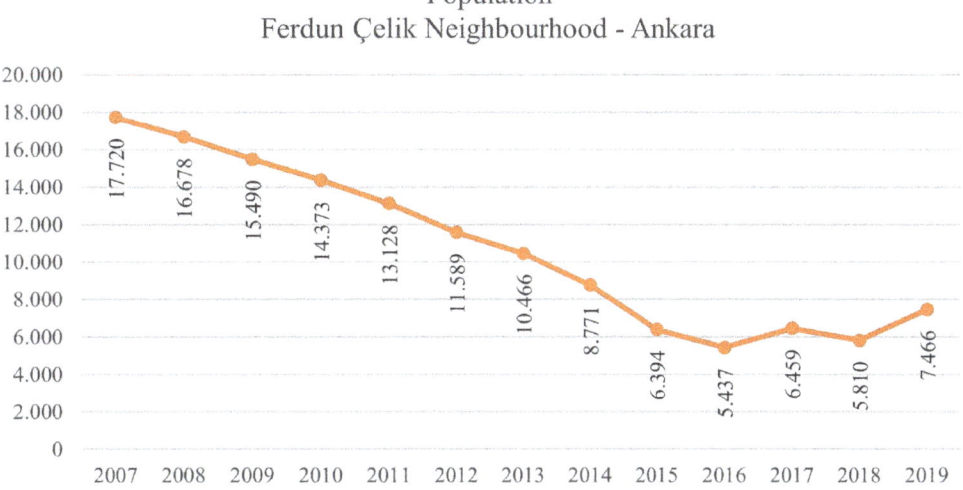

Fig. 6 The boundary of neighbourhood area of Feridun Çelik. Prepared by the author

Fig. 7 Population of Feridun Çelik Neigbourhood between 2007 and 2019 (*Source* TUIK)

in-come social groups. Transforming these "backstage" squats into apartment blocks or high-rise building blocks has produced an image that inhabitants belong to the middle class. Therefore, typomorphological change to the site creates a new image of the city, both socio-politically and socio-economically. The typological change in the urban architecture enables the transformation of a lack of law and order.

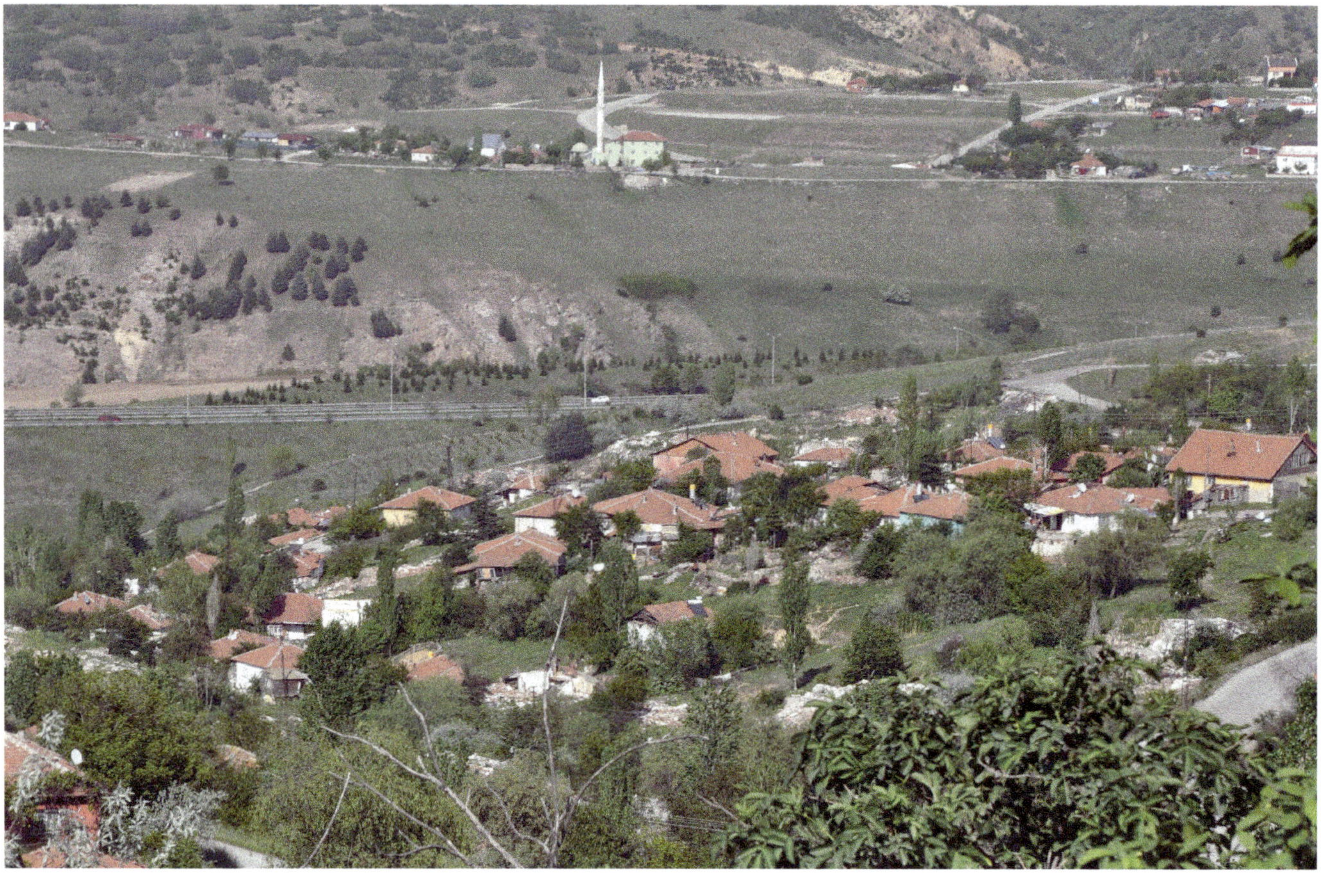

Fig. 8 Squatter area of Feridun Çelik Neighbourhood. Photograph taken by the author

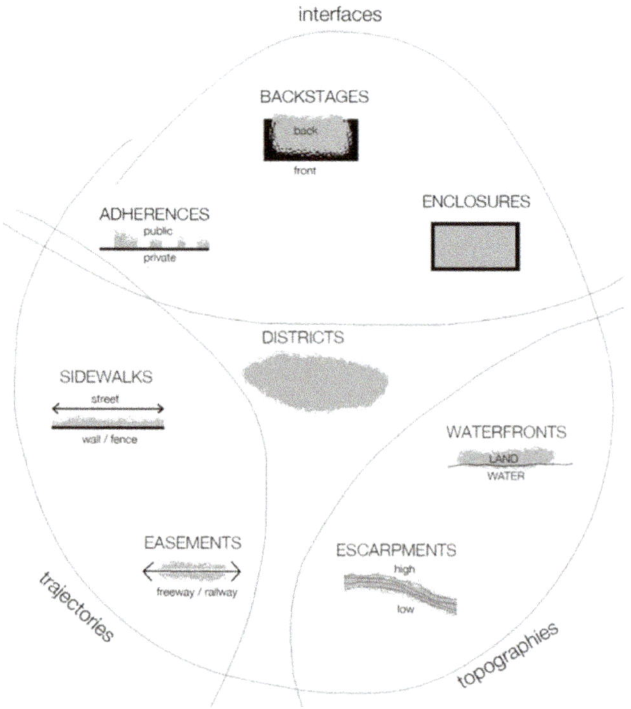

Fig. 9 Types of Squatter Neighbourhoods (*Source* Dovey & King, 2011)

4.1 Site Analysis

4.1.1 Urban Form of the Squatter Settlement

Urban form determines the three-dimensional built form of city by outlining the built and un-built environment. Within the framework of self-urbanity as a morphological process in the neighbourhood, the land uses are not straightforward. Urban form and its individual housing units have an anti-homogeneous character (Fig. 11).

This organic form and land division of the settlement is not a result of formal contract but eventuated by the negotiations between inhabitants. Hence, soft boundaries are observed between the private and public space (Fig. 12). Symbolic boundaries result on the zoning of private, semi-private, semi-public and public spaces. The degree of privatization is provided by architectural barriers or garden walls in the urban space. The creation of zones of transition informs people about the range of possible activity and orients them in the existing living pattern. The lived experience and behaviour of an individual changes according to these symbolic barriers as a matter of course. On the other hand, the regenerated part of the area as a planned place does not include any perceptible zones of transition from the public space to private or semi-private spaces (Fig. 23).

Fig. 10 The intersection of squats and formal-sector buildings (*Source* https://www.altindag.bel.tr/#!haberler/2019)

Fig. 11 Urban form of the squatter settlement in Feridun Çelik Neighbourhood. Prepared by the author

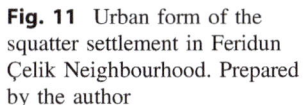

Fig. 12 Soft boundaries within the urban space in the squatter area of Feridun Çelik Neighbourhood. Photograph taken by the author

The change from organic urban form to planned warped parallel structure can also be evaluated according to the associated building blocks. The complicated part of the regeneration project has monotype apartment blocks. On the contrary, the squatter housing units are not identical and have smaller forms shaped according to local topography (Fig. 13).

In the comparison of land use before and after the regeneration, the new settlement after the regeneration includes zoned land use and a flow of inhabitants, whilst the existing pattern of squatter settlement in the form of self-urbanity has the mixed and dynamic making character

of land use. However, this absence of land use let the inhabitants meet different needs over time.

The other important point of the morphological analysis of self-urbanity and its regeneration as a process is the associated access and connectivity. Morphological analysis includes the circulation of space configuration in role of movement and accessibility. By using Habraken's definition, there could be two main patterns of settlement that can be defined in terms of access and connectivity. These are the regularized grid and the organic tree form. Both include different levels of connectivity and control (McCartney & Krishnamurthy, 2018). According to his definition, the

Fig. 13 Morphology as a process: morphological changes between 2004 and 2020 in Feridun Çelik. Prepared by the author from Google images

Fig. 14 Urban form of the squatter area. Prepared by the author

Fig. 15 Urban form of the regenerated area. Prepared by the author

regularized grid form of the regeneration project in Feridun Çelik neighbourhood creates a sense of security by allowing the free filtering of traffic, multiple linkages and options for people. On the other hand, the organic form of squatter areas in the neighbourhood provide less control over access to the community space due to the branching structure of the tree form (Figs. 14 and 15).

4.1.2 Plan Types

Typological process is a key concept in urban morphology regarding to the analysis of type changes as a result of the transformational evaluation of a site (Gokce & Chen, 2019). Within the framework of this study, plan types are defined as a living pattern of inhabitants which reflect their forms of life in a cultural aspect. Therefore, the typological process is used as a tool to reveal the robustness or infirmity of plan types of self-urbanity in the study. The spatial arrangement of the housing unit plans was analysed during the author's site visits and four different typical plans were produced with different plan schemas and spatial configurations. However, it is observed that all squatter housing units are transformed to apartment blocks which have same spatial arrangements

and less flexibility in use with regard to the priorities and needs of different social groups (Figs. 16 and 17).

In self-urbanity, the inhabitants are the decision makers with regard to their housing design in terms of the determining what their needs and priorities are. However, the planned formal housing unit plans have no respect the communities' preferences in Feridun Çelik neighbourhood. In linking placemaking theory and urban regeneration practice, there are some regeneration model examples that allow participatory designing with the community while transforming or upgrading the existing housing conditions. The UN-Habitat A Practical Guide to Designing, Planning and Executing Citywide Squatter Upgrading Programmes states that residents participation improves housing design and achieves a more suitable design result in squatter regeneration programmes if the plans do not remain academic exercises but are implemented with the support of participating residents (UN-Habitat, 2014). Accordingly, the appropriation of a place and spatial comfort can be addressed through taking into consideration any existing plan types and analysing the place's morphological change as a process.

4.1.3 The Housing Unit-Street Relationship

The relationship of the housing unit with the street is important in terms of the ability to analyse the orientation of mass and its entrance, spatial arrangement of indoor–outdoor places and visual communication of public spaces. These factors affect the behavioural patterns of dwellers, neighbouring relationships, security of neighbourhood places, creation of defensible spaces and the sense of belonging in the public realm. The forms of housing units and their relationship with the topography and street network determines the properties of this relationship. The squatter area of Feridun Çelik neighbourhood consists of one-to-two floor housing units including garden spaces (front or back gardens, sometimes both) and balcony spaces in relation to the street spaces (Fig. 18). While the masses are placed in order not to see each other's gardens to provide intimacy between them, they establish a visual communication with the street. In contrast to this, the regeneration project area consists of six-to-ten floor apartment blocks which lack semi-private garden spaces and have less visual communication with the street. Therefore, the squatter housing units create more defensible spaces and control over the public spaces compared to the apartment blocks. This gives an indication that the sense of belonging in squatter settlements is actually stronger than in the regenerated area.

During the site visits, it was also observed that the streets were used as community or socializing space for dwellers and playgrounds for children. However, there was no children on the streets in the regenerated area due to the security and traffic problems (Fig. 19).

Fig. 16 Typical plan schemas of squatter housing-units in Feridun Çelik. Prepared by the author

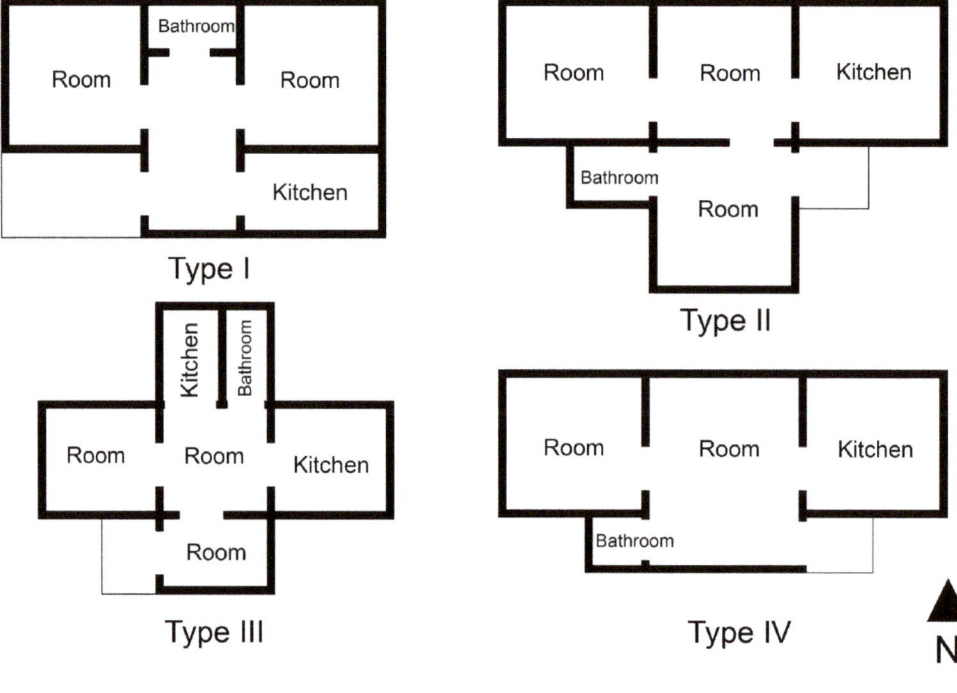

Fig. 17 Spatial configuration of squatter housing units in Feridun Çelik. Prepared by the author

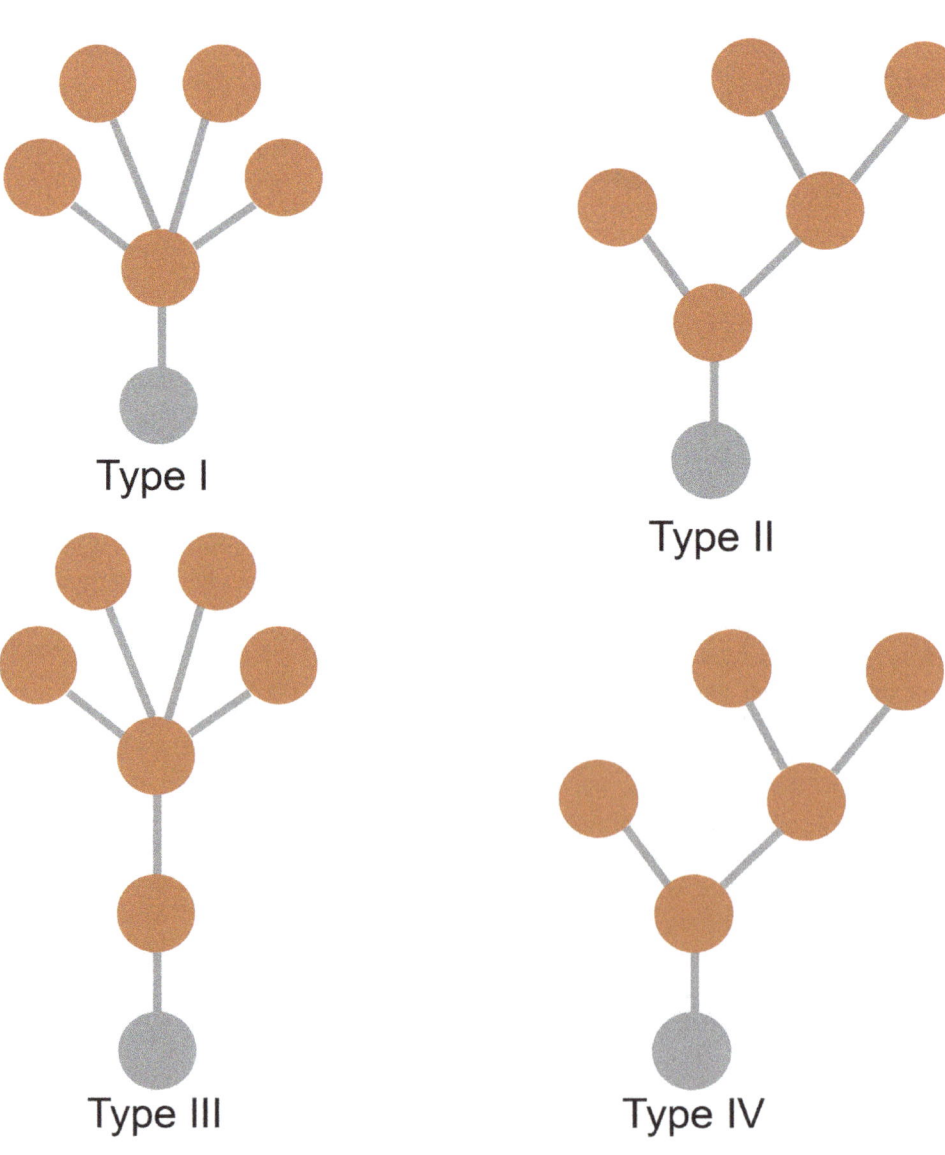

Fig. 18 A squatter housing-unit in Feridun Çelik. Photograph taken by the author

Fig. 19 Street as a community space in the squatter settlement of Feridun Çelik. Photograph taken by the author

Gardens can be defined as the result of the endeavours of inhabitants and they are sites of history in terms of landscape properties. From this perspective, the semi-private garden spaces of the housing units contain architectural values pertaining to the lived experiences of inhabitants.

The other important element in the housing unit-street relationship is that of "wall". During site visits, it was observed that the squatter neighbourhood community used urban walls as a way of communicating. Therefore, the wall,

as a communicative tool, can be defined as a placemaking object of self-urbanity (Figs. 20 and 21).

4.1.4 Climate-Responsive Design and Building Materials

Each stage of the design process is important for the appropriation of the place throughout the lives of the structures. Hence, climate-responsive design is crucial for place-making theory in the sense of attachment to the place

Fig. 20 Wall as a
communicative tool. Photograph
taken by the author

Fig. 21 Wall as a
communicative tool. Photograph
taken by the author

and socio-economic sustainability of a settlement. According to the site analysis, it was observed that housing units were placed a certain distance from each other to properly benefit from the available sunlight. In addition, deciduous trees, especially at the south-west and west of the housing site, have been used to block the summer sun, and traditional pitched roofs further provide the essential shade needed to allow for the everyday activities of the inhabitants (Fig. 18). Also, trees were planted to provide shading and cooling to ensure comfortable retreats, especially during the hot summertime. In this context, the trees on the site can be defined as a memory of place. However, the apartment blocks built for the regeneration projects were not placed with any due consideration for climate, energy efficiency and to receive adequate daylight. The relationship between greenery and apartments was not designed to create appropriate shading (Figs. 22 and 23).

On the other hand, the squatter settlement is a poor quality area in terms of climate change conditions like flooding and extreme weather events. The infrastructure of

Fig. 22 The use of shading element and trees in a squat. Photograph taken by the author

Fig. 23 The new urban regeneration building. Photograph taken from Google Map

the area does not provide for any real resilience to the high risks of emergency services, flooding, heavy storms, etc. Hence, effective policies to improve the poor quality of urban and building infrastructure need to be developed to build such resilience.

Morphological analysis also addresses the material composition of the buildings. In the subject of material qualities of housing units, housing units in the squatter area of the neighbourhood are not built from durable materials or have secure foundations in comparison to the new types. Hence,

Fig. 24 Material quality of
demolished squatter unit.
Photograph taken by the author

Fig. 25 Material quality of a
squatter unit. Photograph taken
by the author.

squats can be defined as temporal housing units in the context of place as a process. Also, the durability of materials and the structure of housing units are ill-suited to the high risk presented by earthquakes in Turkey (Figs. 24 and 25).

4.2 Discussion

The urban form of a squatter settlement as a placemaking process of inhabitants reflects the relations between space, geography and community organizations in the study. The analysis of the building type and urban form of the squatter settlement shows that the housing units randomly scattered on the mountain foothills amongst natural vegetation have a definite rural character. This habitation resulted in different forms of life than the planned urban grid offers. Streets are not only used purely for the purpose of movement but also as socializing spaces for the inhabitants. It is a fact that the spatial standards and material quality of the housing needs to be developed. However, this process should be prepared with due consideration for the current landscape properties and urban symbols of social identity.

The organic urban form of the squatter settlement provides a hierarchy of use, and the low-density scattered layout is related to the privacy, movement and space sharing process of inhabitants in Feridun Çelik. Hence, in-between space, such as semi-private gardens, urban parks, etc., regulates the balance in terms of sharing common space between different social groups in the settlements—e.g., children, women, elderly people, etc. However, an increase in the density of building blocks and population will have certain negative impacts on the social bonds between inhabitants. Also, the linear regulized grid arrangement and wider street layout that fails to include a hierarchy do not provide in-between spaces for different community organizations, and thus results in a loss of social cohesion. The other common building type seen in the squatter settlement is the additional structures that have storage, car parking or woodshed functions. The sizes and forms of these structures as representative forms of flexible construction processes are mostly determined by the needs of the inhabitants over time. On the other hand, the regeneration projects do not include the flexibility to allow inhabitants to determine the form of their buildings or make additions or reductions as part of a placemaking process.

From the perspective of accessibility and connectivity, a regularized grid form has advantages according to the organic tree form. However, the organic form of the squatter area creates more defensible streets, especially in terms of the relationship between housing units and streets. According to the climate-responsive design principles, trees and garden spaces provide energy efficiency and daylighting to the squatter housing units. On the other hand, the apartment blocks that typify the regenerated part were not designed with due consideration for climate type, sunlight or passive energy systems. However, it should be emphasized that the squatter settlements do not create environments that are resilient to the risks inherent to climate change and earthquakes in Turkey.

5 Conclusion

According to the UN-Habitat Report, one third of the global urban population lives in squatter settlements. Although the percentage of squatter areas has decreased, the number of people living in squats has been increasing gradually (UN-Habitat, 2004). Hence, both analysing the living patterns of squatter settlements and developing upgrade and regeneration strategies for them are important, especially in developing countries. In the Turkish context, the study of squatters has previously focused on policy, land tenure and urban poor perspectives. However, this study has analysed

their morphological characteristics and site-specific design responsiveness as pertaining to placemaking theory as a process. The analysis of function and form, connectivity and access and public and private spaces introduced a set of dynamics to the existing morphological approaches. The study approached the squatter neighbourhood as self-urbanity from the perspective of placemaking in the context of sustainability. Accordingly, the analysis method was determined to be one of a typomorphological process to reveal the physical and socio-spatial characteristics of the areas which are squats and the new regenerated settlement in the same neighbourhood. The main limitation of the study was gaining access to the plan drawings of housing units and site plan of the regeneration projects intended for the area. Although the author applied to the local council to gain access to the architectural documents of the regeneration project, the sharing of these documents was rejected by the local planning authority. Hence, the drawings and diagrams pertaining to the project were produced directly by the author during site visits and through internet accessible visuals.

The main question posed by the research—namely that of what changes occurred during the regeneration process of the squatter neighbourhood in the context of Ankara—has been answered from the perspective of socio-spatial characteristics of a city environment as a place. Accordingly, the analysis was conducted according to four different categories, those of "urban form", "plan type", "housing unit-street relationship" and "climate responsive design and building materials". The main finding of the research was the huge change in the urban form over regeneration process of Feridun Çelik. Concerns about the social sustainability of the urban regeneration experience in Turkey have arisen with regard to the consequences of density, building scale and landscape form. The study contributes a new perspective to the scientific field on how spatial arrangements determine social relationships, privacy and community organization in an urban environment. Rather than analysing the main differences of squatter environment and planned urban form, the study offers a social focus on urban habitation, and reveals urban symbols of social identity in the context of the urban regeneration process in Ankara. Suggested future research in this area include research into how to prepare design guidelines or frameworks to gradually improve the physical standards via improved regeneration modelling by conserving the socio-spatial identity and place properties of the environment. The study reveals the spatial properties of a squatter neighbourhood and its social aspects as a place; the next step in this research is to formalize design codes addressing multiple dynamics and factors that shape the growth of squatter settlements. To conclude, one of the challenges of the study of squatter areas was the lack of methodological approach to combining spatial and social inquiries. The main gap in this research field is the lack of data on squatter areas and their inhabitants. These areas are ignored in terms of the collection of detailed data and systematic study by the authorities. This results in inappropriate design solutions and unsustainable regeneration projects in terms of the creation of qualified environments through conserving local culture and social relations. Also, analysing squatter areas with their regeneration processes has been neglected within architecture and urban design literature. However, with the growing visibility of squatter neighbourhoods, especially through morphological studies and geospatial theories, the typomorphological approach has started to provide detailed insights into the analysis of squatter settlements and their regeneration processes and addresses the dynamics of urbanization in the context of different cities.

References

Agnew, J. A. (2014). *Place and politics: The geographical mediation of state and society*. Routledge.

Chen, F., & Thwaites, K. (2018). *Chinese urban design: The typomorphological approach*. Routledge.

Cresswell, T. (2004). *Place: A short introduction* (Vol. 12). Blackwell Ltd.

Dovey, K., & King, R. (2011). Forms of informality: Morphology and visibility of informal settlements. *Built Environment, 37*(1), 11–29.

Duarte, P. B. (2009). Informal settlements: A neglected aspect of morphological analysis. *Urban Morphology, 13*(2), 138.

Dündar, Ö. (2001). Models of urban transformation: Informal housing in Ankara. *Cities, 18*(6), 391–401.

Eren, İ. Ö. (2014). What is the threshold in urban regeneration projects in the context of urban identity? The case of Turkey. *Spatium*, 14–21.

Friedmann, J. (2007). Reflections on place and place-making in the cities of China. *International Journal of Urban and Regional Research, 31*(2), 257–279.

Gokce, D., & Chen, F. (2018). Sense of place in the changing process of house form: Case studies from Ankara, Turkey. *Environment and Planning b: Urban Analytics and City Science, 45*(4), 772–796.

Gokce, D., & Chen, F. (2019). A methodological framework for defining 'typological process': The transformation of the residential environment in Ankara, Turkey. *Journal of Urban Design, 24*(3), 469–493.

Habitat, U. N. (2014). A practical guide to designing, planning, and executing citywide squatter upgrading programmes.

Harvey, D. (1996). *Justice, nature and the geography of difference*. Blackwell.

Heidegger, M. (1971). *Building dwelling thinking. Poetry, language, thought* (trans. A. Hofstadter). Harper Colophon.

Holloway, L., & Hubbard, P. (2001). *People and place: The extraordinary geographies of everyday life*. Pearson Education.

Kap Yucel, S. D., & Aksumer, G. (2019). Urban morphological change in the case of Selcuk, Turkey: A mixed-methods approach. *European Planning Studies, 27*(1), 126–159.

Kellett, P., & Napier, M. (1995). Squatter architecture? A critical examination of vernacular theory and spontaneous settlement with reference to South America and South Africa. *Traditional Dwellings and Settlements Review*, 7–24.

Korkmaz, C., & Balaban, O. (2020). Sustainability of urban regeneration in Turkey: Assessing the performance of the North Ankara Urban Regeneration Project. *Habitat International, 95*, 102081.

Kubat, A. S. (1997). The morphological characteristics of Anatolian fortified towns. *Environment and Planning b: Planning and Design, 24*(1), 95–123.

Kubat, A. S. (1999). The morphological history of Istanbul. *Urban Morphology, 3*(1), 28–41.

Lombard, M. B. (2010). *Making a place in the city: Place-making in urban informal settlements in Mexico* (Doctoral dissertation, University of Sheffield).

McCartney, S., & Krishnamurthy, S. (2018). Neglected? Strengthening the morphological study of informal settlements. *SAGE Open, 8*(1), 2158244018760375.

Norberg-Schulz, C. (1971). *Existence, space & architecture*. Praeger.

Oliveira, V. (2019). An historico-geographical theory of urban form. *Journal of Urbanism: International Research on Placemaking and Urban Sustainability, 12*(4), 412–432.

Pred, A. (1984). Place as historically contingent process: Structuration and the time-geography of becoming places. *Annals of the Association of American Geographers, 74*(2), 279–297.

Rapoport, A. (1988). Spontaneous settlements as vernacular design. *Spontaneous Shelter: International Perspectives and Prospects, 51–77*, 52.

Relph, E. (1976). *Place and placelessness* (Vol. 1). Pion.

Seamon, D. (1980). Body-subject, time-space routines, and place-ballets. *The Human Experience of Space and Place, 148*, 65.

Shields, R. (2002). *Places on the margin: Alternative geographies of modernity*. Routledge.

Toplu Konut İdaresi Başkanlığı. (n.d.). https://www.toki.gov.tr/sosyal-konutlar

Tuan, Y. F. (1977). *Space and place: The perspective of experience*. U of Minnesota Press.

UN-Habitat. (2004). The challenge of squatters: Global report on human settlements 2003. *Management of Environmental Quality: An International Journal, 15*(3), 337–338.

Unlu, T. (2019). Managing the urban change: A morphological perspective for planning. *ICONARP International Journal of Architecture and Planning, 7*, 55–72.

Unlu, T., & Bas, Y. (2017). Morphological processes and the making of residential forms: Morphogenetic types in Turkish cities. *Urban Morphology, 21*(2), 105–122.

Uzun, B., & Simsek, N. C. (2015). Upgrading of illegal settlements in Turkey; the case of North Ankara entrance urban regeneration project. *Habitat International, 49*, 157–164.

Varley, A. (2008). A place like this? Stories of dementia, home, and the self. *Environment and Planning d: Society and Space, 26*(1), 47–67.

Whitehand, J. W. R. (1977). The basis for an historico-geographical theory of urban form. *Transactions of the Institute of British Geographers*, 400–416.

Whitehand, J. W. R. (Ed.). (1981). *The urban landscape: Historical development and management; Papers by MRG Conzen*. Academic Press.

Yaldız, E., Aydın, D., & Sıramkaya, S. B. (2014). Loss of city identities in the process of change: The city of Konya-Turkey. *Procedia-Social and Behavioral Sciences, 140*, 221–233.

Research on the Relationship Between Informal Learning and the Regeneration of Public Spaces on Chinese University Campuses: Taking Xian Jiaotong-Liverpool University as an Example

Jierui Wang

Abstract

As a significant part of urban areas, university campuses take the responsibility for providing students with spaces for being educated, improving the city's aesthetics, and sometimes even serving as a symbol of a curtain place. This research takes China's university campuses as study subjects in order to explore the possible developments of informal learning and public spaces on university campuses. Because of the lack of engagement with informal learning spaces and university campuses since the beginning of tertiary education in China, students' on-campus livelyhoods are often limited to three areas, which are the classroom, the canteen and the dormitory. Therefore, their necessary creative thinking skills for both study and work are insufficient generally. This study focuses on the current condition of informal learning in China's practice of higher education and the relationship between informal learning and the public spaces on university campuses, in order to summarise the strategies for the regeneration of public spaces on university campuses around China. In the final part of this article, Xi'an Jiaotong-Liverpool University (XJTLU), located in Suzhou, China, is taken as an example to assess the strategies this research adopts and to determine the proper relationship between informal learning and the public spaces on university campuses, concluding references to other university campus regeneration in China and elsewhere.

Keywords

Higher education • Campus public spaces • Informal learning • XJTLU

1 Introduction

According to the theory of educational psychology, learning can be divided into formal learning and informal learning, which was firstly put forward by Malcolm S Knowles in 1950 (Wang, 2020b). Formal learning refers to learning which occurs during one's academic education period and the subsequent adult education period after graduation from universities. It is accomplished through teacher-centric educational methods, such as courses, lectures, and seminars. In contrast, informal learning refers to a learner's autonomous learning on informal occasions, which takes place with the help of non-learning social communications; that is, it does not require specialised classrooms and does not represent the distinct essence of an organisation or institution (Chen & Liu, 2011). This type of learning can be viewed as student-centric learning and can take place in atria, corridors, terraces, staircases, squares, etc. (Chen et al., 2010). Therefore, in higher education institutions, campus public spaces are potential venues for implementing informal learning. However, due to the traditional Chinese university pedagogical approach which has been in place since the end of the nineteenth century, educators pay too much attention to formal learning and largely ignore informal learning. Hence, at present, most of the public spaces on Chinese university campuses are short of the design and construction to promote informal learning, leading to limited creation and use of public spaces (Wang, 2018). The attributes of informal learning and its relationship with Chinese university campuses will be outlined in the following sections in order to further explore the regeneration strategies for campus public spaces to develop good informal learning environments.

J. Wang (✉)
Welsh School of Architecture, Cardiff University, Cardiff, C10 3BE, UK
e-mail: wangjierui3@gmail.com

© The Author(s), under exclusive license to Springer Nature Switzerland AG 2022
C. Piselli et al. (eds.), *Innovating Strategies and Solutions for Urban Performance and Regeneration*,
Advances in Science, Technology & Innovation, https://doi.org/10.1007/978-3-030-98187-7_21

2 Attributes of Informal Learning

2.1 Active Knowledge Acquisition of Informal Learning

According to educational psychology and the constructivist learning theory, learning is a process of knowledge construction by learners on their own, rather than a process of absorbing knowledge from the outside, aligning it closely with informal learning, which is the process of autonomous knowledge construction by learners (Chen & Liu, 2011). Based on informal learning, an individual student establishes their own knowledge schema on the ground of their existing knowledge and experiences. Compared with traditional teaching, which demands students' memory and application of learning tasks, informal learning expects them to learn by thinking intelligently. Knowledge acquired in this way may be enhanced and become comprehensible, leading to internalisation (Chen & Liu, 2011).

2.2 Social Interactivity of Informal Learning

Traditional teaching normally regards students' learning as their individual mental activities, irrespective of the social aspect of learning. In contrast, informal learning can help students to internalise relevant knowledge and skills through their participation in certain social and cultural events. Such a learning process is usually carried out in a learning community with learners' cooperation and interaction (Chen & Liu, 2011).

2.3 Situationality of Informal Learning

Traditional teaching insists on 'de-contextualisation', regarding generalisation as the core content of learning tasks. However, the abstract concepts and rules of learning may not help learners to adapt themselves to specific situations flexibly, so it is hard for students to apply what they have learned to solve practical problems in reality and to take part in social practice. Unlike traditional teaching, informal learning focuses on learning in a multi-context manner, putting more value on practical experiences instead of infusing knowledge, which demonstrates that this method can help students to learn more effectively and efficiently (Chen & Liu, 2011).

As shown in Fig. 1, these three attributes show the process of the transformation from traditional to informal learning. It is clear that informal learning has different requirements from traditional teaching and can facilitate a more flexible teaching programme and learning time for students to arrange their out-of-class collaborative and individual learning. The next section will introduce the application of informal learning in Chinese education and the construction of campus public spaces in Chinese universities respectively, to determine if they satisfy the attributes of informal learning as stated above.

3 Literature Review of the Relationship Between Informal Learning and the Public Spaces on Chinese University Campuses

3.1 The Application of Informal Learning in Chinese Education

Exam-oriented education is universally over-valued in China due to its special historical, political, economic, and social factors (Wang, 2013). Since most of Chinese schools and universities lay more emphasis on the mode of classroom-based knowledge transmission, teachers provide students with fixed knowledge in class according to the teaching programme, and the same contents are delivered to students with different characteristics and knowledge reserves on the same platform (He, 2018). Although it

Fig. 1 Informal learning (Wang, 2018)

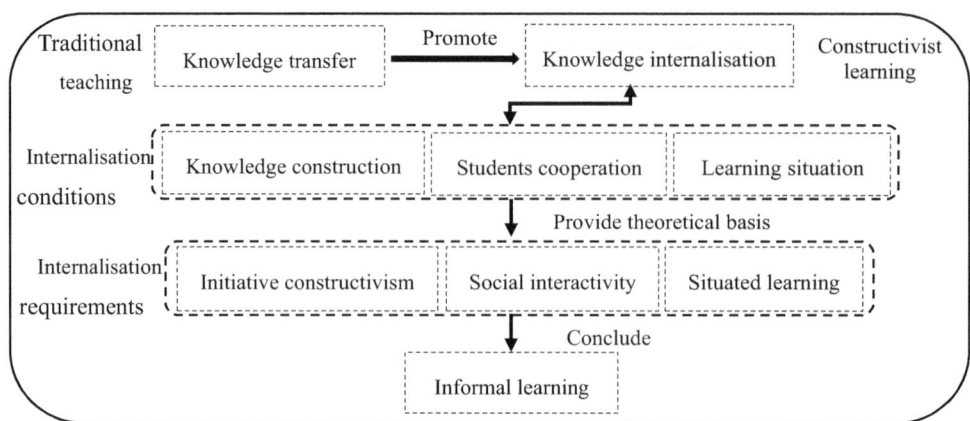

appears as a form of equality for each individual learner, there is no time for them to rethink and internalise their new knowledge, and after-class time is regarded as the period for the assignment of homework, which means that informal learning is seriously neglected within this approach (Li, 2020b). Additionally, the number of public spaces for individual and collaborative learning is severely limited (Liu, 2019). For example, Shao and Zhang (2020) indicated that students largely choose to either stay in classroom or go outside for fun during their after-class periods in pre-university education; thus, it becomes a common scene for the corridors to be crowded with students after class in primary and high schools in China; however, they claimed that corridors are not a suitable venue for students to have discussions. Consequently, after entering universities, spaces for students are widely expanded, but this is based on 'the three points on one line'model, that is to say the classroom, the canteen, and the dormitory. This is therefore the normal style of university students' campus lives (Wang, 2015). Time is restricted for students, thus informal learning is still unrealisable in higher education (Wu, 2020).

3.2 Embodiment of Informal Learning in the Public Spaces on Chinese University Campuses

The future of university education not only relies on data and the internet but also on educational wisdom for employing the rapid development of technology in a more intelligent way (Marshalsey & Sclater, 2020). This means that educational wisdom serves as the focus of campus construction as well as education quality development, and in the stage of campus development, all design and construction must be developed based on teaching, studying, logistics services, the realisation of hardware networks, digital contents, personalised learning, research collaboration, and intelligent services (Cai et al., 2018). It is essential to create a learning environment for out-of class space and time so as to enhance educational wisdom within campus design and construction. Therefore, public spaces on campuses are an essential focus point (Wu, 2020); another reason is that campus public spaces are the best places for implementing informal learning (Zheng et al., 2020). Students can stay in these spaces embedded with learning-support facilities that create an informal learning environment on campus (Wang, 2020a). When ensuring that the implementation of informal learning is smooth, good lighting, a quiet environment, proper heating, etc., may help to contribute to the creation of a campus public space which is also a suitable reasonable informal learning space (Wang, 2018).

Students are the main users of campus facilities, rather than the campus designers and other university stakeholders (Chen et al., 2010; Wang, 2018). This is the reason why campus public spaces ought to be built in a human-oriented manner, a point which most Chinese universities ignore. However, such Chinese university campus design, as based on traditional campus planning and design methods with a bird's eye view, neglects consideration of a human perspective or a pleasant dimension of space (Wang, 2020a); although the design is apparently reasonable in land relation, architectural planning, campus streamlining and function division, it is not suitable for variable students' lives and processing informal learning; the reason is that it does not include an understanding of what students really want and need in such spaces (He, 2018).

3.3 Informal Learning in Chinese University Campus Planning

Chinese university campuses, currently, can be classified into two kinds: city campuses and suburban campuses (Chen, 2014). City campuses are usually older and located in noisy but convenient places for easy access to neighborhoods, while suburban campuses are normally found on the edges of cities and are usually near other universities in the 'university town' whose surrounding environment is natural (Liu, 2019). Chen and Shi (2011) articulated that, at the end of the last century, universities in China were affected by the change of the concept of education, wherein elite education was replaced by universal education, leading to the number of students increasing and campuses expanding. Hence, the restrictions around old campuses have prevented extensive expansion and renovation, giving birth to new campuses on city edges or in rural places (Chen & Shi, 2011). In the first ten years of the twenty-first century, Chinese universities were constructed in increasing numbers, with new campuses and 'university cities' both growing in number.

At present, the future development of university campuses mainly lies in the renovation of old campuses and the expansion of new campuses (Coulson et al., 2011, 2018). However, the public space development for most of Chinese new university campuses has not given full consideration to the construction of informal learning spaces (Wu, 2020). For example, Wu (2020) claimed that connections between functions on campus are seriously weakened, and vacant land begins to emerge which has caused a large amount of waste (as shown in Fig. 2). Furthermore, campus construction and university leadership achievements are linked, which means that the construction process cannot be consistent and future development considerations are ignored (Wang, 2018).

Specifically, Tu (2007) argued that one of the elements of campus planning is the design of university campus road system, which consists of roads, non-car routes, and

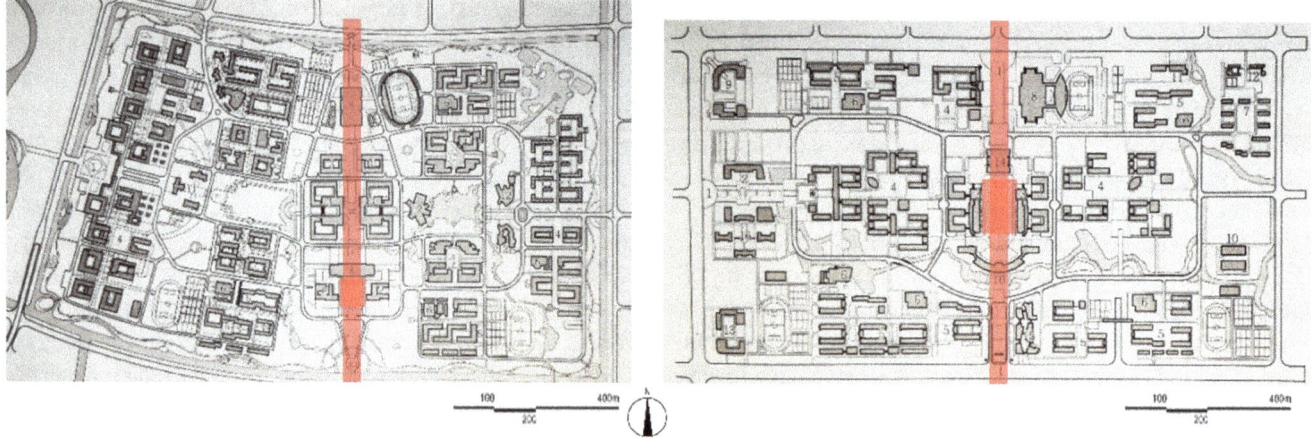

Fig. 2 Typical planning of Chinese university campus (one main axis with a giant square)

pedestrian paths. The design of the system can reveal whether campus planning considers the engagement of informal learning (Zhu & Wang, 2020). Although the current large number of motor vehicles has increased the demand for road system, campus planning should give priority to non-car modes and pedestrian paths; separate systems for roads and pedestrians have been widely adopted, among which the ring road is the most suitable strategy for universities to carry out multi-direction development and to adjust measures to local conditions for further development; polycentric system and multi-loop systems or semi-ring systems derive from this concept (Tu, 2007). Pedestrian spaces can be regarded as valuable places for informal learning, because of their high levels of usage (Li, 2020b). They connect learning and living districts, but, as shown in Fig. 3, they often serve simply as commuting routes without any other functions in most of Chinese university campuses, so students just pass them by rather than stay to learn (Wang, 2018).

Campus squares, as another element of campus planning, are usually the main part and the most significant symbol for universities, so they also need to be considered in campus planning (Tu, 2007). The campus square is an open and iconic place in the campus environment. It can be equipped with plants, sculptures, and small elements to enrich the

landscape and guide the view. Due to the lack of certain atmospheric elements, most campus squares are desolate, with people passing by in haste (Wang, 2020a). Zheng et al. (2020) pointed out that reasonable squares, embedded with additional learning support facilities, can serve as good informal learning spaces where students can gather together to engage in discussion-based learning. However, at present, most of the campus squares in Chinese universities are 'idle' (He, 2018), showing a magnificent presence at the campus entrance or beside the main building, but being essentially non-functional (as shown in Fig. 4).

3.4 Informal Learning in Chinese University Campus Landscape

Generally, the campus landscape is divided into the natural landscape and cultural landscape (Ye, 2014). Just like roads and squares, the natural landscape in Chinese university campuses is generally idle and is not well-connected with students' university activities, which leads to a sense of distance (Wang, 2018). The natural landscape in university campuses is usually divided into two kinds. First, there is the placement of artificial elements including border trees, tree

Fig. 3 Representative university campus roads in China's universities (author's photographs in 2015)

Fig. 4 Representative university campus squares in China's universities (author's photographs in 2016)

pools in squares and landscaping around the building (Zheng et al., 2020). Alongside this, there is the natural environment, including wetlands and vegetation of the original site. The first natural landscape shows too much artificial carving, and the rigid image has become deeply impressed on people (Li, 2020b), while the second landscape presents an excessively natural feeling and fails to integrate with its surroundings (He, 2018) (as shown in Figs. 5 and 6).

On the campuses of Chinese universities, cultural landscaping is very common, including monuments, courtyards, statues, and squares with historical significance (Wang, 2020b). These cultural landscapes are too formalised to be considered as natural, though they have some value as markers on campuses or as essentially souvenirs (Wang, 2020a). Landscape planning has not given full play to the role of these cultural landscapes including developing and stimulating students' thinking and learning, and creating informal learning places. The vast majority of cultural landscaping is now seriously 'idle' and is largely devoid of students (Wang, 2018). Accordingly, whatever cultural or natural landscape within Chinese university campuses, students and the staff can just regard them as normal views rather than informal learning spaces, due to lack of the construction of learning supported facilities and circumstances.

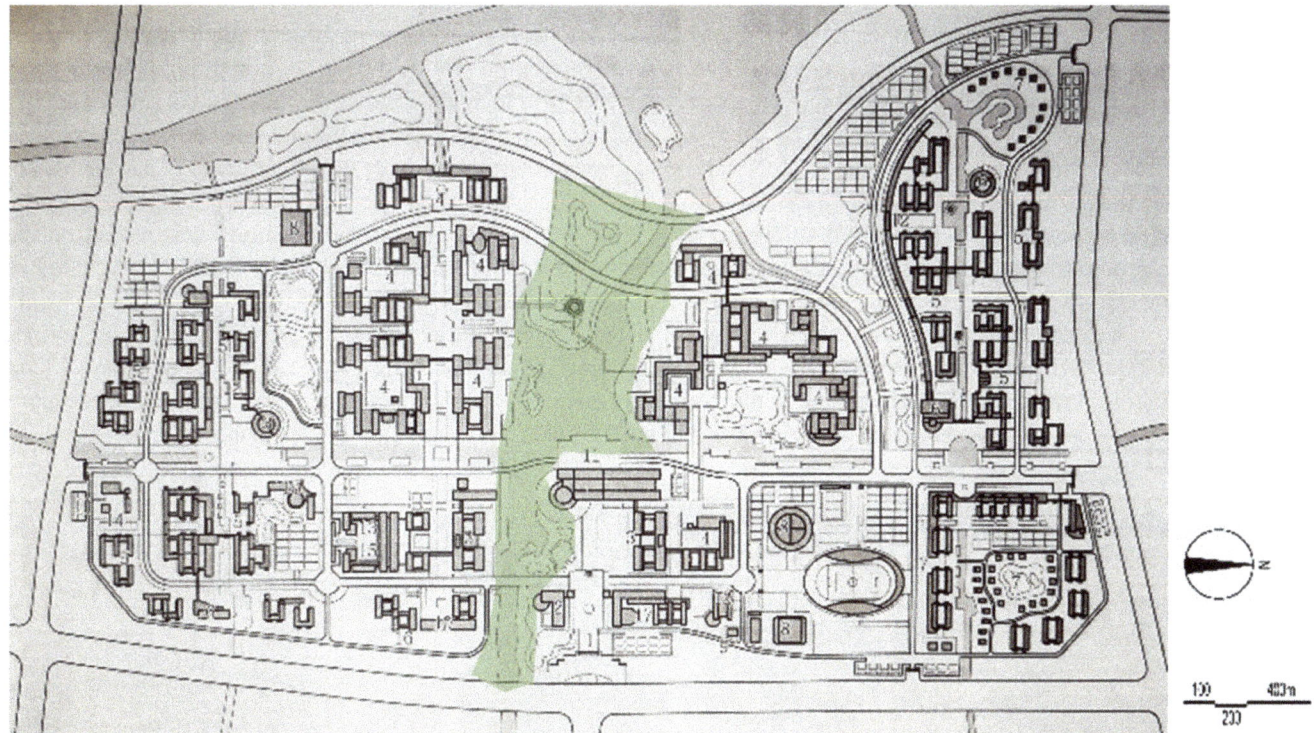

Fig. 5 Typical landscape in Chinese university campus

Fig. 6 Representative campus landscape in China's universities (author's photographs in 2017)

3.5 Informal Learning in Chinese University Campus Architecture

The campus planning of Chinese universities usually focuses on the plane composition and the intentional functional partition with little consideration for students' requirements and activities (Wang, 2018). There is a lack of interaction between buildings, which look like isolated islands, hence there are few spaces for students to implement informal learning.

Due to the influence of the former Soviet architecture after the founding of modern universities in China, the space vitality in the building is very limited, but the approach to function is rigorous (Chen & Shi, 2011). Through the development of the regeneration and opening period, the requirements of higher education for university students have become more diverse, and the vitality of building space has been gradually improved (Li, 2020a). However, the old campuses of Chinese universities have a long history, and the building functions are very old, while new campus construction is developing too fast (Wang, 2020a). There is insufficient consideration for the diversity of functions in the building, so there are not enough informal learning spaces. Specifically, too few open teaching buildings have enough independent fields, and there are few interactions between buildings, leading to fewer spaces and environments for teachers and students to communicate and discuss during

out-of-class time (Liu, 2019). As shown in Fig. 7, most architectural spaces lack the design of learning facilities and atmosphere, with empty, dark and monotonous views.

4 Regeneration Strategies of Campus Public Spaces

4.1 Universities Dominate Campus Planning and Enhance Campus Environmental Diversity

University campus planning should be a cooperation between the university and the government, and it should take into full consideration of the main users, students; furthermore, the school should be more dominant in its role (He, 2018). In reality, however, this is not the case. For instance, historically, the traditional universities experienced too much interference from the government departments, not only in the passive choice of location but also because excessive emphasis in campus planning was placed on the layout based on formalism and symbolism, so the quality of the campus environment could not be guaranteed (Wang, 2018). In the future, the dominant determinant of campus planning should be the university itself (Zhu & Wang, 2020). The government can instead make instructional suggestions and provide funding, and the planning should focus on

Fig. 7 Representative architectural entrance, atrium, and corridor in Chinese universities from left to right (author's photographs in 2017)

improving the quality of students' study, research, and life (Chen, 2018). The functions of study and living areas should be planned clearly and distinctly, but there should be appropriate transitional space between them, such as waterfronts, forests, grasslands, or man-made landscape structures. Campus planning should also place emphasis on the function of space diversity and increasing quality. For example, a square is not only regarded as useless 'open land' but should be able to accommodate many functions as 'a gathering place' for learners; meanwhile, a road can serve as a commuting path and learning space at the same time.

4.2 Improving the Principles of Campus Landscape Design

At present, the campus landscape is usually the most neglected link in campus planning and design within Chinese universities. As stated above, due to the instability of landscape systems and the lack of attention paid to landscape design, the landscapes of university campuses are generally fixed and do not meet the utility needs of universities. To improve this situation, it is necessary to improve the principles of campus landscape design, which will be analysed based on two categories: namely, natural and cultural landscapes.

Firstly, the core principle of natural landscape design is to create a natural environment for teachers and students to satisfy their needs for learning, working and leisure after class (Zheng et al., 2020). The design is restricted by the following three factors: the climate, the site environment, and the vegetation condition. As for the climatic factor, Li (2020b) indicated that the landscape design should follow the local climate conditions to design the best site and structure and to create a microclimate to reduce the impact of climate factors on the landscape environment. As for the site environment, He (2018) articulated that the landscape design should take the advantage of the relationship between the geography of the campus and the surrounding areas and the topographic features of the area itself. The vegetation condition refers to the fact that the vegetation on campus is an important natural condition and should be preserved and utilised as much as possible (Ye, 2014).

By comparison, the principles of cultural landscape, which is the reflection of the history and culture as well as the souls of universities, should cover the following three aspects: the campus cultural background, design concepts and styles, and students' behavioural characteristics (Zhu & Wang, 2020). As for the first, from the ancient Chinese tradition of reverence for 'etiquette' to modern America's 'liberal democracy', and from 'high, big and full' to the more contemporary 'open, efficient and intensive', the concepts formed in different cultural backgrounds have become the frame of reference for the orientation of campus environments (He, 2018). Design concepts and styles indicate that the landscape design cannot have the advanced and practical guiding significance without keeping up with the trend of the times. For the last, the main users of university campuses are students, so the landscape design should be centred around students' behaviours, and consciously guide them to maintain positive and healthy behaviours and lifestyles (Ye, 2014).

4.3 Making Full Use of Spaces in and Among Campus Buildings

As part of the campus environment, there should be a series of interactions and connections with other function areas to strengthen communication in and out of campus buildings, especially teaching buildings. On Chinese university campuses, buildings only have their own rooms for specific functions, such as teaching and learning, meeting, which are supported by service and transport spaces, but they should be open to the public as well to some extent (Wang, 2018). This feature does not apply to all types of campus buildings. Generally, in terms of the degree of openness, public activity centres, student centres, and libraries are the most open buildings, followed by those of departments and public teaching buildings, while students' dormitory buildings come the last (Tu, 2007). At the core part of the campus, the buildings should present a layout of 'architectural complexity', with cross-penetration among functions providing a good place for informal learning. In general, at the entrance, atrium or terrace of these buildings, people may assemble, and informal learning areas can be formed (Chen, 2018). Thus, these spaces could be embedded with learning support facilities and devices to construct an environment for learners to implement informal learning (Zheng et al., 2020). Once the spaces in and among campus buildings, especially the ones with the highest degree of openness, are fully utilised, students' lives would not only be restricted to 'the three points on one line' mode in Chinese universities, because they would have alternative places to undertake individual and collaborative learning which they need to do during their out-of-class time.

5 Case Study: Campus Public Spaces of XJTLU

5.1 A Brief Introduction to XJTLU's Campus and Pedagogy

XJTLU is an international joint university based in Suzhou, Jiangsu, China. Founded in 2006 and an outcome of the partnership between the University of Liverpool and Xi'an

Jiaotong University, it is the first Sino-British joint venture between world-class research-led universities. The University primarily focuses on science, technology, engineering, architecture and business, with a secondary focus in English. Students are rewarded with degrees from both the University of Liverpool and XJTLU. President Xi of China stated that the goal of XJTLU is to combine the best practices of China's universities with those of Western universities to create China's own unique pedagogical and management model (Yi, 2013).

The campus of XJTLU is located at No.111 Renai Road, Suzhou Dushu Lake Science and Education Innovation District in Suzhou Industrial Park, 12 km east of the city center of Suzhou, and 90 km west of Shanghai. Its campus has been developed in two phases. The first phase focused on the North Campus (as shown in Fig. 8a), which was completed in Autumn 2013. The overall campus plan was designed by Perkins + Will, an American architecture firm. The University's iconic Central Building, which is also the university library, was completed in Autumn 2013 and was designed by the British-Asian architecture firm Aedas. It was nominated in the 2014 World Architecture Festival. The second phase focused on the South Campus (as shown in Fig. 8b), which was designed by British architects BDP, and was started in the Summer of 2013. The first phase of the South Campus was opened on 26 July 2016. Architecturally it contrasts the North Campus and features modern designs that complement the surrounding area, as shown in Fig. 6. More than those, there are several facilities and business zones located around the campus, enriching the normal lives of students and the staff within out-of-class time.

As for XJTLU's pedagogy, the university is close to the borrowing/localisation model by creating its own identity through the joint strengths of the two parent universities (Yi, 2013). There are three aspects playing very important roles

in this area: the teaching management system relies on highly efficient but simple functions; the research-led training or teaching process can guide teachers, students and teaching itself to stimulate their potentials; and the teaching quality assurance system can effectively guide and control the teaching quality of higher education in the right way (Cai et al., 2018). These elements all reveal that the implementation of pedagogy is different from that of China's traditional higher education institutions because they show that the education and learning is more reliant on teachers' and students' own perspectives, which can help to make the teaching and research more efficient and of higher quality. Thus, the campus design and construction, which need to cooperate with the design of its teaching system, are totally unique and distinct from those of other university campuses in China. The following content will be the introduction of the differences based on the three aspects.

By contrast, the majority of Chinese suburban university campuses not only lack the assisted facilities around, their main pedagogy is mostly determined or influenced by Chinese educational authorities lacking of flexibility and concerns of student needs.

5.2 Superior Location and Campus Environment

The campus of XJTLU is embedded with complete surrounding facilities and convenient transportation. The campus stands out for its modern architectural image, distinctive landscape, and diverse forms of space. The campus is a pleasant place with no large, idle squares. The campus roadways are clear and are surrounded by other campus elements. The internal road system is highly walkable. Buildings, roads, and squares are all interwoven, forming a multi-dimensional space network. Especially the campus

Fig. 8 **a** XJTLU North Campus (Saieh, 2009) **b** XJTLU South Campus (XJTLU, 2020)

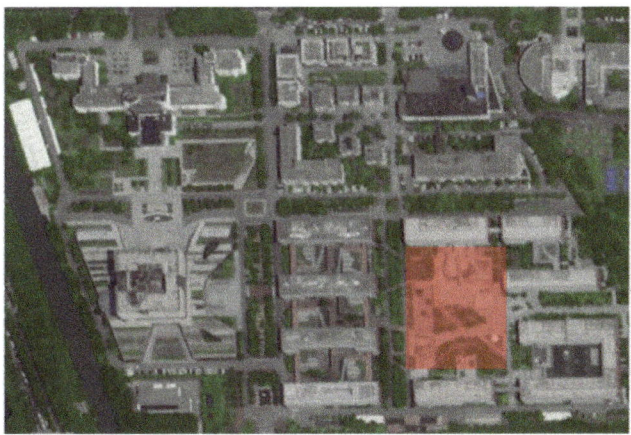 

Fig. 9 **a** Campus square in North Campus **b** Campus square in South Campus

Fig. 10 **a** North Campus central square (Xi'an Jiaotong-Liverpool University, 2021) **b** South Campus central square (BDP, 2018)

squares, different from those on most Chinese university campuses, are not located on the central axis and can generate spatial interactions between different campus buildings (as shown in Fig. 9a, b). The environmental space on-campus is very reasonable, and the public open space is rich in content, as shown in Fig. 10a, b. From almost every angle, there is a good view, and the public teaching space converges a lot, meaning that not only is the atmosphere quiet, but also interactions with the outside spaces can be formed. The good location and campus environment provide good places for students in both a material and spiritual way, making them suitable to implement informal learning.

On general Chinese university campuses, the environmental space on-campus is full of idle squares, landscapes, and other spaces, which students and the staff normally seldom prefer to enjoy or stop to watch. Thereby, the percentage of spatial usage in such space is comparatively lower than that in XJTLU.

5.3 Good Connections Between Campus Buildings

In the North Campus, to serve as 'bridges', clusters of polytechnic teaching buildings are linked with outdoor corridors and galleries, and the inner corridor of the building also opens and communicates with the aisles in the vertical direction. These run through every building (as shown in Fig. 11). Some features are placed in the corridors connected with the outdoor space, such as tables, chairs and sun shades. The outdoor corridors, which also function as building entrances, are located inside the outward galleries. The central square surrounded by the building cluster is the centre of the campus, including the ground and underground spaces. The underground spaces are for commercial shops, providing entrepreneurial places for students, while the ground is reserved as a square and restaurant area. The abundant architectural relationships between teaching

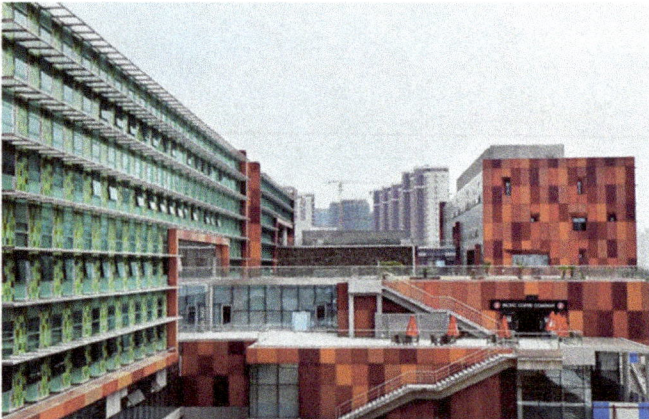

Fig. 11 Connections between clusters of polytechnic teaching buildings (easyuni, 2021)

buildings provide many possibilities for the activities that can occur on campus, which is the essence of informal learning.

The university library is the focus point of the whole campus, and the design is inspired by the Lake Taihu stone in Suzhou (as shown in Fig. 12a, b). The building form is frontally hexahedral, and the two bottom layers are enlarged for the entrance and roof activity platform. On every façade of the building, a hollowed-out design style is used to excavate the hole and converge the building's form into a central atrium in the middle. This design enriches the abundance of the building space and provides a central rest area for people in the library to converse and do other activities to avoid bothering others learning in the library. The library is located close to the southwest corner of the polytechnic teaching buildings, and the two sides of the square provide free access to the campus. This unconstrained architectural layout provides people with possibilities to carry out a variety of activities. Meaningful activities,

including learning, communicating, playing, and taking photographs, can be done anywhere on the campus.

Compared with most traditional Chinese university campuses, whose public spaces are normally idle, various spaces between buildings provide students and the staff with multiple alternative places to spend their out-of-class time on XJTLU campus. Thereby, students can experience informal learning and social activities within such spaces rather than the domitory or the canteen.

5.4 Multipurpose Indoor Spaces

The campus is effective not only as a campus environment for its exterior spaces and the layout of its buildings, but also because the interior spaces of the university are abundant and varied. Take the College of Science and Engineering, in the North Campus, as an example: each building has a similar appearance, but the interior spaces are different. For instance,

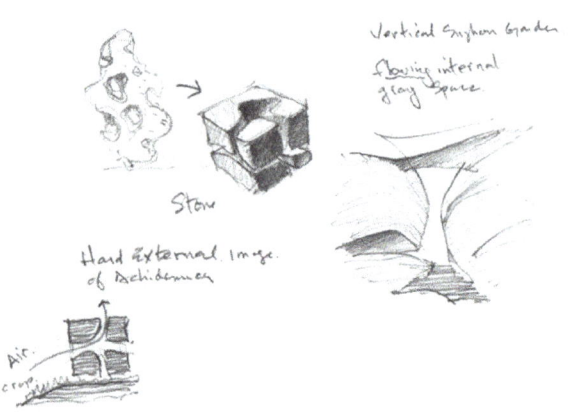

Fig. 12 **a** Design concept of university library (AEDAS, 2013) **b** University library appearance (Aedas, 2013)

Fig. 13 The relationship between the corridor of the teaching building and the external environment (Wang, 2018)

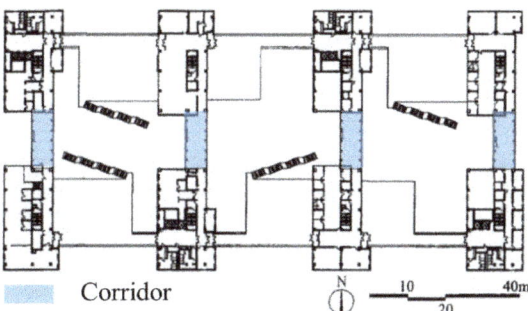

Corridor

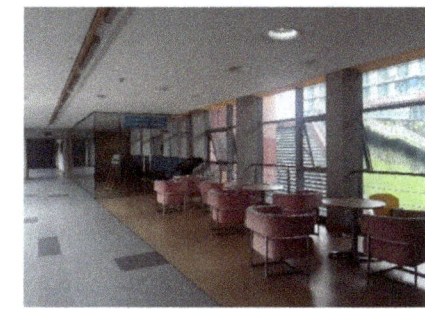

together with different foyer layout and design, as well as the halls and open holes, the varied-width corridors can serve as venues for students to undertake self-directed learning, communication and gathering (as shown in Fig. 13). The design of furniture and colours in the corridor make it fun for students to stay in such spaces doing various activities willingly, with boring spaces entirely absent.

In the South Campus, another type of corridor, located in the new design building, is a wholly inner space, but it utilises soft materials on the corridor walls, which can be regarded as both protection and ornament for the walls. They can create small-scale communication spaces which are used as a convenient place for students and teachers to communicate after class (as shown in Fig. 14). However, this corridor space is limited to specific disciplines, such as design-related ones, so the applicational value is marginally less than that in other areas.

On both the North and South campuses, almost every building, with the exception of the university library, belongs to a specific college or major, so the function and layout requirements for the informal learning spaces are different. Although the appearance of the buildings is similar, there are several kinds of public learning spaces within them, and the relationships between corridors, open spaces, lecture rooms, and classrooms can seem free rather than discursive (as shown in Fig. 15).

In comparison, the indoor spaces, mainly corridors and atria, are just empty and monotonous spaces without learning facilities and contexts on traditional Chinese university campuses; whereas the indoor spaces are various and are facilited with tables, chairs, exhibition walls, etc., presenting several attractive places where students and the staff can communicate and interact after class.

5.5 Impacts of the Campus on the Space Users

The spatial usage on most conventional Chinese university campuses lacks variety, while that on XJTLU campus is diverse. For example, on the campus squares, students and the staff have multiple choices to use the space, since there are several layers in an unit area, which can create many functionally and specifically private or public places for learning and teaching; compared with most Chinese university squares, which are normally pretty big but just used for adding the sense of ceremony; as a result, merely few students and staff use them for learning or teaching after class.

Similarly, except from normal classrooms and lecture rooms in XJTLU, the spaces within campus landscapes and architecture are almost designed for students and the staff who have a great deal of choices to learn and teach outside

Fig. 14 Exhibition space in corridors of new design building (Ke & Shang, 2017)

Fig. 15 Different kinds of public indoor learning spaces (QS, 2021)

Table 1 The comparison between informal learning spaces on campuses of the majority of Chinese universities and XJTLU

	The majority of Chinese universities	XJTLU
Campus planning	Campus construction and university leadership achievements are linked, which means that the construction process cannot be consistent and future development considerations are ignored; connections between functions on campus are seriously weakened, and vacant land begins to emerge which has caused a large amount of waste, which means that space users normally do not spend their out-of-class time within public campus spaces	Campus planning is leded by professional planning and architectural companies; the connections between different campus buildings are various, and space users can organise and experience different activities outside classes or during out-of-class time
Campus landscape	Cultural and natural landscape within the majority of Chinese university campuses lacks the facilities that can facilitate students' normal lives, which means that the landscape is just used for viewing rather than alternative places for students' informal learning outside classes or during the out-of-class time	Campus landscapes within XJTLU provide students with several informal learning spaces with multiple learning supported facilities, such as chairs and tables, which means that students and the staff have alternative places to extend learning outside classes or during the out-of-class time
Campus architecture	Architectural spaces, whatever within or outside buildings, lack variety, which means that students and the staff hardly use such spaces outside classrooms or during out-of-class time	There are abundant architectural spaces which are equipped with learning supported facilities, such as chairs, tables, and exhibition walls, enabling students' informal learning with great convenience

classes or within out-of-class time, such as the 'bridges' and self-learning spaces in North Campus and the corridor-walled areas in South Campus; different from the space users in traditional Chinese university campuses, their lives are normally restricted in three places, which are the classroom, the canteen, and the domitory.

Therefore, in summary, the majority of general Chinese university campus space users have fewer choices to experience informal learning outside formal classrooms and within out-of-class time; by comparison, space users of XJTLU campus can not only study within normal course time but also experience ample informal learning within campus squares, roads, landscapes, corridors, and other public spaces.

Overall, the summarised differences of informal learning spaces on campuses within the majority of Chinese universities and XJTLU are listed in Table 1.

6 Conclusion

It can be seen from the above that informal learning is more suitable for higher education at present, and it is already a trend worldwide. However, China's higher education system still makes formal learning the main learning method for students because traditional pedagogy and rigid campus design limit the implementation of informal learning. Thus, this paper presents the strategies to embed informal learning into campus public spaces from three dimensions, which are campus planning, campus landscapes and campus architecture, in order to provide prospective university campus public space regeneration projects with suggestions to prepare for engagement with informal learning. Finally, the case study, which takes the campus public spaces of XJTLU as an example, demonstrates that innovative campus and school buildings are not only suitable for informal learning but also acceptable for China's higher education institutions and educational environment. In addition, the campus design and construction of XJTLU are good examples for others to learn from when regenerating old campuses or constructing new campuses in China.

References

AEDAS. (2013). *Xi'an Jiaotong-Liverpool University Administration Information Building*. ARCHI TONIC. https://www.architonic.com/en/project/aedas-xi-an-jiaotong-liverpool-university-administration-information-building/5103730

Aedas. (2013). *Xi'an Jiaotong-Liverpool University Administration Information Building/Aedas*. ArchDaily. https://www.archdaily.com/511602/xi-an-jiaotong-liverpool-university-administration-information-building-aedas

BDP. (2018). *XI'AN Jiaotong Liverpool University South Campus*. BDP. http://www.bdp.com/en/projects/p-z/xian-jiaotong-liverpool-university/

Cai, L., et al. (2018). Research on the teaching quality structure of Sino Foreign Joint University and its inspiration—A survey from Xjtlu University. *International Journal of Information, Business and Management, 10*(3), 178–190.

Chen, J. (2014). *A university campus in Beijing*. Tsinghua University Press.

Chen, J. (2018). A case study on the use of informal learning space on campus. *Interior Architecture of China, 09*, 115.

Chen, Q., & Liu, D. (2011). *Educational psychology* (2nd ed.). Higher Education Press.

Chen, X., & Shi, L. (2011). *A brief history of Chinese university campus morphology*. Southeast University Press.

Chen, X., et al. (2010). From classroom to lawn—Campus learning space continium construction. *China Educational Technology, 286*, 1–6.

Coulson, J., et al. (2011). *University planning and campus architecture—The search for perfection*. Routledge.

Coulson, J., et al. (2018). *University trends: Contemporary campus design*. Routledge.

easyuni. (2021). *Xi'an Jiaotong-Liverpool University*. Easyuni. https://www.easyuni.com/china/xian-jiaotong-liverpool-university-11464/photos/

He, T. (2018). *Research of informal learning space design in college and university education buildings*. South China University of Technology.

Hu, J., et al. (2020). On the upgrading of students apartments in universities by equipping them with informal learning space. *Research in Higher Education of Engineering, 04*, 118–123.

Ke, S., & Shang, J. (2017). The third teacher—The new design building at XJTLU: How students become architects. *World Architecture, 7*, 48–57. http://hdl.handle.net/11311/1062199

Li, J. (2020a). *Research on the design of informal learning space in the Department of Architecture*. Hebei University of Architecture.

Li, S. (2020b). *Research on the renewal design of informal learning space in colleges and universities based on multi-source data analysis*. Xi'an University of Architecture and Technology.

Liu, Y. (2019). *A survey on the current status of informal learning space: A case study of old campus of S University*. Sichuan Normal University.

Marshalsey, L., & Sclater, M. (2020). Together but apart: Creating and supporting online learning communities in an era of distributed studio education. *iJADE, 39*(4), 826–840. https://doi.org/10.1111/jade.12331

QS. (2021). *Xi'an Jiaotong Liverpool University*. QS. https://www.topuniversities.com/universities/xian-jiaotong-liverpool-university

Saieh N. (2009, November 05). *Xi'an Jiaotong-Liverpool University Administration & Information Centre/Aedas*. ArchDaily. https://www.archdaily.com/39697/xi%e2%80%99an-jiaotong-liverpool-university-administration-information-centre-aedas

Shao, X., & Zhang, J. (2020). New learning space in primary and secondary schools: Dimensions and methods of informal learning space construction. *Research in Educational Development, 40*(10), 66–72.

Tu, H. (2007). *Integral design for campus-planning, landscape architecture*. China Architectural Industry Press.

Wang, C. (2013). On the problems, causes and countermeasures of exam-oriented education in China. *Kaoshi Zhoukan, 47*, 1–2.

Wang, Y. (2015). Analysis of students' weariness psychology and its countermeasures. *Science and Technology Innovation Herald, 28*, 227–228.

Wang, J. (2018). *Study on the construction of informal learning places for public spaces on university campuses*. Anhui Jianzhu University.

Wang, J. (2020a). A study on the construction of informal learning spaces (ILSs) on University Campuses in China. *IOP Conference Series: Materials Science and Engineering (WMCAUS)*. https://doi.org/10.1088/1757-899x/960/2/022022

Wang, X. (2020b). Design principles for informal learning Spaces on campus. *Artistic Sea, 08*, 116–117.

Wu J. (2020). *Research on the informal learning use of open space on campus: With case study of Beijing Xueyuanlu College Campuses*. Beijing Forestry University.

Xi'an Jiaotong-Liverpool University. (2021). *Xi'an Jiaotong-Liverpool University*. Jobs. https://www.jobs.ac.uk/enhanced/employer/xian-jiaotong-liverpool-university/

XJTLU. (2020). *XJTLU Xi'an Jiaotong-Liverpool University*. https://www.jobs.ac.uk/enhanced/campaign-site/xjtlu/index.html

Ye, X. (2014). *University campus landscape planning and design*. Chemical Industry Press.

Yi, F. (2013). University of Nottingham Ningbo China and Xi'an Jiaotong-Liverpool University: Globalization of higher education in China. *Higher Education, 65*, 471–485. https://doi.org/10.1007/s10734-012-9558-8

Zheng, T. (2018). *Study on informal learning space in university teaching buildings in Southern China*. Shenzhen University.

Zheng, Y., et al. (2020). Practice and exploration of campus informal learning space construction. *Vocational Education, 19*(01), 93–96.

Zhu, Q., & Wang, W. (2020). Reflections on the architectural space of universities in China: Taking informal learning space as an example. *Beauty & times, 03*, 20–21.

Flora of Archaeological Landscape: Case Study of Arslantepe Mound and Its Territory

Şükrü Karakuş and Aysun Tuna

Abstract

Floristic research plays a key role in understanding archaeological landscapes. Floristic studies provide important information for archaeological research, for understanding the diet of societies to determining agricultural activities of the periods. With the data obtained as a result of archaeobotanical studies in recent years, important findings regarding the general flora character of archaeological landscapes have been reached. As of today, the floristic balance of current archaeological landscape is changing as the existing vegetation is destroyed in archaeological sites where excavations are continuing. Based on this, in this paper, current flora of Arslantepe Mound and its territory was examined. The aim of this paper is to determine the plant biodiversity of the study area and to identify the sensitive areas with endemic species. Within the boundaries of the study area, 440 samples were collected between August 2018 and August 2019. As a result of identification of collected samples and evaluation of plants registered in the literature, 704 taxa (681 species, 13 subspecies, 10 variate) belonging to 384 genera and 90 families were determined. 65 of the total species are endemic in the area. The phytogeographic regions of only 233 species out of the collected material have been identified; Irano-Turanian 184, Mediterranean 38, Euro-Siberian 11. The rest 471 species of the total are either pluriregional or phytogeographically unknown. Two species belong to Pteridophyte whereas 702 species belong to Spermatophyta. Within the area, gymnosperms have 16 species and angiosperms have 686 species. Dicotyledons and monocotyledons have 586 and 100 species, respectively, in the Angiosperms. The largest families identified in the study area are as follows: Fabaceae 67, Brassicaceae 65, Asteraceae 58, Poaceae 54, and Lamiaceae 53. The largest genera in the study area are as follows: Alyssum 12, Euphorbia 12, Astragalus 10, Medicago 10, and Trifolium 7. Endemic species located within the study area within the scope of the findings obtained as a result of the floristic research, deployment areas, and endangered species have been identified. Suggestions have been developed to enable the archaeological sites to appear within the urban landscape as alternative green spaces.

Keywords

Endemic • Flora • Taxonomy • Archaeological landscape • Arslantepe mound • Malatya

1 Introduction

Turkey is a country occupying a meeting point of the Asian and European continents between the 35–42° N latitudes and 25–45° E longitudes, and it covers an area of 783,562 km^2. Having extensive variety of climate, topography, main rocks, soil, and aquatic habitats with running water systems, the country has magnificent biodiversity. Furthermore, glacial periods had only a limited destructive effect on the country at higher altitudes, and thus its biodiversity was kept from natural destruction, unlike European biodiversity. Hence, in respect of biodiversity, Turkey is an important center worldwide, especially in the Mediterranean phytogeographical region and it has plenty of biodiversity hotspots for conservation priorities. Moreover, not only the Mediterranean region but also the Irano-Turanian part of Turkey is important biodiversity center and new taxa are continuously described (Dönmez & Yerli, 2019). Based on the floristic studies carried out in Turkey, 9983 species,

Ş. Karakuş (✉)
Faculty of Fine Arts and Design, Department of Landscape Architecture, Inonu University, Malatya, Turkey
e-mail: sukuru.karakus@inonu.edu.tr

A. Tuna
Faculty of Architecture, Department of City and Regional Planning, Bolu Abant Izzet Baysal University, Bolu, Turkey

vascular plant species are known to grow in the country. Turkey has a total of 11,707 taxon with subspecies and varieties (Güner et al., 2012). Among them 3,649 taxa are endemic.

Province of Malatya, hosting the study area, is located in Eastern Anatolia Region of Turkey. The study area is located in southeast region on the B7 according to the grid square system of Henderson (1961), which is one of the main endemism centers in Turkey. It is located in Iran-Turan phytogeographical region (Fig. 1).

In recent years, many new plant species have been identified within the borders of Malatya Province (Ekşi & Yıldırım, 2019; Genç et al., 2012; Karakuş & Mutlu, 2019; Koç & Aksoy, 2013; Koç et al., 2012; Mutlu & Karakuş, 2012, 2015a, b, 2019; Uzunhisarcıklı et al., 2013; Yeşil et al., 2016; Yıldırım, 2015a, b, 2019; Yıldırım & Erol, 2013; Yıldırım & Şenol, 2014a, b, c; Yıldırım et al., 2015). The studies carried out within the borders of Malatya Province are Contributions to Malatya Flora I A Preliminary Working in Sürgü-Çelikhan (Aktoklu, 1996), Floristics Characteristics of Beydağı (Yıldız et al., 2014), A Floristic Study on Poaceae spp. growing naturally in Malatya Province (Arabacı & Yıldız, 2004), Floristic List of İnönü University (Malatya) Main Campus Area (Mutlu & Karakuş, 2015a, b, c) and Floristic List of Tohma (Malatya-Sivas, Turkey) Valley (Karakuş & Mutlu, 2017). As a result of the studies, 2075 species were identified in Malatya Province and 437 of these species were endemic (Karakuş, 2016).

Arslantepe Mound has been an attractive center where the first urbanization process started, due to the region's access to rivers and mineral resources and its dominance over agricultural areas in Malatya Region since 4000 BCE. Today, we learn that the Malatya Plain established close relations with Mesopotamia, Syria, Transcaucasia, and Central Anatolia during this period, thanks to the archaeological findings of Arslantepe Mound which are representing the early statehood and the birth of bureaucracy (Frangipane, 2012).

As a result of excavation work from 1930 to our time, it has been considered to be a "mound" structure with its multi-layered artificial form. The mound, which is 4 hectares and has 30 m of cultural fill, has been home to many civilizations from the 6th millennium B.C. to the Byzantine period as a necropolis (Frangipane, 2012). The first excavations in Arslantepe started in the 1930s. However, the first systematic excavations were initiated in 1961 under the direction of Alba Palmieri from La Sapienza University. Excavations have been continuing since 1990 under the direction of Marcella Frangipane (Frangipane, 2013).

Located within the provincial borders of Malatya and listed on Tentative List of UNESCO World Heritage, Arslantepe Mound is a substantial archaeological site with outstanding universal values.

It has been determined that Arslantepe Mound fulfills three cultural criteria among the criteria measuring its extraordinary universal value determined by the World

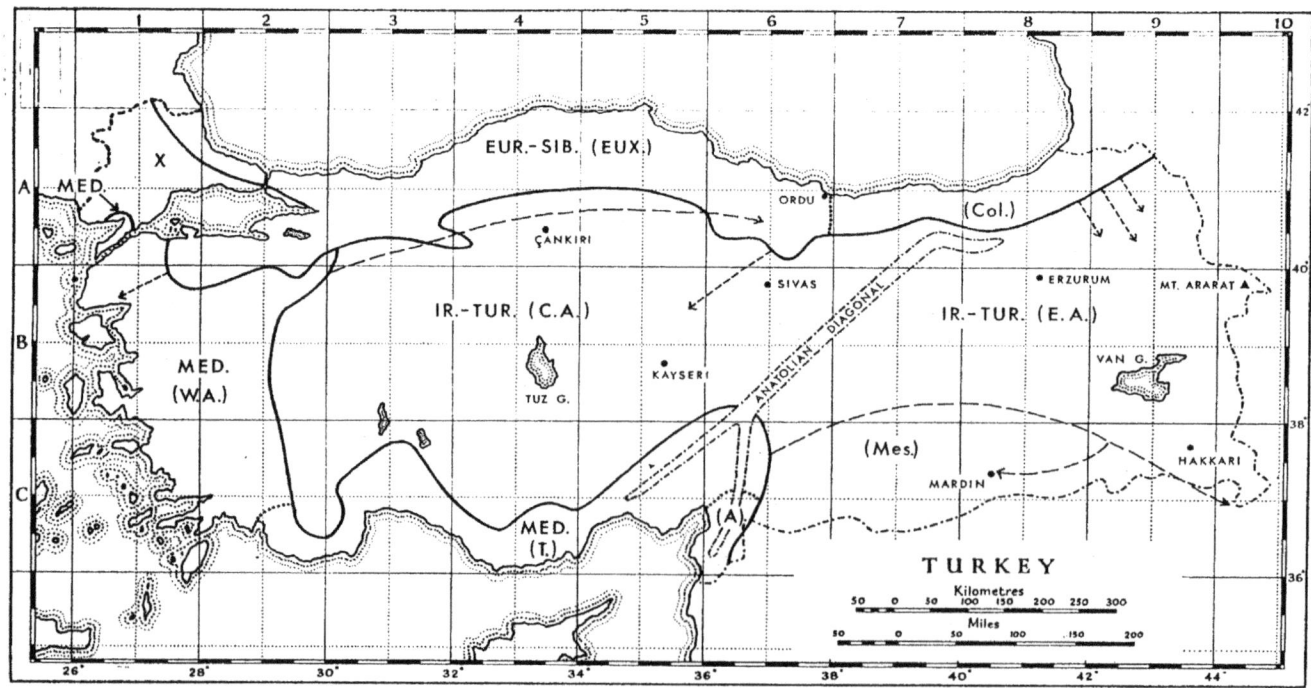

Fig. 1 Phytogeographic regions of Turkey (Davis, 1971)

Heritage Committee, referring to the condition for inclusion in the World Heritage List. These criteria are as follows:

In the context of the criterion (ii): the World Heritage Committee has determined that Arslantepe had a state structure inclusive of the Eastern Anatolia and Mesopotamian civilizations in the 4th Millennium B.C., interchange between such multicultural and social values has introduced a new social and political system based on hierarchy and social stratification and such powerful correlations support new developments such as monumental architecture, administration technologies, and artistic representation power (iconography). Moreover, the Committee further stated that Arslantepe, owing to the palace complex within excavated to the greater extent, reveals the best conserved and advanced example to date within the Upper Mesopotamia and northern periphery (UNESCO, 2014).

In the context of the criterion (iii): the World Heritage Committee has evaluated that association between Arslantepe with the Uruk Civilization presents a proof of its unique value.

Within the scope of the findings obtained as a result of the excavations of the palace complex, the Committee has stated that the entire palace could write a new page in the history of early developments in human society in terms of its anthropological and historical features, such as the existence of a central government controlling the economy, allowing for the explanation of the life and activities of the elite class in detail (UNESCO, 2014).

In the context of the criterion (iv): the World Heritage Committee has stated that there are two reasons for the acknowledgment of the palace complex of Arslantepe as a unique value of the monumental local architecture: first one is that it is the earliest example of the public palaces well known within the Syria-Mesopotamia dating back to 3rd Millennium B.C., revealing the early development of political power regimes in secular order based on economic and administrative control, and the second one is that it has been resolved that the site has a unique and an extraordinary value with its well-conserved structure with the wall with the height of 2–2.5 m., authentic white plaster and murals, revealed as a result of excavation works performed on an area of approx. 2,000 m^2. It is the only unique and single area in the world in this sense, describing a substantial part of human history with its specified features and structures within the Arslantepe Mound and documenting the important stages of human development (UNESCO, 2014).

UNESCO determines that the sustainable conservation of the areas listed as World Heritage can be ensured through a qualified mode of rule. Within this scope, UNESCO has introduced a preliminary requirement in order to prepare a site management plan for the areas on the Tentative List. In accordance with this decision, the requirement of having an archaeological site management plan is raised on the agenda in the process of inclusion on the permanent list of Arslantepe Mound, which is on the Tentative List for the World Heritage. In the content of the Site Management Plan, it is of utmost importance to comprehend not only the archaeological heritage, but also the archaeological landscape it has. In this context, floristic research comes to the forefront in defining the alteration of the archaeological landscape from the past to the present.

2 Materials and Methods

Floristic research covers the entire flowering plants grown naturally within the boundaries of the study area (Fig. 2), mephrolepis exaltata and perennial cultivated plants. This research has been conducted in two stages being as land and herbarium.

Field Studies: Vascular plants have been collected upon performing land studies in various time periods within the boundaries of the study area. As Colchicum falcifolium (February) being the early-flowering type of the year and Sternbergia clusiana (November) being the late-flowering type in Malatya Province, land studies have been performed between February and November.

Herbarium Studies: Plant samples collected as a result of the land studies have been rendered into herbarium material in accordance with the herbarium techniques (Bridson & Forman, 1999).

2.1 Selection of Study Area and Natural Landscape Characteristics of the Study Area

Arslantepe Mound archaeological area located within the boundaries of Orduzu Quarter close to west shore of the Euphrates River (Karakaya Dam Reservoir), 7 km northeast of Malatya and immediate environment thereof have been selected as the study area.

Natural vegetation has been destroyed in Arslantepe Mound as excavations have continued for many years. Therefore, the interaction area of Arslantepe has been determined as the study area for floristic research. Arslantepe Mound interaction area was chosen as the area where flora studies were conducted around Arslantepe Mound. In 2004, the archaeological survey project which were carried out under the direction of Gian Maria DI NOCERA of the Italian Excavation Team was accepted as the boundaries of the interaction area. The floristic research area is surrounded by plains at the foothills of Haroğlu Mountain and Hacı Mustafa Mountains in the north, within the borders of Baskil District of Elazığ City; Beydağları in the south; Şakşak Mountains in the east; and Battalgazi, Yazıhan, and

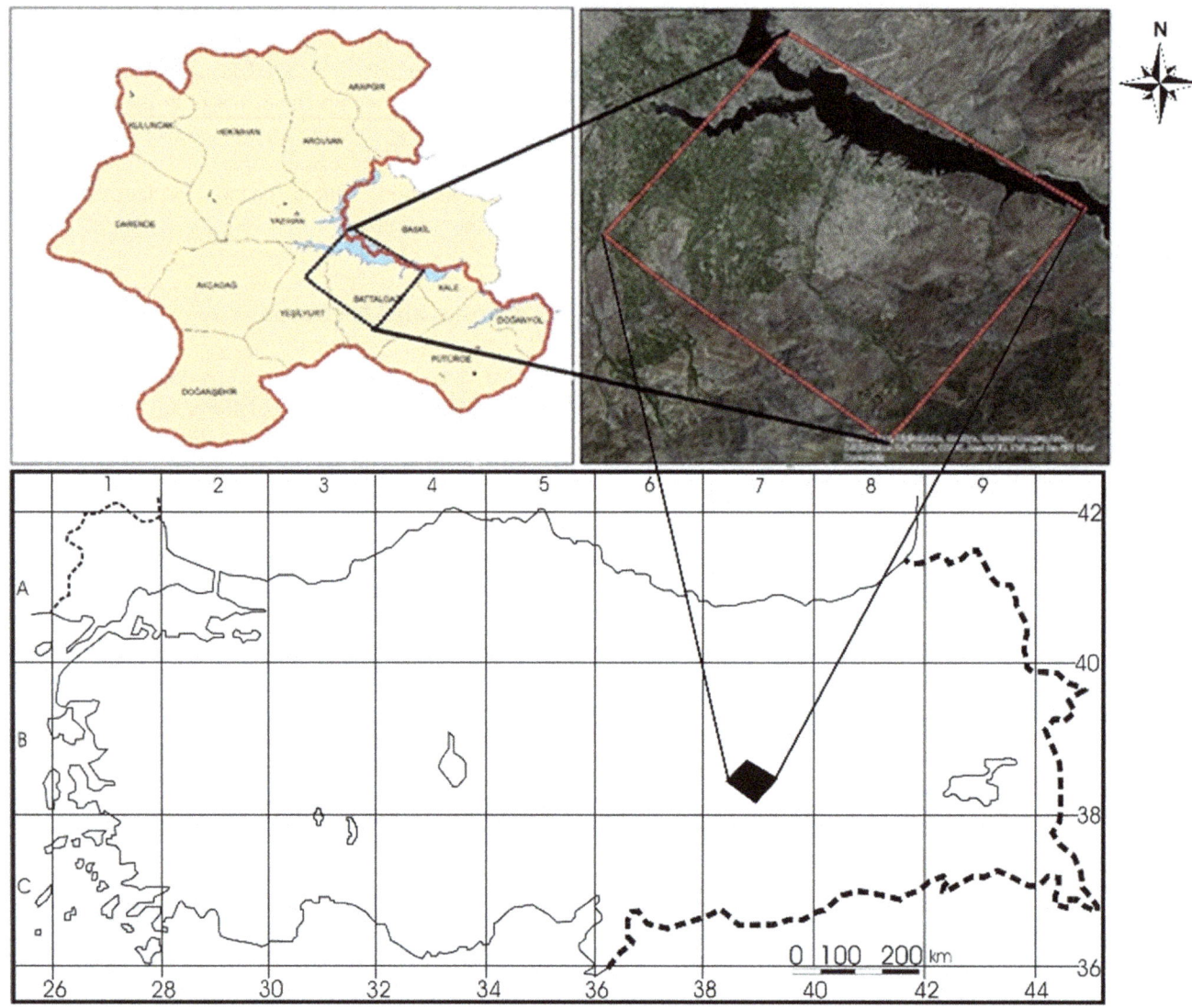

Fig. 2 Geographical location of the study area (Tuna, 2019)

Yeşilyurt Plains in the west. Geographical location of the study area is shown in Fig. 2.

The Malatya Basin, covering the study area, varies within in terms of hydrographic properties. Streams fed from karstic springs in the south and west are high flow and the flow thereof is constant throughout the entire seasons. The streams, having their spring from mountainous areas, are poured into Karakaya Dam Reservoir, which is the lowest part of the basin. Majority of the basin waters are drained by Tohman streamlet, consisting of the substantial bayou of the Euphrates River. The Euphrates River, which was the sole stream located at the north of the study area prior to the year 1985, today, four major streamlets are available poured into Karakaya Dam Reservoir. Those are Hatunsuyu, Orduzu, Ballıdere, and Şişman.

The climate of Malatya Basin, located within the study area is semi-arid D B'2 d b'2, mesothermal, with the absence of or very little excess water, with the climate type close to continental climate (Sunkar et al., 2013). Avcı and Esen (2019) have stated that the average annual temperature of the basin was 11.9 °C based on the 42-year, mean, minimum, maximum temperature, and precipitation meteorological record data covering the period of 1975–2017 of Malatya Station of the Malatya Basin in the Upper Euphrates Section of the Eastern Anatolia Region.

The same study indicates that the highest average temperature is in June 23.5 °C, while the lowest is in January (0.2 °C) and the long-term maximum average temperature is 17.8 °C, and the highest maximum average temperature in August is 30.3 °C, with the lowest in January (4.1 °C), and the minimum average temperature is 6.2 °C, while minimum average temperature drops below 0 °C in winter season.

The climate in Malatya, topography, and the parent material differences occurring have caused differences in soil structure. Such differences reveal themselves in and around the Arslantepe Mound, particularly with the increase in the water surface as a result of the construction of Karakaya Dam. Arslantepe Mound and surrounding soil groups are consisted of brown soils, lime-free brown forest soil, alluvial soils, and colluvial soils. Brown forest soil is predominant at the study area (Tuna, 2019).

2.2 Methodology

Within the boundaries of the study area, 440 samples were collected between August 2018 and August 2019. At least one sample for each taxon was deposited at in the herbarium of İnönü University. Specimens were identified basically using the Flora of Turkey (Davis, 1965–1985; Davis et al., 1988; Güner et al., 2000). Identified species, and other studies used in the writing of the floristic list (Karakuş, 2016; Mutlu & Karakuş, 2015c; Yıldız et al., 2014). In this study, endemic plant and floristic list are given. In the endemic plant list, the following details are stated: the species, authors locality number, phytogeographical region, and IUCN risk categories. In these studies, "Red Book Plant of Turkey" was used for the threatening categories of species (Ekim et al., 2000).

In the floristic list (presented in appendix), the following details are stated: family, genus, species (cultivated plants are indicated an asterisk), authors, locality number, altitude, collection and observed date, collector(s), collector number, phytogeographical region, endemism, and IUCN risk categories.

Collecting localities in the study area are as follows: (1) Malatya: Battalgazi; Arslantepe Mound surroundings; (2) Malatya: Battalgazi; İnönü University campus surroundings; (3) Malatya: Battalgazi; İspendere spirits; (4) Malatya: Battalgazi; Karakaya dam edge; (5) Malatya: Battalgazi; Karaköy surroundings; (6) Malatya: Battalgazi; Karamildan Hill; (7) Malatya: Battalgazi; Meydancık Village surroundings; (8) Malatya: Battalgazi; MÜSİAD Malatya Memorial Forest; (9) Malatya: Battalgazi; Orduzu pond area; (10) Malatya: Battalgazi; Orduzu, Gelincik Hill; (11) Malatya: Battalgazi; Şişmanhan Village surroundings; 8129 Malatya: Battalgazi; Venk Village surroundings; (13) Malatya: Battalgazi; Yıkılgan Hill; (14) Malatya: Yazıhan; Sinanlı Bozburun surroundings; (15) Elazığ: Baskil; Bilaluşağı Village surroundings; (16) Elazığ: Baskil; Ferry pier area.

The abbreviations used in the text and endemic list are as follows: Ir.-Tur.: Irano-Turanian, Medit.: Mediterranean, CR: Critically EN: Endangered, VU: Vulnerable, LR: Lower risk, (cd): Requires protection, (nt): May be threatened, (lc): The least worrying, and INU: İnönü University Department of Biology Herbarium in Malatya.

3 Findings

As a result of identification of collected samples and evaluation of plants registered in the literature, 704 taxa (681 species, 13 subspecies, 10 variate) belonging to 384 genera and 90 families were determined. Two species belong to Pteridophyte whereas 702 species belong to Spermatophyta. Within the area, gymnosperms have 16 species and angiosperms have 686 species. Dicotyledons and monocotyledons have 586 and 100 species, respectively, in the Angiosperms. 65 of the total species are endemic in the area.

The ten largest families according to their species in the area are as follows: Fabaceae 67, Brassicaceae 65, Asteraceae 58, Poaceae 54, Lamiaceae 53, Rosaceae 35, Caryophyllaceae 31, Apiaceae 30, Boraginaceae 25, and Ranunculaceae 20 and the ten largest genera according to their species in the area as follows: Alyssum 12, Euphorbia 12, Astragalus 10, Medicago 10, Trifolium 7, Salvia 7, Hypericum 7, Aegilops 7, Onosma 6, and Papaver 6.

The phytogeographic regions of only 233 species out of the collected material have been determined: Irano-Turanian 184, Mediterranean 38, and Euro-Siberian 11. The rest 471 species of the total are either pluriregional or phytogeographically unknown (Table 1).

Certain varieties identified during the land studies and coming to the forefront with calligraphic properties are shown in Fig. 3.

The deployment map of 65 endemic taxon identified within the project area is indicated in Fig. 4. When the figure is reviewed, four deployment areas of endemic taxon stand out in relief. First area is İnönü University and its surrounding, second area is Venk Village and its surrounding, third area is Sinanlı Village and its surrounding; and the fourth area is Gelinciktepe-Tüllüktepe and its surrounding.

The threat categories of endemic plants within the area are given in Table 2. Distribution to the threatened category of endemic taxa is as follows: 1 taxa CR, 1 taxa EN, 6 taxa

Table 1 Distribution of species determined in the research area according to phytogeographic regions

Phytogeographic regions	Species number
Irano-Turanian	184
Mediterranean	38
Euro-Siberian	11
Pluriregional	471

Fig. 3 (First line) *Crocus cancellatus* Herb. subsp. *cancellatus*, *Eminium spiculatum* (Blume) Schott., *Eremostachys moluccelloides* Bunge, *Lamium garganicum* L. subsp. *striacum* (Sm.) var. *striacum*, *Salvia euphratica* Montbret and Aucher var. *euphratica*. (Second line) *Euphorbia denticulata* Lam., Centaurea urvillei DC. subsp. urvillei, *Arabis alpina* L., Aubrieta canescens (Boiss.) Bornm. (Third line) *Tulipa armena* Boiss. var. *armena*, *Fritillaria imperialis* L., *Genista albida* Willd., *Onosma alborosea* Fisch & C.A. Mey subsp. *alborosea* var. *alborosea*

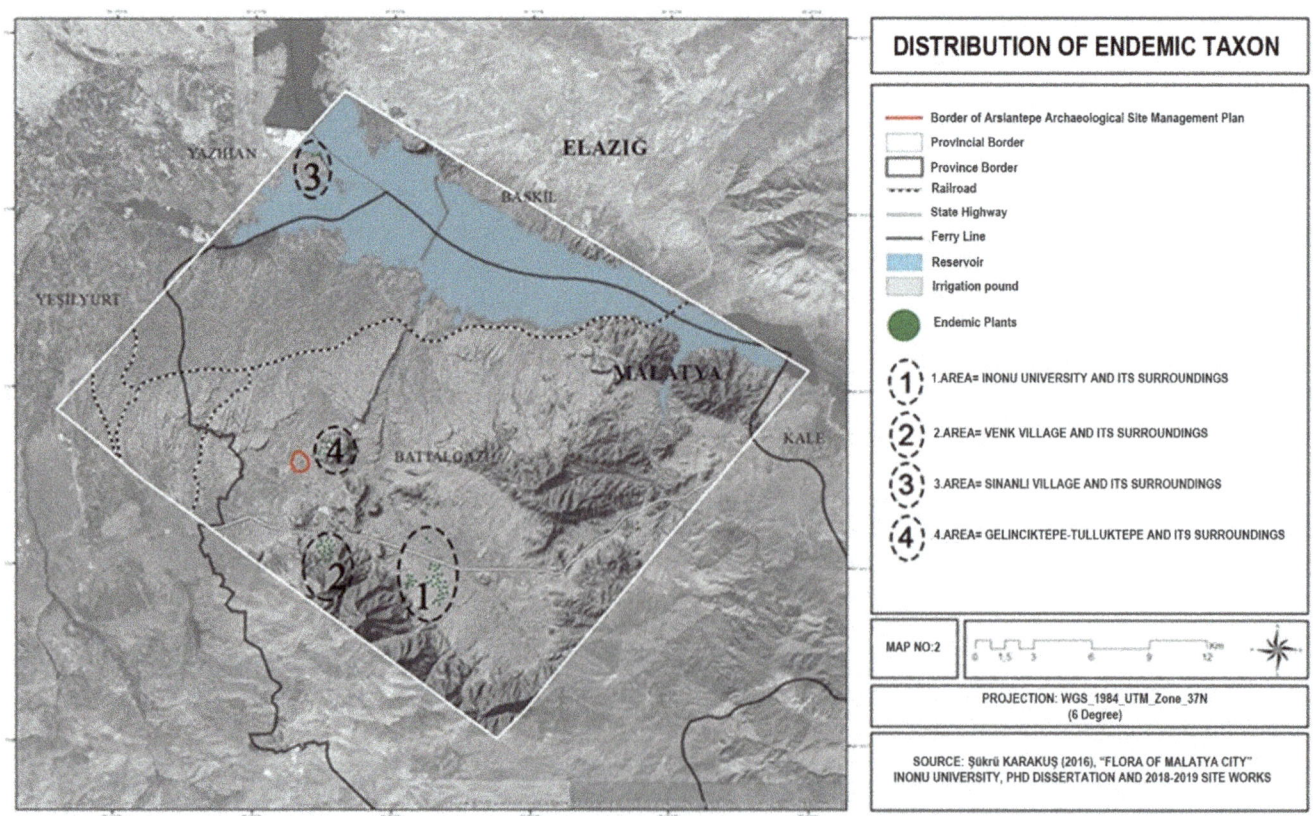

Fig. 4 The distribution of endemic plants areas

Table 2 The threat categories of endemic plants within the area

Threat categories[*]		Species number
CR		1
EN		1
VU		6
LR	9	9
	6	6
	40	40
DD		2

[*] CR: Critically EN: Endangered; VU: Vulnerable; LR: Lower risk; (cd): Requires protection; (nt): May be threatened; (lc): The least worrying

VU, 9 taxa LR(cd), 6 taxa LR(nt), 40 taxa LR(lc), and 2 taxa DD. Knowledge of abundance and spread of *Gundelia tournefortii* var. *armata* and *Tordylium cappadocicum* and are inadequate. Because of this, they included in DD (data deficiency) threat category. *Ornithogalum malatyanum* is included in CR (Critically). *Elymus longearistatus* subsp. *sintenisii* is included in EN (Endangered). *Astragalus macrouroides, Hedysarum aucheri, Hypericum capitatum* Choisy var. *luteum, Phlomis sintenisii, Sterigmostemum sulphureum* subsp. glandulosum, and *Verbascum euphraticum* are included in VU (Vulnerable). Identified endemic taxa in study area are given in Table 3.

Table 3 Endemic taxa identified in the study area

İnönü University and its surroundings		Venk Village and its surroundings		Sinanlı Village and its surroundings	Gelinciktepe-Tüllüktepe and its surroundings
Latin names of plants		Latin names of plants		Latin names of plants	Latin names of plants
Achillea pseudoaleppica Hausskn. ex Hub.-Mor.	*Isatis aucheri* Boiss.	*Achillea lycaonica* Boiss. & Heldr.	*Verbascum varians* Freyn & Sint. subsp. stepporum Hub.-Mor.	*Achillea magnifica* Heimerl & ex Hub.-Mor.	*Achillea pseudoaleppica* Hausskn. ex Hub.-Mor.
Alkanna trichophila Hub.-Mor. var mardinensis Hub.- Mor.	*Lamium orientale* (Fisch. & Mey)	*Allium scabriflorum* Boiss.	*Salvia euphratica* Montbret & Auchervar. leiocalycina (Rech. f.) Hedge	*Haplophyllum cappadocicum* Spach	*Astragalus macrouroides* Hub.-Mor.
Anchusa leptophylla Roem. & Schult. subsp. Incana (Ledeb.) D.F. Chamb.	*Linaria iconia* Boiss. & Heldr.	*Arum rupicola* Boiss. var. rupicola	*Astragalus Leporinus* Boiss. var. leporinus	*Hedysarum aucheri* Boiss.	*Astragalus melitenensis* Boiss.
Anthemis pauciloba Boiss. var. pauciloba	*Marrubium globosum* Montbret & Aucher ex Benth. subsp. Globosum	*Asphodeline damascena* (Boiss.) Baker subsp. rugosa E. Tuzlaci	Ebenus laguroides Boiss		*Cota wiedemanniana* (Fisch. & C.A. Mey.) Holub
Asperula stricta Boiss. subsp. latibracteata (Boiss.) Ehrend.	*Noccaea violascens* (Schott & Kotschy) F. K. Mey	*Astragalus decurrens* Boiss.	*Scorzonera tomentosa* L.		*Elymus lazicus* (Boiss.) Melderis subsp.
Astragalus melitenensis Boiss.	*Onobrychis cappadocica* Boiss.	*Astragalus globosus* Vahl	Tanacetum densum (Labill.) Sch. subsp. amani Heywood		*Maarrubium globosum* Montbret & Aucher ex Benth. subsp. globosum
Consolida glandulosa (Boiss. & Huet) Bornm.	*Onobrychis fallax* Freyn & Sint. ex Freyn var. fallax	*Astragalus macrouroides* Hub.-Mor.			*Paronychia kurdica* Boiss. subsp. haussknechtii Chaudhri
Convolvulus galaticus Rost. Ex Choisy	*Onopordum anatolicum* Boiss. & Heldr. ex Eig	*Cousinia cataonica* Boiss & Hausskn.			*Salvia Euphratica* Montbret & Aucher var. euphratica
Cota wiedemanniana (Fisch. & C.A. Mey.) Holub	*Onosma briquettii* Czeczott.	*Elymus longearistatus* (Boiss.) Tzelev subsp. sintenisii Melderis			*Verbascum Euphraticum* Benth.

(continued)

Table 3 (continued)

İnönü University and its surroundings		Venk Village and its surroundings	Sinanlı Village and its surroundings	Gelinciktepe-Tüllüktepe and its surroundings
Latin names of plants		Latin names of plants	Latin names of plants	Latin names of plants
Elymus lazicus (Boiss.) Melderis subsp.	Paronychia kurdica Boiss. subsp. haussknechtii Chaudhri	Eremogone drypidea (Boiss.) Ikonn.		Verbascum splendidum Boiss.
Eremogone ledeboriana (Fenzl) Ikonn	Phlomis oppositifolia Boiss. & Hausskn	Iris sari Schott ex Baker		Allium scabriflorum Boiss.
Erysimum kotschyanum J. Gay	Phlomis sieheana Rech f.	Johrenia berytea (Boiss. & Hausskn. ex Boiss.)		Asphodeline damascena (Boiss.) Baker subsp. rugosa E. Tuzlaci
Festuca anatolica Markgr.-Dann. subsp. Anatolica	Phlomis sintenisii Rech. f	Linaria corifolia Desf.		Gundelia tournefortii L. var. armata Freyn & Sint
Gundelia tournefortii (L. var. armata Freyn & Sint.)	Plantago euphratica Decne. ex Barnéoud	Onosma mutabilis Boiss & Hausskn		Iris sari Schott ex Baker
Hypericum capitatum Choisy var. luteum N. Robson	Salvia absconditiflora (Montbret & Aucher ex Benth.) Greuter & Burdet	Ornithogalum malatyanum Mutlu		Scorzonera tomentosa L.
Hypericum uniglandulosum Hausskn. ex Bornm	Salvia euphratica Montbret & Aucher var. euphratica	Quercus petraea (Matt.) Liebl. subsp. pinnatiloba (K. Koch) Menitsky		
Sterigmostemum sulphureum (Banks & Soland.) Bornm. subsp. Glandulosum	Stachys cretica subsp. Anatolica Rech. f.			
Tordylium cappadocicum Boiss.	Stachys ramosissima Montbret & Aucher ex Benth.			
Trigonella kotschyi Fenzl	Verbascum euphraticum Benth.			
Veronica multifida L.	Verbascum splendidum Boiss.			

4 Conclusion

Flora studies performed in the archaeological landscapes have a key role not only for the purpose of defining flora in the area where today's archaeological heritage is located, but also defining the alteration of the landscape from the past to the present in light of information obtained from the archaeobotanical findings. The alteration in flora reveals the evolution of climatic changes and therefore natural landscape features such as soil and hydrology. This alteration provides substantial information in understanding the land use patterns and access to resources of communities and determining the socio-cultural properties of societies.

On the other hand, considering today, making use of archaeological sites as an indicator of the natural texture of the city, and as it has a considerable substantial potential of visitors, it raises the issue of considering these areas as alternative public green spaces. Considering the educational function of the archaeological landscape, floristic research is of utmost importance for the planting endeavors to be performed in these areas. In addition to identifying endemic and endangered species and introducing them to the science world, qualitative and quantitative properties of floristic composition, characteristics of plant tissue, and importance of plant tissue in the urban landscape are required to be noted in order to raise academic and social awareness in these areas as alternative educational-recreational green space.

Acknowledgements This paper was supported by the Scientific and Technological Research Council of Turkey (TUBITAK project number 217O290).

Appendix: Floristic List

PTERIDOPHYTA

ASPLENIACEAE

1. **Asplenium. Ceterach** L., (2), 950 m, 28 Iv 2019, Ş.K. Observ.

CYSTOPTERIDACEAE

2. **Cystopteris Fragilis** (L.) Bernh., (12), 1650 m, 10 V 2019, Ş.K. 6620.

PINOPHYTA

CUPRESSACEAE

3. *****Cupressus arizonica** Gren, (2), 10 v 2019, Ş.K. Observ.
4. *****Cupressus sempervirens** L., (2), 10 v 2019, Ş.K. Observ.
5. **Juniperus excelsa** M.Bieb. var. **excelsa**, (2), 950 m, 28 iv 2019, Ş.K. Observ.
6. *****Juniperus rigida** Siebold & Zucc. (2), 930 m, 10 v 2019, Ş.K. Observ.
7. **Juniperus sabina** L., (2), 950 m, 10 v 2019, Ş.K. Observ.
8. *****Platycladus orientalis** L., (2), 950 m, 10 v 2019, Ş. K. Observ.

GINKGOACEAE

9. *****Ginkgo. biloba** L., (2), 950 m, 10 v 2019, Ş.K. Observ.

PINACEAE

10. *****Abies nordmanniana** (Steven) Spach var. **nordmanniana**, (3) 850 m, 31 v 2019, Ş.K. Observ.
11. **Cedrus libani** A.Rich var. **libani**, (8), 1050 m, 28 iv 2019, Ş.K. Observ.
12. *****Picea orientalis** (L.) Link., (2), 950 m, 28 iv 2019, Ş. K. Observ.
13. *****Picea pungens** Engelm, (2), 950 m, 10 v 2019, Ş.K. Observ.
14. **Pinus brutia** Ten. (2), 950 m, 10 v 2019, Ş.K. Observ.
15. *****Pinus nigra** L. subsp. **pallasiana** (Lamb.) Holmboe. var. **pallasiana**, (2), 950 m, 28 iv 2019, Ş.K. Observ. var. *****pyramidata** (Açatay) Yalt., (2), 950 m, 10 v 2019, Ş.K. Observ.
16. *****Pinus sylvestris** L. var. **hamata** Steven, (2), 950 m, 28 iv 2019, Ş.K. Observ.

TAXACEAE

17. *****Taxus baccata** L., (2), 950 m, 28 iv 2019, Ş.K. Observ.

MAGNOLIOPHYTA

ALTINGIACEAE

18. *****Liquıdambar styraciflua** L., (2), 950 m, 12 iv 2019, Ş.K. Observ.

AMARANTHACEAE

19. **Amaranthus albus** L., (2), 950 m, 12 iv 2019, Ş.K. Observ.
20. **Atriplex nitens** Schkuhr, (2), 950 m, 16 × 2011 B. M.11641.
21. **Chenopodıum album** L. subsp. **album** var. **album**, (10), 900 m, Ş.K. Observ.
22. **Noaea mucronata** (Forssk.) Asch. & Schweinf, (10), 900 m, Ş.K. Observ.
23. **Salsola tragus** L. subsp. **tragus**, (2), 950 m, 16 × 2011 B.M.11639.
24. **Spinacia tetrandra** M.Bieb, (2), 900 m, 28 vi 1991, B. Y. 8448a.

ANACARDIACEAE

25. **Pistacia palaestina** Boiss., (12), 1400–1600 m, 07 vii 1996, B.Y. 13736.
26. **Rhus coriaria** L., (2), 950 m, 16 vii 2011, B.M.11635.

APIACEAE

27. **Actinolema macrolema** Boiss., (2), 09 v 1986, B.Y. 7280., Ir-Tur.
28. **Artedia squamata** L., (8), 1050 m, 22 v 2019, Ş.K. 6656.

29. **Astrodaucus orientalis** (L.) Drude, (1), 850 m, 22 v 2019, Ş.K. Observ., Ir-Tur.
30. **Bunium paucifolium** DC., (2), 900 m, 11 vi 2011, B.M.11489.
31. **Bupleurum aleppicum** Boiss., (2), 950 m, 11 vi 2011, B.M.11463, Ir-Tur.
32. **Bupleurum croceum** Fenzl, (2), 900 m, 04 v 1987, G.T. 1019b, Ir-Tur.
33. **Bupleurum gerardii** All., (2), 950 m, 18 vi 2011, B.M.11545.
34. **Bupleurum papillosum** DC., (5), 950 m, 03 viii 2018, Ş.K. 6367.
35. **Caucalis platycarpos** L., (12), 1200 m, 08 vi 1995, B.Y. 12528.
36. **Chaerophyllum macrospermum** Fisch. & C.A.Mey ex Hohen. (12), 15 vi 2013, Ş.K. 4368.
37. **Daucus guttatus** Sibth. & Sm., (14), 700 m, 07 vii 2019, Ş.K. 2911.
38. **Echinophora tenuifolia** L., (2), 950 m, 18 vi 2011, B.M.11648.
39. **Eryngium billardieri** F.Delaroche, (12) 1650 m, 05 vi 2019, Ş.K. 6763, Ir-Tur.
40. **Grammosciadium platycarpum** Boiss. & Hausskn, (1), 850 m, 22 v 2019, Ş.K. Observ.
41. **Ferula rigidula** Fisch. ex DC., (12) 1650 m, 05 vi 2019, Ş.K. 6764, Ir-Tur.
42. **Johrenia berytea** Boiss. & Hausskn., (12), 07 vii 1996, B.Y. 13,675, End., LR(nt)].
43. **Lisaea heterocarpa** (DC.) Boiss., (2), 900 m, 14 vi 1986, B.Y. 7694, Ir-Tur.
44. **Malabaila lasiocarpa** Boiss., (12) 1650 m, 05 vi 2019, Ş.K. 6785.
45. **Opopanax hispidus** (Friv.) Griseb., (12) 1650 m, 05 vi 2019, Ş.K. 6789.
46. **Ormosciadium aucheri** Boiss., (12), 1200 m, 08 vi 1995, B.Y. 12,540.
47. ***Petroselinium crispum** (Mill.) A.W.Hill, (1), 850 m, 22 v 2019, Ş.K. Observ.
48. **Prangos peucedanifolia** Fenzl, (12), 1650 m, 25 v 2013, Ir-Tur.
49. **Scandix aucheri** Boiss., (1), 850 m, 22 v 2019, Ş.K. Observ., Ir-Tur.
50. **S. iberica** M.Bieb., (2), 930 m, 04 vi 2011, B.M.11358.
51. **S. pecten-veneris** L., (1), 850 m, 22 v 2019, Ş.K. Observ.
52. **S. stellata** Banks. & Sol., (16), 700 m, 24 v 2019, Ş.K. 6713.
53. **Tordylium cappadocicum** Boiss., (2), 17 vi 2011, B.M.11305, Ir-Tur., End., [DD].
54. **Torilis leptophylla** (L.) Reich. f., (1), 850 m, 22 v 2019, Ş.K. Observ.
55. **Turgenia latifolia** (L.) Hoffm., (1), 850 m, 22 v 2019, Ş.K. Observ.

APOCYNACEAE

56. **Vinca herbacea** Waldst. & Kit. (2), 1400 m, 07 vii 1996, B.Y. 13715.
57. **Vincetoxicum tmoleum** Boiss., (16), 700 m, 24 v 2019, Ş.K. 6720, Ir-Tur.

AQUIFOLIACEAE

58. ***Ilex aquifolium** L./(2), 900 m, 22 v 2019, Ş.K. Observ.

ARISTOLOCHIACEAE

59. **Aristolochia Maurorum** L., (2), 950 m, 08 V 2005, B.M. Observ., Ir-Tur.

ASTERACEAE

60. **Achillea arabica** Kotschy., (1), 850 m, 22 v 2019, Ş.K. Observ.
61. **Achillea lycaonica** Boiss. & Heldr., (12), 07 vii 1996 B.Y. 13721, Ir-Tur., End., [LR(lc)].
62. **Achillea magnifica** Heimerl & ex Hub.-Mor., (14), 850 m, 30 v 2014, Ş.K. 4887, Ir-Tur., End., [LR(nt)].
63. **Achillea pseudoaleppica** Hausskn. ex Hub.-Mor., (10), 900 m, 11 iv 2019, Ş.K. 6461, Ir-Tur., End., [LR (cd)].
64. **Achillea santolinoides** Lag. subssp. **wilhelmsii** (K. Koch.) Greuter, (16), 650 m, 24 v 2019, Ş.K. 6693, Ir-Tur.
65. **Anthemis hyalina** DC., (16), 650 m, 24 v 2019, Ş.K. 6692.
66. **Anthemis pauciloba** Boiss. var. **pauciloba**, (2), 04 vi 2011, B.M.11385, End., [LR(lc)].
67. ***Calendula officinalis** L., (2), 930 m, 18 vi 2019, Ş.K. Observ.
68. **Carduus nutans** L., subsp. **nutans**, (10), 900 m, 18 vi 2019, Ş.K. Observ.
69. **Carduus pycnocephalus** L., (2), 900 m, 04 vi 2011, B.M.11254.
70. **Carthamus dentatus** (Forssk.) Vahl, (2), 04 vi 1986, E.Aktoklu 229.
71. **Carthamus persicus** Desf. ex Willd., (5), 950 m, 03 viii 2018, Ş.K. 6375, Ir-Tur.
72. **Centaurea aggregata** Fisch. & C.A.Mey., (12), 1000 m, 25 vi 1995, B.Y. 12929.
73. **Centaurea balsamita** Lam., (5), 950 m, 03 viii 2018, Ş.K. 6374, Ir-Tur.

74. **Centaurea urvillei** DC., subsp. **armata** Wagenitz, (10), 900 m, Ş.K. Observ, E.Medit.

75. **Chardinia orientalis** (L.) Kuntze, (2), 930 m, 04 vi 2011, B.M.11263, Ir-Tur.

76. **Cichorium intybus** L., (6), 930 m, 19 iv 2019, Ş.K. 6487.

77. **Cirsium haussknechtii** Boiss., (6), 930 m, 19 iv 2019, Ş.K. 6513, Ir-Tur.

78. **Cnicus benedictus** L., (6), 930 m, 19 iv 2019, Ş.K. 6491.

79. **Conyza canadensis** (L.) Cronquist, (2), 900 m, 11 v 1986, E.Aktoklu 246.

80. **Cota coelopoda** (Boiss.) Boiss., (16), 700 m, 24 v 2019, Ş.K. 6727.

81. **Cota wiedemanniana** Holub, (10), 900 m, 11 iv 2019, Ş.K. 6461, End., [LR(lc)].

82. **Cousinia cataonica** Boiss. & Haussкn., (5), 03 viii 2018, Ş.K. 6361, Ir-Tur., End., [LR(nt)].

83. **Crepis foetida** L., subsp. **foetida**, (13), 1050 m, 26 iv 2019, Ş.K. 6556.
    subsp. **rhoeadifolia** (M.Bieb.) Čelak., (6), 930 m, 19 iv 2019, Ş.K. 6561.

84. **Crepis pulchra** L. subsp. **pulchra**, (2, 950 m, 07 vi 2011, B.M.11433.

85. **Crepis sancta** (L.) Bornm. (1), 850 m, 22 v 2019, Ş.K. Observ.

86. **Crepis syriaca** (Bornm.) Babc.& Navashin, (2), 950 m, 04 vi 2011, B.M.11257.

87. **Crupina crupinastrum** (Moris) Vis., (9), 850 m, 22 v 2019, Ş.K. 6667.

88. **Cyanus depressus** (M.Bieb.) Soják, (1), 850 m, 22 v 2019, Ş.K. Observ.

89. **C. triumfettii** (All.) Dostál ex Á.Löve & D.Löve (2), 900 m, 25 vi 1986, E.Aktoklu 103.

90. **Cymboleana griffithii** (A.Gray) Wagenitz, (2), 950 m, 04 vi 2011, B.M.11532, Ir-Tur.

91. **Echinops orientalis** Trautv., (5), 950 m, 03 viii 2018, Ş.K. 6372, Ir-Tur.

92. **Filago pyramidata** L., (16), 650 m, 24 v 2019, Ş.K. 6679.

93. **Garhadiolus hedypnois** Jaub. & Spach, (9), 850 m, 22 v 2019, Ş.K. 6662, Ir-Tur.

94. **Gundelia tournefortii** L. var. **armata** Freyn & Sint. (10), 950 m, 11 iv 2019, Ş.K. 6463, Ir-Tur., End. [LR (lc)].

95. ***Helianthus annuus** L., (1), 850 m, 22 v 2019, Ş.K. Observ.

96. ***H. tuberosus** L., (2), 950 m, 12 iv 2019, Ş.K. Observ.

97. **Helichrysum plicatum** DC. subsp. **plicatum**, (2) 04 vi 2011, B.M.11395.

98. **Inula graveolens** (L.) Desf., (2), 900 m, 03 iii 1986, E. Aktoklu 103.

99. **Leontodon asperrimus** (Willd.) Endl., (16), 700 m, 24 v 2019, Ş.K. 6715.

100. **Onopordum anatolicum** Boiss. & Heldr. ex Eig, (2),, 11 vi 2011, B.M.11492, Ir-Tur., End., [LR(lc)].

101. **Onopordum candidum** Nábělek, (2), 950 m, 16 × 2011, B.M.11662, Ir-Tur.

102. **Picris strigosa** M.Bieb subsp. **strigosa**, (16), 700 m, 24 v 2019, Ş.K. 6736.

103. ***Santolina chamaecyparissus** L., (2), 950 m, 12 iv 2013, Ş.K. Observ.

104. **Scorzonera mollis** M. Bieb., subsp. **mollis**, (6), 930 m, 19 iv 2019, Ş.K. 6513.

105. **Scorzonera suberosa** K. Koch. subsp. **suberosa**, (12) 1650 m, 10 v 2019, Ş.K. 6607, Ir-Tur.

106. **Scorzonera tomentosa** L., (10), 950 m, 11 iv 2019, Ş. K. 6477, Ir-Tur., End., [LR(lc)].

107. **Senecio vernalis** Waldst & Kit., (1), 850 m, 22 v 2019, Ş.K. Observ.

108. **Siebera pungens** (Lam.) J.Gay, (12),1200 m, 01 vi 1995, B.Y. 13,001, Ir-Tur.

109. **Sonchus asper** (L.) Hill subsp. **asper**, (2), 950 m, 02 iv 2013, B.M.11674.

110. **Tanacetum densum** (Labill.) Sch. subsp. **amani** Heywood, (12) 1650 m, 05 vi 2019, Ş.K. 6761. Ir-Tur., End., [LR(nt)].

111. **Taraxacum buttleri** Soest, (12), 1650 m, 10 v 2019, Ş.K. 6606.

112. **Taraxacum syriacum** Boiss., (10), 900 m, 11 iv 2019, Ş.K. 6455.

113. **Tragopogon coloratus** C.A.Mey., (16), 650 m, 24 v 2019, Ş.K. 6701.

114. **Tussilago farfara** L., (2), 900 m, 03 iii 1987, E. Aktoklu 254, Av-Sib.

115. **Xanthium orientale** L. subsp. **italicum** (Moretti) Greuter, (1), 22 v 2019, Ş.K. Observ.

116. **Xeranthemum longipapposum** Fisch. & C.A.Mey. (5), 03 viii 2018, Ş.K. 6380, Ir-Tur.

117. **Zogea leptaurea** L., (12), 1200 m, 08 vi 1995, B.Y. 12,546, Ir-Tur.

BERBERIDACEAE

118. ***Berberis thunbergii** cv. **atropurpurea**, (2), 950 m, 12 iv 2013, Ş.K. Observ.

BETULACEAE

119. ***Betula pendula** Roth., (2), 950 m, 12 iv 2019, Ş.K. Observ.

*BIGNONIACEAE

120. ***Catalpa bignonioides** Walter, (2), 950 m, 12 iv 2013, Ş.K. Observ.

BORAGINACEAE

121. **Alkanna megacarpa** A.DC., (6), 930 m, 28 iv 2019, Ş.K. 6599.
122. **Alkanna trichophila** Hub.-Mor. var. **mardinensis** Hub.-Mor., (2), 950 m, 18 vi 2011, B.M.11475, Ir-Tur., End., [LR(lc)].
123. **Anchusa azurea** Mill. var. **azurea**, (9), 900 m, 27 iv 2013, Ş.K. 3762.
124. **Anchusa leptophylla** Roem. & Schult. subsp. **incana** (Ledeb.) D.F.Chamb., (2), 1350 m, 15 v 2009, SKö 32, Medit., End., [LR(lc)].
125. **Anchusa pusilla** Guşul, (2), 900 m, 14 v 1987, B.Y. & G.Taş 1051.
126. **Brunnera orientalis** I.M.Johnst., (12), 1650 m, 25 v 2013, Ş.K. 4123.
127. **Buglossoides arvensis** (L.) I.M.Johnst., subsp. **sibthorpiana** (Griseb.) R.Fern., (10), 900 m, 28 iv 2019, Ş.K. Observ.
128. **Echium angustifolium** Mill. (1), 850 m, 22 v 2019, Ş. K. Observ. E.Medit.
129. **Echium glomeratum** Poir., (2), 900 m, 16 vi 1986, B. Y. 7629.
130. **Heliotropium dolosum** De Not., (1), 850 m, 22 v 2019, Ş.K. Observ.
131. **Lappula patula** (Lehm). Asch. ex Gürke, (2), 900 m, 11 v 1987, HÇs.n.
132. **Moltkia coerulea** (Willd.) Lehm., (10), 900 m, 28 iv 2019 Ş.K. Observ., Ir-Tur.
133. **Myosotis refracta** Boiss. subsp. **refracta**, (2), 900 m, 04 v 1987, G.Taş 1042, Medit.
134. **Neatostema apulum** I.M.Johnst., (2), 900 m, 09 v 1986, B.Y. 7243, Medit.
135. **Nonea melanocarpa** Boiss., (12), 1650 m, 10 v 2019, Ş.K. 6640, Ir-Tur.
136. **Onosma alborosea** var. **alborosea** Fisch & C.A.Mey, (12) 1650 m, 10 v 2019, Ş.K. 6641.
137. **Onosma briquettii** Czeczott., (2), 04 vi 2011, B. M.11304, Ir-Tur., End., [LR(cd)].
138. **Onosma bulbotricha** D.C., (16), 700 m, 24 v 2019, Ş. K. 6743, Ir-Tur.
139. **Onosma mollis** D.C. (10), 850 m, 22 v 2019, Ş.K. 6665, Ir-Tur.
140. **Onosma mutabilis** Boiss & Hausskn., (12), 1400 m, 07 vii 1996, B.Y. 13690, Ir-Tur., End., [LR(lc)].
141. **Onosma sericea** Willd. (16), 24 v 2019, Ş.K. 6740; (10), 900 m, Ş.K. Observ., Ir-Tur.
142. **Paracaryum strictum** (K.Koch) Boiss., (12) 1650 m, 05 vi 2019, Ş.K. 6786, Ir-Tur.
143. ***Phacelia tanacetifolia** Benth., (2), 950 m, 05 vi 2014, Ş.K. 5088.
144. **Rochelia disperma** (L.f.) Koch var. **disperma**, (12) 1650 m, 05 vi 2019, Ş.K. 6770.
145. **Symphytum brachycalyx** Boiss., (2), 900 m, 27 iv 1987, G.Taş 1007, E.Medit.

BRASSICACEAE

146. **Aethionema arabicum** (L.) Andrz. ex DC. (6), 930 m, 19 iv 2019, Ş.K. 6531.
147. **Aethionema armenum** Boiss., (12) 1650 m, 10 v 2019, Ş.K. 6638.
148. **Aethionema carneum** (Banks & Sol.) B.Fedtsch., (2), 11 iv 1987, G.Taş 1068, Ir-Tur.
149. **Aethionema grandiflorum** Boiss. var. **grandiflorum**, (12) 1650 m, 05 vi 2019, Ş.K. 6797.
150. **Aethionema heterocarpum** J.Gay, (2), 900 m, 04 vii 1986, E.Aktoklu 192a.
151. **Aethionema iberideum** (Boiss.) Boiss., (12), 1650 m, 10 v 2019, Ş.K. 6636.
152. **Alliaria petiolata** Andrz. ex M.Bieb., (2), 930 m, 30 iv 2019, Ş.K. 6604.
153. **Alyssum armenum** Boiss., (8), 1050 m, 28 iv 2019, Ş. K. 6577.
154. **Alyssum aureum** (Fenzl) Boiss., (8), 1050 m, 28 iv 2019, Ş.K. 6586, Ir-Tur.
155. **Alyssum callichroum** Boiss. & Balansa, (10), 900 m, 11 iv 2019, Ş.K. 6448.
156. **Alyssum condensatum** Boiss. & Hausskn. subsp. **flexibile** (Nyár.) T.R.Dudley, (8), 1050 m, 28 iv 2019, Ş.K. 6581.
157. **Alyssum desertorum** Stapf., (10), 900 m, 11 iv 2019, Ş.K. 6444.
158. **Alyssum filiforme** Nyár., (10), 900 m, 11 iv 2019, Ş. K. 6545.
159. **Alyssum hirsitum** M.Bieb. subsp. **hirsitum**, (6), 930 m, 19 iv 2019, Ş.K. 6530.
160. **Alyssum linifolium** Stephan ex Willd. var. **linifolium**, (6), 930 m, 19 iv 2019, Ş.K. 6546.
161. **Alyssum sibiricum** Willd., (6), 930 m, 19 iv 2019, Ş. K. 6509.
162. **Alyssum simplex** Rudolph, (6), 930 m, 19 iv 2019, Ş. K. 6538.
163. **Alyssum strigosum** Banks & Sol. subsp. **strigosum** (3), 900 m, 13 iii 2013, Ş.K. 3401.

164. **Alyssum szovitsianum** Fisch. & C.A.Mey., (6), 930 m, 19 iv 2019, Ş.K. 6499.

165. **Arabis alpina** L. subsp. **alpina**, (12) 1650 m, 10 v 2019, Ş.K. 6610.

166. **Arabis aucheri** Boiss., (10), 900 m, 11 iv 2019, Ş.K. 6461.

167. **Arabis montbretiana** Boiss., (2), 900 m, GT 1039, Ir-Tur.

168. **Arabis nova** Vill., (6), 930 m, 19 iv 2019, Ş.K. 6502.

169. **Aubrieta libanotica** Boiss. & Hohen., (12) 1650 m, 10 v 2019, Ş.K. 6611.

170. **Boreava orientalis** Jaub. & Spach, (8), 1050 m, 28 iv 2019, Ş.K. 6572.

171. **Brassica elongata** Ehrh. (5), 850 m, 22 v 2019, Ş.K. 6673.

172. ***B. oleracea** L. cv. **asepala** (1), 16 iii 2018, Ş.K. Observ.

173. **Camelina hispida** Boiss., (6), 930 m, 26 iv 2019, Ş.K. 6559.

174. **Camelina rumelica** Velen., (10), 900 m, 11 iv 2019, Ş.K. 6467.

175. **Capsella bursa-pastoris** (L.) Medik., (10), 900 m, 11 iv 2019, Ş.K. 6480.

176. **Chorispora purpurascens** (Boiss. & Sol.) Eig, (10), 900 m, 11 iv 2019, Ş.K. 6480.

177. **Clypeola jonthlapsi** L., (8), 1050 m, 28 iv 2019, Ş.K. 6590.

178. **Conringia orientalis** (L.) Dumort. (2), 900 m, 04 vi 2011, B.M.11295.

179. **Conringia planisiliqua** Fisch. & C.A.Mey., (6), 930 m, 19 iv 2019, Ş.K. 6562.

180. **Crambe orientalis** L. var. **orientalis**, (2), 900 m, 04 vi 2011, B.M.11321.

181. **Descurainia sophia** (L.) Webb ex Prantl., (10), 900 m, 11 iv 2019, Ş.K. 6470.

182. **Draba minima** (C.A.Mey.) Steud. (10), 900 m, 16 iii 2018, Ş.K. Observ.

183. **D. verna** L., (10), 900 m, 11 iv 2019, Ş.K. 6476, Ir-Tur.

184. **Eruca sativa** L., (6), 930 m, 28 iv 2019, Ş.K. 6597.

185. ***Erysimum cherii** (L.) Crantz, (2), 950 m, 08 vi 2008, B.M.10606.

186. **Erysimum crassipes** Fisch. & C.A.Mey., (8), 1050 m, 22 v 2019, Ş.K. 6653.

187. **Erysimum kotschyanum** J.Gay, (2), 900 m, 04 v 1987. G.Taş 1026, End., [LR(lc)].

188. **Erysimum repandum** L., (12) 1650 m, 10 v 2019, Ş.K. 6625.

189. **Erysimum smyrnaeum** Boiss. & Balansa, (16), 650 m, 24 v 2019, Ş.K. 6703.

190. **Fibigia clypeata** (L.) Medik. subsp. **clypeata** var. **clypeata** (12), 10 v 2019, Ş.K. 6619.
var. **eriocarpa** (DC.) Post, (12) 1650 m, 10 v 2019, Ş.K. 6614.

191. **Glastaria glastifolia** (DC.) (7), 800 m, 16 iii 2013, Ş.K. 3450, Ir-Tur.

192. **Hesperis pendula** DC. subsp. **pendula**, (12) 1650 m, 10 v 2019, Ş.K. 6608.

193. **Isatis aucheri** Boiss., (2), 900 m, 04 vi 2011, B.M.11477, Ir-Tur., End., [LR(lc)].

194. **Isatis glauca** Auch. ex Boiss. subsp. **glauca**, (2), 900 m, 04 vii 1986, E.Aktoklu 194.

195. **Lepidium perfoliatum** L., (16), 650 m, 24 v 2019, Ş.K. 6688.

196. **Matthiola longipetala** (Vent.) DC., subsp. **bicornis** (Sibth. & Sm.) P.W.Ball, (10), 900 m, 11 iv 2019, Ş.K. 6482.

197. **Microthlapsi perfoliatum** (L.) F.K.Mey. (6), 930 m, 19 iv 2019, Ş.K. 6511.

198. **Myagrum perfoliatum** L. (8), 1050 m, 22 v 2019, Ş.K. 6657.

199. **Neslia paniculata** (L.) Desv. subsp. **thracica** (Velen.) Bornm., (8), 28 iv 2019, Ş.K. 6576.

200. **Rapistrum rugosum** (L.) All., (16), 700 m, 24 v 2019, Ş.K. 6723.

201. **Ricotia aucheri** (Boiss.) B.L.Burtt, (15), 700 m, 24 v 2019, Ş.K. 6723, Ir-Tur.

202. **Noccaea violascens** (Schott & Kotschy) F.K.Mey., (12) 1650 m, 10 v 2019, Ş.K. 6616, End., [LR(lc)].

203. **S.inapis alba** L. subsp. **alba**, (12) 1650 m, 05 vi 2019, Ş.K. 6763.

204. **Sisymbrium altissimum** L., (6), 930 m, 19 iv 2019, Ş.K. 6514.

205. **Sisymbrium loselii** L., (6), 930 m, 19 iv 2019, Ş.K. 6515.

206. **Sisymbrium orientate** L., (8), 1050 m, 28 iv 2019, Ş.K. 6580.

207. **Sisymbrium septulatum** DC., (2), 950 m, 09 v 2005, B.M.9561.

208. **Sterigmostemum sulphureum** (Banks & Soland.) Bornm. subsp. **glandulosum**, (5), 900 m, Hub.-Mor. 9255., Ir-Tur., End., [VU].

209. **Strigosella africana** (L.) Botsch., (3) 850 m, 31 v 2019, Ş.K. 6750.

BUXACEAE

210. *Buxus sempervirens L., (2), 950 m, 12 iv 2013, Ş.K. Observ.

CAMPANULACEAE

211. Campanula reuterana Boiss. & Balansa, (2), 950 m, 18 vi 2011, B.M.11565, Ir-Tur.
212. Legousia pentagonia (L.) Thell., (12) 1650 m, 05 vi 2019, Ş.K. 6791, E.Medit.
213. Legousia speculum-veneris (L.) Durande ex Vill., (12) 1650 m, 05 vi 2019, Ş.K. 6795.

CANNACEAE

214. Canna indica L., (2), 930 m, 12 iv 2013, Ş.K. Observ.

CAPPARACEAE

215. Capparis sicula Veil. subsp. sicula, (14), 700 m, 30 v 2012, Ş.K. 2041.

CAPRIFOLIACEAE

216. Cephalaria procera Fisch. & Avé-Lall., (14), 700 m, 07 vii 2012, Ş.K. 2910.
217. Cephalaria syriaca (L.) Schrad., (2), 900 m, 04 vi 1991, B.Y. 1986.
218. *Lonicera etrusca Santi var. etrusca, (2), 950 m, 12 iv 2019, Ş.K. Observ.
219. *Lonicera japonica Thunb, (2), 950 m, 12 iv 2013, Ş.K. Observ.
220. Pterocephalus plumosus (L.) Coulter,(16), 700 m, 24 v 2019, Ş.K. 6706.
221. Scabiosa argentea L., (14), 850 m, 14 vi 2019, Ş.K. 6172.
222. Scabiosa calocephala Boiss., (14), 700 m, 30 v 2012, Ş.K. 2044, Ir-Tur.
223. Scabiosa micrantha Desf., (12), 1650 m, 25 v 2013, Ş.K. 4121.
224. Scabiosa rotata M.Bieb., (16), 650 m, 24 v 2019, Ş.K. 6679, Ir-Tur.
225. *Simphoricarpus albus Blake, (2), 950 m, 12 iv 2019, Ş.K. Observ.
226. Valeriana dioscoridis Sm., (13), 1050 m, 26 iv 2019, Ş.K. 6564.
227. Valeriana sisymbriifolia Vahl., (13), 1050 m, 26 iv 2019, Ş.K. 6570, Ir-Tur.
228. Valerianella coronata (L.) DC., (12), 1650 m, 05 vi 2019.
229. Valerianella cymbicarpa C.A.Mey., (12), 1600 m, 23 v 2015, Ş.K. 6079, Ir-Tur.

230. Valerianella locusta (L.) Laterr., (6), 930 m, 19 iv 2019, Ş.K. 6533, Av-Sib.
231. Valerianella pumila (L.) DC., (16), 650 m, 24 v 2019, Ş.K. 6681.
232. Valerianella vesicaria (L.) Moench, (10), 900 m, 11 iv 2019, Ş.K. 6453.

CARYOPHYLLACEAE

233. Arenaria serpyllifolia L., subsp. leptoclados (Rchb.) Nyman, (2), 07 vi 2011, B.M.11423.
    subsp. serpyllifolia, (2), 950 m, 04 vi 2011, B.M.11381.
234. Cerastium dichotomum L. subsp. dichotomum, (12) 1650 m, 10 v 2019, Ş.K. 6648.
235. Cerastium dubium (Bastard) O. Schwarz, (12) 1650 m, 10 v 2019, Ş.K. 6615.
236. Dianthus floribundus Boiss., (2), 950 m, 18 vi 2011, B.M.11556, Ir-Tur.
237. Eremogone drypidea (Boiss.) Ikonn., (12), 15 vi 2013, Ş.K. 4353, Ir-Tur., End., [LR(lc)].
238. Eremogone ledeboriana (Fenzl) Ikonn., (12), 05 vi 2019, Ş.K. 6793, End., [LR(lc)].
239. Gypsophila pallida Stapf, (16), 700 m, 24 v 2019, Ş.K. 6745.
240. Gypsophila perfoliata L. var. perfoliata, 950 m, 18 vi 2011, B.M.11514.
241. Gypsophila pilosa Hudson, (16), 650 m, 24 v 2019, Ş.K. 6702, Ir-Tur.
242. Habrosia spinuliflora (Ser.) Fenzl, (2), 930 m, 18 v 2019, Ş.K. Observ.
243. Herniaria incana Lam., (2), 950 m, 11 vi 2011, B.M.11460; (10), 900 m, Ş.K. Observ.
244. Holosteum tenerrimum Boiss. (2), 900 m, 03 iii 1987, E.Aktoklu 258.
245. Holosteum umbellatum L. var. glutinosum (M.Bieb.) Gay, (1), 22 v 2019, Ş.K. Observ.
246. Minuartia hamata (Hausskn.) Mattf., (2), 900 m, 04 vi 2011, B.M.11314.
247. Minuartia hybrida (Vill.) Schischk. subsp. turcica McNeill, (2), 04 vi 2011, B.M.11292.
248. Minuartia meyeri (Boiss.) Bornm., (5), 850 m, 22 v 2019, Ş.K. 6669, Ir-Tur.
249. Minuartia montana L. subsp. wiesneri (Stapf) McNeill, (6), 19 iv 2019, Ş.K. 6493, Ir-Tur.
250. Paronychia kurdica Boiss., subsp. haussknechtii Chaudhri, (10), 900 m, 11 iv 2019, Ş.K. 6479, End., [LR(lc)].
251. Petrorhagia cretica (L.) P.W.Ball & Heywood, (2), 950 m, 14 vi 1986, B.Y. 7666.
252. Saponaria officinalis L., (2), 950 m, 18 vi 2011, B.M.11628.

253. **Silene argentea** Ledeb., (2), 950 m, 18 vi 2011, B.M.11546, Ir-Tur.
254. **Silene chaetodonta** Boiss., (2), 950 m, 08 vi 2011, B.M.11557.
255. **Silene conoidea** L., (12), 1650 m, 23 v 2015, Ş.K. 6081.
256. **Silene kotschyi** Boiss. var. **kotschyi**, (12), 1200 m, 08 vi 1995, B.Y. 12518.
257. **Silene spergulifolia** (Desf.) M.Bieb., (12) 1650 m, 05 vi 2019, Ş.K. 6760, Ir-Tur.
258. **Silene stenobotrys** Boiss. & Hausskn., (2), 950 m, 18 vi 2011, B.M.11551, Ir-Tur.
259. **Silene supina** M.Bieb. subsp. **pruinosa** Chowdhuri, (2), 950 m, 28 v 1988, G.T 1054d.
260. **Stellaria media** (L.) Vill., (12) 1650 m, 10 v 2019, Ş.K. 6631.
261. **Telephium imperati** L. subsp. **orientale** (Boiss.) Nyman, (12), 05 vi 2019, Ş.K. 6791.
262. **Vaccaria hispanica** (Mill.) Rauschert, (2), 930 m,16 × 2011, B.M.11642.
263. **Velezia rigida** L., (16), 650 m, 24 v 2019, Ş.K. 6684.

CELASTRACEAE

264. ***Euonymus europaeus** L., (2), 950 m, 12 iv 2019, Ş. K. Observ.
265. ***Euonymus fortunei** (Turcz) Hand.-Mazz., (2), 950 m, 12 iv 2019, Ş.K. Observ.

CISTACEAE

266. **Helianthemum microcarpum** Coss. ex Boiss., (16), 700 m, 24 v 2019, Ş.K. 6721.
267. **Helianthemum salicifolium** (L.) Mill., (5), 850 m, 22 v 2019, Ş.K. 6668.

CLEOMACEAE

268. **Cleome ornithopodioides** L., (14), 700 m, 29 ix 2018, Ş.K. Observ., E.Medit.

CONVOLVULACEAE

269. **Calystegia sepium** (L.) R.Br. subsp. **sepium**, (1), 850 m, 22 v 2019, Ş.K. Observ.
270. **Convolvulus arvensis** L., (1), 850 m, 22 v 2019, Ş.K. Observ.
271. **Convolvulus betonicifolius** Mill., subsp. **betonici-folius**, (2), 950 m, 11 vi 2011, B.M.11445. subsp. **peduncularis** (Boiss.) Paris, (14),, 850 m, 30 v 2014, Ş.K. 4886, Ir-Tur.

272. **Convolvulus dorycnium** L. subsp. **oxysepalus** (Boiss.) Rech.f., (12), 1200 m, 25 vi 1995, B.Y. 12,919, E.Medit.
273. **Convolvulus galaticus** Rost. ex Choisy, (2), 950 m, B.M.Observ. Ir-Tur., End., [LR(lc)].
274. **Convolvulus reticulatus** Choisy subsp. **reticulatus**, (2), 16 × 2011, B.M.11637, Ir-Tur.
275. **Cuscuta globularis** Bertol., (12), 1650 m, 15 vi 2013, Ş.K. 4376.
276. **Cuscuta planiflora** Ten., (2), 950 m, 04 vi 2011, B.M.11334.

CRASSULACEAE

277. **Sedum album** L., (12) 1650 m, 05 vi 2019. 22 v 2019, Ş.K. Observ.
278. **Sedum hispanicum** L., (13), 1050 m, 26 iv 2019, Ş.K. 6557, Ir-Tur.
279. **Umbilicus luteus** (Huds.) Webb & Berthel., (12) 1650 m, 05 vi 2019, Ş.K. 6778.

CUCURBITACEAE

280. ***Cucumis sativus** L., (1), 850 m, 22 v 2019, Ş.K. Observ.
281. ***Lagenaria siceraria** (Molina) Standl., (14) 750 m, 30 vi 2012, Ş.K. Observ.

ELAEAGNACEAE

282. ***Elaeagnus angustifolia** L. var. **angustifolia**, (1), 850 m, 22 v 2019, Ş.K. Observ.

EUPHORBIACEAE/SÜTLEĞENGİLLER

283. **Chrozophora tinctoria** (l.) A. Juss., (2), 950 m, 16 × 2011, B.M.11650.
284. **Euphorbia altissima** Boiss. var. **glabrescens** Boiss. ex Khan, (3), 31 v 2019, Ş.K. 6753.
285. **Euphorbia cheiradenia** Boiss. & Hohen., (8), 1050 m, 22 v 2019, Ş.K. 6652.
286. **Euphorbia denticulata** Lam., (12), 1600 m, 23 v 2015, Ir-Tur.
287. **Euphorbia esula** L. subsp. **tommasiniana** Kuzmanov, (1), 03 viii 2018, Ş.K. Observ.
288. **Euphorbia falcata** L. subsp. **falcata**, (2), 950 m, 14 vi 2012, B.Y. 7643.
289. **Euphorbia glareosa** Pall ex M.Bieb., (2), 950 m, 16 × 2011, B.M.11646.
290. * **Euphorbia maculata** L., (2), 980 m, 24 vii 2012, B.M.11653.

291. **Euphorbia peplus** L. var. **peplus**, (2), 950 m, 01 iv 2011, B.M.11672.
292. **Euphorbia petrophila** C.A.Mey., (12) 1650 m, 05 vi 2019, Ş.K. 6792.
293. **Euphorbia phymatosperma** Boiss., (2), 950 m, 04 vi 2011, B.M.11345, Ir-Tur.
294. **Euphorbia physocaulos** Mouterde, (2), 900 m, 04 vi 2011, B.M.11444, E.Medit.
295. **Euphorbia taurinensis** All., (2), 950 m, 14 vi 1986, B.Y. 7642.

FABACEAE

296. **Alhagi maurorum** Medik, var. **maurorum**, (5), 950 m, 03 viii 2018, Ş.K. 6359, Ir-Tur.
297. **Astragalus decurrens** Boiss., (12) 1650 m, 05 vi 2019, Ş.K. 6780, Ir-Tur.
298. **Astragalus globosus** Vahl, (12), 1600 m, 07 vii 1996, B.Y. 13,700, Ir-Tur., End., [LR(lc)].
299. **Astragalus gummifer** Labill., (2), 950 m, 18 vi 2011, B.M.11562, Ir-Tur.
300. **Astragalus guttatus** Banks & Sol., (9), 850 m, 22 v 2019, Ş.K. 6663, Ir-Tur.
301. **Astragalus hamosus** L., subsp. **hamosus** (2), 900 m, 04 vi 1986, B.Y. 7647.
     subsp. **finitimus** (Bunge) D.F.Chamb, (12),, 1650 m, 15 vi 2013, Ş.K. 4357, Ir-Tur.
302. **Astragalus leporinus** Boiss. (12), 1650 m, 05 vi 2019, Ş.K. 6778, Ir-Tur., End., [LR(lc)].
303. **Astragalus macrouroides** Hub.-Mor.,(10), 11 iv 2019, Ş.K. 6445, Ir-Tur., End., [VU].
304. **Astragalus melitenensis** Boiss., (10), 900 m, 11 iv 2019, Ş.K. 6490, Ir-Tur., End., [LR(cd)].
305. **Astragalus tigridis** Boiss., (12) 1650 m, 10 v 2019, Ş.K. 6643, Ir-Tur.
306. **Astragalus xylobasis** Freyn & Bornm., (16), 650 m, 24 v 2019, Ş.K. 6698, Ir-Tur.
307. ***Caragana arborescens** Lam., (2), 950 m, 12 iv 2013, Ş.K. Observ.
308. **Cercis sliquastrum** L. subsp. **sliquastrum**, (2), 950 m, 12 iv 2013, Ş.K. Observ.
309. **Chesneya rytidosperma** Jaub. & Spach, (16), 700 m, 24 v 2019, Ş.K. 6714.
310. **Cicer bijugum** Rech. f., (2), 950 m, 04 vi 2011, B.M.11407, Ir-Tur.
311. **Cicer pinnatifidum** Jaub. & Spach, (1), 850 m, 03 viii 2018, Ş.K. Observ.
312. **Coronilla scorpioides** (L.) D.J.Koch, (2), 950 m, 18 vi 2011, B.M.11525.
313. **Dorycnium hirsutum** (L.) Ser., (16), 700 m, 24 v 2019, Ş.K. 6737, Medit.

314. **Ebenus laguroides** Boiss., (12), 1650 m, 05 vi 2019, Ş.K. 6782, Ir-Tur., End., [LR(lc)].
315. **Genista albida** Willd., (16), 650 m, 24 v 2019, Ş.K. 6690.
316. ***Gleditsia triacanthos** L., (2), 950 m, 04 vi 2011, B.M.11364.
317. **Glycyrrhiza glabra** L., var. **glabra**, (14), 700 m, 07 vii 2012, Ş.K. 2910.
318. **Hedysarum aucheri** Boiss., (14), 850 m, 30 v 2014, Ş.K. 4883, Ir-Tur., End., [VU].
319. **Hedysarum pannosum** (Boiss.) Boiss., (16), 650 m, 24 v 2019, Ş.K. 6683, Ir-Tur., [VU].
320. **Hedysarum pogonocarpum** Boiss., (12) 1650 m, 05 vi 2019, Ş.K. 6779, End., [LR(lc)].
321. **Hedysarum varium** Willd. var. **varium**, (16), 650 m, 24 v 2019, Ş.K. 6697.
322. ***Laburnum anagyroides** Medik, (2), 950 m, 12 iv 2018, Ş.K. Observ.
323. **Lathyrus cicera** L., (2), 900 m, 11 v 1987, İ.Erkuş 1041, Medit.
324. **Lathyrus sativus** L., (2), 950 m, 04 vi 2011, B.M.11357, Medit.
325. **Lens culinaris** Medik. subsp. **orientalis** (Boiss.) Ponert, (14), 850 m, 14 vi 2015, Ş.K. 6173.
326. **Lotus corniculatus** L., (1), 850 m, 03 viii 2018, Ş.K. Observ.
327. **Lotus gebelia** Vent., var. **gebelia**, (2), 950 m, 04 v 1986, E.Aktoklu 119.
328. **Medicago biflora** (Griseb.) E.Small, (12), 1400 m, 15 vi 2013, Ş.K. 4408, Ir-Tur.
329. **Medicago lupilina** L., (10), 900 m, 11 iv 2019, Ş.K. Observ.
330. **Medicago minima** (L.) Bartal. var. **minima**, (10), 900 m, 11 iv 2019, Ş.K. 6486.
331. **Medicago monantha** (C.A.Mey.) Trautv., (2), 950 m, 15 v 1990, B.Y. 7685b, Ir-Tur.
332. **Medicago monspeliaca** (L.) Trautv., (2), 950 m, 04 vi 2011, B.M.11250, Medit.
333. **Medicago noeana** Boiss., (16), 650 m, 24 v 2019, Ş.K. 6700, Ir-Tur.
334. **Medicago orthoceras** (Kar.& Kir.) Trautv., (2), 900 m, 16 vi 1986, B.Y. 7645, Ir-Tur.
335. **Medicago radiata** L., (16), 700 m, 24 v 2019, Ş.K. 6717, Ir-Tur.
336. **Medicago rigidula** (L.) All. var. **rigidula**, (12), 01 vi 1996, B.Y. 13,348-b.
     var. **agrestis** Burniat, (2), 980 m, 24 vii 2012, B.M.11248.
337. **Medicago sativa** L. subsp. **sativa**, (2), 980 m, 24 vii 2012, B.M.11430.
338. **Melilotus officinalis** (L.) Desr., (10), 900 m, Ş.K. Observ.

339. **Onobrychis argyrea** Boiss. subsp. **argyrea**, (12), 1650 m, 05 vi 2019, Ş.K. 6787, Ir-Tur.

340. **Onobrychis cappadocica** Boiss., (2), 950 m, 18 vi 2011, B.M.11548, Ir-Tur., End., [LR(lc)].

341. **Onobrychis fallax** Freyn & Sint. ex Freyn var. **fallax**, (2), 900 m, 09 v 1986, B.Y. 7655, Ir-Tur., End., [LR(lc)].

342. **Onobrychis hypargyrae** Boiss., (2) 950 m, 11 vi 2011, B.M.11447.

343. **Onobrychis kotschyana** Fenzl, (2), 980 m, 09 vi 2011, B.M.11261, Ir-Tur.

344. **Pisum sativum** L., var. **pumilio** Meikle, (2), 980 m, 11 vi 2011, B.M.11400.

345. **Robinia pseudocacia** L. cv. **umbraculifera**, (2), 930 m, 30 iv 2019, Ş.K. Observ.

346. **Sophora alopecuroides** L., var. **alopecuroides**, (6), 930 m, 19 iv 2019, Ş.K. 6558.

347. *** Sophora japonica** L., (2), 950 m, 11 vi 2011, B. M.11479.

348. **Trifolium campestre** Schreb. subsp. **campestre**, (10), 900 m, 11 iv 2019, Ş.K. 6485.

349. **Trifolium nigrescens** Viv. subsp. **petrisavii** Holmboe, (6), 930 m, 28 iv 2019, Ş.K. 6598.

350. **Trigonella coelesyriaca** Boiss., (12) 1650 m, 05 vi 2019, Ş.K. 6781, Ir-Tur.

351. **Trigonella filipes** Boiss., (2), 950 m, 07 vi 2011, B. M.11428.

352. **Trigonella kotschyi** Fenzl, (2) Kampüsü, 900 m, 09 v 1986, B.Y. 7268, End., [LR(lc)].

353. **Trigonella monspeliaca** L., (6), 930 m, 19 iv 2019, Ş. K. 6560, Medit.

354. **Trigonella spicata** Sibth. & Sm., (2), 980 m, 07 vi 2011, B.M.11427, E.Medit.

355. **Trigonella spruneriana** Boiss., (2), 950 m, 09 v 2011, B.M.11294, Ir-Tur.

356. **Trigonella velutina** Boiss., (2), 950 m, 04 vi 2011, B. M.11380.

357. **Vicia narbonensis** L. var. **narbonensis**, (2), 980 m, 04 vi 2011, B.M.11368.

358. **Vicia noeana** Boiss. & Reut. ex Boiss. var. **noeana**, (2), 07 vi 2011, B.M.11429, Ir-Tur.

359. **Vicia peregrina** L., (12) 1650 m, 10 v 2019, Ş.K. 6646.

360. **Vicia sativa** L., subsp. **sativa**, (2), 900 m, 09 v 1986, B.Y. 7286.
  subsp. **nigra** (L.) Ehrh., var. **nigra**, (2), 980 m, 07 vi 2011, B.M.11431.
  var. **segetalis** (Thuill.) Ser. ex DC., (2), 900 m, H.Eren s.n.

361. **Vicia sericocarpa** Fenzl var. **sericocarpa**, (2),, 950 m, 11 vi 2011, B.M.11356.

362. *** Wisteria sinensis** (Sims.) DC., (2), 950 m, 12 iv 2018, Ş.K. Observ.

FAGACEAE

363. **Quercus infectoria** Oliv., subsp. **infectoria**, (8), 1050 m, 28 iv 2019, Ş.K. 6574, Av-Sib.
  subsp. **veneris**, (12) 1650 m, 10 v 2019, Ş.K. 6650.

364. **Quercus libani** Oliv., (12), ky, kayalık, 1650 m, 15 vi 2013, Ş.K. Observ., Ir-Tur.

365. **Quercus petraea** (Matt.) Liebl. subsp. **pinnatiloba** (K. Koch) Menitsky, (12), 1650 m, 15 vi 2013, Ş.K. 4365, End., [LR(lc)].

366. **Quercus robur** L./subsp. **robur**, (2), 950 m, 03 viii 2013, B.M.11692, Av-Sib.

GENTIANACEAE

367. **Centaurıum erythraea** Rafn. subsp. **erythraea**, (2), 900 m, 08/07/1986, E.Aktoklu 217.

GERANIACEA

368. **Erodium acaule** (L.) Becherer & Thell., (10), 900 m, 11 iv 2019, Ş.K. 6490, Medit.

369. **Erodium ciconium** L'Hér., (10), 900 m, 11 iv 2019, Ş. K. 6489.

370. **Erodium malacoides** (L.) L'Hér., (2), 900 m, 27 v 1987, G.Taş 1008, Medit.

371. **Geranium lucidum** L., (12) 1650 m, 10 v 2019, Ş.K. 6628.

372. **Geranium pusillum** Burm.f., (12) 1650 m, 10 v 2019, Ş.K. 6612.

373. **Geranium tuberosum** L., (6), 930 m, 19 iv 2019, Ş.K. 6526, Ir-Tur.

374. **Pelargonium endlicherianum** Fenzl, (12) 1650 m, 10 v 2019, Ş.K. Observ.

HYDRANGEACEAE

375. *** Hydrangea macrophylla** (Thunb.) Ser., (2), 950 m, 12 iv 2018, Ş.K. Observ.

376. **Philadelphus coronarius** L., (2), 950 m, 12 iv 2013, Ş.K. Observ.

HYPERICACEAE

377. **Hypericum amblysepalum** Hocst., (2), 950 m, E. Aktoklu 116, Ir-Tur.

378. **Hypericum capitatum** Choisy, var. **capitatum**, (2), 980 m, 04 vi 2011, B.M.11339, Ir-Tur.
  var. **luteum** N.Robson, (2), 950 m, 04 vi 2011, B. M.11397, Ir-Tur., End. [VU].

379. **Hypericum lydium** Boiss., (2), 950 m, 18 vi 2011, B.M.11543.
380. **Hypericum scabrum** L., (1), 850 m, 03 viii 2018, Ş.K. Observ., Ir-Tur.
381. **Hypericum thymifolium** Banks & Sol., (2), 900 m, 16 × 2011, B.Y. 9759, E.Medit.
382. **Hypericum triquetrifolium** Tura, (2), 900 m, 16 × 2011, B.M.11640.
383. **Hypericum uniglandulosum** Hausskn. ex Bornm., (2), 950 m, 18 vi 2011, B.M.11534, Ir-Tur., End., [LR(nt)].

JUGLANDACEAE

384. ***Juglans regia** L., (1), 850 m, 22 v 2019, Ş.K. Observ.

LAMIACEAE

385. **Ajuga chamaepitys** (L.) Schreb., subsp. **chia** (Schreber), (16), 700 m, 24 v 2019, Ş.K. 6718.
subsp. **laevigata** (Banks & Sol.) P.H.Davis, (12) 1650 m, 10 v 2019, Ş.K. 6645, Ir-Tur.
386. **Clinopodium graveolens** (M.Bieb.) Kuntze subsp. **graveolens**, (2), 04 vi 2011, B.M.11342.
387. **Eremostachys moluccelloides** Bunge, (8), 1050 m, 28 iv 2019, Ş.K. Observ., Ir-Tur.
388. **Lallemantia iberica** (M.Bieb.) Fisch & C.A.Mey., (12) 1650 m, 10 v 2019, Ş.K. 6647.
389. **Lamium amplexicaule** L., (6), 930 m, 28 iv 2019, Ş.K. 6602.
390. **Lamium garganicum** L. subsp. **striacum** (Sm.), (12) 1650 m, 10 v 2019, Ş.K. 6621, Medit.
391. **Lamium macrodon** Boiss. & Huett, (12) 1650 m, 10 v 2019, Ş.K. 6621 Ir-Tur.
392. **Lamium orientale** Fisch. & Mey, (2), 950 m, 24 iv 1988, M.D. 1049, Ir-Tur., End., [LR(lc)].
393. **Marrubium cuneatum** Banks & Sol., (2), 950 m, 11 vi 2011, B.M.11469, Ir-Tur.
394. **Marrubium globosum** Montbret & Aucher ex Benth. var. **globosum**, (10), 900 m, 11 iv 2019, Ir-Tur., End., [LR(lc)].
395. **Mentha spicata** L. subsp. **spicata**, (1), 850 m, 22 v 2019, Ş.K. Observ.
396. **Phlomis oppositiflora** Boiss. & Hausskn (2), 18 vi 2011, B.M.11538, Ir-Tur., End., [LR(lc)].
397. **Phlomis sieheana** Rech f., (2), 950 m, 04 vi 2011, B.M.11396, Ir-Tur., End., [LR(lc)].
398. **Phlomis sintenisii** Rech. f., (12), 1500 m, 20 v 1995, B.Y. 12,697, Ir-Tur., End., [VU].
399. **Salvia absconditiflora** (Montbret & Aucher ex Benth.) Greuter & Burdet, (2), 950 m, 04 vi 2011, B.M.11325, Ir-Tur., End., [LR(lc)].
400. **Salvia ceratophylla** L., (2), 950 m, 04 vi 2011, B.M. Observ., Ir-Tur.
401. **Salvia euphratica** Montbret & Aucher, var. **euphratica**, (10), 900 m, 11 iv 2019, Ş.K. 6447, Ir-Tur., End., [LR(cd)].
var. **leiocalycina** (Rech. f.) Hedge, (12), 20 v 1995, B.Y. 12,208, Ir-Tur., End., [LR(cd)].
402. **Salvia multicaulis** Vahl., (6), 930 m, 19 iv 2019, Ş.K. 6521, Ir-Tur.
403. **Salvia palaestina** Benth., (10), 650 m, 24 v 2019, Ş.K. 6691, Ir-Tur.
404. **Salvia staminea** Montbret & Aucher ex Benth., (12) 1650 m, 05 vi 2019, Ş.K. 6758., r-Tur.
405. **Salvia suffruticosa** Montbret & Aucher ex Benth., (2), 950 m, 15 v 1991, B.Y. 8821, Ir-Tur.
406. **Satureja hortensis** L.,(2) (14),, 700 m, 15 ix 2018, Ş.K. 6426.
407. **Scutellaria orientalis** L., subsp. **cretacea** (Boiss. & Hausskn) J.R.Edm., (16), 650 m, 24 v 2019, Ş.K. 6705, Ir-Tur. [EN].
408. **Sideritis montana** L., subsp. **montana**, (12),1600 m, 20 v 1995, B.Y. 12,193, E.Medit.
subsp. **remota** (d'Urv.) P.W. Ball, (2), 950 m, 18 vi 2011, B.M.11524, E.Medit.
409. **Stachys annua** L. subsp. **annua**, var. **lycaonica** R. Bhattacharjee, (16), 700 m, 24 v 2019, Ş.K. 6716, E. Medit.
410. **Stachys cretica** L., subsp. **anatolica** Rech.f., (2), 18 vi 2011, B.M.11535, End., [LR(lc)].
411. **Stachys ramosissima** Montbret & Aucher ex Benth., (2), 900 m, 25 vi 1986. E.Aktoklu 97, Ir-Tur., End., [LR(cd)].
412. **Teucrium chamaedrys** L., subsp. **syspirense** (K. Koch) Rech.f., (16), 24 v 2019, Ş.K. 6680.
413. **Teucrium orientale** L., var. **puberulens** Ekim, (2), 950 m, 14 vi 1995, B.Y. 8404b, Ir-Tur.
414. **Teucrium polium** L. subsp. **polium**, (5), 950 m, 03 viii 2018, Ş.K. 6360, Ir-Tur.
415. **Thymus sipyleus** Boiss., (2), 950 m, 18 vi 2011, B.M.11533.
416. **Ziziphora capitata** L., (2), 950 m, 04 vi 2011, B.M.11299.
417. **Ziziphora taurica** M.Bieb. subsp. **taurica**, (2), 950 m, 04 vi 2011, B.M.11403.

LINACEAE

418. **Linum mucronatum** Bertol., subsp. **mucronatum**, (16), 24 v 2019, Ş.K. 6687, Ir-Tur.
419. **Linum nodiflorum** L., (2), 950 m, 18 vi 2011, B.M.11550, Medit.
420. ***Punica granatum** L., (1), 850 m, 22 v 2019, Ş.K. Observ.

MAGNOLIACEAE

421. ***Magnolia grandiflora** L., (2), 950 m, 12 iv 2019, Ş. K. Observ.

MALVACEAE

422. **Alcea hohenackeri** (Boiss. & Huet) Boiss., (1), 850 m, 22 v 2019, Ş.K. Observ., Ir-Tur.
423. ***Hibiscus syriacus** L., (2), 950 m, 12 iv 2013, Ş.K. Observ.
424. **Malva neglecta** Wallr., (1), 850 m, 03 viii 2018, Ş.K. Observ.
425. ***Tilia argentea** Desf., (2), 950 m, 03 viii 2018, Ş.K. Observ.
426. ***Tilia plathphyllos** Scop. subsp. **plathphyllos**, (2), 950 m, 03 viii 2018, Ş.K. Observ.

MELIACEAE

427. ***Melia azedarach** L., (2), 950 m, 14 vi 2011, B. M.11493.

MORACEAE

428. **Ficus carica** L., subsp. **carica**, (1), 850 m, 22 v 2019, Ş.K. Observ.
429. ***Maclura pomifera** (Raf.) C.K.Schneid, (2), 950 m, 15 vi 2011, B.M.11499.
430. **Morus alba** L., (2), (1), 850 m, 03 viii 2018, Ş.K. Observ.
431. **Morus nigra** L.,(1), 850 m, 03 viii 2018, Ş.K. Observ.
432. **Morus rubra** L., (2), 950 m, 12 iv 2018, Ş.K. Observ.

NITRARIACEAE

433. **Peganum harmala** L., (14), 700 m, 30 v 2012, Ş.K. 2039.

OLEACEAE

434. ***Forsythia x intermedia** Zab., (2) Kampüsü, 950 m, 12 iv 2018, Ş.K. Observ.
435. ***Fraxinus excelsior** L. subsp. **excelsior** L., (1), 850 m, 03 viii 2018, Ş.K. Observ.
436. **Jasminum fruticans** L., (3), 850 m, 31 v 2019, Ş.K. Observ.
437. **Syringa vulgaris** L., (2), 950 m, 12 iv 2018, Ş.K. Observ.

OROBANCHACEAE

438. **Orobanche elatior** Sutton, (12), 1600 m, 08 vi 1995, B.Y. 12,618.
439. **O. ramosa** L., (5), 850 m, 22 v 2019, Ş.K. 6661.
440. **O. sintenisii** Beck ex Bornm., (2) 950 m, 18 vi 2011, B.M.11538.
441. **Parentucellia latifolia** (L.) Caruel subsp. **flaviflora** (Boiss.) Hand.-Mazz., (6), 930 m, 19 iv 2019, Ş.K. 6540.
442. **Phelipanche coelestis** (Reut) Soják, (12), 1650 m, 15 vi 2013, Ş.K. 4354.

OXALIDACEAE

443. **Oxalis corniculata** L., (2), 980 m, 01 iv 2013, B. M.11673.

PAPAVERACEAE

444. **Fumaria officinalis** L., subsp. **officinalis**, (10), 900 m, 11 iv 2019, Ş.K. Observ.
445. **Fumaria vaillantii** Loisel, (10), 900 m, 11 iv 2019, Ş. K. 6492.
446. **Glaucium corniculatum** (L.) Rudolph var. **corniculatum** (16) 700 m, 24 v 2019, Ş.K 6738.
447. **Glaucium flavum** Crantz, (16), 700 m, 24 v 2019, Ş. K. 6739.
448. **Hypecoum dimitiatum** Delile, (10), 900 m, 11 iv 2019, Ş.K. 6483.
449. **Hypecoum pendulum** L., (10), 900 m, 11 iv 2019, Ş. K.6662.
450. **Papaver argemone** L. subsp. **argemone**, (3) 850 m, 31 v 2019, Ş.K. 6755.
451. **Papaver dubium** L. subsp. **dubium**, (6), 930 m, 19 iv 2019, Ş.K. 6336.
452. **Papaver macrostomum** Boiss. & A. Huet, (10), 900 m, 11 iv 2019, Ş.K. 6481, Ir-Tur.
453. **Papaver persicum** Lindl. subsp. **persicum**, (6), 930 m, 19 iv 2019, Ş.K. 6541.
454. **Papaver rhoeas** L., (10), 900 m, 11 iv 2019, Ş.K. 6461.
455. **Papaver somniferum** L. var. **somniferum**, (16), 650 m, 24 v 2019, Ş.K. 6674.
456. **Roemera hybrida** (L.) DC. subsp. **hybrida**, (1), 850 m, 03 viii 2018, Ş.K. Observ.

PAULOWNIACEAE

457. ***Paulownia tomentosa** (Thunb) Steud., (2), 950 m, 12
iv 2018, Ş.K. Observ.

PEDALIACEAE

458. ***Sesamum indicum** L., (1), 850 m, 22 v 2019, Ş.K.
Observ.

PHYLLANTHACEAE

459. **Andrachne telephioides** L., (2), 950 m, 09 v 1986, B.
Y. 7247a.

PITTOSPORACEAE

460. ***Pittosporum tobira** (Thumb.) W.T.Aiton, (2),
950 m, 12 iv 2013, Ş.K. Observ.

PLANTAGINACEAE

461. **Anarrhinum orientale** Benth.,(1), 850 m, 22 v 2019,
Ş.K. Observ., Ir-Tur.
462. **Kickxia spuria** (L.) Dumort. subsp. **integrifolia** (Brot.)
R.Fernnandes, (2), 950 m, 16 × 2011, B.M.11644.
463. **Linaria chalapensis** (L.) Mill. subsp. **chalapensis**,
(10), 11 iv 2019, Ş.K. 6463, E.Medit.
464. **Linaria corifolia** Desf., (12), 1650 m, 15 vi 2013, Ş.K.
4374, Ir-Tur., End., [LR(lc)].
465. **Linaria iconia** Boiss. & Heldr., (2), 950 m, 04 vi 2011,
B.M.11353, Ir-Tur., End., [LR(lc)].
466. **Linaria simplex** DC., (6), 930 m, 19 iv 2019, Ş.K.
6543, Medit.
467. **Plantago euphratica** Decne ex Barnéoud (2) 18 vi
2011, B.M.11564, Ir-Tur. End., [LR(cd)].
468. **Plantago lanceolata** L., (3) 850 m, 31 v 2019, Ş.K.
6754.
469. **Plantago major** L. subsp. **major** (Glib.) Lange, (2),
950 m, 07 vii 2011, B.M.11621.
470. **Veronica multifida** L., (12) 1650 m, 05 vi 2019, Ş.K.
6794, Ir-Tur., End., [LR(lc)].
471. **Veronica polita** Fr., (12) 1650 m, 05 vi 2019, Ş.K.
6788.
472. **Veronica praecox** All., (2), 950 m, 24 iv 1986, M.D.
1062a.

PLATANACEAE

473. **Platanus orientalis** L., (1), 850 m, 22 v 2019, Ş.K.
Observ.

PLUMBAGINACEAE

474. **Acantholimon armenum** Boiss. & Huet, var. **armenum** (16), 24 v 2019, Ş.K. 6690, Ir-Tur.
var. **balansae** Boiss. & Huet., (2), 950 m, 04 vi 2011,
B.M.11419, Ir-Tur.
475. **Acantholimon caryophyllaceum** Boiss., (2), 950 m,
18 vi 2011, B.M.11563, Ir-Tur.

POLYGONACEAE

476. **Atrophaxis billardieri** Jaub.& Spach var. **billardieri**,
(12), 20 v 1995, B.Y. 12383, Ir-Tur.
477. **Polygonum aviculare** L., (2), 950 m, 11 vi 2011, B.
M.11471.
478. **Rumex scutatus** L., (12) 1650 m, 05 vi 2019, Ş.K.
6784.
479. **Rumex tuberosus** L. subsp. **horizontalis** Rech.f., (12),
1650 m, 15 vi 2013, Ş.K. 4364.

PORTULACEAE

480. **Portulaca oleracea** L., (1), 850 m, 03 viii 2018, Ş.K.
Observ.

PRIMULACEAE

481. **Anagallis arvensis** L. subsp. **arvensis**, (2), 900 m, 04
vii 1986, E.Aktoklu 199.
482. **Anagallis foemina** Mill., (2), 950 m, 18 vi 2011, B.
M.11516, Medit.
483. **Androsace maxima** L., (6), 930 m, 19 iv 2019, Ş.K.
6536.

RANUNCULACEAE

484. **Adonis aestivalis** L., subsp. **aestivalis**, (6), 930 m, 19
iv 2019, Ş.K. 6532.
subsp. **parviflora** (Fisch ex DC.) N.Busch., (7), 800 m,
16 iii 2013, Ş.K. 3460.
485. ***Aquilegia olympica** Boiss., (2), 900 m, 04 v 2014, Ş.
K. 4623.
486. **Ceratocephala falcata** (L.), (10), 900 m, 11 iv 2019,
Ş.K. 6443.
487. **Clematis orientalis** L., (11), 850 m, 01 xi 2018, Ş.K.
6430.
488. **Consolida glandulosa** (Boiss. & Huet) Bornm.,
(2) Kampüsü, 950 m, B.Y. 7663, Medit., Ir-Tur., End.,
LR(lc)].
489. **Consolida orientalis** (J.Gay) Schrödinger, (2), 950 m,
03 iii 1987, E.Aktoklu 265.

490. **Delphinium albiflorum** DC., (12), 1650 m, 15 vi 2013, Ş.K. 4367.

491. **Delphinium peregrinum** L., (2), 950 m, 11 vi 2011, B.M.11626, Medit.

492. **Nigella arvensis** L., var. **caudata** Boiss., (2), 950 m, 18 vi 2011, B.M.11529.

493. **Nigella latisecta** P.H.Davis, 950 m, 04 vi 2011, B. M.11286, Ir-Tur.

494. **Nigella nigellastrum** (L.) Willk., (2), 950 m, 04 vi 2011, B.M.11387.

495. **Nigella orientalis** L., (14), 850 m, 30 v 2012, Ş.K. 2042.

496. **Nigella oxypetala** Boiss., (2), 950 m, 04 vi 2011, B. M.11359, Ir-Tur.

497. **Ranunculus arvensis** L., (12) 1650 m, 05 vi 2019, Ş. K. 6762.

498. **Ranunculus constantinopolitanus** (DC.) d'Urv., (12) 1650 m, 10 v 2019, Ş.K. 6613.

499. **Ranunculus cuneatus** Boiss., (13), 1050 m, 26 iv 2019, Ş.K. 6568.

500. **Ranunculus isthmicus** Boiss. subsp. **stepporum** P.H. Davis, (10), 11 iv 2019, Ş.K. 6460.

501. **Ranunculus kochii** Ledeb., (6), 930 m, 28 iv 2019, Ş. K. 6603.

502. **Ranunculus muricatus** L., (6), 930 m, 19 iv 2019, Ş. K. 6550.

503. **Thalictrum isopyroides** C.A.Mey., (12), 1650 m, 10 v 2019, Ş.K. 6623.

RESEDACEAE

504. **Reseda lutea** L. var. **lutea**, (6), 930 m, 19 iv 2019, Ş. K. 6517.

ROSACEAE

505. **Alchemilla lithophila** Juz., (12), 1650 m, 05 vi 2019, Ş.K. 6772.

506. **Amygdalus communis** L., (12), 1650 m, 10 v 2019, Ş. K. 6630.

507. **Amygdalus orientalis** Mill., (6), 930 m, 19 iv 2019, Ş. K. 6520, Ir-Tur.

508. **Amygdalus trichamygdalus** (Hand.-Mazz.) Woronow, (5), 16 iii 2018, Ş.K. 3423, Ir-Tur.

509. *****Armeniaca vulgaris** Lam., (1), 850 m, 03 viii 2018, Ş.K. Observ.

510. *****Cerasus avium** (L.) Moench., (1), 850 m, 22 v 2019, Ş.K. Observ.

511. **Cerasus mahaleb** Mill. var. **mahaleb**, (10), 900 m, 11 iv 2019, Ş.K. 6442.

512. **Cerasus prostrata** (Labill.) Ser. var. **prostrata**, (12) 1650 m, 10 v 2019, Ş.K. 6635, Medit.

513. **Cerasus vulgaris** Mill., (1), 850 m, 22 v 2019, Ş.K. Observ.

514. *****Chaenomeles speciosus** (Sweet) Nakai, (2), 950 m, 12 iv 2018, Ş.K. Observ.

515. *****Chaenomeles japonica** (Thunb.) Lindl. ex Spach, (2), 950 m, 12 iv 2018, Ş.K. Observ.

516. *****Cotoneaster franchetii** Boiss., (2), 950 m, 12 iv 2018, Ş.K. Observ.

517. **Cotoneaster integerrimus** L., (12), 1650 m, 15 vi 2018, Ş.K. 4360.

518. **Cotoneaster melanocarpus** Lodd, (2) Kampüsü, 950 m, 12 iv 2018, Ş.K. Observ.

519. **Cotoneaster nummularius** Fisch. C.A.Mey., (12), 1650 m, 25 v 2018, Ş.K. 4114.

520. **Crataegus azarolus** L. var. azarolus, (10), 900 m, 11 iv 2019, Ş.K. 6443.

521. **Crataegus** monogyna Jacq. subsp. **monogyna**, (11), 850 m, 01 xi 2018, Ş.K. 6427.

522. **Crataegus orientalis** Pall. ex M.Bieb., subsp. **orientalis** (12) 1650 m, 15 vi 2018, Ş.K. 4361.

523. *****C. persimilis** Sarg. cv. **prunifolia**, (2), 950 m, 12 iv 2018, Ş.K. Observ.

524. *****C. oblonga** Mill., (1), 850 m, 22 v 2019, Ş.K. Observ.

525. *****Malus floribunda** Siebold ex Van Houtte, (2), 950 m, 12 iv 2018, Ş.K. Observ.

526. *****Malus pumila** Mill., (14), 750 m, 20 v 2018, Ş.K. Observ.

527. *****Malus x purpurea** Rehd., (2), 950 m, 12 iv 2018, Ş. K. Observ.

528. *****Persica vulgaris** Mill., (14), 750 m, 20 v 2018, Ş.K. Observ.

529. **Potentilla recta** L., (6), 930 m, 19 iv 2019, Ş.K. 6520.

530. *****Prunus cerasifera** Ehrh., (1), 850 m, 22 v 2019, Ş.K. Observ.

531. *****Prunus laurocerasus** L., (2), 950 m, 11 vi 2011, B. M.11732.

532. *****Prunus subhirtella** Miq. cv. **snofozam**, (2), 950 m, 12 iv 2018, Ş.K. Observ.

533. *****Pyracantha coccinea** M.Roem., (2), 950 m, 12 iv 2018, Ş.K. Observ., Av-Sib.

534. *****Pyrus communis** L. subsp. **communis**, (1), 850 m, 22 v 2019, Ş.K. Observ.

535. **Rosa canina** L., (11), 850 m, 01 xi 2018, Ş.K. 6429.

536. **Rosa foetida** J.Herrm., (2), 950 m, 12 iv 2018, Ş.K. Observ.

537. **Rubus sanctus** Schreb., (11), 850 m, 01 xi 2018, Ş.K. 6428.

538. **Sanguisorba minor** L., subsp. **minor**, (1), 850 m, 22 v 2019, Ş.K. Observ.

539. *****Spiraea x vanhouttei** (Briot) Carriére, (2), 950 m, 12 iv 2013, Ş.K. Observ.

RUBIACEAE

540. **Asperula affinis** Boiss. & A.Huet, (10), 900 m, 11 iv 2019, Ş.K. 6457.
541. **Asperula arvensis** L., (2), 04 iv 1987, İ.Erkuş 1026.
542. **Asperula glomerata** (M.Bieb.) Griseb., subsp. **glomerata**, (12), 05 vi 2019, Ş.K. 6767.
543. **Asperula stricta** Boiss., subsp. **stricta**, (16), 700 m, 24 v 2019, Ş.K. 6729.
subsp. **latibracteata** (Boiss.) Ehrend., (14) 30 v 2014, Ş.K. 4884, Ir-Tur., End., [LR(lc)].
544. **Callipeltis cucullaria** (L.) Steven, (16), 700 m, 24 v 2019, Ş.K. 6735, Ir-Tur.
545. **Crucianella macrostachya** Boiss., (12), 1600 m, 08 ix 1995, B.Y. 12607, E.Medit.
546. **Cruciata articulata** (L.) Ehrend., (1), 850 m, 22 v 2019, Ş.K. Observ.
547. **Galium floribundum** Sm. subsp. **floribundum**, (2), 950 m, 04 vi 2011, B.M.11340.
548. **Galium setaceum** Lam., (2), 950 m, 11 vi 2011, B.M.11281.
549. **Galium tenuissimum** M. Bieb. subsp. **trichophorum**, (12), 05 vi 2019, Ş.K. 6757, Ir-Tur.
550. **Galium tricornutum** Dandy, (12), 1500 m, 20 v 1995, B.Y. 12715, Medit.

RUTACEAE

551. **Haplophyllum cappadocicum** Spach, (14), 30 v 2012, Ş.K. 2037, Ir-Tur., End., [LR(nt)].

SALICACEAE

552. **Populus alba** L. var. **alba**, (1), 850 m, 22 v 2019, Ş.K. Observ.
553. **Populus nigra** L., subsp. **caudina** (Ten.) Bugala, (2), 950 m, 12 iv 2018, Ş.K. Observ.
554. **Salix alba** L. subsp. **alba**, (11), 850 m, 01 xi 2018, Ş.K. 6433, Av-Sib.
555. ***Salix babylonica** L. var. **babylonica**, (1), 850 m, 03 viii 2018, Ş.K. Observ.
556. **Salix cinerea** L. var. **cinerea**, (2), 950 m, 12 iv 2013, Ş.K. Observ., Av.-Sib.

SANTALACEAE

557. **Chrysothesium aureum** (Jaub. & Spach) Hendrych, (2), 18 vi 2011, B.M.11511, Ir-Tur.
558. **Thesium macranthum** Fenzl., (14),, 850 m, 30 v 2014, Ş.K. 4879, Ir-Tur.

SAPINDACEAE

559. ***Acer negundo** L., (2), 950 m, 12 iv 2018, Ş.K. Observ.
cv. **flamingo**, (2), 950 m, 12 iv 2018, Ş.K. Observ.
560. ***Aesculus hippocastanum** L., (2), 950 m, 12 iv 2018, Ş.K. Observ.
561. ***Koelreuteia paniculata** Laxm., (2), 950 m, 05 vii 2011, B.M.11622.

SCROPHULARIACEAE

562. ***Buddleja davidii** Franch., (2), 950 m, 05 vii 2011, B.M.11620.
563. **Scrophularia xanthoglossa** Boiss. var. **decipiens** Boiss., (6), 19 iv 2019, Ş.K. 6500.
564. **Verbascum euphraticum** Benth., (10), 900 m, 11 iv 2019, Ş.K. 6488, Ir-Tur., End., [VU].
565. **Verbascum pyramidatum** M. Bieb., (8), 1050 m, 22 v 2019, Ş.K. 6658.
566. **Verbascum splendidum** Boiss., (10), 11 iv 2019, Ş.K. 6446, E.Medit., End., [LR(lc)].
567. **Verbascum varians** Freyn & Sint. subsp. **stepporum** Hub.-Mor., (12),, 1650 m, 25 v 2013, Ş.K. 4111, End., [LR(cd)].

SIMAROUBACEAE

568. **Ailanthus altissima** (Mill.) Swingle, (1), 850 m, 03 viii 2018, Ş.K. Observ.

SOLANACEAE

569. **Hyoscyamus niger** L., (2), 950 m, 09 vi, 2018, Ş.K. Observ.
570. **Hyoscyamus reticulatus** L., (12) 1650 m, 05 vi 2019, Ş.K. 6768.
571. ***Lycopersicon esculentum** Mill., (14), 750 m, 20 v 2018, Ş.K. Observ.
572. **Solanum dulcamara** L., (1), 850 m, 03 viii 2018, Ş.K. Observ, Av-Sib.
573. **Solanum luteum** Mill., (1), 850 m, 03 viii 2018, Ş.K. Observ.
574. ***Solanum melongena** L., (1), 850 m, 22 v 2019, Ş.K. Observ.

TAMARICACEAE

575. **Tamarix smyrnensis** Bunge, (16), 700 m, 24 v 2019, Ş.K. 6733.

THYMELAEACEAE

576. **Dapne oleoides** Schreb. subsp. **kurdica** (Bornm.) Bornm., (12), 10 v 2019, Ş.K. Observ.

577. **Thymelaea passerina** (L.) Coss. & Germ., (5), 950 m, 03 viii 2018, Ş.K. 6363.

ULMACEAE

578. **Ulmus glabra** Huds., (1), 850 m, 03 viii 2018, Ş.K. Observ., Av-Sib.

URTICACEAE

579. **Parietaria judaica** L., (8), 1050 m, 28 iv 2019, Ş.K. 6580.

580. **Urtica dioica** L. subsp. **dioica**, (1), 850 m, 22 v 2019, Ş.K.Observ.

581. **Urtica pilulifera** L., (10), 9000 m, 11 iv 2019, Ş.K. Observ.

VERBENACEAE

582. **Verbana officinalis** var. **officinalis**, (2), 950 m, 15 vii 2011, B.M.11630.

VIOLACEAE

583. **Viola kitaibeliana** Roem. & Schult., (6), 930 m, 19 iv 2019, Ş.K. 6494.

584. **Viola occulta** Lehm., (13), 1050 m, 26 iv 2019, Ş.K. 6557.

VITACEAE

585. *****Vitis vinifera** L., (1), 850 m, 22 v 2019, Ş.K. Observ.

ZYGOPHYLLACEAE

586. **Trıbulus terrestris** L., (2), 900 m, 04 v 1987, G.Taş 1019c.

MONOCOTYLEDONS

AMARYLLIDACEAE

587. **Allium atroviolaceum** Boiss., (12), 1650 m, 05 vi 2019, Ş.K. 6776.

588. **Allium callidictyon** C.A.Mey. ex Kunth, (13), 1050 m, 26 iv 2019, Ş.K. 6660, Ir-Tur.

589. *****Allium cepa** L., (13), 1050 m, 26 iv 2019, Ş.K. 6565.

590. **Allium dictyoprasum** C.A.Mey. ex Kunth, (16), 700 m, 24 v 2019, Ş.K. 6744, Ir-Tur.

591. **Allium chrysantherum** Boiss. & Reut., (14), 700 m, 09 vi 2017, Ş.K. 6322, Ir-Tur.

592. **Allium flavum** L. subsp. **tauricum** (Beser ex Rchb.) Stearn var. **tauricum**, (2), 950 m, 18 vi 2011, B. M.11504b, Medit.

593. **Allium kharputense** Freyn & Sint., (12) 1650 m, 05 vi 2019, Ş.K.6765, Ir-Tur.

594. **Allium scabriflorum** Boiss., (10), 950 m, 11 iv 2019, Ş.K. 6474, Ir-Tur., End., [LR(lc)].

595. **Sternbergia clusiana** Ker Gawl. ex Spreng., (2), 1350 m, 14 × 2006, B.M.10170, Ir-Tur.

596. **Arum elongatum** Steven, (6), 930 m, 19 iv 2019, Ş.K. 6553.

597. **Arum rupicola** Boiss. var. **rupicola**, (12), 08 vi 1996, B.Y. 13,376, Ir-Tur., End., [LR(nt)].

598. **Eminium rauwolffii** (Blume) Schott var. **rauwolffii**, (2), 950 m, 10 iv 1989, B.Y. 8411.

599. **Eminium spiculatum** (Blume) Schott, (8), 1050 m, 28 iv 2019, Ş.K. 6587, Ir-Tur.

ASPARAGACEAE

600. **Bellevalia longipes** Post, (12) 1650 m, 10 v 2019, Ş.K. 6634, Ir-Tur.

601. **Muscari armeniacum** Leichtlin ex Baker, (6), 930 m, 19 iv 2019, Ş.K. 6512.

602. **Muscari comosum** (L.) Mill., (9), 850 m, 22 v 2019, Ş.K. 6665.

603. **Muscari longipes** Boiss., (12) 1650 m, 05 vi 2019, Ş.K. 6766, Ir-Tur.

604. **Ornithogalum malatyanum** Mutlu, (12), 1650 m, 25 v 2013, Ş.K. 4112, Ir-Tur., End., [CR].

605. **Ornithogalum oligophyllum** E.D.Clarke, (13), 1050 m, 26 iv 2019, Ş.K. 6653.

606. **Ornithogalum sphaerocarpum** A.Kern., (16), 700 m, 24 v 2019, Ş.K. 6707.

607. **Scilla melaina** Speta, (2), 950 m, 23 v 1994, B.Y. 11,427-b, E.Medit.

COLCHICACEAE

608. **Colchicum szovitsii** Fisch. C.A.Mey., (2), 950 m, 09 v 1986, B.Y. 7274, Ir-Tur.

609. **Colchicum triphyllum** Kunze, (10), 900 m, 11 iv 2019 Ş.K. Observ, Medit.

CYPERICACEAE

610. **Carex distans** L. subsp. **distans**, (4), 650 m, 01 xi 2018, Ş.K. Observ.

IRIDACEAE

611. **Crocus biflorus** Mill., subsp. **tauri** (Maw) B.Mathew, (2), 24 iv 1988, H.Ç. 1010, Ir-Tur.

612. **Crocus cancellatus** Herb. subsp. **damescanus** (Herb.) B.Mathew, (7), 850 m, 01 xi 2018, Ş.K. 6434, Ir-Tur.

613. **Crocus pallasii** Goldb.subsp. **turcicus** B.Mathew, (7), 650 m, 01 xi 2018, Ş.K. 6436.

614. **Gladiolus atroviolaceus** Boiss., (10), 900 m, Ş.K. Observ, Ir-Tur.

615. **Iris x germanica** L., (2), 950 m, 06 xi 2018, Ş.K. Observ.

616. **Iris persica** L., (7), 800 m, 16 iii 2018, Ş.K. 3424, Ir-Tur.

617. **Iris sari** Schott ex Baker, (12) 1650 m, 10 v 2019, Ş.K. Observ, Ir-Tur., End., [LR(lc)].

IXIOLIRIACEAE

618. **Ixıolırıon tataricum** (Pallas) Herbent, (13), 1050 m, 26 iv 2019, Ş.K. 6659, Ir-Tur.

LILIACEAE

619. **Fritillaria imperialis** L., (12), ky, kayalık, 1500 m, 06.05.2012, Ş.K. 1766, Ir-Tur., [VU].

620. **Gagea bulbifera** (Pall.) Salisb., (12) 1650 m, 10 v 2019, Ş.K. 6622, Av-Sib.

621. **Gagea fibrosa** (Desf.) Schult. & Schult.f., (2), 950 m, 24 iv 1998., M.D. 1032.

622. **Gagea granatelli** (Parl.) Parl., (6), 930 m, 19 iv 2019, Ş.K. 6510.

623. **Tulipa armena** Boiss. var. **armena**, (12) 1650 m, 24 iv 2019, Ş.K. 6639.

POACEAE

624. **Aegilops biuncialis** Vis., (16), 700 m, 24 v 2019, Ş.K. 6731.

625. **Aegilops columnaris** Zhukovsky, (2), 950 m, 18 vi 2011, B.M.11365, Ir-Tur.

626. **Aegilops cylindrica** Host., (9), 850 m, 22 v 2019, Ş.K. 6671, Ir-Tur.

627. **Aegilops geniculata** Roth., (16), 650 m, 24 v 2019, Ş.K. 6696, Medit.

628. **Aegilops speltoides** Tausch, var. **speltoides**, (16), 650 m, 24 v 2019, Ş.K. 6676.
var. **ligustica** (Savign.) Bornm, (16), 650 m, 24 v 2019, Ş.K. 6695.

629. **Aegilops triuncialis** L. subsp. **triuncialis**, (2), 980 m, 11 vi 2011, B.M.11457.

630. **Aegilops umbellulata** Zhuk., (8), 1050 m, 22 v 2019, Ş.K. 6654, Ir-Tur.

631. **Amblyopyrum muticum** (Boiss.) Eig, var. **loliaceum** (Jaub. & Spach) Eig, (2), 950 m, 16 vi 1987, B.Y. 7687.

632. **Avena eriantha** Durieu, (16), 700 m, 24 v 2019, Ş.K. 6726.

633. ***Avena sativa** L., (8), 1050 m, 22 v 2019, Ş.K. 6655.

634. **Bothrıochloa ischaemum** (L.) Keng, (2), 980 m, 15 vii 2011, B.M.11624.

635. **Briza humilis** M.Bieb., (16), 700 m, 24 v 2019, Ş.K. 6711.

636. **Briza maxima** L., (16), 650 m, 24 v 2019, Ş.K. 6675.

637. **Bromus danthoniae** Trin. subsp. **danthoniae**, (16), 650 m, 24 v 2019, Ş.K. 6677.

638. **Bromus japonicus** Thunb., subsp. **anatolicus** (Boiss. & Heldr.) Pelzes, (7) 800 m, 11 v 2018, Ş.K. Observ. subsp. **japonicus**, (16), 650 m, 24 v 2019, Ş.K. 6674.

639. **Bromus riparius** Rehm., (12),, 1650 m, 15 vi 2013, Ş. K. 4351.

640. **Bromus sterilis** L., (2), 950 m, 15 v 1990., B.Y. 8770.

641. **Bromus tectorum** L., (2), 950 m, 04 vi 2011, B. M.11312.

642. **Chrysopogon gryllus** (L.) Trin. subsp. **gryllus**, (2), 950 m, 04 vi 2011, B.M.11317.

643. **Crithopsis delileana** (Schult.) Roshev., (2), 950 m, 09 v 1986, B.Y. 7239, Ir-Tur.

644. **Cynodon dactylon** (L.) Pers. var. **villosus** Regel, (2), 980 m, 11 vi 2011, B.M.11464.

645. **Dactylis glomerata** L., subsp. **glomerata**, (1), 850 m, 03 viii 2018, Ş.K. Observ.

646. **Echinaria capitata** (L.) Desf., (12) 1650 m, 10 v 2019, Ş.K. 6627.

647. **Echinochloa colonum** (L.) Link, B7, 900 m, 15 × 2000, T.Arabacı 1177.

648. **Echinochloa crus-galli** (L.) P.Beauv., (2), 950 m, 15 × 2000, T.Arabacı 1176.

649. **Elymus hispidus** (Opis) Melderis, subsp. **barbulatus** (Schur) Melderis, (14), 850 m, 10 v 2014, Ş.K. 4812. subsp. **hispidus**, (16), 700 m, 24 v 2019, Ş.K. 6710.

650. **Elymus lazicus** (Boiss.) Melderis, (10), 900 m, 11 iv 2019, Ş.K. 6479, End., [LR(lc)].

651. **Elymus longearistatus** (Boiss.) Tzelev subsp. **sintenisii** Melderis, (12),, 1650 m, 15 vi 2013, Ş.K. 4350, Ir-Tur., End., [EN].

652. **Eremopoa altaica** (Trin.) Roshev., (2), 950 m, 10 v 1990, B.Y. 8772, Ir-Tur.

653. **Festuca anatolica** Markgr.-Dann. subsp. **anatolica**, (2), 950 m, 04 vi 2011, B.M.11291, End., [LR(lc)].

654. **Gaudınıopsis macra** M.Bieb., (2), 950 m, 04 vi 2011, B.M.11306, Ir-Tur.

655. **Hordeum bulbosum** L., (2), 950 m, 28 vi 1986, E. Aktoklu 169.

656. **Hordeum murinum** L., subsp. **glaucum** (Steud.) Tzvelev, (2), 950 m, 04 v 1987, HEs.n
subsp. **leporium** (Link) Arcang, (2), 950 m, 04 vi 2011, B.M.11352, Ir-Tur.

657. **Hordeum spontaneum** K.Koch, (1), 850 m, 03 viii 2018.

658. **Loliolum subulatum** (Banks & Sol.) Eig., 950 m, 04 vi 2011, B.M.11309, Ir-Tur.

659. **Melica persica** Kunth subsp. **persica**, (16), 700 m, 24 v 2019, Ş.K. 6728.

660. **Parapholis incurva** (L.) C.E.Hubb., (2), 900 m, 14 vi 1986, B.Y. 7696.

661. **Pennisetum orientale** Rich., (2), 950 m, 15 v 1991, B.Y. 8825, Ir-Tur.

662. **Phleum boissieri** Bornm., (16), 700 m, 24 v 2019, Ş.K. 6730, Ir-Tur.

663. **Phragmites australis** (Cav.) Trin. ex Steud., (11) 850 m, 01 xi 2018, Ş.K. 6431, Av-Sib.

664. **Poa bulbosa** L., (1), 850 m, 03 viii 2018, Ş.K. Observ.

665. **Psilurus incurvus** (Gouan) Schinz & Thell., (2), 950 m, 04 vi 2011, B.M.11310.

666. **Rostraria cristata** (L.) Tzvelev var. **cristata**, (2), 950 m, 04 vi 2011, B.M.11307.

667. **Setaria glauca** (L.) P.Beauv., (2), 980 m, 15 vii 2011, B.M.11623.

668. **Setaria viridis** (L.) P.Beauv., (2), 980 m, 14 vi 2000, T.Arabacı 1169.

669. **Sorghum halepense** (L.) Pers., var. **halepense**, (2), 980 m, 14 vi 2000, T.Arabacı 1145.

670. **Stipa arabica** Trin. & Rupr., (9), 850 m, 22 v 2019, Ş.K. 6670, Ir-Tur.

671. **Stipa holosericea** Trin., (1), 850 m, 22 v 2019, Ş.K. Observ.

672. **Taeniatherum caput-medusae** (L.) Nevski, (2), 980 m, 07 vi 2011, B.M.11303, Ir-Tur.

673. **Triticum baeoticum** Boiss., (2), 950 m, 27 vi 1986, E. Aktoklu 139.

674. **Triticum durum** Desf., (2), 980 m, 11 vi 2011, B.M.11470.

675. **Vulpia ciliata** Dumort. subsp. **ciliata**, (2), 950 m, 04 vi 2011, B.M.11416.

676. **Vulpia persica** (Boiss & Buhse) Krecz. & Bobrov, (2), 04 vi 2011, B.M.11566, Ir-Tur.

677. **Zea mays** L. subsp. **mays**, (1), 850 m, 22 v 2019, Ş.K. Observ.

XANTHORRHOEACEAE

678. **Asphodeline damascena** (Boiss.) Baker, subsp. **damascena**, (8), 1050 m, 22 v 2019, Ş.K. 6651, Ir-Tur.
subsp. **rugosa** E. Tuzlaci, (10), 950 m, 11 iv 2019, Ş.K. 6471, E.Medit. End., [LR(lc)].

679. **Eremurus cappadocicus** J.Gay ex Baker, (16), 700 m, 24 v 2019, Ş.K. 6746.

680. **E. spectabilis** M.Bieb., (12) 1650 m, 05 vi 2019, Ş.K. Observ.

681. **Hemerocallis fulva** (L.) L., (2), 950 m, 20 vi 2016, B.M.11657, Ir-Tur.

References

Aktoklu, E. (1996). Malatya Florasına Katkılar I Sürgü-Çelikhan Yöresinde bir Ön Çalışma. *Turkish Journal of Botany, 20*, 257–278.

Arabacı, T., & Yıldız, B. (2004). A floristical study on Poaceae spp. growing naturally in Malatya Province. *Turkish Journal of Botany, 28*, 361–368.

Avcı, A., & Esen, F. (2019). Malatya Havzası'nda Sıcaklık ve Yağışın Trend Analizi. *İnönü University International Journal of Social Sciences, (INIJOSS), 8*(1), 230–246.

Bridson, D., & Forman, L. (1999). The herbarium handbook: Royal botanic gardens. Kew Press.

Davis, P. H. (Ed.). (1965–1985). Flora of Turkey and the East Aegean Islands. Edinburgh University Press.

Davis, P. H. (Ed.). (1971). *Distribution patterns is Anatolia particular reference to Endemizm, in D. Harper and Hedge Plant Life of SW Asia*. Edinburgh University Press.

Davis, P. H., Mill, R. R., & Tan, K. (Eds.). (1988). Flora of Turkey and the East Aegean Islands. Edinburgh University Press.

Dönmez, A. A., & Yerli, S. V. (2019). Biodiversity in Turkey: Global biodiversity: Selected countries in Europe. Apple Academic Press.

Genç, İ, Özhatay, N., & Koyuncu, M. (2012). Allium purpureoviride sp. nov. (Alliaceae) from east Anatolia, Turkey. *Nordic Journal of Botany, 30*, 333–336.

Ekim, T., Koyuncu, M., Vural, M., Duman, H., Aytaç, Z. & Adıgüzel, N. (2000). *Red data book of Turkish plants (Pteridophyta ve Spermatophyta)*. TTKD and Van Centennial University.

Ekşi, G., & Yıldırım, H. (2019). Allium yamadagensis (Amaryllidaceae) a new species from Turkey. *Phytotaxa, 400*(1), 31–36.

Frangipane, M. (2012). The Collapse of the 4th millennium centralised system at Arslantepe and the far-reaching changes in 3rd millennium societies. *Origini, XXXIV*, 237–260

Frangipane, M. (2013). *The origins of power (Arslantepe Gücün Kökeni)*. Malatya Provincial Directorate of Culture Press.

Henderson, D. M. (1961). Contribution to the Bryophyte flora of Turkey: IV. *Notes from Royal Botanic Garden Edinburg, 23*, 263–278.

Güner, A., Özhatay, N., Ekim, T., & Başer K.H.C. (Eds.). (2000). Flora of Turkey and the East Aegean Islands. Edinburg University Press.

Güner, A., Aslan, S., Ekim, T., Vural, M., Babaç, M. T. (Eds.). (2012). Türkiye Bitkileri Listesi (Damarlı Bitkiler), Nezahat Gökyiğit Botanik Bahçesi ve Flora Araştırmaları Derneği Yayını.

Karakuş, Ş. (2016). Malatya ili Florası, (Publication No: 424336) (Doctoral dissertation, İnönü University), InuResearch Information. https://avesis.inonu.edu.tr/yonetilen-tez/c054647c-5f34-4116-a12a-894fb7a9c5e2/malatya-florasi

Karakuş, Ş, & Mutlu, B. (2017). Floristic list of Tohma (Malatya-Sivas, Turkey) Valley. *Hacettepe Journal of Biology and Chemistry, 1* (45), 95–116.

Karakuş Ş., & Mutlu, B. (2019). Allium dönmezii, a new species of Allium sect. Melanocrommyum (Amaryllidaceae) from Turkey: Morphological and molecular evidence. *Phytotaxa, 411*(3), 194–204.

Koç, M., Hamzaoğlu, E., & Budak, Ü. (2012). Minuartia aksoyi sp. nov. and M. buschiana subsp. artvinica subsp. nov. (Caryophyllaceae) from Turkey. *Nordic Journal of Botany, 30*(3), 337–342.

Koç, M., & Aksoy, A. (2013). Minuartia hamzaoglui (Caryophyllaceae), a new species from Turkey. *Turkish Journal of Botany, 37*, 428–433.

Mutlu, B., & Karakuş, Ş. (2012). A new species of Ornithogalum (Hyacinthaceae) from East Anatolia, Turkey. *Turkish Journal of Botany, 36*, 125–133.

Mutlu, B., & Karakuş, Ş. (2015a). A new species of Sisymbrium (Brassicaceae) from Turkey: Morphological and molecular evidence. *Turkish Journal of Botany, 39*, 325–333.

Mutlu, B., & Karakuş, Ş. (2015b). A new species of Campanula (Campanulaceae) from Turkey. *Phytotaxa, 234*(3), 287–293.

Mutlu, B., & Karakuş, Ş. (2015c). Floristic list of İnönü University (Malatya) main campus area. *Hacettepe Journal of Biology & Chemistry, 43*(2), 73–89.

Mutlu and Karakuş, 2019Mutlu, B. & Karakuş Ş. (2019). Poskıllı ketentere [Camelina hispida Boiss. var. lasiocarpa (Boiss. & C.I. Blanche) Post (Brassicaceae)] taksonunun Türkiye'deki Varlığı ve Taksonomik Durumu. *Bağ Bahçe Bilim Dergisi, 6*(3), 9–16.

Sunkar, M., Hatun Ü., & Toprak, A. 2(013). Malatya Havzası ve Çevresinde İklim Özelliklerinin Meyveciliğe Etkisi, 3rd International Geography Symposium—GEOMED 2013 Symposium Proceedings. ISBN: 978-605-62253-8-3.

Tuna, A. (2019). Project Report of TUBITAK (Project No:217O290). Development of Archaeological Landscape Restoration and Management Strategy in Arslantepe Mound and Its Territory.

Yeşil, Y., Yıldırım, H., Akalın, E., & Altıoğlu, Y. (2016). Pimpinella enguezekensis (Apiaceae), a new species from East Anatolia Region (Turkey). *Phytotaxa, 289*(3), 237–246.

Yıldırım, H., & Erol, O. (2013). Crocus yakarianus sp. nov. from eastern Turkey. *Nordic Journal of Botany, 31*(4), 426–429.

Yıldırım, H., & Şenol, S. G. (2014a). Alkanna malatyana (Boraginaceae), a new species from East Anatolia, Turkey. *Phytotaxa, 64* (2), 124–132.

Yıldırım, H., & Şenol, S. G. (2014b). Campanula alisan-kilincii (Campanulaceae), a new species from eastern Anatolia, Turkey. *Turkish Journal of Botany, 38*, 22–30.

Yıldırım, H., & Şenol, S.G. (2014c). Reseda malatyana (Resedaceae), a new chasmophytic species from eastern Anatolia, Turkey. *Turkish Journal of Botany, 38*, 1013–1021.

Yıldırım, H., Altıoğlu, Y., Şahin, B., & Aslan, S. (2015). Bellevalia chrisii sp. nov. (Asparagaceae) from eastern Anatolia, Turkey. *Nordic Journal of Botany, 33*(1), 45–49.

Yıldırım, H. (2015a). Parietaria semispeluncaria (Urticaceae), a new species from eastern Turkey. *Phytotaxa, 226*(3), 281–287.

Yıldırım, H. (2015b). Taraxacum rupicolum (Asteraceae) Doğu Anadolu'dan yeni bir Karahindiba (Taraxacum F.H.Wigg.) türü. *Bağ Bahçe Bil. Dergisi, 1*(3), 72–81.

Yıldırım, H. (2019). Allium sultanae-ismailii (Amaryllidaceae), a new species from eastern Turkey. *Phytotaxa, 403*(1), 39–46.

Yıldız, B., Bahçecioğlu, Z., & Arabacı, T. (2014). Floristics characteristics of Beydağı (Malatya). *Turkish Journal of Botany, 28*, 391–419.

UNESCO. (2014). Archaeological Site of Arslantepe. Retrieved February 29, 2021, from https://whc.unesco.org/en/tentativelists/5908/

Uzunhisarcıklı, M. E., Duman, H., & Yılmaz, S. (2013). A new species of Bellevalia (Hyacinthaceae) from Turkey. *Turkish Journal of Botany, 37*, 651–655.

Disused Urban Cemeteries: Unearthing User Experiences of Abney Park Cemetery

Corianne Rice

Abstract

Disused urban cemeteries (DUCs) represent complex landscapes that have long been recognised as sites of social, environmental and recreational fulfilment. Despite this, they are still marginally understood and rarely incorporated into urban planning conversations. This paper investigates perceptions of DUCs and how they can help bridge the gap in planning knowledge. Through an explorative case study of London's Abney Park Cemetery, user perceptions were elicited through 26 semi-structured interviews. Thematic analyses revealed the DUC is still a highly valued land use because it provides urban inhabitants with a break from the monotony of man-made. This function and its associated features are then considered within a framework of urban planning implications.

Keywords

Urbanisation • Public perceptions • Cemeteries • London • Extemporaneity • Man-made

1 Introduction

When the Rio Summit was held in 2012, the world's population had just surpassed seven billion (UN-DESA, 2012). At the same time, the Summit highlighted an additional dynamic of population growth: urbanisation. More people than ever were living in cities and will continue to do so. As populations grow and densification increases, demand for urban space becomes more competitive (Basmajian & Coutts, 2010). New urban forms are generated mainly through choices to build up or tear down, pave over or replant, renovate or abandon. Yet one of the very few urban landscapes to persist as is, is the cemetery.

Cemeteries are more than cities of the dead; they also meet the needs of the living. A wealth of evidence suggests that communities have appreciated the amenity value of cemeteries for some time (Dunk & Rugg, 1994; Worpole, 1997). However, land scarcity is beginning to raise questions about the viability of these spaces going forward (Allam, 2019; Rugg, 1998). This should warrant considerable attention, yet cemeteries remain surprisingly overlooked in city policy and planning (Basmajian & Coutts, 2010; Harvey, 2006). The knowledge held by historians, planners or even cemetery managers is likely to differ from intimate visitor experiences, a fact commonly overlooked in planning discourse (Beebeejaun, 2017; Hofmann et al., 2012). This paper maintains that in order to make informed planning decisions, more work is needed to explore urban cemeteries from a user perspective.

This paper focuses explicitly on disused urban cemeteries ('DUCs') in London and defines the DUC as *cemetery land, no longer used for new interments, within a metropolitan environment*. Whilst it is recognised that other types of burial spaces exist, such as modern municipal cemeteries, churchyards and crematoria, those sites, along with their related cultural and religious connotations, shall remain outside the scope of discussion. They are worth briefly defining, however, as a lack of clarity of these terms may stand in the way of planning considerations (Dunk & Rugg, 1994).

The terms 'cemetery,' 'graveyard' and 'churchyard' are often mistakenly interchanged (Dunk & Rugg, 1994). 'Churchyard' and 'graveyard' refer to small burial lots (one to two acres) of specialised religious affiliation. In the U.K., 'churchyards' are situated around church parishes (usually Anglican) and 'graveyards' serve denominations outside the established Church (i.e., Baptists or Quakers) (Dunk & Rugg, 1994). The cemetery, on the other hand, is typically larger in size and owned by a secular institution (Rugg, 2000). The primary role of a cemetery is to bury and

C. Rice (✉)
King's College London, London, UK
e-mail: corianne.rice@kcl.ac.uk

© The Author(s), under exclusive license to Springer Nature Switzerland AG 2022
C. Piselli et al. (eds.), *Innovating Strategies and Solutions for Urban Performance and Regeneration*,
Advances in Science, Technology & Innovation, https://doi.org/10.1007/978-3-030-98187-7_23

memorialise the deceased (White & Hodson, 2007). How-ever, many urban cemeteries, especially those in major metropolitan areas, reached capacity and therefore the end of their functional purpose long ago. Today, it is rare for these grounds to see new interments and for bereaved visitors to pay respects to those interred. Hence, this paper applies Dunk and Rugg's (1994, p. 14) definition of disused burial space as that '...in which there is no room for new burial, but where interments might still be taking place in existing family plots,' to form the definition of DUCs. It is DUC land in London that this paper will specifically focus on.

The U.K. has an uncommon policy that does not allow grave reuse after a certain time (Rugg, 1998). As a conse-quence, some of these spaces have persisted unchanged for decades, if not centuries (Barrett & Barrett, 2001). The impacts of this rule have generated questions regarding the 'wastefulness' of underutilised cemetery land (Lehrer, 1974, p. 197; Rugg, 1998). Apprehensions have also been raised over 'spatial justice' and whether the living are being unfairly affected by the choice positioning and permanence of cemetery space (Allam, 2019; Klaufus, 2016). As such, it is not uncommon for disused burial land to be cleared or redeveloped (Al-Akla et al., 2018; Bennett & Davies, 2015; Brown, 2013; Capels & Senville, 2006). In London, The Metropolitan Public Gardens Association (MPGA) con-verted almost 60 former burial grounds into parks and gar-dens by the turn of the twentieth century (Brown, 2013). Other, more recent examples of repurposed cemetery land include playgrounds in Berlin (Berlin Tourismus & Kon-gress GmbH, 2021), a metro station in Washington, DC (Schneider, 2020), a mega energy project in Johor (Borneo Post, 2012) and a mixed-use residential development in Perth (Warriner, 2019). Whilst these decisions can allow cities greater amenity space, revenue and oversight (Allam, 2019; Capels & Senville, 2006; Skår et al., 2018), such action can be viewed as authoritarian (Brown, 2013), impertinent (Allam, 2019; Cloke & Jones, 2004; Harvey, 2006) or prejudiced (Rivers, 2020; Schneider, 2020).

Public perceptions on these matters are not entirely clear; however, studies have shown that experts and the public often perceive landscapes differently (Al-Akla et al., 2018; Hofmann et al., 2012). Whilst U.K. planning authorities generally acknowledge the role of a cemetery as a heritage site and green asset (White & Hodson, 2007), I believe there are other interpretations that remain largely ignored, espe-cially of disused cemeteries within rapidly changing cities. The aim of this research is to (1) contribute to a growing body of knowledge on DUCs by investigating perceptions beyond the realm of grief, yet within the realm of urban change and (2) connect the findings with urban planning policies and practices through the following research questions:

– What function does the DUC provide for urban residents?
– What do users value about DUC experiences?
– What are DUC's defining features, as ascribed by those who use them?

This paper begins with a brief overview of the history of cemeteries in London, specifically Abney Park and its unique characteristics. It then analyses the cemetery and spatial theory, mainly through the works of Lefebvre, de Certeau and Foucault. Given the limited available literature on DUCs, this section will centre on cemeteries more broadly. The paper then introduces the conducted research methodology, analysis and limitations. Findings are then presented and discussed within the framework of the research objectives and implications for planning. Lastly, it concludes with a summary of thoughts and suggestions for future research.

2 Where People Become Places

London presents a compelling location in which to study DUCs. Ethnic, religious and socio-economic diversity cre-ates an appropriate backdrop for a multidimensional study of DUC users. There are over 120 cemeteries within a ten-mile radius of central London alone, covering an estimated 3,500 acres, or one-fifth of the total cemetery acreage in all of England (Meller & Parsons, 2011; White & Hodson, 2007). Seven of these are DUCs that have risen to such prominence; they have their own moniker, the 'Magnificent Seven' (Table 1). To my knowledge, no other city in the world has this type of DUC designation, making London a suitable place for their study.

Moreover, London has played a principal role in ceme-tery evolution, seeing how people have lived—and died—in this city for at least 5,000 years (Dunning, 2017). By the nineteenth century, rapid population growth and epidemics had placed enormous stress on the churchyard burial system, which had been in place since medieval times. These one-to-two-acre plots were unable to tolerate the increasing number of London's dead, resulting in unsightly conditions, vandalism, intrusive odour and contamination (Francis et al., 2005; Rugg, 2000; Scholz, 2017). As a result, the govern-ment banned burials within city limits, leading to the 'garden cemetery' movement of the 1820s–1850s (Brooks et al., 1989; Tarlow, 2000). In a time before large municipal parks, these sites were relocated outside the city center and designed as lavish gardens meant to attract and benefit the public. Some, such as Kensal Green and Highgate, served as blueprints for cemeteries elsewhere across England and overseas (Francis et al., 2005). Nonetheless, the grounds eventually fell victim to disrepair due to overcrowding,

Table 1 The magnificent seven

Cemetery name	Post code	Acreage	Founded	Active	Owner	'Friends' group
Abney Park	N16	32	1840	FP	London Borough of Hackney	Yes
Brompton	SW10	39	1840	Yes	The Crown (managed by The Royal Parks)	Yes
Highgate	N6	37	1839	Yes	The Friends of Highgate Cemetery Trust	Yes
Kensal Green	W10	72	1833	Yes	The General Cemetery Company	Yes
Nunhead	SE15	52	1840	No	London Borough of Southwark	Yes
Tower Hamlets	E3	31	1841	No	London Borough of Tower Hamlets	Yes
West Norwood	SE27	40	1836	C & FP	London Borough of Lambeth	Yes

Note C = Still operating for cremations only. FP = Current burial is available in existing family plots only
Source Dunk and Rugg (1994), Meller and Parsons (2011), The Royal Parks (2021)

reduced funding, vandalism, the aftermath of two world wars, bombings from one, and a growing predilection for cremation (Meller & Parsons, 2011). If not for 'friends' groups (conservation groups for individual cemeteries), many of them would have been demolished by their local councils for being unkempt and unsafe (Meller & Parsons, 2011).

Abney Park Cemetery, located in Hackney, East London, is one of the 'Magnificent Seven.' Having seen the better part of its interments occur before the turn of the twentieth century, a dwindling number of burials take place in its existing family plots today (Dunk & Rugg, 1994; Rugg & Pleace, 2011). The site is home to notable structures (Fig. 1) and a vast array of biodiversity, in addition to its 200,000-plus human remains (Abney Park Trust, 2019). As urban ecologist Matthew Gandy describes, the 'labyrinthine space' welcomes a wide array of visitors who 'peacefully coexist: artists, cruisers, dog walkers, drinkers, ecologists, joggers, lovers, mourners, photographers, poets, writers, and many others,' making the site fit to meet the outlined research objectives (Gandy, 2012, p. 727).

Located on 32 acres in Stoke Newington, in the borough of Hackney, Abney Park is distinct from the other Magnificent Seven in two ways. First, its grounds were never consecrated, so it did not have to pay the parish of the interred individual (Joyce, 1994). Second, Abney Park was not only built as a cemetery but also as an arboretum. It is located on the site of what was once two prominent estates, Abney House[1] and Fleetwood House.[2] The two great homes

boasted exceptional gardens, planned in part by the poet and hymn-writer Dr. Isaac Watts (born 1674), who lived in Abney House from 1734 until his death in 1748. Features of these two gardens were kept in the cemetery design and complemented with plantings from a nearby nursery, Loddiges of Hackney (Joyce, 1994).

William Hosking (1800–1861) built Abney Park's chapel, which is believed to be the first non-denominational chapel in all of Europe (Fig. 2) (Abney Park Trust, 2019; White & Stamper, 2018). Meanwhile, Egyptologist Joseph Bonomi Junior (1796–1878) designed the Egyptian style columns, office and gate piers at the east-facing entrance on Stoke Newington High Street (Meller & Parsons, 2011). The cemetery's only other entrance, a south-facing entry on Church Street, is where the gate to Abney House still stands (Fig. 3).

The cemetery opened to much success on 20 May 1840, interring over 5,000 people in its first 10 years of operation (Joyce, 1994). In fact, the grounds were effectively full by 1855, yet burials continued, surpassing 113,000 by 1908 (Abney Park Historical Tour, 2019, 4 August). By the mid-1900s, Abney Park, like its sibling cemeteries, began to struggle under increased management demands, cremation competition, diminishing space for new burials and dwindling revenue (Cloke & Jones, 2004). As pressures mounted, the site was effectively abandoned by its owners in 1972. The grounds fell victim to deterioration and vandalism until concerned members of the public formed the Save Abney Park Cemetery (SAPC) association in 1974 to raise awareness of the space's unique history and lobby for its preservation (Joyce, 1994). Then, in 1979, The London Borough of Hackney purchased the site for £1.00 (Miller, 2018). In the 1980s, Hackney and SAPC were able to afford the site certain protections by securing English Heritage designation for the grounds and Grade II listing for the architecturally and historically significant chapel and Egyptian entrance. Abney Park reached another milestone in 1993 when it was finally designated as the first Local Nature Reserve in Hackney (Gandy, 2012).

---

[1] Named after Mary Abney, wife to Thomas Abney, Lord Mayor of London (1700-01) and a founder of the Bank of England. Erected in 1700 and demolished in 1843 (The Conservation Studio, 2004).
[2] Named after Charles Fleetwood, son-in-law to Oliver Cromwell. Erected around 1634, and demolished in 1872 (Joyce, 1994; The Conservation Studio, 2004).

Fig. 1 Three of Abney Park's 12 listed monuments: Henry Richard (top left), Frank and Susannah Bostock (bottom left), William and Catherine Booth (bottom centre). London's only public statue of Dr. Isaac Watts (right). Photographs by the author

Fig. 2 Hosking's Gothic-style chapel. The chapel was never consecrated and operated only for funerals. It is Hosking's lone surviving creation (*Source* Abney Park Trust, 2019). Photographs by the author

Fig. 3 Abney Park's east main entrance, on Stoke Newington High Street, is flanked by the Egyptian columns (left), and the southern entrance on Church Street still secured by the original gate of Abney House (right). Photographs by the author

3 Nature Unleashed

Although Abney Park was the only one of the Magnificent Seven to boast an arboretum, it is not unique as an urban green space. As these sites were once created on town peripheries, DUCs have become fragments of countryside trapped by urban expansion (Fig. 4) (Cloke & Jones, 2004; Meller & Parsons, 2011; White & Hodson, 2007). Consequently, DUCs have not only turned into biodiversity hotspots (Barrett & Barrett, 2001; Harvey, 2006), but have also been linked to urban green effects such as decreased air pollution, noise reduction and temperature regulation (Hofmann et al., 2012).

However, what makes the DUC so unique, indeed striking to most, is its 'wild' natural atmosphere. Due to a multitude of factors mentioned previously, cemeteries from this era are known today for their 'romantic decay' (Joyce, 1994, p. 63; see also Brooks et al., 1989; Dunk & Rugg, 1994; Worpole, 1997; Cloke & Jones, 2004; Meller & Parsons, 2011). Abney Park is no different, almost completely overgrown by vegetation, resulting in a rural city space which Gandy describes as a hybrid 'urban nature' (2012, p. 730). Disorder is enhanced by the deterioration of the Victorian monuments, whose elaborate designs prove challenging to maintain and are further threatened by dense overgrowth, unchecked root systems and falling trees (Fig. 5).

This disorder has similarities to the marginal urban space concept of 'terrain vague' (Mariani & Barron, 2013), urban 'wastelands' (Mabey, 2010), interstitial space (Lévesque, 2009; Wright, 1991) and informal urban greenspace (Rupprecht & Byrne, 2014). Despite arguments that this type of unstructured urban vegetation provides important ecological

benefits, it is not always seen as a positive social or aesthetic contribution (Del Tredici, 2010). This was echoed in a 2012 study in Kuala Lumpur, where researchers found many Malays avoided public cemeteries due to a melancholy atmosphere they felt was the consequence of neglect (Afla & Reza, 2012). Similarly, Al-Akla et al. (2018) reported cemeteries in Beirut were perceived more positively when they exhibited the organised approach of planned, decorative landscapes. Alternatively, some visitors in the U.K. prefer the unpredictability of the 'alive' and 'wild' to the 'sterile' and 'boring' attributes of formal parks and cemeteries (Francis et al., 2005, p. 208). Differing interpretations are likely influenced by cultural and religious idiosyncrasies. For example, poltergeist superstitions heighten the wariness of overgrown cemeteries in Malaysia (Afla & Reza, 2012). Whilst differences across religions and cultures are outside the scope of this discussion, they would certainly be worth exploring in the future as they are likely to impact cemetery perceptions.

The preferred level of order may also reflect how some understand humankind's relationship to nature. In the eighteenth century, the English developed a theory of spatial relationships that depicted humans and nature as two opposing forces; contending for power in the sublime, the picturesque and the pastoral (Sloane, 1991). The sublime was wild nature, left unmarked by human influence; the picturesque represented a balance of nature and art; and the pastoral subjected nature to humankind (Sloane, 1991). Public parks and even modern cemeteries are generally pastoral, with shrubs, grass and trees representing a predetermined framework (Francaviglia, 1971). De Certeau (1984) extended this argument, claiming the evident planning of the pastoral demands a deliberate mapping of space that can only be undertaken by those in the position of

Fig. 4 Aerial view of three London DUCs: Abney Park Cemetery N16 (left), Tower Hamlets Cemetery E3 (upper right) and Nunhead Cemetery SE15 (lower right) (*Source* Google Maps, 2019)

Fig. 5 Overgrown vegetation and felled trees within Abney Park which can damage and destabilise memorials. Photographs by the author

power. Contemporary green spaces may thus reveal a range of social responses to urban nature, ranging from the unstructured appropriation of the sublime to the controlling discourse of the pastoral (Gandy, 2012).

Of course, this paper does not argue that all overgrown DUCs represent a repudiation of institutional control. In fact, deterioration is a natural expectation from a site that has been exposed to the elements for almost two centuries. However, the extent of unregulated urban space is an unusual experience (Carmona, 2014) that would be considered unacceptable for someone's garden, a park or other green space (Francis et al., 2005). It is because of this that we must consider Gandy's (2012) claim that the exception for disorder conjures interesting parallels with how the modern city is regulated.

4 Cemeteries and Spatial Theory

The presence of wild urban nature lies in tension with many other spaces in the contemporary city, such as brightly lit open squares, shopping centres and manicured parks (Deering, 2010; Gandy, 2012). Concern that these spaces are becoming increasingly segregated has resurfaced in Lefebvre's debates about the 'rights to the city' (Beebeejaun, 2017) and has evoked criticism for 'privatised, exclusionary, architecturally deterministic, over-designed, and…cheap,' places in London (Carmona, 2015, p. 376) that may restrict non-normative behaviour (Gandy, 2012). As a result, some subcultures might turn away from public squares, surveilled and stripped of vegetative undergrowth for example, to less-governed sites such as the DUC. Gandy (2012) reported such findings when he discovered some visitors use Abney Park as a cruising ground. Meanwhile, Deering (2010) uncovered that teenagers in Paris and southern England utilise cemetery space to engage in 'unofficial' behaviour such as 'drinking, having sex and creating general disturbance.' In a city full of overly administered, uniform space, these acts of 'purple recreation,'[3] and the places in which they can occur may offer a sense of sovereignty and belonging (Deering, 2010).

This concept again resonates with de Certeau (1984), to whom regular urban citizens wield the power to resist discursive patterns of authority. As people move throughout everyday space, particularly through the act of walking, they can imprint their own personal meaning onto their surroundings, allowing each individual a temporary sense of control (Beebeejaun, 2017; de Certeau, 1984). Thus, public space like the DUC can be a site of contested meaning,

constantly redefined and appropriated (Skår et al., 2018). Both de Certeau and Lefebvre's theories are challenged, however, by a narrow Marxist, anti-capitalist agenda (Purcell, 2002). The prevailing literature on these two concepts overlooks spatial strategy for some underrepresented groups who may lack safe and equal access to public space (Beebeejaun, 2017; Purcell, 2002). There are also questions related to the role of other non-human agents, such as the technology-enabled city (Amin, 2015) or plants and wildlife (Cloke & Jones, 2004). Consideration of these factors results in a more complicated nexus of how urban rights are negotiated or exercised.

The complex nature of spatial actors and equity conjures the concept of heterotopia, a term coined by Foucault (1967), and defined as space that exists across all societies, but in a realm beyond what one might call 'normal' (Lee, 2015). As cemeteries, particularly DUCs, have distinguishing features that often defy 'normal' societal presumptions, it is easy to see why some researchers (Åsdam, 1995; Clements, 2017; Gandy, 2012; Johnson, 2008; Lee, 2015) and Foucault himself (1967), chose to address the cemetery through a heterotopic lens (Table 2). It both reflects established norms like respect for the dead (Francis et al., 2005; Nordh et al., 2017; Rugg, 2000; Skår et al., 2018), but embodies opposition in the form of escape (Deering, 2010; Gandy, 2012; Clements, 2017). It breaks the traditional sense of time, offering a sense of permanence whilst the city is in constant flux (Davies & Bennett, 2016; Rugg, 2000; Woodthorpe, 2011; Worpole, 1997). The cemetery is a microcosm within the city that offers a public place for solitude (Mytum, 2000; Clements, 2017). Moreover, it accommodates an abundance of life amidst death (Barrett & Barrett, 2001; Gandy, 2018; Miller, 2018).

However, I argue the obscure nature of the heterotopia which perpetuates the cemetery's liminal status that often goes undetected by planning experts. Furthermore, the concept is so broad that it can be, and has been, applied to many types of social space—brothels, asylums, even ships (Johnson, 2013). Whilst the term is a creative way of capturing its multifaceted sentiments, I believe clarity is needed, so DUCs can be better represented in the more concrete discourse of planning.

That is not to say there have never been concerted efforts to bring Britain's historic cemeteries into planning conversations (see Jones, 1979; Brooks et al., 1989; Ernst, 1991; Curl, 2000; Mytum, 2000). However, the study of death spaces remains a growing discipline with very little research specific to DUCs. The main exception is the work by Julie Dunk and Julie Rugg (1994), who were the first to define disused burial land and conduct an extensive survey of its prevalence in the early 1990s. What they found by speaking to cemetery managers and authorities was that individuals visited cemeteries for four main purposes: (1) history;

---

[3] This term was coined by Joseph E. Curtis (1979) and refers to forms of recreation that may be seen as deviant (i.e., sex, drinking, drugs).

Table 2 Foucault's six principles of the heterotopia

No.	Description
1	Exist across all cultures but in multiple variations
2	Are not static and provide different functions as cultures evolve
3	Possess concurrent elements that would seemingly not coexist
4	Operate outside normal spatio-temporal regularities
5	Enforce their own unique conventions
6	Embody illusory or compensatory reflections to other space

Note Adopted from Foucault (1967), Lee (2015)

(2) nature; (3) education and (4) passive recreation (Dunk & Rugg, 1994). However, their research, whilst certainly additive, was generally limited in its approach by relying upon indirect accounts of visitation habits and not investigating visitor experiences directly. Considering the inactive status of the DUC, the spatial critiques of the urban built environment and the scrutiny on cemeteries due to land scarcity, I believe research is needed to see if Dunk and Rugg's (1994) reasons still hold true.

Therefore, I employed an exploratory case study by conducting semi-structured interviews with Abney Park visitors. Abney Park Cemetery was chosen for its categorisation as one of the Magnificent Seven. It was also selected because of its location, which reflects a rapidly changing urban environment. Hackney, on the east side of London, is a borough with socio-economic indicators of poverty and deprivation (Francis et al., 2005). However, the borough is also experiencing accelerating gentrification, as evidenced in Abney Park's neighbourhood of Stoke Newington (Moran, 2007). In this way, the case study site was relevant in achieving the research objectives of examining how people interpret a multifaceted DUC within the realm of precipitous urban change.

5 Unearthing DUC Experiences

Fieldwork was conducted over a six-week period from 22 June to 4 August 2019. During this time span, I performed participant and site observations, conducted interviews and participated in three site-sponsored events. A breakdown of activities is shown in Table 3. Seventeen interviews were conducted with individuals and groups, resulting in a sample size of 26 people. Responses have been anonymised with numbers in place of names throughout the remainder of this paper (R1 = respondent 1). The selection criteria for participants were that they were (1) within the case study site, (2) over the age of 16 and (3) did not appear to be grieving or otherwise preoccupied.

Semi-structured interviews enabled participants to recount their own detailed narratives whilst granting the flexibility to modify questions if needed. I conducted

interviews within the study site to allow for an immersive active recount of surroundings (Rupprecht & Byrne, 2014), using an interview guide to establish the commonality of responses. Participants were generally asked a series of 18 questions, but at times probes were inserted, if appropriate (see Bernard & Ryan, 2010, pp. 31–33). Questions were purposefully selected to avoid sensitive information such as income, political beliefs and thoughts specific to 'purple recreation' (Curtis, 1979). The omission of these topics was to adhere to the minimal ethical risk approval granted for this research. However, if the participant initiated any of those topics, they were captured in the dialogue. Interviews were recorded, so more attention could be given to the nature of the conversation, non-verbal cues and overall demeanour, whilst capturing verbatim quotes via transcription. Participants were assured of their anonymity and consent to be recorded was sought verbally with the assurance the interview could be terminated at any point.

An important goal for analysis was to search for the essence of a pattern, guided by the objective of identifying DUC functions and features that affected how urban users valued the space. Primary data were analysed using a combination of thematic analysis and systematic text condensation as outlined by Braun and Clarke (2006) and Malterud (2012), respectively. In line with the objectives of this study, this approach was chosen for its flexibility and ability to provide rich data from which thematic findings could be shared (Braun & Clarke, 2006; Raimbault & Dubois, 2005). The analysis encompassed three steps: (1) acclimating to the data, (2) identifying and organising codes and themes and (3) recontextualising and defining themes.

Whilst I made every attempt at following the steps above, I was constrained in part by some limitations. Most site visits occurred on weekends and a convenience sampling technique may have decreased the reliability and representativeness of findings (Bernard & Ryan, 2010). A questionnaire would have allowed for a more representative sample. However, I felt the richness in data acquired through semi-structured interviews outweighed representativeness. Lastly, it is hard to deny the active role I played in identifying codes and themes (Braun & Clarke, 2006). As there was no intercoder reliability, the analysis would have

Table 3 On-site fieldwork activity

Date	Day	Activity	Participants
21 June	Friday	User group meeting	APUG
22 June	Saturday	Interviews	R1, R2, R3, R4a, R4b, R5a, R5b
23 June	Sunday	Observation	N/A
24 June	Monday	Interviews	R6, R7
27 June	Thursday	Observation	N/A
28 June	Friday	Interviews	R8a, R8b, R9, R10a, R10b, R11
12 July	Friday	Interviews	R12a, R12b
13 July	Saturday	Litter pick, Interviews	APUG, R13, R14, R15
20 July	Saturday	Interview	R16
3 August	Saturday	Observation	N/A
4 August	Sunday	Historical tour, Interview	R17a, R17b, R17c, R17d, R17e

Note APUG = Abney Park User Group. In instances when an interview involved more than one person, these individuals were catalogued as R#a, R#b, and so on

benefited by having an additional researcher to see if data would have been coded in a similar way (Bernard & Ryan, 2010; Malterud, 2012).

Of the 26 participants, most were White (76%) and 25–34 years of age (42%) (Table 4). The sample included slightly more males (16) than females (10). Ten individuals were visiting the site for the first time, whereas 16 had visited previously. Of these 16 individuals, 5 were classified as intermittent visitors (reported more than one but less than ten unique visits) and 11 were classified as regular users (more than ten unique visits). Of the sample, six were lone visitors (R2, R9, R11, R14, R15 and R16) and four were

Table 4 Participant profile summary (N = 26)

Variable	Category	Number	Total %
Age group	16–24 Years	6	23.07
	25–34 Years	11	42.30
	35–44 Years	3	11.54
	45–54 Years	1	3.85
	55–64 Years	1	3.85
	65–74 Years	3	11.54
	>75 Years	0	0
	Unknown	1	3.85
Gender	Male	16	61.54
	Female	10	38.46
Religion	None	11	42.31
	Christian	8	30.76
	Spiritual	2	7.69
	Atheist	1	3.85
	Agnostic	2	7.69
	Hindu	1	3.85
	Unknown	1	3.85
Visitation	First Time	10	38.46
Frequency	Intermittent	5	19.23
	Regular	11	42.31

lone visitors with dog(s) (R1, R3, R6 and R7). Five interviews were of pairs (R4a&b, R5a&b, R8a&b, R10a&b and R12a&b), one interview (R13) was of a parent with two children (not interviewed) and another was an interview of five peers (R17a–e).

Whilst some mentioned they would like to see certain structures or monuments within the site, none said the purpose of their visit was to see one specific grave or pay respects to a family plot. When initially asked what brought them to the site, many gave reasons such as 'get out of the house' (R13), 'walk the dog' (R7) or 'go for a wander' (R5a).

6   Extemporaneity

The primary theme garnered from user accounts is that the DUC is valued for its ability to provide a break from the monotony of the surrounding cityscape. I labelled this theme as extemporaneity, which in some ways supports, and in other ways, contradicts existing theories on how cemetery space is interpreted. Defined as a departure from the artificiality and predictability of the man-made, the extemporaneity function was made up of three features: wild nature, intrigue and authenticity. A common thread is a repudiation of the controlled or artificial with the untouched nature of Abney Park; a topic of repeated focus.

6.1   Wild Nature

This feature was evident in almost every interview, depicted as nature's ability to grow unimpeded, which was always portrayed in a positive light. Thirty-six codes such as 'wild,' 'wilderness,' 'overgrown,' 'native' and 'dishevelled' were used to describe the atmosphere. This sentiment was embodied by both first-time users, such as R5b who expressed, 'I like the wild and the nature,' as well as regular users, like R1, who stated, 'I want it to be like this, it feels wild and that's why I enjoy it.'

It was generally acknowledged that the site was not entirely abandoned, but cared for within a commonly understood, yet unspoken margin of acceptability; for instance, 'it's obviously tended to a certain extent, but I think it's quite nice, letting things be how they are' (R8b). In addition, 'I'm sure it's being maintained, but in a way that lets it be a bit wild and I think that's about perfect' (R10b). The figurative line would be crossed, it seemed, if the place was tidied too much that it began to resemble a modern municipal cemetery or other public parks, as expressed here:

As compared to Hyde Park, that is so manicured, it feels artificial, and here, it's well kept, but there are things that seem more wild, which is cool. (R17a)

This implied that users preferred the unruly atmosphere but did not necessarily attribute it to a lack of oversight. In these instances, participants seemed aware that management existed, but respected that it did not exert too much human manipulation. An important distinction here is that users were careful to not confuse wild with messy or careless, often citing that rubbish, dog poo or vandalism were unwelcome. This supports previous research, which found users preferred more natural settings but also want to see evidence of upkeep (Özgüner & Kendle, 2006). Although there is some allusion to a reversal of authority, there is not enough evidence to presume it was the underlying cause for this preference. In general, participants seemed to enjoy the naturalised landscape for its own sake and appreciated the contrast to the urban surroundings. Findings that untamed nature could be interpreted as menacing or scary (Özgüner & Kendle, 2006) were not fully supported in this study. The one participant who found the site 'gloomy' (R11) seemed to associate that feeling with the fact that the site was a cemetery, not because it was overgrown.

6.2   Intrigue

Many would mention how 'weird' (R1, R10a, R13 and R14) or 'strange' (R3, R5b and R8a) it was that they enjoyed the space. Whilst most attributed their own predilections more towards 'the inexplicable' rather than 'the pathological' (Taylor & Ussher, 2001, p. 310), they still alluded to a feeling outside cultural norms. These comments reflect Young and Light's (2016) observation that despite the many ways in which people utilise cemeteries, they are still categorised as 'alternative' spaces in academic and common discourse. To foster acceptance as commonplace, the authors recommend addressing, if not entirely removing 'the stigma of freakiness' from cemeteries and other death spaces (Young & Light, 2016, p. 64). Whilst this may be said for people who choose to avoid such sites, this sentiment was not entirely echoed by participants in this study. Many visitors acknowledged the DUC's appeal despite the associated stigma. However, an interesting pattern emerged with visitation frequency, suggesting that as the DUC becomes less of a novelty, the more it fosters a sense of connection between users and community. For example, first and intermittent users would speak of a shared sentimentality, like R10b, who brought first-time visitor R10a to the site in order to experience something special together. In other examples, participants would speak of shared stories derived from monument inscriptions, like R4b, who thought it was 'beautiful' that she could 'be a little part' of the community by reading about those buried there.

On the other hand, regular users would express a deeper sense of connectivity through networks and belonging. R12a and R12b claimed the DUC had been serving as their 'meeting place' for the last 25 years. On that day, it was the two of them, but at other times 'there might be four or five' (R12b). Unlike other regular users who live nearby, they reach the DUC by bus. Their commute takes about 20 min, but is worth it, as they 'prefer this to open parks' (R12a).

Other regular users continued to reference community. R1 had moved to London six years earlier and, like R4b, felt she belonged to the 'area by reading the names of the people who used to live here.' From R15's designated bench, he could look out for his friends he visited daily, declaring, 'we know everyone.' R6 enjoyed interacting with the 'whole network of dogwalkers within.' R7 appreciated meeting 'interesting people,' like a birdwatcher willing to share knowledge. Similarly, R16 prized the 'good friendly environment' and claimed to have met 'the most amazing people' within the site. These accounts demonstrate that whilst users were aware of the heterotopic relationship of the DUC, it did not seem to weaken its appeal. Participants also did not express experiencing this type of captivation in other city spaces. These findings support studies that suggest burial

grounds can promote the formation of a shared cultural memory, which can strengthen identity and belonging, especially in diverse urban communities (Bennett & Davies, 2015; Francis et al., 2005).

To others, the site was 'unlike anything else nearby' (R10b) with many 'surprising' and 'random' qualities you 'rarely find,' such as the growth of wild garlic (R13). Fundamental to this fascination were the site's tall trees, mentioned by half of the participants. R16 gave an illustrative example of how trees dominate the space, as in: 'What I especially like about it, these trees are so tall and high, it's almost like being in nature's cathedral.'

In many narratives, the trees took on lives of their own, helped in part by the informative placards located throughout the site (Fig. 6). The placards spoke of the trees as veterans, victims and survivors, painting them as active personalities within the space. This supports Cloke and Jones' (2004) assertion that like humans, trees can play an active role in ecosystems, thereby helping construct a social place. Previous literature has also shown that tree height is a significant consideration in human fascination (Hofmann et al., 2012). Thus, it appears that the veteran trees in Abney Park not only captivated visitors because of their contrast to the surrounding cityscape but also helped build a connection between the users and the site.

'Fascinating' (R5a) monuments were also mentioned as features that one does not usually find in a 'regular

Fig. 6 Light shining through the trees in Abney Park (left). Example of one of the many placards throughout the site to provide information about plant life (middle). Face carved into one of the trees (right) (*Photos* Author)

cemetery,' (R9). Those too were employed as tools to maintain users' interest, inspire imagination or stave off boredom:

> Sometimes I just walk around and read the things. If I'm bored sometimes, I walk around and find someone who has my date of birth. I like that. I will literally find someone who matches my birthday. (R2)

Like R2 above, R5a was engaging with the monuments in an arguably playful way. She had been told of the Bostock family monument by an artist a few days earlier. During our interview, she admitted she had enlisted her friend on her 'mission' to find it. Past literature has mentioned the role that monuments play in fostering legacy and connection (The Builder, 1843; Mytum, 2000; Rugg, 2000; Burk, 2003); however, it was evident here that they played a notable part in users' curiosity and fascination, as well.

6.3 Authenticity

The third extemporaneity feature was authenticity, which I define as local and timeless. Users appreciated the DUC for being true to its history and setting, for example, 'It's got a real Hackney vibe to it' (R6), whilst remaining apart from sites that are over-engineered. Here it was less about nature, but more about the regularity of the DUC, even though as we have seen, the site is far from ordinary. Still, there was an appreciation for locals moving in and out as they conducted their daily routines. The site was occasionally referred to as a shortcut, suggesting that those familiar with the area, consciously engage with the space. As one participant admitted, she uses the site whilst running errands as 'a way to give my kids a bit of free space before I go' (R13). She imagined a lot of other people in the area did the same. This was echoed by R11 who stated, 'I think what some people do, is they use it as a shortcut to get from A to B if they know the area' (R11).

Authenticity was also expressed as 'keeping with the style' (R8b). Typically, this was articulated when considering modern amenities, which were met with mixed reactions. R16, for example, thought it would be reasonable to add a café around the chapel to capitalise on its beauty and bring people together. Similarly, R7 thought 'even if [the chapel] had a little café and some chairs outside…that would be nice,' making sure to clarify 'nothing much more, and certainly nothing exclusive,' given the site's nondenominational history. After some deliberation, R8b decided the site could benefit from a few more benches or seating areas, as long as they fit the 'style' of the grounds. Whilst these statements draw parallels to cheap and commercial public space (Carmona, 2015), it appears users were more concerned with staying true to the site's legacy, where privatisation, commercialism and segregation were absent.

6.4 Extemporaneity Summary

Overgrown vegetation is the core component of the extemporaneity function, with almost all users referring to it in a positive light. This renunciation of the controlled seems to provide a visual and mental break from the monotony of the urban landscape (Harvey, 2006). This supports previous studies in that untouched nature can not only act as a sanctuary for many uncommon species of plants and animals, but also for the person seeking respite from the hectic pace of the city (Hartig & Staats, 2006; Shahhoseini et al., 2015).

We must also consider ecological and psychological theories behind visual landscape preference, and how they may account for DUC perceptions. In the city of Tabriz, Iran, Shahhoseini et al. (2015) examined almost 400 responses of visual preference in small urban parks using components from information-processing and prospect-refuge theories. They found that the most preferred spatial arrangements met the criteria of a mystery first, followed by coherence, refuge and complexity (Shahhoseini et al., 2015). This meant that visitors in their study preferred small urban parks that piqued their curiosity with partially concealed areas, a variety of stimuli, winding paths, and densely vegetated, layered spaces, much like the DUC. They also found that visitors preferred wide open, easily legible spaces the least. Whilst there are many parallels with their findings and those I gathered from the DUC, there are also some obvious differences in terms of location, population size and cultural heterogeneity. As such, it would be interesting in the future to explore the extent that once again culture has in landscape preference, but also the role that psychological and ecological theories like prospect and refuge play in DUC perceptions.

Regardless, more consideration should be given to plant diversity and the allowance for reseeding and growth with minimal maintenance. Natural diversity has been associated with human captivation (Ratcliffe et al., 2013). As Del Tredici (2010) has shown, this has important ecological benefits for the surrounding city. The presence of veteran trees can offer intrigue as well as a home for wildlife (Gandy, 2018). Even dead or dying trees, normally cleared in other spaces, should be left in situ, for the shade, shelter and nutrients deadwood can provide to plant and animal life (Gandy, 2018; Lehvävirta & Rita, 2002). Along with monuments, wooden debris can also act as barriers to human wear, aiding natural plant regeneration (Lehvävirta & Rita, 2002). The only challenge foreseen here is potential safety hazards posed by unstable monuments or weakened trees. Abney Park, for example, closes during inclement weather to reduce the risk of injury from wayward branches or destabilised monuments (Abney Park Trust, 2019). If planners wish to follow a similar approach, they would need to determine the suitability and trade-offs of doing so.

The fact that users also employed creative ways of site interaction (i.e., seeking out specific monuments) should urge planners to adopt the informational tactics employed by Abney Park. It should also call into question suggestions for greater use of promotional material (Dunk & Rugg, 1994; Revelle, 1967), as these might take away some of the site's enigmatic character. More attention should be paid to authenticity by 'friends' groups and others responsible for promoting such places, as there seems to be a fine balance as to when a space feels genuine and when it feels manufactured. Insight can likely be gleaned from studies of 'dark tourism,' which was not a focus of this research, as such instances typically represent a one-time interaction between visitors and spaces (Tomašević, 2018). Still, future studies could benefit from comparing experiences of Abney Park with other DUCs that charge for entry like Highgate or the historic cemeteries that have become major tourist attractions like in Savannah, Georgia or Père la Chaise in Paris (Capels & Senville, 2006). In instances where cemetery redevelopment may be unavoidable, planners should consider ways to save or reuse monuments, and contemplate more subtle ways of user engagement. However, part of what DUC users enjoy is the air of mystery, hence there seems room to include less overt engagement tactics in public space than for instance, practices of the playful city (Donoff & Bridgman, 2017).

7 Disused Not Disregarded

DUCs represent complex landscapes that have long been recognised as providers of social, environmental and recreational fulfilment. Despite this, they are still marginally understood and rarely incorporated into urban planning conversations. Furthermore, users' direct experiences are not always included in planning decisions. This study bridges this gap by investigating perceptions of DUCs and identifying the qualities that impact how urban users value this space.

I found that the DUC is still a highly appreciated land use that mainly provides urban dwellers with the ability to break away from the scripted. As expected, DUC perceptions reflect and expand prior research on cemetery space. Dunk and Rugg's (1994) report on the management of old cemeteries listed four reasons people visit the cemetery: historical, ecological, education and leisure. Over a decade later, their findings were reflected by White and Hodson (2007), who contended historic cemeteries are valued for their architecture, landscaping, ecology and leisure. Remnants of these functions are echoed in my findings; however, my research

suggests that today's DUC users value these sites for more active purposes. Features such as history, landscape and architecture, whilst important, are inherent to old cemetery sites, by virtue of when, where and how they were built. The findings of my research suggest that users do not perceive these features as primary reasons for visiting the DUC per se, rather they view parts of them, such as the overgrown, as a meaningful reflection of the built environment in which they are situated.

In addition to extemporaneity, other themes that emerged in this research were that users enjoyed the DUC for control, connection and respite. Specifically, participants expressed enjoyment in deciding with whom to speak, where to walk and for how long, reflecting the de Certeau-esque concept that DUC users look to reclaim some level of autonomy. The DUC should also be considered as a space that invites users to connect to others with whom they relate, both living and dead. This is notable since rapidly changing urban environments can isolate inhabitants from their surroundings as well as from each other (Sloane, 1991). Furthermore, DUCs should be studied in association with mental health, given the evidence that natural environments (Hartig, 2004; Nordh, et al., 2017), biological diversity (Fuller et al., 2007) and landscape mystery (Shahhoseini et al., 2015) can encourage mental restoration.

In terms of additional research, I argue there is more to learn about the relationship between property values and DUCs. Whilst residents commonly object to new cemetery development over concerns of falling property prices, there is evidence that some gentrified areas in London, such as Highgate, Chelsea and Hampstead, happen to have sizeable amounts of burial land (Allam, 2019; Anderson & West, 2006; Capels & Senville, 2006; Moran, 2007; Steinmetz-Wood et al., 2017). I suspect there is further delineation between urban property values near active cemeteries and those near older DUCs.

This research suggests urban dwellers interpret DUC space differently than other public spaces, and in densely populated cities, there is something to be said for having as many types of public spaces as possible. Cemeteries should therefore be brought into mainstream planning conversations, with DUCs—and their users—given particular attention. Despite pressures on burial practices and cemetery land use, this study implies the DUC is becoming more appreciated for the experiences users can achieve within their grounds that arguably cannot be instilled elsewhere. Therefore, I urge authorities to pay closer attention to DUCs, and who visits them, so we continue to learn as much as we can about these unique places, use that knowledge to shape the future of urban planning and prevent disused urban cemeteries from becoming disregarded.

Acknowledgements Many thanks are owed to Professor Phil Hubbard, the Right Reverend Dr. Stephen Venner, Ed Venner and Alphege Bell. I would also like to thank Seeley Jennings, Superintendent of New Haven's Grove Street Cemetery, as well as the Abney Park User Group, The Abney Park Trust, and the visitors who participated in this research. I am also grateful to Julia Sequeira and the anonymous reviewers of IEREK Press, whose comments were invaluable in helping improve the quality of this report. Lastly, thanks to Andrew Carr for introducing me to Abney Park Cemetery.

References

Abney Park Trust. (2019). *History of Abney Park* [online]. Retrieved April 18, 2021, from https://abneypark.org/

Afla, M., & Reza, M. (2012). Sustainability of urban cemeteries and the transformation of Malay burial practices in Kuala Lumpur metropolitan region, World Academy of Science. *Engineering and Technology, 71*, 808-829.

Al-Akla, N., Nasser Karaan, E., Al-Zein, M., & Assaad, S. (2018). The landscape of urban cemeteries in Beirut: Perceptions and preferences. *Urban Forestry & Urban Greening, 33*, 66–74.

Allam, Z. (2019). The city of the living or the dead: On the ethics and morality of land use for graveyards in a rapidly urbanised world. *Land Use Policy, 87*, In Press.

Amin, A. (2015). Animated space. *Public Culture, 27*(2), 239–258 [Online]. Retrieved August 03, 2019, from https://doi.org/10.1215/08992363-2841844

Anderson, S., & West, S. (2006). Open space, residential property values, and spatial context. *Regional Science and Urban Economics, 36*, 773–789.

Åsdam, K. (1995) *Heterotopia: Art, pornography and cemeteries*, translated into English by D. A. Marmorstein [Online]. Retrieved June 11, 2019, from http://www.heterotopiastudies.com/wp-content/uploads/2015/06/knut-asdam-article-pdf1.pdf

Barrett, G. W., & Barrett, T. L. (2001). Cemeteries as repositories of natural and cultural diversity. *Conservation Biology, 15*, 1820–1824.

Basmajian, C., & Coutts, C. (2010). Planning for the disposal of the dead. *Journal of the American Planning Association, 76*(3), 305–317.

Beebeejaun, Y. (2017). Gender, urban space, and the right to everyday life. *Journal of Urban Affairs, 39*(3), 323–334.

Bennett, G., & Davies, P. J. (2015). Urban cemetery planning and the conflicting role of local and regional interests. *Land Use Policy, 42*, 450–459.

Berlin Tourismus & Kongress GmbH. (2021). *Leise Park*. Visit Berlin [Online]. Retrieved April 11, 2021, from https://www.visitberlin.de/en/leise-park

Bernard, H., & Ryan, G. (2010). *Analyzing qualitative data: Systematic approaches*. SAGE Publications Inc.

Borneo Post Online. (2012). Over 1,500 Muslim graves in Pengerang to be relocated due to RAPID project. *The Borneo Post* [Online]. Retrieved April 11, 2021, from https://www.theborneopost.com/2012/09/12/over-1500-muslim-graves-in-pengerang-to-be-relocated-due-to-rapid-project/

Braun, V., & Clarke, V. (2006). Using thematic analysis in psychology. *Qualitative Research in Psychology, 3*(2), 77–101.

Brooks, C., Elliot, B., Litten, J., Robinson, E., Robinson, R., & Temple, P. (1989). *Mortal remains: The history and present state of the Victorian and Edwardian cemetery*. Wheaton Publishers.

Brown, T. (2013). The making of urban 'Healtheries': The transformation of cemeteries and burial grounds in Late-Victorian East London. *Journal of Historical Geography, 42*, 12–23.

Builder, The. (1843). Miscellanea. *The Builder, 4*, 51.

Burk, A. L. (2003). Private griefs, public places. *Political Geography, 22*(3), 317–333.

Capels, V., & Senville, W. (2006). Planning for cemeteries. *Planning Commissioners Journal, 54*, 3–10.

Carmona, M. (2014). The place-shaping continuum: A theory of urban design process. *Journal of Urban Design, 19*(1), 2–36.

Carmona, M. (2015). Re-theorising contemporary public space: A new narrative and a new normative. *Journal of Urbanism: International Research on Placemaking and Urban Sustainability, 8*(4), 373–405.

Clements, P. (2017). Highgate cemetery Heterotopia: A creative counterpublic space. *Space and Culture, 20*(4), 470–484.

Cloke, P., & Jones, O. (2004). Turning in the graveyard: Trees and the hybrid geographies of dwelling, monitoring and resistance in a Bristol cemetery. *Cultural Geographies, 11*(3), 313–341.

The Conservation Studio. (2004). *Stoke Newington conservation area appraisal*. The Conservation and Design Team [Online]. Retrieved May 26, 2021, from https://drive.google.com/file/d/120j3jfzP35IksWsTEcJmgRnRxJVBizGy/view

Curl, J. S. (2000). *The Victorian celebration of death*. Sutton Publishing.

Curtis, J. E. (1979). *Recreation: Theory and practice*. The C.V. Mosby Company.

Davies, P. J., & Bennett, G. (2016). Planning, provision and perpetuity of deathscapes—Past and future trends and the impact for city planners. *Land Use Policy, 55*, 98–107.

de Certeau, M. (1984). *The practice of everyday life*. The University of California Press [Online]. ProQuest Ebook Central. Retrieved August 24, 2019, from https://ebookcentral.proquest.com/lib/kcl/detail.action?docID=922939

Deering, B. (2010) From anti-social behaviour to X-rated: Exploring social diversity and conflict in the cemetery. In A. Maddrell & J. D. Sidaway (Eds.), *Deathscapes: Spaces for death, dying, mourning and remembrance* (pp. 75–93). Routledge, 2016.

Del Tredici, P. (2010). Spontaneous urban vegetation: Reflections of change in a globalized world. *Nature and Culture, 5*(3), 299–315.

Donoff, G., & Bridgman, R. (2017). The playful city: Constructing a typology for urban design interventions. *International Journal of Play, 6*(3), 294–307.

Dunk, J., & Rugg, J. (1994). *The management of old cemetery land: Now and the future*. Shaw & Sons.

Dunning, H. (2017). A history of burial in London. *Natural History Museum* [online]. Retrieved April 19, 2021, from http://www.nhm.ac.uk/discover/a-history-of-burial-in-london.html

Ernst, J. (1991). Land for the living?—The land use and conservation of urban cemeteries and churchyards. *Local Government Policy Making, 17*, 14–21.

Foucault, M. (1967). *Of other spaces: Utopias and Heterotopias, Architecture/Mouvement/Continuité*, translated into English by J. Miskowiec [online]. Retrieved May 26, 2021, from http://web.mit.edu/allanmc/www/foucault1.pdf

Francaviglia, R. (1971). The cemetery as an evolving cultural landscape. *Annals of the Association of American Geographers, 61*(3), 501–509.

Francis, D., Kellaher, L., & Neophytou, G. (2005). *The secret cemetery*. Berg Publishers.

Fuller, R., Irvine, K., Devine-Wright, P., Warren, P., & Gaston, K. (2007). Psychological benefits of greenspace increase with biodiversity. *Biology Letters, 3*(4), 390–394.

Gandy, M. (2012). Queer ecology: Nature, sexuality, and heterotopic alliances. *Environment and Planning D: Society and Space, 30*(4), 727–747.

Gandy, M. (2018). The fly that tried to save the World: Saproxylic geographies and other-than-human ecologies. *Transactions of the Institute of British Geographers, 44*(2), 392–406.

Hartig, T. (2004). Restorative environments. In C. D. Spielberger (Ed.), *Encyclopedia of applied psychology* (Vol. 3, pp. 273–279). Elsevier.

Hartig, T., & Staats, H. (2006). The need for psychological restoration as a determinant of environmental preferences. *Journal of Environmental Psychology, 26*(3), 215–226.

Harvey, T. (2006). Sacred places, common places: The cemetery in the contemporary American city. *The Geographical Review, 96*(1), 295–312.

Hofmann, M., Westermann, J. R., Kowarik, I., & van der Meer, E. (2012). Perceptions of parks and urban derelict land by landscape planners and residents. *Urban Forestry & Urban Greening, 11*(3), 303–312.

Johnson, P. (2008). The modern cemetery: A design for life. *Social & Cultural Geography, 9*(7), 777–790.

Johnson, P. (2013). The geographies of heterotopia. *Geography Compass, 7*, 790–803.

Jones, J. (1979). *How to record graveyards*. C.B.A. and Rescue.

Joyce, P. (1994). *A guide to Abney Park Cemetery* (2nd ed.). Abney Park Cemetery Trust.

Klaufus, C. (2016). The dead are killing the living: Spatial justice, funerary services, and cemetery land use in urban Colombia. *Habitat International, 54*, 74–79.

Lee, T. (2015). Place and identity: What can we learn from the dead? In K. Lawrence (Ed.), *Landscape, Place and Identity in Craft and Design, Craft + Design Enquiry [Discontinued], 7*, 99–112 [online]. Retrieved May 26, 2021, from http://press-files.anu.edu.au/downloads/press/p328141/pdf/ch072.pdf

Lehrer, J. (1974). Cemetery Land Use and the Urban Planner. *Urban Law Annual; Journal of Urban and Contemporary Law, 7*(1), 181–197 [Online]. Retrieved May 26, 2021, from https://openscholarship.wustl.edu/cgi/viewcontent.cgi?article=1745&context=law_urbanlaw

Lehvävirta, S., & Rita, H. (2002). Natural regeneration of trees in urban Woodlands. *Journal of Vegetation Science, 13*(1), 57–66.

Lévesque, L. (2009). Towards an interstitial approach to urban Landscape. *Territorio, 48*, 77–82.

Mabey, R. (2010). *Weeds: The story of outlaw plants*. Profile Books.

Malterud, K. (2012). Systematic text condensation: A strategy for qualitative analysis. *Scandinavian Journal of Public Health, 40*, 795–805.

Mariani, M., & Barron, P. (2013). *Terrain vague: Interstices at the edge of the pale*. Routledge.

Meller, H., & Parsons, B. (2011). *London cemeteries: An illustrated guide and gazetteer* (5th ed.). The History Press.

Miller, R. (2018) *The trees and woodland of Abney Park Cemetery* (3rd ed.). Swallowtail Print. Reprinted from *The London Naturalist, 87*, 2008.

Moran, J. (2007). Early cultures of gentrification in London, 1955–1980. *Journal of Urban History, 34*(1), 101–121.

Mytum, H. C., & Council for British Archaeology and English Heritage (2000). *Recording and analysing graveyards*. Council for British Archaeology in association with English Heritage.

Nordh, H., Evensen, K., & Skår, M. (2017). A peaceful place in the City—A qualitative study of restorative components of the cemetery. *Landscape and Urban Planning, 167*, 108–117.

Özgüner, H., & Kendle, A. D. (2006). Public attitudes towards naturalistic versus designed landscapes in the city of Sheffield (UK). *Landscape and Urban Planning, 74*(2), 139–157.

Purcell, M. (2002). Excavating Lefebvre: The right to the city and its urban politics of the inhabitant. *GeoJournal, 58*, 99–108.

Raimbault, M., & Dubois, D. (2005). Urban soundscapes: Experiences and knowledge. *Cities, 22*(5), 339–350.

Ratcliffe, E., Gatersleben, B., & Sowden, P. (2013). Bird sounds and their contributions to perceived attention restoration and stress recovery. *Journal of Environmental Psychology, 36*, 221–228.

Revelle, R. (1967). Outdoor recreation in a hyper-productive society. *Daedalus, 96*(4), 1172–1191 [Online]. Retrieved May 26, 2021, from http://www.jstor.org/stable/20027111

Rivers, M. (2020). More than 100 Uyghur graveyards demolished by Chinese authorities, satellite images show. *CNN* [Online]. Retrieved April 11, 2021, from https://www.cnn.com/2020/01/02/asia/xinjiang-uyghur-graveyards-china-intl-hnk/index.html

Royal Parks, The. (2021). *About Brompton cemetery* [Online]. Retrieved May 26, 2021, from https://www.royalparks.org.uk/parks/brompton-cemetery/about-brompton-cemetery

Rugg, J. (1998). A few remarks on modern sepulture: Current trends and new directions in cemetery research. *Mortality, 3*(2), 111–128.

Rugg, J. (2000). Defining the place of burial: What makes a cemetery a cemetery? *Mortality, 5*(3), 259–275.

Rugg, J., & Pleace, N. (2011). *An audit of London burial provision*. Greater London Authority [Online]. Retrieved May 26, 2021, from https://www.london.gov.uk/file/5284/download?token=sLOljOSB

Rupprecht, C., & Byrne, J. (2014). Informal urban green space: A typology and trilingual systematic review of its role for urban residents and trends in the literature. *Urban Forestry & Urban Greening, 13*(4), 597–611.

Schneider, G. S. (2020). A Virginia state senator found headstones on his property. It brought to light a historic injustice in D.C. *The Washington Post* [Online]. Retrieved April 11, 2021, from https://www.washingtonpost.com/local/virginia-politics/headstones-black-cemetery-potomac-river/2020/10/25/3586f0d4-0d7a-11eb-8074-0e943a91bf08_story.html

Scholz, M. (2017). Over our dead bodies: The fight over cemetery construction in nineteenth-century London. *Journal of Urban History, 43*(3), 445–457.

Shahhoseini, H., Bin, M. K., & Bin Maulan, S. (2015). Visual preferences of small urban parks based on spatial configuration of place. *International Journal of Architectural Engineering and Urban Planning, 25*(2), 84–93.

Skår, M., Nordh, H., & Swensen, G. (2018). Green urban cemeteries: More than just parks. *Journal of Urbanism: International Research on Placemaking and Urban Sustainability, 11*(3), 362–382.

Sloane, D. C. (1991). *The last great necessity: Cemeteries in American history*. The Johns Hopkins University Press.

Steinmetz-Wood, M., Wasfi, R., Parker, G., Bornstein, L., Caron, J., & Kestens, Y. (2017). Is gentrification all bad? Positive association between gentrifcation and individual's perceived neighborhood collective efficacy in Montreal, Canada. *International Journal of Health Geographics, 16*(24), 24–33.

Tarlow, S. (2000). Landscapes of memory: The nineteenth-century garden cemetery. *European Journal of Archaeology, 3*(2), 217–239.

Taylor, G., & Ussher, J. (2001). Making sense of S&M: A discourse analytic account. *Sexualities, 4*(3), 293–314.

Tomašević, A. (2018). Cemeteries as tourist attraction. *Broj, 21*, 13–24.

UN-DESA. (2012). *Rio 2012 Issues Brief 14—Population dynamics and sustainable development* [Online]. UNCSD Secretariat. Retrieved May 26, 2021, from. https://sustainabledevelopment.un.org/content/documents/543brief14.pdf

Warriner, J. (2019) East Perth historical cemetery to be excavated, revealing mysteries of early Perth settlement. *ABC News Australia* [Online]. Retrieved April 11, 2021, from https://www.abc.net.au/news/2019-11-28/historic-east-perth-cemetery-site-excavated-for-development/11743560

White, J., & Hodson, J. (2007). *Paradise preserved: An introduction to the assessment, evaluation, conservation and management of historic cemeteries*. English Heritage [Online]. Retrieved May 26,

2021, from https://thegardenstrust.org/wp-content/uploads/2016/11/EH-Paradise-Preserved-2007-1.pdf

White, J., & Stamper, P. (2018). *List of registered cemeteries*. Historic England, 51684 and 51685 [Online]. Retrieved May 26, 2021, from https://historicengland.org.uk/images-books/publications/list-of-registered-cemeteries/

Woodthorpe, K. (2011). Sustaining the contemporary cemetery: Implementing policy alongside conflicting perspectives and purpose. *Mortality, 16*(3), 259–276.

Worpole, K. (1997). *The cemetery in the city: A report by Ken Worpole for the Gulbenkian Foundation*. Comedia.

Wright, P. (1991). *A journey through ruins: The last days of London*. Radius Books.

Young, C., & Light, D. (2016). Interrogating spaces of and for the dead as 'Alternative Space': Cemeteries, corpses and sites of dark tourism. *International Review of Social Research, 6*(2), 61–72.

Correction to: Multiplying Effects of Urban Innovation Districts. Geospatial Analysis Framework for Evaluating Innovation Performance Within Urban Environments

Jeremy Burke, Ramon Gras Alomà, and Fernando Yu

Chapter 15 in: C. Piselli et al. (eds.), Innovating Strategies and Solutions for Urban Performance and Regeneration, Advances in Science, Technology & Innovation, https://doi.org/10.1007/978-3-030-98187-7_15

In the original version of the book, the author's name "Ramon Gras (Given name) Alomà (Family name) has been changed to "Ramon (Given name) Gras Alomà (Family name) in the Frontmatter, Backmatter and in Chapter 15.

The updated version of this chapter can be found at
https://doi.org/10.1007/978-3-030-98187-7_15

Ingram Content Group UK Ltd.
Milton Keynes UK
UKHW052301050723
424587UK00001B/1